Image Processing in Radiation Therapy

IMAGING IN MEDICAL DIAGNOSIS AND THERAPY

William R. Hendee, Series Editor

Quality and Safety in Radiotherapy

Todd Pawlicki, Peter B. Dunscombe, Arno J. Mundt, and Pierre Scalliet, Editors
ISBN: 978-1-4398-0436-0

Adaptive Radiation Therapy

X. Allen Li, Editor
ISBN: 978-1-4398-1634-9

Quantitative MRI in Cancer

Thomas E. Yankeelov, David R. Pickens, and Ronald R. Price, Editors
ISBN: 978-1-4398-2057-5

Informatics in Medical Imaging

George C. Kagadis and Steve G. Langer, Editors
ISBN: 978-1-4398-3124-3

Adaptive Motion Compensation in Radiotherapy

Martin J. Murphy, Editor
ISBN: 978-1-4398-2193-0

Image-Guided Radiation Therapy

Daniel J. Bourland, Editor
ISBN: 978-1-4398-0273-1

Targeted Molecular Imaging

Michael J. Welch and William C. Eckelman, Editors
ISBN: 978-1-4398-4195-0

Proton and Carbon Ion Therapy

C.-M. Charlie Ma and Tony Lomax, Editors
ISBN: 978-1-4398-1607-3

Comprehensive Brachytherapy: Physical and Clinical Aspects

Jack Venselaar, Dimos Baltas, Peter J. Hoskin, and Ali Soleimani-Meigooni, Editors
ISBN: 978-1-4398-4498-4

Physics of Mammographic Imaging

Mia K. Markey, Editor
ISBN: 978-1-4398-7544-5

Physics of Thermal Therapy: Fundamentals and Clinical Applications

Eduardo Moros, Editor
ISBN: 978-1-4398-4890-6

Emerging Imaging Technologies in Medicine

Mark A. Anastasio and Patrick La Riviere, Editors
ISBN: 978-1-4398-8041-8

Cancer Nanotechnology: Principles and Applications in Radiation Oncology

Sang Hyun Cho and Sunil Krishnan, Editors
ISBN: 978-1-4398-7875-0

Monte Carlo Techniques in Radiation Therapy

Joao Seco and Frank Verhaegen, Editors
ISBN: 978-1-4665-0792-0

Image Processing in Radiation Therapy

Kristy Kay Brock, Editor
ISBN: 978-1-4398-3017-8

Forthcoming titles in the series

Stereotactic Radiosurgery and Radiotherapy

Stanley H. Benedict, Brian D. Kavanagh, and David J. Schlesinger, Editors
ISBN: 978-1-4398-4197-6

Informatics in Radiation Oncology

Bruce H. Curran and George Starkschall, Editors
ISBN: 978-1-4398-2582-2

Tomosynthesis Imaging

Ingrid Reiser and Stephen Glick, Editors
ISBN: 978-1-4398-7870-5

Cone Beam Computed Tomography

Chris C. Shaw, Editor
ISBN: 978-1-4398-4626-1

Image Processing
in Radiation Therapy

Edited by

Kristy K. Brock

CRC Press
Taylor & Francis Group
Boca Raton London New York

CRC Press is an imprint of the
Taylor & Francis Group, an **informa** business

A TAYLOR & FRANCIS BOOK

MATLAB® is a trademark of The MathWorks, Inc. and is used with permission. The MathWorks does not warrant the accuracy of the text or exercises in this book. This book's use or discussion of MATLAB® software or related products does not constitute endorsement or sponsorship by The MathWorks of a particular pedagogical approach or particular use of the MATLAB® software.

CRC Press
Taylor & Francis Group
6000 Broken Sound Parkway NW, Suite 300
Boca Raton, FL 33487-2742

© 2014 by Taylor & Francis Group, LLC
CRC Press is an imprint of Taylor & Francis Group, an Informa business

No claim to original U.S. Government works

Printed on acid-free paper
Version Date: 20130709

International Standard Book Number-13: 978-1-4398-3017-8 (Hardback)

Library of Congress Cataloging-in-Publication Data

Image processing in radiation therapy / editor, Kristy Kay Brock.
   p. ; cm. -- (Imaging in medical diagnosis and therapy)
   Includes bibliographical references and index.
   ISBN 978-1-4398-3017-8 (hardcover : alk. paper)
   I. Brock, Kristy Kay, editor of compilation. II. Series: Imaging in medical diagnosis and therapy.
   [DNLM: 1. Radiotherapy, Computer-Assisted--methods. 2. Image Processing, Computer-Assisted--methods. 3. Radiation Oncology--methods. 4. Radiographic Image Interpretation, Computer-Assisted--methods. WN 250.5.R2]
   RC78.7.R38
   616.07'57--dc23
      2013026187

Visit the Taylor & Francis Web site at
http://www.taylorandfrancis.com

and the CRC Press Web site at
http://www.crcpress.com

To my family—my husband Rob,

my daughter Olivia,

and my son Lucas

Contents

SECTION I   Image Processing in Radiotherapy Applications

SECTION II   Registration

SECTION III   Segmentation

SECTION IV   Advanced Imaging Techniques

Foreword

Over the past three decades, radiation therapy has evolved into the most complex and intricate form of cancer therapy. These advances have made significant contributions to cancer management—increasing disease control rates and/or reducing toxicity, as well as defining new treatment paradigms. In parallel with this advancing complexity, the field has maintained a remarkable degree of safety and control in the application of this powerful therapeutic. Despite this period of rapid advancement, the field is poised to pursue yet another escalation in complexity—one that seeks to maximize the use of the imaging information collected before and during therapy to guide the design and dynamic redesign of an individual's radiation treatment course. This is enabled by the accelerated deployment of imaging systems into radiation oncology, including those in the simulation process (e.g., CT, PET, MR), as well as those in the treatment room (e.g., CT, MVCT, cone-beam CT). These imaging systems demonstrate more and more clearly the dynamic nature of the normal and diseased anatomy in the context of therapy. It is these observations that have led to an emerging "4D hypothesis"—that is, adapting a patient's treatment to observed changes in anatomical or functional imaging during the course of therapy can improve the therapeutic ratio. While simple enough to state, the realistic progression of these concepts to robust and safe clinical application remains a major technical and scientific challenge in the field of medical physics. Furthermore, the ultimate clinical evaluation of the benefit of these approaches will hinge on the integrity and performance of these tools.

This book provides a unique and valuable compilation of the many technologies, algorithms, and implementations that are needed to fully leverage the use of imaging in radiation therapy. While elements of the algorithms can be found in the published literature, this work draws them together in a dedicated volume with consistency in presentation and illustrated application to radiation therapy. Linking the clinical motivation, algorithm descriptions, adaptations to radiation therapy, and validation results makes this a truly valuable resource to the clinical medical physicist. There are several highlights in the book. Chapters 1 and 2 establish the context with motivation and examples for the development of adaptive and on-line corrections. Chapter 3 includes a unique and comprehensive review of deformable image registration performance that allows the reader to appreciate the capabilities of these algorithms while also inspiring further improvements. Chapters 4, 5, 7, and 8 nicely localize the fundamentals of image processing to the radiation therapy problem, which will be appreciated by the novice and expert reader. Chapter 6 is a pleasure to read and leaves the readers feeling they have strengthened their understanding of mechanics and given them the capability to relate these concepts to the challenge of biomechanical models of deformation. Chapters 9 thru 12 tackle the problem of segmentation from the most basic to the most complex—the expert authors provide an excellent overview of the fundamentals as well as details of the implementation in clinical systems—this is one of the many highlights of the textbook. A deep dive into the topics of cone-beam CT reconstruction, respiratory sorting, and the technology of image-guided radiation therapy is beautifully presented by leaders in the field. Finally, Chapter 16 gives us a glimpse into the "fast future" of image acquisition, reconstruction, segmentation, and post-processing using GPU technologies. Overall, an outstanding text in which experts have carefully localized complex algorithms to the exciting and important challenges found in image-guided radiation therapy. My compliments to the authors and to the editor! Such a text will not only advance the degree of understanding in the field, but also inspire many young minds to participate in the exciting future of image-guided radiation therapy.

David A. Jaffray, Ph.D.
Head, Radiation Physics, Radiation Medicine Program
Princess Margaret Cancer Centre
Director, Techna Institute, University Health Network
Toronto, Ontario, Canada

Series Preface

Since their inception over a century ago, advances in the science and technology of medical imaging and radiation therapy are more profound and rapid than ever before. Further, the disciplines are increasingly cross-linked because imaging methods become more widely used to plan, guide, monitor, and assess treatments in radiation therapy. Today, the technologies of medical imaging and radiation therapy are so complex and so computer driven that it is difficult for the persons (physicians and technologists) responsible for their clinical use to know exactly what is happening at the point of care, when a patient is being examined or treated. The persons best equipped to understand the technologies and their applications are medical physicists, and these individuals are assuming greater responsibilities in the clinical arena to ensure that what is intended for the patient is actually delivered in a safe and effective manner.

The growing responsibilities of medical physicists in the clinical arenas of medical imaging and radiation therapy are not without their challenges, however. Most medical physicists are knowledgeable in either radiation therapy or medical imaging and are experts in one or a small number of areas within their discipline. They sustain their expertise in these areas by reading scientific articles and attending scientific talks at meetings. In contrast, their responsibilities increasingly extend beyond their specific areas of expertise. To meet these responsibilities, medical physicists periodically must refresh their knowledge of advances in medical imaging or radiation therapy, and they must be prepared to function at the intersection of these two fields. How to accomplish these objectives is a challenge.

At the 2007 annual meeting of the American Association of Physicists in Medicine in Minneapolis, this challenge was the topic of conversation during a lunch hosted by Taylor & Francis Publishers and involving a group of senior medical physicists (Arthur L. Boyer, Joseph O. Deasy, C.-M. Charlie Ma, Todd A. Pawlicki, Ervin B. Podgorsak, Elke Reitzel, Anthony B. Wolbarst, and Ellen D. Yorke). The conclusion of this discussion was that a book series should be launched under the Taylor & Francis banner, with each volume in the series addressing a rapidly advancing area of medical imaging or radiation therapy of importance to medical physicists. The aim would be for each volume to provide medical physicists with the information needed to understand technologies driving a rapid advance and their applications to safe and effective delivery of patient care.

Each volume in the series is edited by one or more individuals with recognized expertise in the technological area encompassed by the book. The editors are responsible for selecting the authors of individual chapters and ensuring that the chapters are comprehensive and intelligible to someone without such expertise. The enthusiasm of volume editors and chapter authors has been gratifying and reinforces the conclusion of the Minneapolis luncheon that this series of books addresses a major need of medical physicists.

Imaging in Medical Diagnosis and Therapy would not have been possible without the encouragement and support of the series manager, Luna Han of Taylor & Francis Publishers. The editors and authors, and most of all I, are indebted to her steady guidance of the entire project.

William Hendee
Series Editor
Rochester, Minnesota

Preface

The demand for image processing tools to provide integration of data into the radiotherapy process has led to exciting collaborations among medical physicists, computer scientists, and mathematicians. Image processing addresses the reconstruction, adaptation, integration, and evaluation of imaging data into the radiation therapy process to ensure an accurate, efficient, and effective use of the information for radiation planning, treatment delivery, and outcomes assessment. Image processing includes the mathematical manipulation of images for integration at the reconstruction and evaluation stage as well as the incorporation of this information into the increasingly complex radiation therapy workflow. This book provides a description of the techniques used as well as their applications in radiation therapy.

Over the past two decades, the influx of images into the radiotherapy process has dramatically increased. The field quickly moved from the acquisition of a single computed tomography at the start of treatment to the inclusion of multiphase computed tomography scans, magnetic resonance, positron emission tomography, and single-photon emission computed tomography for planning, volumetric imaging at the time of delivery, and the additional acquisition of diagnostic-quality images throughout the treatment course for reassessment. The addition of these images enables clinicians to identify the tumor and critical normal structures as well as the impact of physiological motion on the position of these structures, with increasing precision. In addition to the ability to identify these structures, advances in treatment planning (e.g., intensity-modulated radiation therapy) have enabled sophisticated treatment plans to be developed to precisely avoid these critical structures and target the tumor. The escalation of the number of images, the number of structures to be contoured, and the precision of treatment planning placed an increasing demand on the clinical process to integrate and segment these images. The time spent contouring increased from minutes to hours, introducing the demand for automated segmentation and the need for highly accurate registration of multimodality images. These demands further increased with the introduction of in-room imaging, which provided additional information of the patient, including the complexities that are exhibited over the course of treatment including motion and changes in motion patterns and response to treatment. This wealth of new imaging devices demanded novel reconstruction techniques and methods to handle uncertainties in the data acquisition, such as breathing motion. The acquisition of in-room volumetric images placed additional demands on image registration, requiring high efficiency to allow a streamlined integration of these technologies into a busy clinical workflow. With the ability to view the 3D anatomy of the patient on a daily basis, the detection of tumor and normal tissue response over the course of treatment created the demand for additional diagnostic imaging along with its associated registration and segmentation. This overall patient treatment can easily result in more than 40 3D image volumes and more than 20 structures to be segmented per image. With this trajectory, the demand for image processing, including reconstruction, segmentation, and registration, became clear.

This work spans the topics of deformable registration, segmentation, image reconstruction, and integration of these practices into the radiation therapy environment. Section I of this book motivates the need for advances in image processing in radiotherapy applications, including adaptive radiotherapy, online monitoring and tracking, dose accumulation, and accuracy assessment. Section II describes the mathematical approach to deformable registration. The description of similarity metrics used for registration techniques is described, including their effectiveness and applicability in the radiation therapy environment. In addition, Section II includes a detailed description and evaluation of parametric and nonparametric image registration techniques and their applications in radiation therapy processes. Section III addresses the techniques for image segmentation. This section addresses atlas-based, level set, and registration-based techniques, assessing their efficiency, robustness, and breadth of application. Section IV of this book focuses on advanced imaging techniques for radiotherapy. Advancements in 3D image reconstruction, respiration sorting, in-room imaging techniques, and advanced applications of image registration using a graphics processor unit are described.

The goal of this book is to provide a comprehensive description of techniques and algorithms for image segmentation and registration, in-room imaging techniques and advanced reconstruction techniques, and clinical rationale and implementation. The target audience includes medical physicists, clinicians, dosimetrists, radiation therapists, and trainees in medical physics and related fields. This book benefits from the contributions of nationally and internationally recognized authors who have developed state-of-the-art image processing techniques and pursued their integration into the clinical environment. Their extensive knowledge is evident from their comprehensive descriptions of the algorithms and techniques, practical examples, and resulting clinical translation. The editor is sincerely thankful for their excellent contributions.

MATLAB® is a registered trademark of The MathWorks, Inc. For product information, please contact:

The MathWorks, Inc.
3 Apple Hill Drive
Natick, MA 01760-2098 USA
Tel: 508 647 7000
Fax: 508-647-7001
E-mail: info@mathworks.com
Web: www.mathworks.com

Acknowledgment

Dr. Brock would like to like to acknowledge and express her sincere gratitude for the expert editorial assistance of Luna Han, Taylor & Francis Group, without whom this work would not have been possible.

About the Editor

Kristy K. Brock earned her Ph.D. in nuclear engineering and radiological sciences from the University of Michigan. She was on the radiation oncology faculty at the University of Toronto (Princess Margaret Hospital) and is now an associate professor at the Department of Radiation Oncology, University of Michigan. She is currently a member of the board for the American Association of Physicists in Medicine and the editorial board for *Medical Physics*, and serves as the 2013 Therapy Science Director for the American Association of Physicists in Medicine annual meeting. Dr. Brock is a diplomat of the American Board of Radiology in therapeutic radiological physics. She is known for her research in deformable image registration, validation techniques, and dose accumulation. Her current research interests focus on deformable registration for dose accumulation and adaptive radiotherapy, the application of biomechanical modeling in correlative pathology, and therapy response assessment.

Contributors

Daniel H. Adler
Department of Bioengineering
University of Pennsylvania
Philadelphia, Pennsylvania

Adil Al-Mayah
Department of Civil and Environmental
    Engineering
University of Waterloo
Waterloo, Ontario, Canada

Kristy K. Brock
Department of Radiation Oncology
University of Michigan
Ann Arbor, Michigan

Sonny Chan
Department of Computer Science
Stanford University
Stanford, California

Edward L. Chaney
Department of Radiation Oncology
University of North Carolina
Chapel Hill, North Carolina

Lei Dong
Scripps Proton Therapy Center
San Diego, California

Issam El Naqa
Medical Physics Unit
Department of Oncology
McGill University
Montreal, Quebec, Canada

Timothy H. Fox
Department of Radiation Oncology
and
Winship Cancer Institute
Emory University School of Medicine
Atlanta, Georgia

Arun Gopal
New York Presbyterian Hospital
New York, New York

Geoffrey D. Hugo
Virginia Commonwealth University
Richmond, Virginia

Jenny H. M. Lee
Princess Margaret Hospital
University Health Network
Toronto, Ontario, Canada

Daniel A. Low
Department of Radiation Oncology
University of California at Los Angeles
Los Angeles, California

Todd R. McNutt
Department of Radiation Oncology and
    Molecular Radiation Sciences
Johns Hopkins University School of
    Medicine
Baltimore, Maryland

J. Ross Mitchell
Diagnostic Radiology
The Mayo Clinic
Scottsdale, Arizona

Frédéric Noo
Department of Radiology
University of Utah
Salt Lake City, Utah

Stephen M. Pizer
Department of Computer Science
University of North Carolina at Chapel
    Hill
Chapel Hill, North Carolina

William Plishker
Department of Electrical and Computer
    Engineering
University of Maryland
College Park, Maryland

Sanjiv S. Samant
Department of Radiation Oncology
University of Florida
Gainesville, Florida

David Sarrut
Université de lyon
CREATIS, CNRS, Léon Bérard
    Cancer Center
Lyon, France

Eduard Schreibmann
Department of Radiation Oncology
and
Winship Cancer Institute
Emory University School of Medicine
Atlanta, Georgia

Raj Shekhar
Sheikh Zayed Institute for Pediatric
    Surgical Innovation
Children's National Medical Center
Washington, District of Columbia

Jeffrey H. Siewerdsen
Department of Biomedical Engineering
Johns Hopkins University
Baltimore, Maryland

Jan-Jakob Sonke
Department of Radiation Oncology
The Netherlands Cancer Institute-Antoni
    van Leeuwenhoek Hospital
Amsterdam, the Netherlands

J. Webster Stayman
Department of Biomedical Engineering
Johns Hopkins University
Baltimore, Maryland

Jef Vandemeulebroucke
Department of Electronics and
    Informatics
Vrije Universiteit Brussel
and
Department of Future Media and
    Imaging
iMinds
Ghent, Belgium

Michael Velec
Princess Margaret Hospital
University Health Network
Toronto, Ontario, Canada

He Wang
Department of Radiation Physics
The University of Texas M. D. Anderson
    Cancer Center
Houston, Texas

Jian Wu
Proton Therapy Institute
University of Florida
Gainesville, Florida

Junyi Xia
Department of Radiation Oncology
University of Iowa Hospitals and Clinics
Iowa City, Iowa

Jinzhong Yang
The University of Texas M. D. Anderson
    Cancer Center
Houston, Texas

Yongbin Zhang
The University of Texas M. D. Anderson
    Cancer Center
Houston, Texas

I

Image Processing in Radiotherapy Applications

1

Image Processing in Adaptive Radiotherapy

Lei Dong
Scripps Proton Therapy Center

Jinzhong Yang
*The University of Texas M. D.
Anderson Cancer Center*

Yongbin Zhang
*The University of Texas M. D.
Anderson Cancer Center*

1.1 Radiotherapy Process

1.1.1 Introduction

Radiotherapy is an effective technique for treating localized cancer (Leibel et al. 2003). For radiation therapy, there are two primary treatment goals to achieve the effectiveness of therapy: target dose escalation and normal tissue sparing. The modern radiation therapy of cancers uses computerized treatment planning systems to design treatment delivery parameters that are patient specific. Individual treatment parameters are designed to collimate and direct radiation beams toward the target volume to avoid excessive radiation to organs at risk (OARs).

With intensity-modulated radiation therapy (IMRT) (Wu et al. 2000; Purdy 2001), it is possible to treat very complicated target volumes at variable dose levels while sparing the adjacent normal structures. The implementation of IMRT requires an adequate selection of target volumes, an appropriate specification and dose prescription, dose-volume constraints for normal organs,

and proper knowledge of setup uncertainties (Gregoire et al. 2003). Target delineation often uses different imaging modalities, such as computed tomography (CT), magnetic resonance imaging (MRI), magnetic resonance spectroscopic imaging, positron emission tomography (PET), single-photon emission tomography (SPECT), and ultrasound, to assess the extent of disease and delineate the target regions to be irradiated and the normal tissues to be protected. Therefore, imaging is not a stranger to radiotherapy. It is not surprising that image processing is an important part of modern radiation therapy (Apisarnthanarax and Chao 2005).

This is particularly true when image-guided radiotherapy (IGRT) is introduced recently. IGRT uses various in-room imaging techniques to detect target position, relative to the planned radiation beams, and to apply geometric corrections to improve the accuracy of treatment delivery (Jaffray et al. 2002; Jaffray 2005). IGRT allows the online correction of target deviation before the start of radiation treatment, which often results in a more accurate treatment and reduction of treatment margins.

The frequent use of in-room imaging gives clinicians an opportunity to evaluate the treatment progress during a treatment course.

Treatment deviations can be generally classified in two categories: (1) setup errors, which describe the target deviation from the planned treatment beam, and (2) target volume changes, which imply that the original target determined in the treatment simulation process is changed due to the patient's nonrigid anatomic changes caused by either the treatment effects (weight loss or tumor shrinkage) or the normal physiologic processes (such as organ motion and day-to-day variations in bladder filling). Image-guided intervention can be performed by aligning the beam with the target before radiation to correct for setup errors. However, a simple repositioning of the patient's anatomy relative to the treatment beam is usually insufficient to correct for shape variations in target or nearby critical structures. The more complex correction is usually done by replanning: a process called adaptive radiotherapy, which includes the modification of an initial plan to adapt to the changes in target volume or normal organs.

1.2 Image-Guided Adaptive Radiotherapy

The process of image-guided adaptive radiotherapy is illustrated in Figure 1.1. It may be useful to define each step in more detail, so that the imaging needs and the image processing tools can be better understood. The following terminology and explanation will be used to describe each substep:

- **Treatment simulation:** In this step, the patient will be first immobilized to the tabletop and a CT scan is performed as if the patient is in a treatment position. CT imaging is important for radiation therapy not only because it provides excellent geometric representation of the patient's anatomy but also because it can be used for

radiation dose calculation. CT numbers are calibrated to represent the relative electron density of human tissue at each voxel, which can be used for the accurate dose calculation. The simulation CT establishes an initial patient anatomic model, which describes the geometric relationship between radiation beam and the patient's anatomy.

- **Target delineation:** For modern radiation therapy (IMRT), a clear delineation of target volume in 3D is required for the automatic treatment design using inverse planning techniques (Bortfeld 1999; Purdy 2001). IMRT uses mathematical optimization techniques to derive the radiation patterns and spatial dose distributions that will best match with the shape of the target and avoid normal organs at the same time. The target delineation is perhaps the most critical component for IMRT and for the adaptive radiotherapy process. Various investigators studied site-specific target delineation strategies (Leunens et al. 1993; Tai et al. 1998; Remeijer et al. 1999; Hurkmans et al. 2001; Steenbakkers et al. 2005; Jansen et al. 2010; Louie et al. 2010; Symon et al. 2011). It is worth mentioning that functional imaging, such as PET, SPECT, or diffusion MRI, can help reducing interobserver variations in target delineation and perhaps moving toward the more meaningful biology-based target definition (Ling et al. 2000; Hong et al. 2007; Vesprini et al. 2008; Wang et al. 2009; Krengli et al. 2010). In addition to target volume, the delineation of normal anatomic structures is equally important for conformal radiotherapy.

- **Treatment planning:** The treatment design process uses graphical interface to visualize the target position and critical structures relative to the planned radiation beams. Computerized mathematical optimization will be performed to create radiation intensity patterns that will optimize the 3D dose distribution. The treatment plan design, unfortunately, is typically based on a single snapshot of the patient's anatomy during the CT simulation.

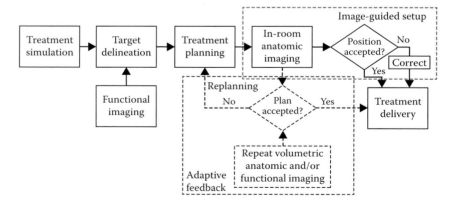

FIGURE 1.1 Image-guided adaptive radiotherapy is a process involving (1) initial planning, (2) image-guided setup, (3) repeat imaging and adaptive planning, and (4) treatment delivery. The initial planning is based on a single simulation CT and optional functional images to establish baseline target volume and functional activity levels. The image guidance process involves only positional interventions, whereas the adaptive replanning corrects for nonrigid changes in anatomy and functional changes. The upper dotted box indicates the components of image-guided setup (IGRT), and the lower dotted box indicates the components of adaptive radiotherapy.

Due to the positional setup errors and the patient's day-to-day and minute-to-minute anatomic changes (which will be discussed later), a large treatment margin is usually used to ensure target coverage, which inevitably irradiates more normal tissues surrounding the target. A treatment plan defines the radiation dose distribution relative to a machine reference point (the rotation center/isocenter, in most situations). If the geometric relationship between the spatial dose distribution and the patient's anatomy is changed (due to setup error, organ motion, or tumor shrinkage), the original treatment plan becomes suboptimal.

- **Anatomic imaging and setup evaluation:** To reduce day-to-day setup error, IGRT is typically used for high-precision radiation therapy. IGRT will be used to guide the patient's position before treatment delivery, as shown in Figure 1.1. Various in-room imaging techniques have been developed recently to detect the target position relative to radiation beams. These IGRT technologies include (1) 2D projection radiographs (Fuss et al. 2007; Chang et al. 2010), (2) ultrasound (Fuss et al. 2004), (3) surface imaging (Bert et al. 2006), and (4) volumetric CT or cone-beam CT (CBCT) (Bissonnette et al. 2009). In particular, CBCT was invented to be an integral part of the accelerator design (Jaffray et al. 2002; Jaffray 2007). CBCT produces 3D CT images to represent the actual position in which the patient received treatment. This will facilitate the treatment plan evaluation (Ding et al. 2007).

- **IGRT:** Although IGRT can be a broad concept, it is usually referred to as the image-guided setup. In this process, the target position will be detected using in-room imaging techniques and compared with the planned target position. The differences in target position will be corrected by a translational shift of the treatment couch, resulting in the alignment of the target with the treatment machine geometry. Some treatment couch can perform six degrees of freedom: translations and rotations. This will require the software capable of performing six-degrees-of-freedom image registration. Although IGRT can minimize the differences between the original plan and the patient's anatomy at the treatment, nonrigid changes of the tumor and normal organs cannot be corrected effectively by a simple couch shift or rotation. The IGRT process is illustrated in the upper dotted box of Figure 1.1.

- **Adaptive radiotherapy:** The target volume sometimes cannot be represented by a single point. When the shape of the target is changed or nonrigid variations occur in the patient's anatomy, the best correction is to replan or to modify the original plan to adapt the new geometric relationship of the tumor and normal anatomy. Repeat volumetric imaging can be performed either online using CBCT or off-line using conventional simulation CT. To reduce the workload in recontouring, deformable image registration is practically necessary to map the original target and normal structures onto the new CT, archiving

autosegmentation of the new anatomy for rapid replanning. Sometimes, treatment protocols can be designed to treat biological targets. Repeat functional images may be required to measure the residual tumor activities or the functional activities of normal organs. The adaptive feedback is illustrated in the lower dotted box of Figure 1.1.

1.3 Morphologic and Physiologic Adaptation

Adaptive radiotherapy is generally described as changing the radiation treatment plan delivered to a patient during a course of radiotherapy to account for the temporal changes in anatomy (e.g., tumor shrinkage, weight loss, or internal motion) or the changes in tumor biology/function (e.g., hypoxia). The former is usually called "anatomy"- or "morphology"-based adaptation, which uses primarily CT images to detect morphologic changes. The latter is called "physiology"- or "function"-based adaptation, which uses biological/functional images, such as PET, diffusion-based MRI, or SPECT, to study the functional changes of the underlining tumor biology.

Due to the treatment effects, many patients experienced weight loss and/or tumor shrinkage during the course of a fractionated radiotherapy. An example of a significant morphologic change in the patient's anatomy is shown in Figure 1.2, in which a head-and-neck patient underwent IMRT treatment. The left picture shows the original treatment plan with target volumes designed on the simulation CT images. The patient had a bulky lymph node, which shrank significantly during treatment. In addition, a 5% body weight loss was also experienced. As a result, the patient's anatomy has significantly deviated from the original plan. The contours of the target volume, after rigid registration between the two CT images, do not match well with the

Planning CT   During treatment

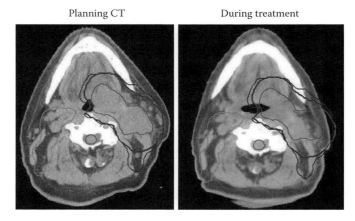

FIGURE 1.2 (See color insert.) Significant anatomic changes can occur during radiation therapy for head-and-neck cancer patients. Left: Treatment planning CT with original target volumes overlaid on the CT images. Right: CT image acquired after 3 weeks of radiotherapy. The original target volume is no longer matching well with the patient's anatomy. The treatment will be suboptimal if a new plan is not derived from the new CT image.

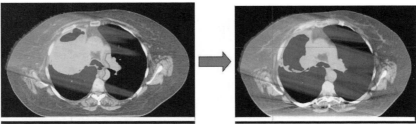

Original proton plan   Dose recalculated on the new anatomy

FIGURE 1.3 (See color insert.) Impact of tumor shrinkage to proton dose distributions. Left: Original proton therapy plan. After about 1 month of treatment, the primary tumor shrank significantly and the originally collapsed lung tissues also expanded. As a result, the proton beam penetrated further into the contralateral (healthy) lung tissue, potentially resulting higher toxicity.

new CT images acquired approximately 3 weeks into the treatment over a 6-week treatment course. As expected, the original highly conformal IMRT plan will be suboptimal when the target volume does not match with the patient's new anatomy.

Similarly, some lung cancers can exhibit substantial anatomy changes as well. Figure 1.3 shows an example of a lung cancer patient who received proton therapy treatment. Due to the significant primary tumor shrinkage, the original proton beam stopped much further into the contralateral lung. Without replanning, the healthy (left) lung could receive higher doses than the original plan indicated. Therefore, it is important to monitor the anatomic changes during the course of treatment for highly conformal treatment.

1.4 Image Registration Methods

Image registration plays a central role in various stages of image-guided adaptive radiotherapy. It is a process of establishing spatial correspondences across two images of the same scene (Maintz and Viergever 1998; Zitova and Flusser 2003). The purpose is to determine a geometric transformation characterizing the relationship of spatial coordinates of corresponding points

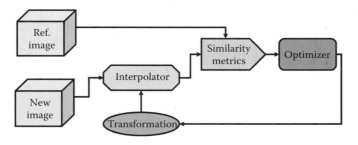

FIGURE 1.4 Generalized image registration process consists of the following components: (1) a pair of images to be registered, (2) a similarity metric to measure the success of registration, (3) an optimization algorithm to drive the direction and the magnitude of transformation, and (4) the transformation and interpolation modules to change one image to match with the other image. The transformation can be either rigid or deformable.

in these two images. The general image registration process can be illustrated in Figure 1.4. In this process, the new image will be transformed to the reference image space iteratively by an optimization process. The success of the registration will be measured by a similarity metric. According to the nature of the geometric transformation (Maintz and Viergever 1998), the image registration methods are normally categorized into rigid image registration methods and deformable image registration methods.

1.5 Rigid Image Registration

Rigid registration between two images only allows rotations and translations. Rigid transformation is a special case of a more general transformation, that is, global or affine transformation. Affine transformation is composed of rotations, translations, scaling, and shearing.

Various rigid image registration algorithms have been developed in the literature (Maintz and Viergever 1998; Hawkes 2001; Hill and Batchelor 2001; Hill et al. 2001; Zitova and Flusser 2003), mostly preceding to the intensive development of nonrigid or deformable registration methods. According to the different registration metrics, these methods can be summarized in the following three categories.

1.5.1 Landmark-Based Rigid Registration

An intuitive registration method is to identify selected corresponding point landmarks, or fiducial markers, from both images and then compute the transformation between them. These landmarks could be internal fiducial markers or external skin markers. Ideally, three pairs of corresponding landmarks are sufficient to compute the rigid transformation for two 3D images, providing they are not on a line. The root-mean-square error is usually referred to as the "target registration error" (TRE) (Fitzpatrick et al. 1998). The TRE sometimes can be used as a metric to minimize the actual distance between two sets of points instead of only the distance between centroids of the fiducial markers.

1.5.2 Feature-Based Rigid Registration

More general features other than landmark points can be used for rigid registration. Examples of such features are lines, curves, point clouds, or surfaces (Goshtasby and Stockman 1985; Besl and McKay 1992; Jiang et al. 1992; Zhang 1994; Meyer et al. 1995; Rangarajan et al. 1997). In general, these features can be extracted by a computer automatically or by a human manually. The correspondence of features in both images is usually not assumed (as unknowns). Normally, it requires the registration to be able to tolerate the feature extraction errors to some extent. The registration algorithms are applied directly to these features. Due to unknown correspondences, iterative optimization is usually required to estimate both the feature correspondence and the transformation simultaneously. The most popular algorithm is the iterative closest point (ICP) algorithm (Besl and McKay 1992; Zhang 1994), which is widely applied to register two sets of points or surface. In some applications, it has been reported to produce accurate and robust results (Sun et al. 2005).

1.5.3 Intensity-Based Rigid Registration

Another thread in the development of rigid registration methods is based on image intensity values. The registration metric is computed from the similarity between two images. This method does not require a presegmentation or predelineation of corresponding structures and thus could be fully automated and generally more robust than feature-based registration methods. Some commonly used similarity metrics for registration include the sum of squared differences (SSD) of intensities (Friston et al. 1995; Hajnal et al. 1995), the cross-correlation coefficient (CC) (Lemieux et al. 1994), the ratio image uniformity (also known as AIR registration) (Woods et al. 1992), and the information theory-based metrics, such as joint entropy (Collignon et al. 1995; Studholme et al. 1995), mutual information (MI) (Maes et al. 1997; Viola and Wells 1997), and Kullback–Leibler (KL) distance (Chung et al. 2002). Most of these metrics are directly applied to the entire image, which is often referred to as the global measures. To achieve a more accurate registration, some researchers also proposed their local measurements using the same metrics as the global ones but characterizing a region of interest (ROI) instead of the entire image domain. Typical examples include template matching (Ding et al. 2001) and salient region feature matching (Huang et al. 2004).

1.6 Rigid Image Registration for IGRT

For image-guided setup, the method of intervention is to reposition the patient and align the radiation beam with the target. In this process, a robust rigid image registration method is usually sufficient to guide the alignment process. For high-precision radiotherapy, such as the hypofractionated radiotherapy, it is a crucial procedure to verify and correct setup errors before the high radiation dose is delivered. Rigid image registration methods also depend on the type of images acquired for patient setup. In the following subsections, we will review the applications in radiation therapy for (1) 2D portal image registration, (2) 2D projections to 3D volumetric image registration, and (3) 3D volumetric registration.

1.6.1 2D–2D Registration

The goal of single portal image registration is to find the 2D transformation (i.e., translations and rotations) that can achieve optimal alignment between the approved reference portal image and the daily portal images. Within the last two decades, numerous methods have been implemented, although their success is rather limited. Feature-based methods rely on geometric primitives extracted from images to calculate registration parameters. To achieve this goal, landmarks (Ding et al. 1993; Lam et al. 1993; Mcparland 1993; Michalski et al. 1993) and other primitives such as lines, curves, and structured patterns (Balter et al. 1992; van Herk and Gilhuijs 1994; Leszczynski et al. 1995, 1998; Mcparland and Kumaradas 1995; Cai et al. 1996, 1998; Kreuder et al. 1998; Pizer et al. 1999; Petrascu et al. 2000; Tang et al. 2000; Matsopoulos et al. 2004) were generally used in the literature. These geometric primitives can either be annotated manually or be defined automatically using feature extraction algorithms. Feature-based approach could be very reliable, as long as the geometric primitives could be accurately defined and associated across the two images. Unfortunately, this is proven to be a challenging problem. In intensity-based approach, the pixel intensities are directly used to find optimal transformation that maximizes a certain similarity measure. Although it is straightforward and fully automatic, the result highly relies on the extent of intensity variations across the two images. Researchers have investigated various similarity functions (Dekker et al. 2003) to address specific problems, such as correlation-based approach (Jones and Boyer 1991; Moseley and Munro 1994; Dong and Boyer 1995; Hristov and Fallone 1996; Dong 1998), moment-based approach (Leszczynski et al. 1993; Dong and Boyer 1996; Shu et al. 2000), and information theoretic-based approach (Alvarez and Sanchiz 2005; Alvarez et al. 2005). Although 2D portal image registration has been used in the clinic, there exist intrinsic drawbacks. First, the portal image quality is very poor, which limits the accuracy of intensity-based automatic registration method. With the new kilovoltage x-ray-based 2D method, this problem is somewhat reduced; however, limited dynamic range in kilovoltage 2D images still presents a challenge for accurate 2D–2D matching. Second, rotations that are out of plane will introduce deformation in the perspective projection, which decreases the robustness of registration.

1.6.2 2D–3D Registration

2D–3D image registration is performed between 3D CT (usually the simulation CT) and daily 2D x-ray projection images.

In contrast to 2D–2D registration, 2D–3D registration calculates the 3D pose (i.e., translations and rotations) of the patient. If the feature correspondences could be accurately established across 3D CT volume and 2D projection x-ray images, the registration can be readily computed using standard stereovision algorithms. However, automatic and accurate feature detection and matching remain an unsolved problem in computer vision. It usually requires implanted fiducial markers within target volumes (Shirato et al. 2004; Nelson et al. 2008; Budiharto et al. 2009). As a result, intensity-based or hybrid approach is still preferred. By ray tracing, simulated radiographs from 3D CT images can be generated to directly compare with the acquired projection x-rays. These simulated x-ray projections are also called the digitally reconstructed radiographs (DRRs). Intensity-based registration can be designed to find the optimal 3D patient pose, at which a simulated DRR can match closely with the projection x-rays. This will lead to an iterative automatic 2D–3D registration algorithm. Numerous applications have been implemented using this method (Bansal et al. 1998, 1999, 2003; Penney et al. 1998; Hill et al. 2001; Birkfellner et al. 2003; Clippe et al. 2003; Muller et al. 2004; Russakoff et al. 2005; Jans et al. 2006; Khamene et al. 2006; Munbodh et al. 2006; Kuenzler et al. 2007; Song et al. 2008; J. Wu et al. 2009). In recent years, graphic processing unit (GPU) has drawn enormous attention to radiation therapy applications. DRRs can be computed in near real time for 2D–3D registration (Khamene et al. 2006; Spoerk et al. 2007).

1.6.3 3D–3D Volumetric Registration

With the advent of in-room volumetric imaging techniques, volumetric (3D) images of the patient can be acquired in the treatment room immediately before the radiation treatment on a daily/weekly basis. These volumetric image data provide explicit information about the patient anatomy in high resolution. The 3D registration is usually performed between daily acquired CBCT images and the reference CT. Indeed, the 3D approach is expected to be more accurate and flexible compared with the previous approaches we have discussed, because it uses a much more accurate representation of the patient's anatomy. Several studies have been carried out to perform the automatic 3D–3D registration for patient setup and to reduce the patient setup errors (Court and Dong 2003; Smitsmans et al. 2005; Zhou et al. 2010).

In general, rigid registration is a global compensation to the setup errors. It should use as much image information as possible. However, clinical studies have demonstrated that the patient's anatomy cannot be assumed a rigid body. Due to the nonrigid changes, such as arm position or neck curvature variations, different regions may not move the same way as the other parts of the body. Different ROIs will have different setup errors, which cannot be compensated by a single global shift and/or rotation (Zhang et al. 2006; Van Karanen et al. 2009). Therefore, for clinical applications, the selection of ROI for 3D–3D rigid registration becomes an important strategy to minimize registration error.

1.7 Deformable Image Registration

Today, the most exciting and challenging research on image registration involves the development of deformable or nonrigid registration algorithms (Goshtasby et al. 2003; Crum et al. 2004; Sarrut 2006; Kaus and Brock 2007). Unlike the rigid or affine transformation, the nonrigid transformation is normally a local free-form mapping. Due to the degeneracy of deformation, it is impossible to perform realistic transformation without proper regularization—the rule of deformation. These regularizations are normally based on the physical properties of the object for which the deformation will occur. Different regularizations will result in different degrees of freedom for the deformation, which can span from a very stiff form (near rigid) to a completely free form. Some commonly used regularization models are described below.

1.7.1 Regularization on Deformable Transformation

An intuitive extension from a global registration to a local one is a piecewise affine registration by applying the linear registration to local image patches and combining the final local affine transformation to an overall deformable transformation (Goshtasby 1987; Flusser 1992). Radial basis functions (RBFs) are a group of global mapping functions that can handle local geometric distortions. The most typical RBFs are the thin-plate splines (TPSs). TPS has been widely applied to image registration (Bookstein 1989; Wahba 1990; Yang et al. 2006). TPS minimizes a spatial integral of the square of the second-order derivatives of the assumed transformation. The optimal solution to TPS is in a closed form, which is generally desirable for fast implementation. In general, a parameter is used to control the rigidity of the TPS (RBF as well), so that the deformation could be adaptive to the different applications. A special case is the registration based on free-form deformations (FFDs) (Rueckert et al. 1999), where the local deformation is modeled by B-spline-based FFD and the smoothness of overall transformation is regularized by TPS. Another group of registration approaches minimizes an energy function defined by a physical model. The most typical one is the elastic image registration (Bajcsy and Kovacic 1989; Shen and Davatzikos 2002). This method achieves the registration by minimizing an energy function that is defined by an external force stretching the image for matching and an internal force restricting the deformation from being unrealistic. Elastic registration is generally localized not allowing big deformations. Fluid registration, or diffeomorphism registration, attempts to solve this problem by modeling the reference image as thick fluid that flows out to match with the other image (Christensen et al. 1996; Morten and Claus 1996; Vercauteren et al. 2009). Based on the different physical models, there are other methods including diffusion-based registration or demons methods (Thirion 1998; Wang et al. 2005a,b), level set-based registration (Vemuri et al. 2003), and optical flow-based registration (DeCarlo and Metaxas 2000).

1.7.2 Registration Metrics

According to registration metrics, deformable image registration methods can be classified into three categories: the feature-based methods, the intensity-based methods, and the hybrid methods that combine the previous two. The feature-based methods sometimes are named as point-based methods or model-based methods by some researchers (Lu et al. 2006a,b; Jaffray et al. 2008). The traditional feature-based methods (Bookstein 1989; Fornefett et al. 1999; Can et al. 2002; Chui and Rangarajan 2003; Brock et al. 2005; Xie et al. 2009) use sparse features extracted from images, such as points, curves, and surface patches. The registration task is to find correspondence and to compute an optimal transformation. The transformation is normally described by the RBFs with the features as control points. These methods are relatively fast but require robust feature extraction, accurate feature correspondence, and sometimes user interaction. The standard intensity-based methods (Thirion 1998; Cachier et al. 1999; Coselmon et al. 2004; Lu et al. 2004; Foskey et al. 2005; Guetter et al. 2005; Wang et al. 2005b) operate directly on the intensity values of the full image content without prior feature extraction. These algorithms always attempt to find a smooth transformation that maximizes intensity-based similarity measure. The transformation is normally described by a physical model. These methods can act without user interaction, but they involve high computational cost and require preregistration to bring the source images close enough for optimal registration. Recent research on nonrigid registration tends to develop sophisticated hybrid methods that integrate the merits of both feature-based and intensity-based methods (Hartkens et al. 2002; Shen and Davatzikos 2002; Cachier et al. 2003; Hellier and Barillot 2003; Yang et al. 2006; Han et al. 2008).

1.7.3 Techniques to Improve Registration

Varied techniques have been proposed to improve the performance of deformable image registration in terms of speed, accuracy, and robustness. The most common and important techniques include the multiresolution/multiscale method, the relaxation method, and the inversed consistency method, as described below.

1.7.3.1 Multiresolution Method

The multiresolution/multiscale method has been widely used in image registration (Bajcsy and Kovacic 1989; Shen and Davatzikos 2002; Wang et al. 2005b). The basic idea of multiresolution method is to obtain an approximate transformation in a lower resolution (coarser scale) and then use this approximated transformation as the guidance for registration in the higher resolution (finer scale). The approximated transformation is then refined to a more accurate level later. The lower-resolution image is normally obtained via down-sampling of the original source image. The number of scale levels is determined by the input image size and applications. Some benefits one may expect from multiresolution include (1) faster convergence, (2) improved robustness by overcoming the local minimum problem during optimization, and (3) the ability to achieve the registration for relatively larger deformations.

1.7.3.2 Relaxation Method

The multiresolution method is applying a hierarchical registration on the image; similarly, the hierarchical method can be applied to extract feature points for registration, that is, the relaxation method (Hummel and Zucker 1983; Price 1985; Rangarajan et al. 1997). The traditional method assumes one-to-one correspondence in searching for corresponding feature points during optimization. However, the relaxation method relaxes the constraint by allowing a fuzzy correspondence, that is, each feature points in one image can be mapped to the feature points in the other image with a probability value between 0 and 1. When the registration algorithm finally converges, the one-to-one correspondence will be ensured by some additional techniques, such as the deterministic annealing (Yuille and Kosowsky 1994). This method has been widely applied to shape and image registration. Promising results have been reported (Chui and Rangarajan 2003; Yang et al. 2006; Zheng and Doermann 2006; Osorio et al. 2009; Shen 2009).

1.7.3.3 Inverse Consistency

Due to the complexity of deformable transformation and the complicated optimization solver, deformable image registration is generally asymmetric. The deformation field obtained by registering image A to image B is not equal to the inverse of deformation field by registering image B to image A. This discrepancy indicates the inconsistency in registration, which is undesirable. To minimize this problem, Christensen and Johnson (2001) first proposed the inversed consistency by formulating the registration in both directions into an overall energy function in the optimization. This ensures the consistency of registration in two directions. Later, different inversed consistency techniques were proposed and improvements were reported (Shen and Davatzikos 2002; Wang et al. 2005b; Shen 2009).

1.8 Deformable Image Registration for Morphology-Based Adaptive Radiotherapy

Deformable image registration is playing a central role in modern radiation therapy (Kessler 2006; Lu et al. 2006; Sarrut 2006; Xing et al. 2006; Kaus and Brock 2007). Deformable image registration is essential to link the anatomy at one time to another while maintaining the desirable one-to-one geographic mapping. In addition, deformable image registration can be used to map secondary images or treatment parameters. We will discuss some of the applications of deformable registration in adaptive radiotherapy settings.

1.8.1 Autosegmentation for Adaptive Planning

When the patient's anatomy is changed and a new adaptive plan is to be developed, the biggest bottleneck is to redefine all contours and target volumes for IMRT planning. Typically, it will take a significant amount of time (from a few hours to a few days) for a radiation oncologist to draw contours from scratch. When the patient's anatomy is changing on a daily basis, it would be impractical to contour the target on a daily basis. Fortunately, deformable image registration provides such a convenience that the patient's anatomy (target volumes and normal structures) can be automatically segmented using the original treatment planning CT and labeled contours as a reference atlas. In this approach, deformable image registration can be performed first between the original planning CT and the new in-room CBCT or the resimulation CT. The same deformable transformation matrix, which maps the original CT to the new CT, can be used to map the structure contours to the new CT, achieving autosegmentation on the new CT images for adaptive planning. Automatic contouring is the most common application of deformable image registration for image-guided adaptive radiotherapy (Gao et al. 2006; Lu et al. 2006; Han et al. 2008; Wang et al. 2008a; Reed et al. 2009). The automatically delineated contours include both the treatment targets and the normal organs, which save the time for physicians to perform the cumbersome contouring. In addition, the geometrically mapped contours using a well-defined deformable image registration method may be more objective by eliminating the interobserver variations.

An example of such application for head-and-neck cancer adaptive radiotherapy is shown in Figure 1.5. In this case, the original treatment plan includes variable target volumes at different risk levels (each target volume receives different radiation

Planning CT

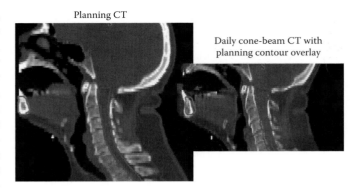

Daily cone-beam CT with planning contour overlay

FIGURE 1.6   (See color insert.) Nonrigid variation of the patient's anatomy cannot be easily corrected by a simple couch shift even when IGRT is used. Neck curvature and chin positional variations are considered nonrigid changes. If the changes are systematic, a replanning may be the best strategy to correct such complicated shape variations.

dose). The primary gross tumor volume (GTV) is shown in purple, the high-risk clinical target volume (CTV) is shown in red, the intermediate-risk CTV is shown in blue, and the low-risk CTV is shown in yellow. The normal organs, which include the parotid glands (left and right) and the spinal cord (red), are also shown. After performing a deformable image registration between the new CT and the original planning CT, all of these contours can be deformably mapped to the new CT for adaptive planning. Typically, 20 to 30 contoured structures are used for head-and-neck IMRT planning. Autosegmentation by deformable image registration saves substantial time in this adaptive radiotherapy process. It can be argued that deformable image registration is a key enabling technology for adaptive radiotherapy.

It should be mentioned that tumor shrinkage is not the only reason for adaptive radiotherapy. Any nonrigid deviations from the original plan can be considered as candidates for adaptive radiotherapy. One example is shown in Figure 1.6, in which the neck curvature change cannot be easily corrected by image-guided setup. If such variations are consistent (which means that they become systematic errors), replanning is perhaps a better corrective approach than daily adjustment of the patient's head rest and chin positions. Therefore, adaptive radiotherapy can be used for correcting nonrigid morphologic changes in the patient's anatomy.

1.8.2 Dose Accumulation

Treatment planning is a prediction process. The dose distribution shown on the simulation CT is the planned dose distribution, which may not be equal to the actual delivered dose distribution. Due to the anatomic changes during the course of treatment, as described previously, the actual dose distribution may be quite different from the planned dose distribution. Figure 1.7 shows an example comparing the planned versus the delivered dose distributions for a head-and-neck case. If daily CT/CBCT images are available to calculate the daily dose

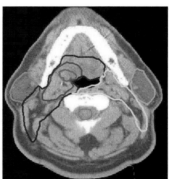

Original planning CT image and contours   Automatically segmented contours on a new CT

FIGURE 1.5   (See color insert.) Example of autosegmentation by deformable image registration is shown for a head-and-neck cancer patient. Left: CT slice with labeled structures. The target volumes include the primary GTV (purple), the high-risk CTV (red), the intermediate-risk target volume (blue), and the low-risk target volume (yellow). The parotid glands (left and right) are also shown in blue and green contours. After deformable image registration between the new CT after 3 weeks of treatment and the original planning CT, the same contours are deformably mapped to the new CT for adaptive planning.

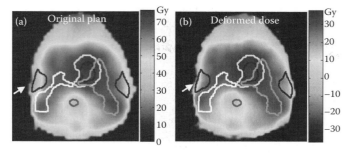

FIGURE 1.7 Example illustrating the difference between (a) the planned dose distribution and (b) the actually realized dose distribution after the entire course of treatment is delivered. Image (b) is the cumulative dose distribution after mapping and adding each daily dose distribution from each daily CT image to the planning CT using deformable image registration method.

distribution, deformable image registration can be used to map the daily delivered dose distribution back to the original planning CT. After adding each treatment fraction voxel-by-voxel, the actually delivered dose distribution can be compared against the planned dose distribution. In this case, due to a combination of setup error (the isocenter was slightly shifted toward the patient's right) and anatomy shrinkage, the right parotid received much higher dose than the original plan. Such information is very valuable to calculate the actual dose response for a critical organ, such as the parotid. It is believed that the actual dose tolerance for the parotid is higher than the published literature, because the parotid volume used for dose response calculation was collected before the radiation treatment (Eisbruch et al. 1999, 2001). The process to use deformable image registration to transfer daily doses into a reference framework was described by Lu et al. (2006), and a study to compare the delivered dose distribution and the planned dose distribution for parotid was described by O'Daniel et al. (2007).

1.8.3 Contour Propagation in 4D CT Applications

4D CT is a series of 3D CT images acquired at different breathing phases. 4D CT is commonly used in lung cancer patients to model organ motion. Similar to the automatic contouring in daily CT images, one can apply deformable image registration to 4D CT and propagates contours delineated on one phase to the other phases (Chang et al. 2008; Wijesooriya et al. 2008). This can dramatically save the time for contouring because 4D CT has a large amount of data, which would take a long time to contour all the phases manually. Another novel application is to improve the image quality of 4D CT in liver cancer patients. By coregistering images from the difference phases and deformably mapping them to a reference phase, one can enhance the quality of 4D CT, which proves to be much more effective than using standard image enhancement filters to reduce noise (Wang et al. 2008b).

1.9 Image Registration in Physiology-Based Adaptive Radiotherapy and Quantitative Outcome Assessment

Metabolic and functional imaging modalities, such as positron emission tomography (PET), dynamic contrast-enhanced magnetic resonance imaging, and diffusion-weighted magnetic resonance imaging, have become increasingly popular in radiation therapy as biomarkers for assessing the treatment response (Graves et al. 2001; Ciernik et al. 2003; Kessler 2006; Steenbakkers et al. 2006). As a result, the fusion of morphologic images with functional images is a prerequisite to integrate functional information into a treatment planning process. Sometimes, rigid fusion may achieve satisfactory results (perhaps, in part, due to the poor spatial resolution in some of these functional images), although deformable fusion remains a challenging and perhaps unsolved problem.

Many previous studies have shown that functional imaging, such as fluorodeoxyglucose (FDG)-PET, provides additional information over the gross anatomic imaging. Voxel-based prescriptions of deliberately nonuniform dose distributions based on functional imaging (the so-called "dose painting") require a particular transformation that maps the image intensities to prescribed doses. However, the functional form of this transformation is currently unknown (Bowen et al. 2009). Nevertheless, nonuniform dose distributions can be created to boost tumor voxels with higher tumor concentration.

An even more appealing use of FDG-PET for "dose painting" can be integrated in the adaptive radiotherapy procedure. Using the imaged residual metabolically active tumor after a part of the treatment course has been delivered has been suggested by the Brussels group (van Baardwijk et al. 2006; Geets et al. 2007; Gregoire et al. 2007; Gregoire and Haustermans 2009). The residual activity may represent parts of the tumor that are more radioresistant than the parts that ceased to be PET avid after some radiation had been delivered. If FDG-PET imaging during therapy indeed represents a resistant subvolume, the FDG-PET avid tumor can be used as target for dose escalation, which theoretically should improve the effectiveness of therapy. Similar to FDG-PET, the tumor subvolumes defined by hypoxia imaging have been proposed as targets for integrated boost as "hypoxia-guided radiotherapy" (Thorwarth et al. 2007; Lin et al. 2008; Sovik et al. 2007, 2009; Bentzen and Gregoire 2011). Compared with FDG-PET, which is spatially stable within the tumor before and throughout radiotherapy, hypoxia imaging shows significant spatial and temporal variability due to reoxygenation and changes in the tumor microenvironment (Lin et al. 2008). The changes in spatial distribution of tumor hypoxia compromised the coverage of hypoxic tumor volumes achievable by dose-painting IMRT. In general, the existing functional imaging is still premature and requires a significant validation before using it for adaptive dose painting.

A realistic application of deformable image registration is to quantify the changes in tumor cell activity. Figure 1.8 shows an

Pre-RT PET/CT   2 months post-RT

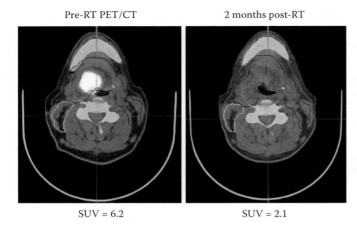

SUV = 6.2   SUV = 2.1

FIGURE 1.8 (See color insert.) Deformable image registration in quantitative assessment of functional outcome. PET/CT images acquired before and 2 months after radiation therapy are shown on the left and right, respectively. To evaluate the changes in the treated target volume, CTVs (shown in colored contours) were mapped to each PET/CT from the planning CT. A reduction of the mean SUV within the GTV (red) was measured.

example of using deformable image registration to map the GTV to each of the preradiation and postradiation therapy PET/CT scans. Taking advantage of deformable image registration, the same ROI can be used to measure the reduction of the standardized uptake value (SUV) in FDG-PET images. In this example, the reduction in the mean SUV inside the GTV was close to 30%.

1.10 Deformable Image Registration for Automatic Treatment Planning

Adaptive radiotherapy usually requires a rapid replanning process. Because the patient's anatomy will continue to change during the course of treatment, any delay in the new (adaptive) plan based on images acquired a few days ago could introduce additional uncertainties. The autosegmentation of target volumes and normal structures can speed up the replanning process; however, IMRT planning itself could be a bottleneck for adaptive radiotherapy.

Various strategies have been employed for replanning. The simplest approach is to keep the same dose-volume histogram (DVH) constraints as in the original plan (Q. Wu et al. 2009). However, due to volume changes, these original DVH constraints may not be achievable or too generous. For example, with a shrinking parotid gland, it is much more difficult to achieve the same DVH constraint than with a large parotid before radiation therapy.

Some investigators studied to use deformable image registration directly mapping and adapting radiation beams to the new CT. A method was proposed by Mohan et al. (2005), who first used deformable image registration method to autodelineate target volumes (prostate and seminal vesicles) and critical structures (rectum and bladder). Then, the IMRT fluence patterns in the original IMRT plan can be deformed for each beam angle based on the projected new anatomy in the beam's eye view, as illustrated in Figure 1.9. They demonstrated that the resulting treatment plan based on deformed IMRT fluence maps is a

(a) Segmented BEV based on pre-tx CT   (b) Segmented BEV based on fx 7 CT

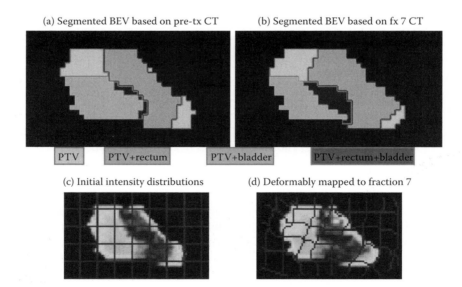

PTV   PTV+rectum   PTV+bladder   PTV+rectum+bladder

(c) Initial intensity distributions   (d) Deformably mapped to fraction 7

FIGURE 1.9 One way to achieve plan adaptation is to deform the IMRT fluence distributions in the beam's eye view (BEV). (a, b) BEV apertures for each beam are segmented into regions of overlap of the PTV with normal critical structures. Then, intensity distributions within each segment from (a) the original pretreatment IMRT plan are mapped to (d) the corresponding segment within the current treatment's BEV. (Reprinted from *International Journal of Radiation Oncology Biology Physics*, 61(4), Mohan, R. et al., Use of deformed intensity distributions for on-line modification of image-guided IMRT to account for interfractional anatomic changes. 1258–1266, Copyright 2005, with permission from Elsevier.)

good approximation to the complete replanning from scratch. The method may be a rapid way to produce new treatment plans online in near real time for adaptive radiotherapy using daily CT images.

Other researchers also developed reoptimization method for online adaptive radiotherapy. Feng et al. (2006) also used deformable image registration method to derive the 3D geometric transformation matrix. Rather than replanning or shifting the patient, they used the deformation matrix to morph the treatment apertures as a potential online correction method. Wu et al. (2008) used the deformable image registration method to map the original dose distribution as the new objective function for optimization. Fluence maps were reoptimized via linear programming. Mestrovic et al. (2007) developed a direct aperture optimization method for online adaptive radiotherapy. They found that the average time needed to adapt the original plan to arrive at a clinically acceptable plan was roughly half of the time needed for complete plan regeneration, although they had some trouble with extremely deformed anatomy. Recently, Ahunbay et al. published a series of papers using a two-step optimization approach. The first step is to use segment aperture morphing by applying the spatial relationship between the apertures and the contours of the planning target and OARs to the new anatomy. The second step uses the segment weight optimization to optimize the plan and account for the interfraction variations of the patient's anatomy (Ahunbay et al. 2008, 2009, 2010).

Currently, online adaptive radiotherapy is still a challenge due to the slowness of deformable image registration, dose calculation, and plan optimization. The use of parallel computing resources, such as the GPU, showed promise to achieve very fast adaptive planning (Gu et al. 2009, 2010; Men et al. 2009, 2010). It was reported that a new IMRT plan could be completed within 3 s (Men et al. 2009). GPU-based deformable image registration takes only 11 s (Gu et al. 2010).

1.11 Challenges in Deformable Image Registration

Although deformable image registration has been used in many clinical applications, it is very challenging to measure the accuracy of the algorithm due to the lack of ground truth in most clinical applications. Deformable image registration has been validated in controlled studies for various situations (Wang et al. 2005b; Kashani et al. 2007; Brock 2010). Typically, the accuracy is on the order of the size of voxel; however, substantial errors, as much as 6 to 10 mm (2 to 5 voxels), were found in some algorithms.

There are many reasons for such large registration errors. For radiation therapy applications, there are three major factors that may impact the accuracy of deformable image registration algorithms: intensity inconsistency between two images of the same object, motion discontinuity, and correspondence ambiguity. CT-to-CBCT registration is a good example of image intensity inconsistency. Due to the scatter and beam hardening effect, CT numbers in CBCT are not well calibrated. As a result, the same object may have different intensity in CBCT and conventional CT. Algorithms based on image intensity value would not work well if image intensity inconsistencies exist between the two images.

Most of existing deformable registration algorithms explicitly or implicitly assumed motion continuity. Deformable registration is an intrinsic ill-posed problem (Tikhonov and Arsenin 1977), where multiple solutions exist due to underconstrained systems: the number of constraints is less than the number of unknowns. The role of the regularization term is to enforce a physical model for the displacement field. Most of existing approaches take advantage of generic priors, such as continuous, differentiable, diffeomorphic, and incompressible constraints. A known problem is the motion discontinuity at the interface of the chest wall and a sliding lung tumor. In this scenario, the homogenous regularization/filter approach tends to result in severe mapping error in the area near the chest wall. An example is shown in Figure 1.4. In this case, the tumor slides against the chest wall with an amplitude as large as 3 cm. The motion field, shown as the red vector field in Figure 1.4, was incorrect due to the isotropic smoothing of motion vectors across the boundary of the lung tissue and the chest wall, leading to unrealistic warping of the tumor in the deformed image. Ideally, the regulator should be anisotropic and adaptive to the stiffness of the tissue type (Staring et al. 2007).

An implicit assumption in any registration algorithm is the existence of one-to-one physical correspondence between the

| Inspiration phase | Expiration phase | Deformed expiration phase |

FIGURE 1.10 (See color insert.) Illustration of deformable registration error caused by isotropic smoothing of displacement field. Right: Deformed expiration phase image to match with the inspiration phase image. The shape of the tumor was dragged unrealistically due to the smoothing requirement inside the deformable image registration algorithm.

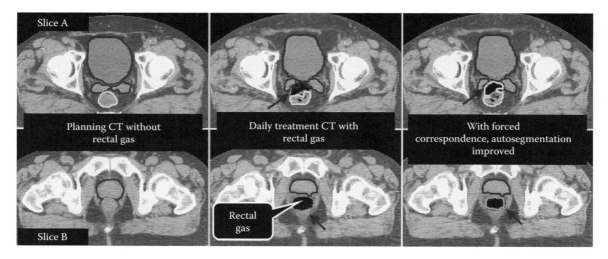

FIGURE 1.11 **(See color insert.)** If the planning CT has an empty rectum (left), registration errors may occur near the area of CT images containing the rectal gas. Incorrect rectal wall is detected by the autosegmentation algorithm (middle column). One method to handle this situation is to artificially modify the planning CT images so that it will contain an air pocket in the center of the contoured (empty) rectum. This will result in a "virtual" correspondence between the gaseous regions. The improvement in autosegmentation can be seen in the CT images on the right column.

voxels in two images. Unfortunately, this assumption is often violated in many clinical situations. For instance, the insertion or removal of brachytherapy applicator will result in images with drastically different image intensities. It will be virtually impossible to register such images because there is no physical correspondence in the voxels of the applicator when the applicator was taken out (of the CT image). Correspondence ambiguity can also happen to prostate cancer patients in which the rectal gas filling may not be consistent in day-to-day treatment. Registration error can occur when an empty rectum is to register the gas-filled rectum, as demonstrated in Figure 1.11. The presence and absence of bowel/rectum gas will strongly affect the accuracy of an image intensity-based deformable image registration algorithm (Gao et al. 2006). In the above scenarios, automatic deformable registration algorithms will create unrealistic deformation, leading to inaccurate autosegmentation, as seen in Figure 1.11 (middle). The solution for this correspondence ambiguity issue is very limited. The true solution will require human input or prior knowledge. Gao et al. (2006) addressed this problem by taking advantage of the planning contours, which indicates the location of the rectum. The authors implemented an image preprocessing approach to synthesize artificial gas patterns within the rectum region in the planning CT, so that the corrected planning image matches better with the daily CT image in presence of rectum gas. Although the virtual gas does not exist in the planning CT, it helped the image intensity-based registration algorithm to better localize the rectal wall. A similar problem was overcome by Foskey et al. (2005) using a variational method to fill in the gas region before registration.

1.12 Summary

Clinical experience has demonstrated that patients who have undergone radiotherapy can exhibit significant anatomic changes due to either physiologic factors or treatment effects. As a result, the initial treatment plan may not reflect the actual delivered dose in patients. To achieve high-precision radiation therapy, these daily variations and time trends should be taken into consideration when designing a treatment strategy. Adaptive radiotherapy is a new image-guided treatment strategy, which corrects for nonrigid changes in the patient's anatomy. It uses advanced image processing techniques, such as deformable image registration, and rapid replanning tool to realize its goal. Physiologic and functional image-guided adaptive radiotherapy is still in its early stage of development. The potential application of such an approach may have a strong clinical impact to radiation therapy. Deformable image registration plays a key role in adaptive radiotherapy. More accurate and robust algorithms are still needed, in particular, for multimodality image registration.

References

Ahunbay, E. E., C. Peng et al. (2008). An on-line replanning scheme for interfractional variations. *Medical Physics, 35*(8), 3607–3615.

Ahunbay, E. E., C. Peng et al. (2009). An on-line replanning method for head and neck adaptive radiotherapy. *Medical Physics, 36*(10), 4776–4790.

Ahunbay, E. E., C. Peng et al. (2010). Online adaptive replanning method for prostate radiotherapy. *International Journal of Radiation Oncology Biology Physics, 77*(5), 1561–1572.

Alvarez, N. A. and J. M. Sanchiz (2005). Image registration from mutual information of edge correspondences. *Progress in Pattern Recognition, Image Analysis and Applications, Proceedings, 3773*, 528–539.

Alvarez, N. A., J. M. Sanchiz et al. (2005). Contour-based image registration using mutual information. *Pattern Recognition and Image Analysis, Part 1, Proceedings, 3522*, 227–234.

Apisarnthanarax, S. and K. S. Chao (2005). Current imaging paradigms in radiation oncology. *Radiation Research, 163*(1), 1–25.

Bajcsy, R. and S. Kovacic (1989). Multiresolution elastic matching. *Computer Vision Graphics and Image Processing, 46*, 1–21.

Balter, J. M., C. A. Pelizzari et al. (1992). Correlation of projection radiographs in radiation-therapy using open curve segments and points. *Medical Physics, 19*(2), 329–334.

Bansal, R., L. H. Staib et al. (1998). A novel approach for the registration of 2D portal and 3D CT images for treatment setup verification in radiotherapy. *Medical Image Computing and Computer-Assisted Intervention, 1496*, 1075–1086.

Bansal, R., L. H. Staib et al. (1999). Entropy-based, multiple-portal-to-3DCT registration for prostate radiotherapy using iteratively estimated segmentation. *Medical Image Computing and Computer-Assisted Intervention, 1679*, 567–578.

Bansal, R., L. H. Staib et al. (2003). Entropy-based dual-portal-to-3-DCT registration incorporating pixel correlation. *IEEE Transactions on Medical Imaging, 22*(1), 29–49.

Bentzen, S. M. and V. Gregoire (2011). Molecular imaging-based dose painting: A novel paradigm for radiation therapy prescription. *Seminars in Radiation Oncology, 21*(2), 101–110.

Bert, C., K. G. Metheany et al. (2006). Clinical experience with a 3D surface patient setup system for alignment of partial-breast irradiation patients. *International Journal of Radiation Oncology Biology Physics, 64*(4), 1265–1274.

Besl, P. J. and N. D. McKay (1992). A method for registration of 3-D shapes. *IEEE Transactions on Pattern Analysis and Machine Intelligence, 14*(2), 239–256.

Birkfellner, W., J. Wirth et al. (2003). A faster method for 3D/2D medical image registration—A simulation study. *Physics in Medicine and Biology, 48*(16), 2665–2679.

Bissonnette, J. P., T. G. Purdie et al. (2009). Cone-beam computed tomographic image guidance for lung cancer radiation therapy. *International Journal of Radiation Oncology Biology Physics, 73*(3), 927–934.

Bookstein, F. L. (1989). Principal warps: Thin-plate splines and the decomposition of deformations. *IEEE Transactions on Pattern Analysis and Machine Intelligence, 11*(6), 567–585.

Bortfeld, T. (1999). Optimized planning using physical objectives and constraints. *Seminars in Radiation Oncology, 9*(1), 20–34.

Bowen, S. R., R. T. Flynn et al. (2009). On the sensitivity of IMRT dose optimization to the mathematical form of a biological imaging-based prescription function. *Physics in Medicine and Biology, 54*(6), 1483–1501.

Brock, K. K. (2010). Results of a multi-institution deformable registration accuracy study (MIDRAS). *International Journal of Radiation Oncology Biology Physics, 76*(2), 583–596.

Brock, K. K., M. B. Sharpe et al. (2005). Accuracy of finite element model-based multi-organ deformable image registration. *Medical Physics, 32*(6), 1647–1659.

Budiharto, T., P. Slagmolen et al. (2009). A semi-automated 2D/3D marker-based registration algorithm modelling prostate shrinkage during radiotherapy for prostate cancer. *Radiotherapy and Oncology, 90*(3), 331–336.

Cachier, P., X. Pennec et al. (1999). Fast non-rigid matching by gradient descent: Study and improvements of the demons algorithm. *INRIA*.

Cachier, P., E. Bardinet et al. (2003). Iconic feature based nonrigid registration: The PASHA algorithm. *Computer Vision and Image Understanding, 89*(2–3), 272–298.

Cai, J. L., S. Q. Zhou et al. (1996). Alignment of multi-segmented anatomical features from radiation therapy images by using least square fitting. *Medical Physics, 23*(12), 2069–2075.

Cai, J. L., J. C. H. Chu et al. (1998). A simple algorithm for planar image registration in radiation therapy. *Medical Physics, 25*(6), 824–829.

Can, A., C. V. Stewart et al. (2002). A feature-based, robust, hierarchical algorithm for registering pairs of images of the curved human retina. *IEEE Transactions on Pattern Analysis and Machine Intelligence, 24*(3), 347–364.

Chang, J. Y., L. Dong et al. (2008). Image-guided radiation therapy for non-small cell lung cancer. *Journal of Thoracic Oncology, 3*(2), 177–186.

Chang, Z., Z. Wang et al. (2010). 6D image guidance for spinal non-invasive stereotactic body radiation therapy: Comparison between ExacTrac X-ray 6D with kilo-voltage cone-beam CT. *Radiotherapy and Oncology, 95*(1), 116–121.

Christensen, G. E. and H. J. Johnson (2001). Consistent image registration. *IEEE Transactions on Medical Imaging, 20*(7), 568–582.

Christensen, G., R. D. Rabbitt et al. (1996). Deformable templates using large deformation kinematics. *IEEE Transactions on Medical Imaging, 5*(10), 1435–1447.

Chui, H. and A. Rangarajan (2003). A new point matching algorithm for non-rigid registration. *Computer Vision and Image Understanding, 89*(2–3), 114–141.

Chung, A. C., W. M. Wells III, A. Norbash and W. E. L. Grimson (2002). Multimodal image registration by minimising kullback-leibler distance. In *Medical Image Computing and Computer-Assisted Intervention—MICCAI 2002*, Springer Berlin Heidelberg, 525–532.

Ciernik, I. F., E. DizenDorf et al. (2003). Radiation treatment planning with an integrated positron emission and computer tomography (PET/CT): A feasibility study. *International Journal of Radiation Oncology Biology Physics, 57*(3), 853–863.

Clippe, S., D. Sarrut et al. (2003). Patient setup error measurement using 3D intensity-based image registration techniques. *International Journal of Radiation Oncology Biology Physics, 56*(1), 259–265.

Collignon, A., F. Maes et al. Automated multimodality image registration using information theory, in Information Processing in Medical Imaging. Kluwer, 1995.

Coselmon, M. M., J. M. Balter et al. (2004). Mutual information based CT registration of the lung at exhale and inhale breathing states using thin-plate splines. *Medical Physics, 31*(11), 2942–2948.

Court, L. E. and L. Dong (2003). Automatic registration of the prostate for computed-tomography-guided radiotherapy. *Medical Physics, 30*(10), 2750–2757.

Crum, W. R., T. Hartkens et al. (2004). Non-rigid image registration: Theory and practice. *British Journal of Radiology, 77*(2), S140–S153.

DeCarlo, D. and D. Metaxas (2000). Optical flow constraints on deformable models with applications to face tracking. *International Journal of Computer Vision, 38*(2), 99–127.

Dekker, N., L. S. Ploeger et al. (2003). Evaluation of cost functions for gray value matching of two-dimensional images in radiotherapy. *Medical Physics, 30*(5), 778–784.

Ding, G. X., S. Shalev et al. (1993). A p-theta technique for treatment verification in radiotherapy and its clinical-applications. *Medical Physics, 20*(4), 1135–1143.

Ding, L., A. Goshtasby et al. (2001). Volume image registration by template matching. *Image and Vision Computing, 19*(12), 821–832.

Ding, G. X., D. M. Duggan et al. (2007). A study on adaptive IMRT treatment planning using kV cone-beam CT. *Radiotherapy and Oncology, 85*(1), 116–125.

Dong, L. (1998). Portal image correlation and analysis. *Imaging in Radiation Therapy* (24), 415–444.

Dong, L. and A. L. Boyer (1995). An image correlation procedure for digitally reconstructed radiographs and electronic portal images. *International Journal of Radiation Oncology Biology Physics, 33*(5), 1053–1060.

Dong, L. and A. L. Boyer (1996). A portal image alignment and patient setup verification procedure using moments and correlation techniques. *Physics in Medicine and Biology, 41*(4), 697–723.

Eisbruch, A., R. K. Ten Haken et al. (1999). Dose, volume, and function relationships in parotid salivary glands following conformal and intensity-modulated irradiation of head and neck cancer. *International Journal of Radiation Oncology Biology Physics, 45*(3), 577–587.

Eisbruch, A., J. A. Ship et al. (2001). Partial irradiation of the parotid gland. *Seminars in Radiation Oncology, 11*(3), 234–239.

Feng, Y., C. Castro-Pareja et al. (2006). Direct aperture deformation: An interfraction image guidance strategy. *Medical Physics, 33*(12), 4490–4498.

Fitzpatrick, J. M., J. B. West et al. (1998). Predicting error in rigid-body point-based registration. *IEEE Transactions on Medical Imaging, 17*(5), 694–702.

Flusser, J. (1992). An adaptive method for image registration. *Pattern Recognition, 25*, 45–54.

Fornefett, M., K. Rohr et al. (1999). Elastic registration of medical images using radial basis functions with compact support. *Proceedings of Computer Vision and Pattern Recognition.*

Foskey, M., B. Davis et al. (2005). Large deformation three-dimensional image registration in image-guided radiation therapy. *Physics in Medicine and Biology, 50*(24), 5869–5892.

Friston, K. J., J. Ashburner et al. (1995). Spatial registration and normalization of images. *Human Brain Mapping, 2*, 165–189.

Fuss, M., B. J. Salter et al. (2004). Daily ultrasound-based image-guided targeting for radiotherapy of upper abdominal malignancies. *International Journal of Radiation Oncology Biology Physics, 59*(4), 1245–1256.

Fuss, M., J. Boda-Heggemann et al. (2007). Image-guidance for stereotactic body radiation therapy. *Medical Dosimetry, 32*(2), 102–110.

Gao, S., L. Zhang et al. (2006). A deformable image registration method to handle distended rectums in prostate cancer radiotherapy. *Medical Physics, 33*(9), 3304–3312.

Geets, X., M. Tomsej et al. (2007). Adaptive biological image-guided IMRT with anatomic and functional imaging in pharyngo-laryngeal tumors: Impact on target volume delineation and dose distribution using helical tomotherapy. *Radiotherapy and Oncology, 85*(1), 105–115.

Goshtasby, A. (1987). Piecewise cubic mapping functions for image registration. *Pattern Recognition, 20*, 525–533.

Goshtasby, A. and G. Stockman (1985). Point pattern matching using convex hull edges. *IEEE Transactions on Systems, Man, and Cybernetics, 15*(5), 631–637.

Goshtasby, A., L. Staib et al. (2003). Nonrigid image registration: Guest editors' introduction. *Computer Vision and Image Understanding, 89*(2–3), 109–113.

Graves, E. E., A. Pirzkall et al. (2001). Registration of magnetic resonance spectroscopic imaging to computed tomography for radiotherapy treatment planning. *Medical Physics, 28*, 2489–2496.

Gregoire, V. and K. Haustermans (2009). Functional image-guided intensity modulated radiation therapy: Integration of the tumour microenvironment in treatment planning. *European Journal of Cancer, 45*(Supplement 1), 459–460.

Gregoire, V., P. Levendag et al. (2003). CT-based delineation of lymph node levels and related CTVs in the node-negative neck: DAHANCA, EORTC, GORTEC, NCIC, RTOG consensus guidelines. *Radiotherapy and Oncology, 69*(3), 227–236.

Gregoire, V., K. Haustermans et al. (2007). PET-based treatment planning in radiotherapy: A new standard? *Journal of Nuclear Medicine, 48*(1 Supplement).

Gu, X., D. Choi et al. (2009). GPU-based ultra-fast dose calculation using a finite size pencil beam model. *Physics in Medicine and Biology, 54*(20), 6287–6297.

Gu, X., H. Pan et al. (2010). Implementation and evaluation of various demons deformable image registration algorithms on a GPU. *Physics in Medicine and Biology, 55*(1), 207–219.

Guetter, C., C. Xu et al. (2005). Learning based non-rigid multi-modal image registration using Kullback-Leibler divergence. *Medical Image Computing and Computer-Assisted Intervention, 8*(Part 2), 255–262.

Hajnal, J. V., N. Saeed et al. (1995). Detection of subtle brain changes using subvoxel registration and subtraction of serial MR images. *Journal of Computer Assisted Tomography, 19*, 677–691.

Han, X., M. Hoogeman et al. (2008). Atlas-based auto-segmentation of head and neck CT images. *Medical Image Computing and Computer-Assisted Intervention, 11*(Part 2), 434–441.

Hartkens, T., D. L. G. Hill et al. (2002). Using points and surfaces to improve voxel-based non-rigid registration. *Medical Image Computing and Computer-Assisted Intervention,* 565–572.

Hawkes, D. J. (2001). Registration methodology: Introduction, in *Medical Image Registration*. Eds. J. V. Hajnal, D. L. G. Hill and D. J. Hawkes, CRC Press LLC, Boca Raton, FL, 11–38.

Hellier, P. and C. Barillot (2003). Coupling dense and landmark-based approaches for non rigid registration. *IEEE Transactions on Medical Imaging, 22*(2), 217–227.

Hill, D. L. G. and P. Batchelor (2001). Registration methodology: Concepts and algorithms, in *Medical Image Registration*. Eds. J. V. Hajnal, D. L. G. Hill and D. J. Hawkes, CRC Press LLC, Boca Raton, FL, 39–70.

Hill, D. L. G., P. G. Batchelor et al. (2001). Medical image registration. *Physics in Medicine and Biology, 46*, R1–R45.

Hong, R., J. Halama et al. (2007). Correlation of PET standard uptake value and CT window-level thresholds for target delineation in CT-based radiation treatment planning. *International Journal of Radiation Oncology Biology Physics, 67*(3), 720–726.

Hristov, D. H. and B. G. Fallone (1996). A grey-level image alignment algorithm for registration of portal images and digitally reconstructed radiographs. *Medical Physics, 23*(1), 75–84.

Huang, X., Y. Sun et al. (2004). Hybrid image registration based on configural matching of scale-invariant salient region features. IEEE CVPR Workshop on Image and Video Registration.

Hummel, R. A. and S. W. Zucker (1983). On the foundations of relaxation labeling processes. *IEEE Transactions on Pattern Analysis and Machine Intelligence, 5*(3), 267–287.

Hurkmans, C. W., J. H. Borger et al. (2001). Variability in target volume delineation on CT scans of the breast. *International Journal of Radiation Oncology Biology Physics, 50*(5), 1366–1372.

Jaffray, D. A. (2005). Emergent technologies for 3-dimensional image-guided radiation delivery. *Seminars in Radiation Oncology, 15*(3), 208–216.

Jaffray, D. A. (2007). Kilovoltage volumetric imaging in the treatment room. *Frontiers of Radiation Therapy and Oncology, 40*, 116–131.

Jaffray, D. A., J. H. Siewerdsen et al. (2002). Flat-panel cone-beam computed tomography for image-guided radiation therapy. *International Journal of Radiation Oncology Biology Physics, 53*(5), 1337–1349.

Jaffray, D. A. et al. (2008). Applications of image processing in image-guided radiation therapy. *Medicamundi, 52*(1), 32–39.

Jans, H. S., A. M. Syme et al. (2006). 3D interfractional patient position verification using 2D-3D registration of orthogonal images. *Medical Physics, 33*(5), 1420–1439.

Jansen, E. P., J. Nijkamp et al. (2010). Interobserver variation of clinical target volume delineation in gastric cancer. *International Journal of Radiation Oncology Biology Physics, 77*(4), 1166–1170.

Jiang, H., R. A. Robb et al. (1992). A new approach to 3-D registration of multimodality medical images by surface matching. SPIE Visualization in Biomedical Computing, Bellingham, Washington.

Jones, S. M. and A. L. Boyer (1991). Investigation of an FFT-based correlation technique for verification of radiation treatment setup. *Medical Physics, 18*(6), 1116–1125.

Kashani, R., M. Hub et al. (2007). Technical note: A physical phantom for assessment of accuracy of deformable alignment algorithms. *Medical Physics, 34*(7), 2785–2788.

Kaus, M. R. and K. K. Brock (2007). Deformable image registration for radiation therapy planning: Algorithms and applications, in *Biomechanical Systems Technology: Computational Methods*. Ed. C. T. Leondes, World Scientific Publishing, Singapore, 1–28.

Kessler, M. L. (2006). Image registration and data fusion in radiation therapy. *British Journal of Psychiatry, 79*, S99–S108.

Khamene, A., P. Bloch et al. (2006). Automatic registration of portal images and volumetric CT for patient positioning in radiation therapy. *Medical Image Analysis, 10*(1), 96–112.

Krengli, M., B. Cannillo et al. (2010). Target volume delineation for preoperative radiotherapy of rectal cancer: Inter-observer variability and potential impact of FDG-PET/CT imaging. *Technology in Cancer Research and Treatment, 9*(4), 393–398.

Kreuder, F., B. Schreiber et al. (1998). A structure-based method for on-line matching of portal images for an optimal patient set-up in radiotherapy. *Philips Journal of Research, 51*(2), 317–337.

Kuenzler, T., J. Grezdo et al. (2007). Registration of DRRs and portal images for verification of stereotactic body radiotherapy: A feasibility study in lung cancer treatment. *Physics in Medicine and Biology, 52*(8), 2157–2170.

Lam, K. L., R. K. Tenhaken et al. (1993). Automated-determination of patient setup errors in radiation-therapy using spherical radio-opaque markers. *Medical Physics, 20*(4), 1145–1152.

Leibel, S. A., Z. Fuks et al. (2003). Technological advances in external-beam radiation therapy for the treatment of localized prostate cancer. *Seminars in Oncology, 30*(5), 596–615.

Lemieux, L., N. D. Kitchen et al. (1994). Voxel based localisation in frame-based and frameless stereotaxy and its accuracy. *Medical Physics, 21*, 1301–1310.

Leszczynski, K., S. Loose et al. (1993). A comparative-study of methods for the registration of pairs of radiation-fields. *Physics in Medicine and Biology, 38*(10), 1493–1502.

Leszczynski, K., S. Loose et al. (1995). Segmented chamfer matching for the registration of field borders in radiotherapy images. *Physics in Medicine and Biology, 40*(1), 83–94.

Leszczynski, K. W., S. Loose et al. (1998). An image registration scheme applied to verification of radiation therapy. *British Journal of Radiology, 71*(844), 413–426.

Leunens, G., J. Menten et al. (1993). Quality assessment of medical decision making in radiation oncology: Variability in target volume delineation for brain tumours. *Radiotherapy and Oncology, 29*(2), 169–175.

Lin, Z., J. Mechalakos et al. (2008). The influence of changes in tumor hypoxia on dose-painting treatment plans based on 18F-FMISO positron emission tomography. *International Journal of Radiation Oncology Biology Physics, 70*(4), 1219–1228.

Ling, C. C., J. Humm et al. (2000). Towards multidimensional radiotherapy (MD-CRT): Biological imaging and biological conformality. *International Journal of Radiation Oncology Biology Physics, 47*(3), 551–560.

Louie, A. V., G. Rodrigues et al. (2010). Inter-observer and intra-observer reliability for lung cancer target volume delineation in the 4D-CT era. *Radiotherapy and Oncology, 95*(2), 166–171.

Lu, W., M.-L. Chen et al. (2004). Fast free-form deformable registration via calculus of variations. *Physics in Medicine and Biology, 49*(14), 3067–3087.

Lu, W., G. H. Olivera et al. (2006a). Deformable registration of the planning image (kVCT) and the daily images (MVCT) for adaptive radiation therapy. *Physics in Medicine and Biology, 51*, 4357–4374.

Lu, W. G., G. H. Olivera et al. (2006b). Automatic re-contouring in 4D radiotherapy. *Physics in Medicine and Biology, 51*(5), 1077–1099.

Maes, F., A. Collignon et al. (1997). Multi-modality image registration by maximization of mutual information. *IEEE Transactions on Medical Imaging, 16*(2), 187–198.

Maintz, J. B. A. and M. A. Viergever (1998). A survey of medical image registration. *Medical Image Analysis, 2*(1), 1–36.

Matsopoulos, G. K., P. A. Asvestas et al. (2004). Registration of electronic portal images for patient set-up verification. *Physics in Medicine and Biology, 49*(14), 3279–3289.

Mcparland, B. J. (1993). Uncertainty analysis of field placement error measurements using digital portal and simulation image correlations. *Medical Physics, 20*(3), 679–685.

Mcparland, B. J. and J. C. Kumaradas (1995). Digital portal image registration by sequential anatomical matchpoint and image correlations for real-time continuous field alignment verification. *Medical Physics, 22*(7), 1063–1075.

Men, C., X. Gu et al. (2009). GPU-based ultrafast IMRT plan optimization. *Physics in Medicine and Biology, 54*(21), 6565–6573.

Men, C., X. Jia et al. (2010). GPU-based ultra-fast direct aperture optimization for online adaptive radiation therapy. *Physics in Medicine and Biology, 55*(15), 4309–4319.

Mestrovic, A., M. P. Milette et al. (2007). Direct aperture optimization for online adaptive radiation therapy. *Medical Physics, 34*(5), 1631–1646.

Meyer, C. R., G. S. Leichtman et al. (1995). Simultaneous usage of homologous points, lines and planes for optimal 3-D linear registration of multimodality imaging data. *IEEE Transactions on Medical Imaging, 14*, 1–11.

Michalski, J. M., J. W. Wong et al. (1993). An evaluation of 2 methods of anatomical alignment of radiotherapy portal images. *International Journal of Radiation Oncology Biology Physics, 27*(5), 1199–1206.

Mohan, R., X. Zhang et al. (2005). Use of deformed intensity distributions for on-line modification of image-guided IMRT to account for interfractional anatomic changes. *International Journal of Radiation Oncology Biology Physics, 61*(4), 1258–1266.

Morten, B.-N. and G. Claus (1996). Fast fluid registration of medical images, in *Proceedings of the 4th International Conference on Visualization in Biomedical Computing*, Springer-Verlag.

Moseley, J. and P. Munro (1994). A semiautomatic method for registration of portal images. *Medical Physics, 21*(4), 551–558.

Muller, U., J. Hesser et al. (2004). Fast rigid 2D-2D multimodal registration. *Medical Image Computing and Computer-Assisted Intervention, 3216*(Part 1), 887–894.

Munbodh, R., D. A. Jaffray et al. (2006). Automated 2D-3D registration of a radiograph and a cone beam CT using line-segment enhancement. *Medical Physics, 33*(5), 1398–1411.

Nelson, C., P. Balter et al. (2008). A technique for reducing patient setup uncertainties by aligning and verifying daily positioning of a moving tumor using implanted fiducials. *Journal of Applied Clinical Medical Physics, 9*(4), 110–122.

O'Daniel, J. C., A. S. Garden et al. (2007). Parotid gland dose in intensity-modulated radiotherapy for head and neck cancer: Is what you plan what you get? *International Journal of Radiation Oncology Biology Physics, 69*(4), 1290–1296.

Osorio, E. M. V., M. S. Hoogeman et al. (2009). A novel flexible framework with automatic feature correspondence optimization for nonrigid registration in radiotherapy. *Medical Physics, 36*(7), 2848–2859.

Penney, G., J. Weese et al. (1998). A comparison of similarity measures for use in 2D-3D medical image registration. *IEEE Transactions on Medical Imaging, 17*, 568–595.

Petrascu, O., A. Bel et al. (2000). Automatic on-line electronic portal image analysis with a wavelet-based edge detector. *Medical Physics, 27*(2), 321–329.

Pizer, S., D. S. Fritsch et al. (1999). Segmentation, registration and measurement of shape variation via image object shape. *IEEE Transactions on Medical Imaging, 18*(10), 851–865.

Price, K. E. (1985). Relaxation matching techniques—A comparison. *IEEE Transactions on Pattern Analysis and Machine Intelligence, 7*(5), 617–623.

Purdy, J. A. (2001). Intensity-modulated radiotherapy: Current status and issues of interest. *International Journal of Radiation Oncology Biology Physics, 51*(4), 880–914.

Rangarajan, A., H. Chui et al. (1997). The Softassign Procrustes Matching algorithm, in *Information Processing in Medical Imaging, 15th International Conference*, Springer, Poultney, VT.

Reed, V. K., W. A. Woodward et al. (2009). Automatic segmentation of whole breast using atlas approach and deformable image registration. *International Journal of Radiation Oncology Biology Physics, 73*(5), 1493–1500.

Remeijer, P., C. Rasch et al. (1999). A general methodology for three-dimensional analysis of variation in target volume delineation. *Medical Physics, 26*(6), 931–940.

Rueckert, D., L. I. Sonoda et al. (1999). Non-rigid registration using free-form deformations: Application to breast MR images. *IEEE Transactions on Medical Imaging, 18*(8), 712–721.

Russakoff, D. B., T. Rohlfing et al. (2005). Intensity-based 2D-3D spine image registration incorporating a single fiducial marker. *Academic Radiology, 12*(1), 37–50.

Sarrut, D. (2006). Deformable registration for image-guided radiation therapy. *Zeitschrift fur Medizinische Physik, 16*(4), 285–297.

Shen, D. (2009). Fast image registration by hierarchical soft correspondence detection. *Pattern Recognition, 42*(5), 954–961.

Shen, D. and C. Davatzikos (2002). HAMMER: Hierarchical attribute matching mechanism for elastic registration. *IEEE Transactions on Medical Imaging, 21*(11), 1421–1439.

Shirato, H., M. Oita et al. (2004). Three-dimensional conformal setup (3D-CSU) of patients using the coordinate system provided by three internal fiducial markers and two orthogonal diagnostic X-ray systems in the treatment room. *International Journal of Radiation Oncology Biology Physics, 60*(2), 607–612.

Shu, H. Z., Y. Ge et al. (2000). An orthogonal moment-based method for automatic verification of radiation field shape. *Physics in Medicine and Biology, 45*(10), 2897–2911.

Smitsmans, M. H., J. de Bois et al. (2005). Automatic prostate localization on cone-beam CT scans for high precision image-guided radiotherapy. *International Journal of Radiation Oncology Biology Physics, 63*(4), 975–984.

Song, Y. L., B. Mueller et al. (2008). A hybrid method for reliable registration of digitally reconstructed radiographs and kV X-ray images for image-guided radiation therapy for prostate cancer. *Medical Imaging 2008: Visualization, Image-Guided Procedures, and Modeling, 6918*(Parts 1 and 2), 69182W.

Sovik, A., E. Malinen et al. (2007). Radiotherapy adapted to spatial and temporal variability in tumor hypoxia. *International Journal of Radiation Oncology Biology Physics, 68*(5), 1496–1504.

Sovik, A., E. Malinen et al. (2009). Strategies for biologic image-guided dose escalation: A review. *International Journal of Radiation Oncology Biology Physics, 73*(3), 650–658.

Spoerk, J., H. Bergmann et al. (2007). Fast DRR splat rendering using common consumer graphics hardware. *Medical Physics, 34*(11), 4302–4308.

Staring, M., S. Klein et al. (2007). Nonrigid registration with tissue-dependent filtering of the deformation field. *Physics in Medicine and Biology, 52*(23), 6879–6892.

Steenbakkers, R. J., J. C. Duppen et al. (2005). Observer variation in target volume delineation of lung cancer related to radiation oncologist-computer interaction: A "Big Brother" evaluation. *Radiotherapy and Oncology, 77*(2), 182–190.

Steenbakkers, R. J., J. C. Duppen et al. (2006). Reduction of observer variation using matched CT-PET for lung cancer delineation: A three-dimensional analysis. *International Journal of Radiation Oncology Biology Physics, 64*(2), 435–448.

Studholme, C., D. L. G. Hill et al. (1995). Multiresolution voxel similarity measures for MR-PET registration, in *Information Processing in Medical Imaging*. Kluwer.

Sun, Y., F. S. Azar et al. (2005). Registration of high-resolution 3D atrial images with electroanatomical cardiac mapping: Evaluation of registration methodology. *SPIE Medical Imaging*.

Symon, Z., L. Tsvang et al. (2011). An interobserver study of prostatic fossa clinical target volume delineation in clinical practice: Are regions of recurrence adequately targeted? *American Journal of Clinical Oncology, 34*(2), 145–149.

Tai, P., J. Van Dyk et al. (1998). Variability of target volume delineation in cervical esophageal cancer. *International Journal of Radiation Oncology Biology Physics, 42*(2), 277–288.

Tang, T. S. Y., R. E. Ellis et al. (2000). Fiducial registration from a single X-ray image: A new technique for fluoroscopic guidance and radiotherapy. *Medical Image Computing and Computer-Assisted Intervention, 1935*, 502–511.

Thirion, J. P. (1998). Image matching as a diffusion process: An analogy with Maxwell's demons. *Medical Image Analysis, 2*(3), 243–260.

Thorwarth, D., S. M. Eschmann et al. (2007). Hypoxia dose painting by numbers: A planning study. *International Journal of Radiation Oncology Biology Physics, 68*(1), 291–300.

Tikhonov, A. N. and V. Y. Arsenin (1977). *Solutions of Ill-Posed Problems*. Winston & Sons, Washington.

van Baardwijk, A., B. G. Baumert et al. (2006). The current status of FDG-PET in tumour volume definition in radiotherapy treatment planning. *Cancer Treatment Reviews, 32*(4), 245–260.

van Herk, M. and K. G. A. Gilhuijs (1994). A quantitative comparison of methods for anatomy matching of portal and simulator images. *Third International Workshop on Electronic Portal Imaging*, San Francisco.

Van Karanen, S., S. Van Beek et al. (2009). Setup uncertainties of anatomical sub-regions in head-and-neck cancer patients after offline CBCT guidance. *International Journal of Radiation Oncology Biology Physics, 73*(5), 1566–1573.

Vemuri, B. C., J. Ye et al. (2003). Image registration via level-set motion: Applications to atlas-based segmentation. *Medical Image Analysis, 7*(1), 1–20.

Vercauteren, T., X. Pennec et al. (2009). Diffeomorphic demons: Efficient non-parametric image registration. *NeuroImage, 45*(1, Supplement 1), S61–S72.

Vesprini, D., Y. Ung et al. (2008). Improving observer variability in target delineation for gastro-oesophageal cancer—The role of (18F)fluoro-2-deoxy-D-glucose positron emission tomography/computed tomography. *Clinical Oncology, 20*(8), 631–638.

Viola, P. and W. M. Wells III (1997). Alignment by maximization of mutual information. *International Journal of Computer Vision, 24*(2), 137–154.

Wahba, G. (1990). *Spline Models for Observational Data*. SIAM, Philadelphia, PA.

Wang, H., L. Dong et al. (2005a). Implementation and validation of a three-dimensional deformable registration algorithm for targeted prostate cancer radiotherapy. *International Journal of Radiation Oncology Biology Physics, 61*(3), 725–735.

Wang, H., L. Dong et al. (2005b). Validation of an accelerated "demons" algorithm for deformable image registration in radiation therapy. *Physics in Medicine and Biology, 50*(12), 2887–2905.

Wang, H., A. S. Garden et al. (2008a). Performance evaluation of automatic anatomy segmentation algorithm on repeat or four-dimensional computed tomography images using deformable image registration method. *International Journal of Radiation Oncology Biology Physics, 72*(1), 210–219.

Wang, H., S. Krishnan et al. (2008b). Improving soft-tissue contrast in four-dimensional computed tomography images of liver cancer patients using a deformable image registration method. *International Journal of Radiation Oncology Biology Physics, 72*(1), 201–209.

Wang, H., H. Vees et al. (2009). 18F-fluorocholine PET-guided target volume delineation techniques for partial prostate reirradiation in local recurrent prostate cancer. *Radiotherapy and Oncology, 93*(2), 220–225.

Wijesooriya, K., E. Weiss et al. (2008). Quantifying the accuracy of automated structure segmentation in 4D CT images using a deformable image registration algorithm. *Medical Physics, 35*(4), 1251–1260.

Woods, R. P., S. R. Cherry et al. (1992). Rapid automated algorithm for aligning and reslicing PET images. *Journal of Computer Assisted Tomography, 16*, 620–633.

Wu, Q., M. Manning et al. (2000). The potential for sparing of parotids and escalation of biologically effective dose with intensity-modulated radiation treatments of head and neck cancers: A treatment design study. *International Journal of Radiation Oncology Biology Physics, 46*(1), 195–205.

Wu, Q. J., D. Thongphiew et al. (2008). On-line re-optimization of prostate IMRT plans for adaptive radiation therapy. *Physics in Medicine and Biology, 53*(3), 673–691.

Wu, J., M. Kim et al. (2009). Evaluation of similarity measures for use in the intensity-based rigid 2D-3D registration for patient positioning in radiotherapy. *Medical Physics, 36*(12), 5391–5403.

Wu, Q., Y. Chi et al. (2009). Adaptive replanning strategies accounting for shrinkage in head and neck IMRT. *International Journal of Radiation Oncology Biology Physics, 75*(3), 924–932.

Xie, Y., M. Chao et al. (2009). Tissue feature-based and segmented deformable image registration for improved modeling of shear movement of lungs. *International Journal of Radiation Oncology Biology Physics, 74*(4), 1256–1265.

Xing, L., B. Thorndyke et al. (2006). Overview of image-guided radiation therapy. *Medical Dosimetry, 31*(2), 91–112.

Yang, J., J. P. Williams et al. (2006). Non-rigid image registration using geometric features and local salient region features. *Proceedings on Computer Vision and Pattern Recognition, 1*, 825–832.

Yuille, A. L. and J. J. Kosowsky (1994). Statistical physics algorithms that converge. *Neural Computation, 6*(3), 341–356.

Zhang, Z. (1994). Iterative point matching for registration of free-form curves and surfaces. *International Journal of Computer Vision, 13*(2), 119–152.

Zhang, L., A. S. Garden et al. (2006). Multiple regions-of-interest analysis of setup uncertainties for head-and-neck cancer radiotherapy. *International Journal of Radiation Oncology Biology Physics, 64*(5), 1559–1569.

Zheng, Y. and D. Doermann (2006). Robust point matching for nonrigid shapes by preserving local neighborhood structures. *IEEE Transactions on Pattern Analysis and Machine Intelligence, 28*(4), 643–649.

Zhou, J., B. Uhl et al. (2010). Analysis of daily setup variation with tomotherapy megavoltage computed tomography. *Medical Dosimetry, 35*(1), 31–37.

Zitova, B. and J. Flusser (2003). Image registration methods: A survey. *Image and Vision Computing, 21*(11), 977–1000.

2

Online Monitoring, Tracking, and Dose Accumulation

Geoffrey D. Hugo
*Virginia Commonwealth
University*

2.1 Tissue Motion during Radiotherapy

2.1.1 Motion during Radiotherapy Delivery

Advances in treatment planning and delivery technology over the past several decades have enabled substantial gains in the available precision of radiotherapy delivery. Higher precision has enabled clinicians to reduce the volume of incidentally irradiated healthy tissue while still achieving proper dose to the target. However, high precision requires an attendant increase in accuracy to prevent missing the target. Sophisticated, integrated systems are now capable of imaging the patient and guiding the treatment beam to the target in a matter of minutes to within an error of a few millimeters and delivering a known dose to this target within an error of a few percent of the prescribed dose.

However, ensuring the patient to be positioned properly before irradiation is not the end of the story. Due to practical issues (personnel exposure, mechanical constraints, and maximum achievable dose rate), the required time to deliver a single treatment ranges from several minutes to several hours for modern external beam or brachytherapy. Radiation beams and implanted sources are often delivered serially, further prolonging the treatment time. During this time, it is expected that the anatomy will remain stationary with respect to the treatment beam or radiation source, as originally localized. As anyone required to sit in one spot for a long period of time knows, remaining still to within a few millimeters is a difficult prospect, particularly in what may be a slightly uncomfortable treatment position. Even if the patient remains "still" by not actively moving their musculature, processes such as peristalsis (digestion), respiration (breathing), or circulation (heartbeat) can cause internal organs or tumor tissue to move during treatment.

2.1.2 Sites and Types of Motion

Before discussing the impact of motion of the target or healthy tissue on treatment delivery, first we must have a general idea of what types of motion might be expected for certain types of cancers and disease sites. For the purpose of this chapter, we are concerned with motion that alters the patient anatomy during a single treatment session. This type of motion is often referred to as "intrafraction motion," as it occurs during a single treatment session or fraction. Change in the anatomy from day to day, or "interfraction motion," is not addressed in this chapter. Four major types of motion are of interest in radiotherapy: musculoskeletal, digestive, respiratory, and circulatory.

"Musculoskeletal motion" refers to gross motion controlled by the patient's musculature, including articulation of the skeleton (e.g., actively shifting of the body on the treatment table or rotating of the head). Because this type of motion is voluntary, it is difficult to predict and follows no standard pattern or magnitude. Musculoskeletal motion may influence any possible treatment site and should therefore be a concern for all treatments. Even with an integrated image guidance and delivery system, the primary means to control this type of motion remains good-quality immobilization. However, it is clear that one should strive to make the patient as comfortable as possible within the limits of good immobilization. Minimizing discomfort, even to the expense of some immobility, may help minimize muscle movement during a long procedure. Many immobilization techniques have been developed and refined for a variety of clinical sites; an excellent source for the interested reader is the text by Bentel (1999).

"Digestive motion" refers to motion due to digestion and excretion. Peristalsis, the muscular contractions of the esophagus and intestines, is a major component of this type of motion. For the purposes of this chapter, we will define all types of digestion- and excretion-related motion (swallowing, peristalsis, bladder, and rectal filling) expected during a treatment session as digestive motion. Digestive motion includes motion of/in the oral cavity, esophagus, stomach, intestines, bladder, and rectum and will influence the surrounding organs and regions such as the neck, mediastinum, upper and lower abdomen (liver, pancreas, and kidneys), and pelvis (prostate, uterus, and rectum). Digestive motion may influence the radiotherapy treatment of cancers involving these organs or areas. For example, digestion-related motion may affect the location of the target volume in head-and-neck cancers (due to patient swallowing during treatment) and prostate cancer (due to peristalsis in the intestines).

In the abdomen and thorax, digestive motion is generally small in relation to respiratory motion, making it difficult to accurately quantify. In the pelvis, bladder filling produces a time-trending motion of bladder wall cancers in the range of several centimeters, which is directly related to observation time. Digestive motion of several centimeters has been observed in prostate and rectal cancer

due mainly to peristalsis and rectal filling. This type of motion is complex and difficult to predict instantaneously (Figure 2.1). Instead, most studies attempt to define a range of likely motion as a function of observation time. For example, Langen et al. (2008) reported the probability of the prostate deviating from the intended position by 3 mm as 13% after 5 min and 25% after 10 min.

"Respiratory motion" refers to motion due to breathing. Respiration is driven by the diaphragm and the intercostal muscles, which generate pressure gradients inside the lungs relative to the atmosphere, forcing exchange of gases through ventilation. Report 91 by the American Association of Physicists in Medicine (AAPM) Task Group 76 (Keall et al. 2006) is an excellent resource for a detailed description of the process of respiration and methods to manage respiratory motion. Respiration will influence the organs and sites in the thorax (esophagus, lung, and breast) and upper abdomen (liver, pancreas, and stomach). Respiration-induced motion of lower abdominal or pelvic structures such as the prostate has been measured but is generally considered to be small in relation to digestive motion of these structures.

The most common treatment sites prone to respiratory motion are lung, liver, and breast. The influence of respiration on whole-breast radiotherapy is generally considered to be small, with the mean range of motion generally reported to be 0.3 cm due to respiration. As such, the effect of respiratory motion on the dose distribution in the breast is small. However, the management of

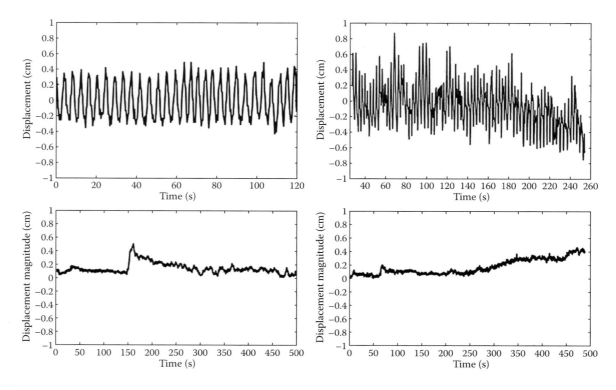

FIGURE 2.1 Example motion patterns observed during radiotherapy treatment. Clockwise, from top left: Regular respiration (superior/inferior motion of a lung tumor); irregular respiration demonstrating cycle-to-cycle variation in superior/inferior position and long-term drift; typical prostate motion (vector magnitude of prostate motion); typical prostate motion, showing drift over time. (Prostate data courtesy of M. D. Anderson Cancer Center, Orlando, Florida.)

respiration may have a role in intensity-modulated and external beam partial breast irradiation. Respiratory maneuvers, such as active breathing control or deep inspiration breath hold, are used to expand the lung volume during treatment to move the treated breast and treatment fields away from the heart to reduce cardiac toxicity. In this case, "motion" consists of the reproducibility of the breast and heart positions from breath hold to breath hold.

For tumors in the lung and liver, the excursion of tumors due to respiration is generally reported to be approximately 1 cm, on average, although the range of motion varies substantially between patients (from less than 0.5 cm of motion up to 3 cm for some tumors). Tumors near the diaphragm have a larger range of motion than those near the apex of the lung (Liu et al. 2007), although the tumor size and stage, and attachment to the chest wall or mediastinum, may outweigh the influence of location on motion range. Respiration-induced tumor motion obeys a semiperiodic pattern that is roughly sinusoidal in shape, with the tumor at minimum velocity at end of inhalation and end of exhalation (Figure 2.1). Some patients demonstrate a prolonged state of rest at the end of exhalation, up to 1 to 2 s. Some tumors may exhibit hysteresis, taking a different path during exhalation than during inhalation. Respiration-driven tumor motion may be very regular and periodic, with the tumor motion pattern having the same shape and the tumor returning to the same location during each cycle. Other patients demonstrate irregular breathing, with the shape of motion and/or position of the tumor at end of inhalation or exhalation varying between cycles. In general, the end exhalation position has been demonstrated to be less variable from cycle to cycle than the end of inhalation position. Finally, there is evidence that breathing may change from periodic and regular to semiperiodic and irregular during a single session (Ozhasoglu and Murphy 2002) or that the position of the tumor in a certain state of breathing (e.g., end of exhalation) may drift from some baseline position over time.

"Circulatory motion" refers to the motion due to heartbeat. This type of motion may influence sites situated near the heart and great vessels, including lung tumors. However, the magnitude of tumor motion due to circulatory motion is small (less than 0.2 cm). Therefore, circulatory motion is generally ignored as a contributor to intrafraction motion in most practices to date.

2.1.3 General Effect of Motion on the Dose Distribution/Rationale

Any type of motion during treatment will affect the delivered dose in relation to the planned dose distribution. The key question is whether this effect has any bearing on the fidelity of the planned treatment. For example, in a large, uniform field of radiation encompassing the entire liver, a small target in the center of the liver will likely receive the same dose whether in motion or not. However, if a small, tight field is used to treat this same tumor, and the target gradually drifts out of the defined beam aperture, the target is likely to receive a lower dose than intended. The effect of different types of motion on the delivered

dose distribution and therefore on the treatment fidelity is a complex function of the magnitude and shape of the motion pattern, what structures are in motion and how the motion patterns of these independent structures are related, the density of the tissue of these structures, and the shape (particularly the gradient) of the dose distribution and delivery methodology [e.g., static beam or intensity-modulated radiotherapy (IMRT) delivery].

Although the effects of motion are complex, there are a few simple observations that can be used to estimate the effect of a given motion pattern on a known dose distribution (Figure 2.2). Periodic or random motion that is stationary (does not change with time, on average) generally has a "blurring" effect on the dose distribution. Areas of the dose distribution with the highest curvature—such as in the penumbral regions or in highly non-uniform IMRT distributions—exhibit the most blurring. The drifting type of motion tends to "shift" the dose distribution relative to the anatomy. Heterogeneous tissues (of varying electron density) will introduce complex changes in the shape of the dose distribution. Dynamic delivery methods, such as IMRT and arc therapy, introduce the effects related to the time at which motion occurred relative to beam delivery. The interaction between collimator motion and tissue motion introduces a complex change in dose distribution shape separate from blurring that has been termed the "interplay effect" (Yu et al. 1998). The magnitude of this effect is generally small for commonly observed motion and IMRT deliveries, particularly as the number of treatment fractions increases (Bortfeld et al. 2002), although there are some situations in which interplay may cause significant delivery errors (Court et al. 2008). Interplay effect due to collimator, gantry, and tissue motion in dynamic arc therapies has not been reported to date. Other more complex effects of motion on dose (changes in tissue shape due to deformation and tissue density variation) are detailed in Section 2.2.

2.2 Surveillance during Radiotherapy

2.2.1 Requirements/Goals

The purpose of surveillance, or monitoring, during treatment is to detect and correct for variations in the patient anatomy that may be detrimental to treatment delivery. This detection and correction is made "online," meaning it is performed while the patient remains on the treatment table, and thus must be made in a quick and efficient manner. The monitoring process consists of three components: a method to detect the anatomical position, a method to assess the error of this position in relation to the anatomical position during treatment planning, and a method to intervene to correct error. Often, this last component may be separate from the monitoring system itself, with the monitoring system providing a binary "yes/no" suggestion on whether treatment should continue, leaving this final decision and intervention up to the operator.

The monitoring process may be the same or completely different from an initial positioning process or use some common methods. For example, one commercial system uses stereoscopic

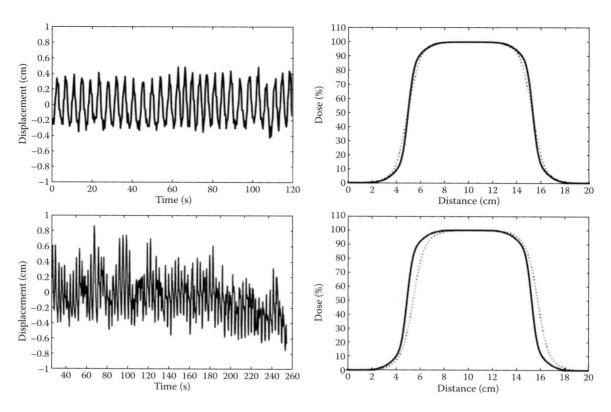

FIGURE 2.2 Effect of motion on the dose distribution. Top and bottom left: Regular and irregular respiration motion patterns, respectively. Top right: Static (solid line) dose profile and motion-influenced (dotted line) dose profiles for regular respiration. The motion pattern has a "blurring" effect on the dose distribution. Bottom right: Static and motion-influenced dose profiles for irregular respiration. The dose is still blurred, but the long-term drift also introduces a systematic shift in the dose distribution.

x-ray radiographs to position the bony anatomy for treatment initially and then relies on an optical infrared system for monitoring position changes during treatment. Another system uses implanted electromagnetic beacons for both initial positioning and in-treatment monitoring. Whether to use the same or different systems for initial positioning and monitoring can be largely based on the answers to two questions: (1) Can the system resolve the types of motion and anatomical change likely to be observed both interfraction and intrafraction with sufficient accuracy and precision? and (2) Is the system capable of detecting and correcting these types of motion at the required frequency? The ability of various systems to perform these tasks will be addressed for the individual systems below. To answer these two questions and select a monitoring system, we must first define the required accuracy and frequency. These parameters can be defined by understanding the types of motion expected to be observed, the planned safety margins and dose distribution, and the fractionation schedule.

2.2.1.1 Influence of Fractionation Schedule

The errors in delivery that occur in a single fraction early in the course of a 35-fraction treatment can be corrected much easier than the same error in a single fraction in a 3-fraction treatment. As the total number of fractions decreases, the required accuracy of delivery during any one fraction increases; thus, the

monitoring requirements increase. As the total number of fractions decreases, there is a tendency for the total treatment time per fraction to increase to deliver a curative dose. Some evidence has been found that there may be greater tendency for larger intrafraction motion in longer fractions (Murphy et al. 2003), although there has been no definitive study showing causation of increased motion with treatment duration.

2.2.1.2 Influence of Required Precision/Margins

As discussed above, a moving structure in a large, uniform dose distribution will be less sensitive to position errors and will require less monitoring. Highly nonuniform dose distributions, such as those produced in IMRT and dynamic arc therapies, will require increased monitoring. As safety margins decrease, the actual target location during treatment becomes closer, on average, to the high-gradient beam edge; thus, monitoring requirements increase.

2.2.1.3 Influence of Type of Motion

The type of possible motion influences mainly the required frequency of motion. Slow, drifting motion requires less frequent monitoring than quick, random changes in position. Large-range motion, such as respiratory motion, may increase the monitoring requirements due to the predilection to try to reduce the effect of such motion. If "tight" margins or 4D planning strategies are used to incorporate a measure of the motion into the treatment plan design, monitoring that the motion remains within the planned range is necessary.

2.2.1.4 Stereotactic Radiosurgery and Body Radiotherapy

Stereotactic radiosurgery and stereotactic body radiotherapy (SBRT) are high-precision treatment modalities characterized by a high per-fraction dose and a limited number of fractions. These modalities are some of the heaviest users of monitoring and online intervention. Often, small margins of 2 to 5 mm are used to limit the probability of normal tissue complications. Compounded with a prescription that places a very high dose gradient at the edge of the target and produces a dose distribution with a high degree of nonuniformity, the confluence of these factors cause the dose distribution in the target to be extremely sensitive to motion and changes in motion. Treatment session durations are often extended relative to conventional fractionated treatments and can range from 30 min to several hours. In SBRT, targets are often located in the lung, liver, and upper abdomen, the site of substantial anatomical motion, as described above. For these reasons, monitoring should be strongly considered for stereotactic and other hypofractionated schedules.

2.2.2 Monitoring Methods

Although radiotherapy surveillance technology has not been as well integrated to date in commercial systems as image-guidance technology has, currently a broad number of exciting options have become available and an interesting research front has opened in this area. Due to the recent explosion of available commercial products for monitoring and the coming development of integrated guidance and monitoring systems, a discussion focused on available commercial products would be quickly out of date. Thus, the discussion herein will focus on the generic modalities and the strengths and drawbacks of each. An effort will be made to note those that are well established in commercial systems and that are not yet available as integrated systems. Finally, because this text is focused on image processing applications in radiotherapy, the mechanical systems (e.g., strain gauges and spirometers) will not be discussed in detail here.

2.2.2.1 Optical

Optical methods are one of the most common and well-established methods for online monitoring for several reasons. An optical system can be developed from existing off-the-shelf components such as economical digital cameras and light sources. Because optical methods use either infrared or visible light, they produce no ionizing radiation and can thus be used continually during a treatment fraction without the need to consider imaging dose. Optical systems are fast and responsive, with operational frequencies of 30 Hz.

The major drawback of optical systems is the inability to image internal anatomy. Optical systems track markers or points on the patient surface or a dense map of points of the patient surface. Thus, such systems are limited to monitoring gross patient musculoskeletal motion or surrogates for internal motion, if they exist. The fidelity of surrogate signals is discussed in Section 2.2.2.1.4.

2.2.2.1.1 Stereoscopic

Because the goal of surveillance is to track a surrogate for the target position during therapy, 3D information is often desirable. For this reason, most of the optical systems have been designed with at least two detectors to enable simultaneous measurement of object location. Each detector provides a 2D measurement of position; without prior knowledge of the object shape and size, depth cannot be measured accurately with a 2D camera system. A set of 2D measurements, coupled with the knowledge of the detector orientation and position (the "pose" of the imaging system), enables the estimation of the 3D position of the object and is termed stereophotogrammetry. The reader is referred to Jahne (2005) for an introduction and mathematical treatment. In some systems, extra detectors are used to ensure that at least two detectors are available to image the object, even if one of the detectors is occluded (e.g., if the treatment head is in between the patient and the detector).

Most stereoscopic optical systems require a high-contrast, well-defined object for tracking and use external objects placed on the patient surface that generate a point-like signal in the detector. If visual light is used, the recognition of a marker in a crowded scene (with other materials, such as the treatment unit and patient, being imaged as well) may be difficult. For this reason, many optical systems use infrared light to enhance object contrast in relation to the background.

Markers function in one of two manners:

- Passive object (reflective markers): These are generally spherical markers coated with a reflective substance. If infrared light is used, a separate infrared light source is required.
- Active object (emissive marker): A light-emitting object, often a light-emitting diode, is used as the marker itself. Although this type of system eliminates the need for the external light source required for passive markers, the light-emitting diode requires a power source that generally means that wiring must be run to the marker for power.

Many systems are capable of resolving multiple markers simultaneously. The use of multiple markers adds the ability to not only determine position but also pose, including the rotation of the patient or marker configuration. Such measurements are most accurate if the marker configuration is rigid (i.e., if the markers cannot move relative to each other), with a reduction in accuracy if markers are allowed to change configuration. Nonrigid marker configurations may be implemented by design to quantify some possibly higher-level deformation of the patient. For example, Baroni et al. (2000) placed markers in locations representative of gross musculoskeletal movement (the suprasternal notch and near the clavicles) and in locations representative of respiration (abdomen and near the xyphoid process), enabling measurement of both quantities simultaneously. Other applications may be in measuring articulation of the patient skeletal anatomy, by placing markers near different bony landmarks such as the skull, jaw, and upper torso.

2.2.2.1.2 Monoscopic

Although a 3D marker position is desirable in many situations, for the measurement of a surrogate signal for respiration, a 1D or 2D signal may suffice, if one wishes to measure only the phase or amplitude of breathing with time. One of the most common systems consists of a single camera and an infrared light source (Kubo and Hill 1996) and was commercialized as the Real-time Position Management (RPM) system (Varian Medical Systems, Palo Alto, California). The RPM system produces a 1D phase or amplitude signal as a function of time by tracking a marker block placed on the patient surface, generally between the umbilicus and the xyphoid process, to maximize anterior/posterior motion of the marker.

2.2.2.1.3 Multiple Point-Based/3D Surface

Recently, systems have been developed and commercialized that enable the monitoring of a large number of arbitrary points, or some portion of the patient surface, without the necessity of external markers. One system (AlignRT, Vision RT Ltd., London, UK) uses a similar setup to the marker-based approaches, with a pair of calibrated cameras and an infrared light source. A key innovation is the use of a pseudo-random "speckle" pattern of light that enables unambiguous detection of the 3D location of a dense grid of surface points. Another advantage is the ability to form these grid points into a connected mesh of vertices, which enables higher-level analysis of patient surface motion such as interaction between points on the surface.

2.2.2.1.4 General Use of Surrogates for Anatomy

For the optical-based methods discussed so far, a key requirement for use in monitoring motion other than musculoskeletal is the validity of the patient surface as a surrogate for such motion. It is generally assumed that digestive motion is not well represented by the measurement of the patient surface, with no reports of optical systems used for this purpose. However, a large amount of work is available in the literature on the use of the patient surface as a surrogate for respiratory motion of internal structures, including tumors.

The use of the patient surface as a surrogate for respiratory-driven tumor motion is predicated on the knowledge that the diaphragm supplies most of the driving force for ventilation. Because the human abdomen is not compressible, the motion of the abdominal surface is well correlated with diaphragmatic motion (Vedam et al. 2003). For tumors in the lung, liver, pancreas, and other thoracic and upper abdominal structures, abdominal surface motion has been studied as a surrogate for tumor motion. However, in the upper abdomen, the addition of peristalsis and gastrointestinal filling may reduce the surrogate accuracy. In the thorax, secondary forces supplied by the intercostal muscles, pressure gradients within different regions of the lung, and heart contraction reduce the direct correlation between the abdominal surface and the lung tumor motion.

The implications of these secondary forces in the lung and abdomen are that (1) there may be error in using the surrogate signal to predict the position of the tumor at a given point in time, and (2) there may be phase differences between a surrogate signal and a tumor position signal. Models have been developed that attempt to incorporate secondary forces into tumor position prediction based on either the diaphragm position (Zhang et al. 2007a) or an external signal such as a spirometer (Low et al. 2005). Furthermore, there is evidence that phase differences between the surrogate and tumor position may change during a single session, as the patient relaxes into a semistationary pattern, and between sessions (Ozhasoglu and Murphy 2002). As discussed below, phase differences between the signals should be calibrated at the beginning of every treatment session.

The correlation between a surrogate signal and tumor position is often calculated to represent the quality of a given surrogate in predicting tumor position. Note that this correlation rarely has a slope of unity. In such cases, it is erroneous to assume that the tumor motion range is equivalent to the range of the surrogate. Often, an abdominal surrogate signal has a substantially larger magnitude than that of the tumor, particularly for lung tumors. Thus, interventional strategies should be designed with an action level based on tumor motion, not surrogate motion range alone.

Another useful method to quantify surrogate quality is to measure the actual tumor position in relation to the expected tumor position based on a surrogate signal. In gated radiotherapy (discussed in Section 2.3.1), this quantity is known as the "residual motion." The residual motion of lung tumors using a solitary surrogate signal from the abdominal surface has been reported to vary between patients, between treatment sessions for the same patient, and between multiple tumors in the same patient. As discussed below, surrogate fidelity should be periodically reassessed at the beginning and throughout a treatment session.

2.2.2.2 X-Ray Radiographic

Radiographic imaging using kilovoltage or megavoltage x-ray is another common modality for intrafraction monitoring, because it enables direct imaging of internal anatomy at a high frequency. In radiographic imaging, an x-ray source and planar detector are used to measure the attenuation along a set of projective ray lines through the patient. For the purposes of this chapter, radiographic imaging will also encompass fluoroscopic acquisition, where the x-ray source and detector are rapidly pulsed to generate a set of time-dependent radiographic frames, capturing the motion of the patient.

One drawback to radiographic imaging is the use of ionizing radiation. Long exposure times—particularly with continuous radiographic imaging for the entire duration of treatment—may result in unacceptable patient dose. One solution to this problem is to pulse the imaging to acquire radiographs at a lower frequency but still at a high enough frequency to appropriately sample the expected motion pattern. For example, if slowly varying or drifting motion is expected, radiographs may only need to be acquired every few seconds or minutes, reducing patient exposure. Imaging dose for monitoring should always be kept to the minimum necessary to achieve the objective and the imaging frequency be controlled to achieve this minimum and no more.

Another detriment is that a single radiograph or set of radiographs from a fixed source/detector geometry provides only 2D information. If some knowledge of the object size and shape exists, it may be possible to estimate 3D information from a single radiograph. In the case of radiotherapy monitoring, this estimation is possible if the object has been previously imaged with 3D techniques previously and if the object size or shape is not expected to vary during treatment. As discussed in Section 2.2.2.4, this method is used extensively for estimating the pose and position of bony anatomy. Due to the requirement of rigidity of the object, monoscopic methods require other information for soft-tissue monitoring. With x-ray imaging, it is often difficult to acquire dense 3D information of the patient anatomy in real time while keeping patient exposure to a minimum. One solution to this problem is to use prior information to estimate 3D position (see Section 2.2.2.3). Another is to acquire stereoscopic or quasistereocopic images and compute the 3D position of a small number of objects (see Section 2.2.1.1).

Radiographic imaging with kilovoltage x-ray provides high-contrast images with low dose. However, these imaging systems generally require additional imaging devices mounted either to the treatment gantry or in the treatment room. Such external imaging systems also expose the patient to radiation dose in regions that may not coincide with the high-dose treatment volumes. Thus, care must be taken to quantify the imaging dose and the location of this exposure as well. The megavoltage treatment beam itself can be used for monitoring, which enables monitoring without an additional imaging dose. However, even with modern electronic portal imaging detectors, inherent contrast of megavoltage portal images is poorer than that from kilovoltage beams, which may impact the accuracy and precision of monitoring. The small, irregular beam apertures present in modern irradiation techniques such as IMRT may limit the visibility of structures at some time points during treatment, making monitoring of such structures difficult. An interesting strategy has been devised to incorporate the estimated object position into the IMRT planning process, optimizing the shape of the collimated beam to minimize occlusion of implanted markers by the collimator (Wiersma et al. 2009).

2.2.2.2.1 Stereoscopic

A common research and commercial setup for radiographic monitoring is to fix a pair of kilovoltage x-ray sources and corresponding detectors in orthogonal or nearly orthogonal geometries in the treatment room. In these systems, the imaging system is independent of the treatment beam delivery system (usually a gantry-mounted linear accelerator or robotic arm-mounted linear accelerator). In a perfect system, measurement

of 3D position can be accomplished by triangulating object positions from the 2D measurements from the separate detectors. However, imaging systems are never perfect; errors such as the calibration of the pose of the detectors, imperfect measurement of the 2D object position, and movement of the object between acquisition with each detector cause reduced accuracy in triangulation. For this reason, stereoscopic 3D position estimation is generally posed as an optimization problem, where the goal is to iteratively determine the 3D position that minimizes the sum of square errors. Because the imaging systems are stationary in the treatment room, calibration of the source/detector geometry is straightforward and accomplished with a calibration phantom of known geometry and pose. However, certain positions of the treatment unit can occlude one of the imaging systems (i.e., block the view of one of the source/detector systems). For this reason, the system developed by Mitsubishi (Shirato et al. 2000) adds a third source/detector system.

2.2.2.2.2 Monoscopic

With the recent commercial availability of gantry-mounted cone-beam computed tomography (CBCT) systems, a number of groups have investigated using the CBCT source/detector system as an intrafraction monitor. With this method, radiographs may be acquired at fixed poses (with the gantry stationary) before, during, or after beam delivery. Radiographs may also be acquired between deliveries of subsequent beams as the gantry is rotating around the patient (or during delivery of rotational therapy). A CBCT consists of a set of radiographs acquired over a partial or complete rotation of the gantry; therefore, operations performed on a single radiograph could also be applied to the set of radiographs constituting the CBCT acquisition. Because the imaging system is moving with the treatment gantry, occlusion with the gantry is not an issue as with fixed-imaging systems. However, due to this motion, mechanical deflection may introduce errors into the source/detector geometry, which must be corrected by calibration over the possible gantry positions (Cho et al. 2005).

For a CBCT acquisition system adapted for monoscopic monitoring, radiographs are acquired over an angular range in sequence rather than simultaneously or nearly simultaneously as with a stereoscopic system. A key question is how to estimate the 3D position of a well-defined object from this set of monoscopic radiographs. Consider a well-defined point, the position of which can be measured accurately in 2D with the monoscopic system (Figure 2.3a). Because the radiographs are not acquired simultaneously in time, it may be possible that this point has moved between acquisitions of subsequent radiographs. In this case, conventional triangulation will not be useful in resolving the actual 3D positions of the point at two separate times (Figure 2.3b). For a large number of radiographs, back-projection can be used to determine the most likely position of the point over the entire time of acquisition. However, this calculation will produce only a single 3D position of a point. For monitoring, we require the instantaneous position of the point as a function of time during acquisition.

(a)

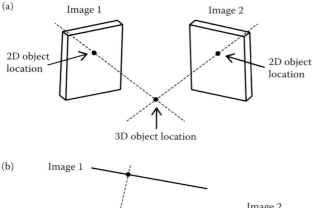

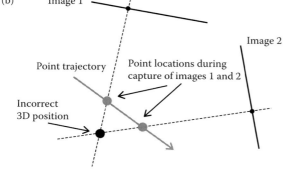

(b)

FIGURE 2.3 (a) Stereoscopic estimation of 3D point object location from simultaneous 2D images. The orientation and position of both images in space must be known, generally through calibration of the camera systems. (b) Error in estimation of the 3D point object location, if the object is in motion and images are not captured simultaneously. The magnitude of the error depends on the time between image capture, the system geometry, and the velocity of the object.

To solve this problem, Adamson and Wu (2008) implemented a shortest-path approximation used to estimate the most likely path of a point object during monoscopic acquisition. Essentially, the shortest possible trajectory that matches the measured 2D object positions in all images is selected as the most likely path. To solve the same problem, Poulsen et al. (2008) used a two-step approach: first, a 3D probability density function of most likely positions of the point object is made using the entire set of radiographs. Second, for each radiograph, the out-of-plane position of the object is sampled from this likelihood distribution and used to estimate the instantaneous 3D position of the object. For real-time operation, an a priori probability distribution would be necessary rather than a distribution estimated from the acquisition itself. Hugo et al. (2007) described a different two-step approach: first, the most likely 3D position of the object is estimated by reconstructing a 3D CT image from the set of radiographs and comparing the 3D object position with that from a reference image. The second step is similar to the Poulsen et al. method, where this estimated 3D position is used to estimate the out-of-plane position and then the instantaneous 3D position. The method developed by Poulsen et al. is more amenable to well-defined point objects such as markers, whereas the method of Hugo et al. is more suited for anatomical objects.

Rather than using separate monoscopic methods with the treatment beam or a CBCT system, several groups have

attempted to design a hybrid stereoscopic method using simultaneous (or near-simultaneous) imaging with both the megavoltage treatment beam and the kilovoltage CBCT system (Cho et al. 2009; Wiersma et al. 2009). This method enables the stereoscopic detection of the instantaneous 3D position of a point object using triangulation rather than the monoscopic estimation methods described above. However, the megavoltage beam presents similar challenges in object detection and recognition as described above for megavoltage monoscopic methods.

2.2.2.2.3 Processing to Extract Anatomical Signals

Up to this point, we have assumed that the detection and recognition of an object in the radiographic image is exact. However, depending on the type of object to be detected and imaging system, this task may be the most difficult in the monitoring process. For the purposes of this discussion, we will separate the methods of extracting the position of an object from an image into two major groups. The first, which we will term "feature detection," involves detecting a feature that represents the object directly from the 2D image. This process may involve filtering to enhance the feature or remove extraneous information from the image. An example would be use of an edge detection filter to enhance a high-contrast marker followed by extraction of the marker boundaries and estimation of the marker centroid. The second position extraction method is "registration." In this method, a 3D model of the object is created, and the pose (and possibly size and shape) of the object is adjusted. A synthetic image of the object is created, mimicking the image formation geometry and physics of the measured image. The synthetic image and measured image are compared with a similarity metric to determine agreement. Finally, the pose (size and shape) of the object that maximizes similarity is selected. The use of templates to find a 2D object in a 2D image blurs the lines of these two categories. For example, one could consider a 2D template to be an enhancement kernel, which would fall into the first category. Conversely, the use of templates can also be considered a 2D registration problem. For the purposes of this discussion, use of a 2D template will be considered a feature detection method if the template is formed based on 2D properties of an object (e.g., a fiducial marker), and a registration method if the template is produced by projecting through a 3D model of the object.

2.2.2.2.4 Feature Detection Methods

Implementing the feature detection method requires high-contrast objects that can be easily and quickly enhanced and extracted. For the most part, this can be difficult for soft-tissue structures. The two main types of objects that can be easily extracted directly from a radiograph are implanted fiducial markers and bony anatomy. Fiducial markers are generally small metal (often gold) objects, often spherically or cylindrically shaped, of small size (generally on the order of a millimeter or so). The small size is required to limit the deformation of the adjacent tissue and to enable easy implantation with existing interstitial or endoscopic equipment. By far, the most commonly reported site of use is the prostate for several reasons. The prostate exhibits large, relatively unpredictable interfraction and intrafraction variation, which makes localization and monitoring important for high-precision targeting. The prostate boundary is generally low contrast and thus is extremely difficult to resolve with radiographic imaging. Finally, markers can be easily implanted through the perineum with minimal morbidity. Recently, fiducial marker implantation has become more common in other sites such as liver, lung, breast, and head and neck as the importance of intrafraction motion and monitoring has become apparent.

To automatically recognize markers, a variety of classic image processing algorithms have been tested, including region growing (Aubin et al. 2003), edge detection, and morphologic operations (Buck et al. 2003; Keall et al. 2004). Due to the existence of spurious signals from other high-contrast structures mimicking a marker such as bony ridges and the treatment table, recent implementations have tended toward template or kernel-based methods (Lam et al. 1993; Nederveen et al. 2000; Mao et al. 2008a). For spherical markers, the image of the marker appears the same regardless of orientation or pose. For such markers, kernel methods only need to consider scale (size of the marker), as perspective projection of the marker changes the size of the marker image, depending on the location of the marker in depth from the imaging plane (Buck et al. 2003). Scale is a key issue in "tuning" a kernel to generate the maximum signal for a particular marker type. "Scale-space" methods have been developed (Lindeberg 1993), which enable the detection of features over a range of scales with a fixed kernel by incorporating an additional kernel for blurring. Generally, however, the range of possible scales is small (because the markers are usually found in a small region in space around the treatment isocenter, limiting the effect of magnification). For cylindrical markers, marker orientation can affect not only the orientation of the image of the marker but also the apparent length due to out-of-plane rotation of the marker. Several current methods use oriented and scaled kernels for optimal extraction, with the added benefit of marker orientation determination (Mao et al. 2008b).

Although image processing with filters can enhance the ability to detect markers in radiographic images, simpler approaches should not be overlooked. Often, it is possible to constrain the search region based on known information about the location of the marker. For example, if 3D imaging is performed for localization, marker positions can be estimated and projected into the radiographic imaging space to serve as a starting point. As discussed above, the knowledge of the marker size and shape can be used to tune an extraction kernel to generate the maximum signal when matching a marker-like object. Such information can also be used to remove candidates that are not "marker like," such as bony ridges or calcifications.

Feature detection methods have been applied to the detection of anatomical objects, primarily consisting of bony anatomy. These methods generally work by extracting features from a treatment image and a reference image and then finding the feature correspondences between the two images to determine an appropriate transform to map the features from one image

to the other. The reference image may be an image from a previous treatment session, or a digitally reconstructed radiograph (DRR), generated by projecting in the imaging geometry through a 3D model of the patient. One prominent method is chamfer matching, which extracts edges and formulates edge map correspondence by minimizing the sum of squares distance between the edge maps (Gilhuijs and van Herk 1993). Several other automated and semiautomated methods have been developed (Balter et al. 1992). Although these methods do not require the use of implanted markers, the detection of bony anatomy as a surrogate for tumor position relies on the assumption of stasis of the tumor in relation to the bone being imaged, so such methods are rarely applied for online monitoring.

2.2.2.2.5 Image Registration Based on Anatomical Objects

With the advent of 3D conformal radiotherapy, 3D patient models, usually generated from CT scans, became routinely available in the clinic. This development, coupled with the necessary computational resources, encouraged the use of 2D/3D registration techniques for object position analysis from monitoring radiographs. The general idea behind 2D/3D registration is that a 3D model is adjusted (either in position or shape), a set of DRRs is generated through this new model, and the DRRs are compared with the measured radiographs with some type of similarity metric. The model is readjusted until the similarity between the DRRs and the measured radiographs is maximized. For a detailed description, the reader is referred to the article by Murphy (1997). The major advancements in 2D/3D registration have involved the development of optimization techniques to minimize the number of required adjustments, the exploration and optimization of similarity metrics, the improved modeling of the physics of the image formation process to improve DRR rendering, and the improved processing to remove background features and enhance target features in both the DRR and the radiographic images. 2D/3D registration methods allowing six degrees-of-freedom transformations (rigid-body translation and rotation) are currently not fast enough for continuous, real-time monitoring, although the use of graphics processing units to speed up DRR rendering and similarity metric calculation is rapidly removing this barrier (Sadowsky et al. 2006).

For real-time monitoring, it is common to try to reduce the scope of the problem. Often, in-plane translation of the object is the objective, reducing the free parameters to two (Berbeco et al. 2005). If a fixed, known view angle is used, a library of "templates" from offline DRR renderings can be precalculated. Templates are simply images of the target object, generally limited to a small region surrounding the target, often with background or occlusive information removed from the image. Templates are often generated using DRR rendering techniques but can also be generated from a previously acquired radiograph or some other model of the target object, if known. The initial position of the target object is often known from a pretreatment localization procedure (e.g., CBCT), so the search region can be limited. The goal of these constraints is to reduce the time from image acquisition to position determination to a few tens or hundreds of milliseconds, enabling fast reaction to anatomical motion.

Berbeco et al. (2005) proposed template registration for monitoring lung tumor position. The moving tumor was emphasized relative to occlusive anatomy by frame averaging over a set time and subtracting this average image from each radiograph. This "motion-enhanced" image increased the contrast of moving structures by removing static structures. A preexisting end-of-exhale motion-enhanced image was used as a template and compared with the motion-enhanced online images with cross-correlation. Rather than estimate the position of the tumor directly, a respiratory-correlated signal was generated from the cross-correlation similarity index. This group then built on this work by using multiple templates (Cui et al. 2007b), incorporating support vector machines for better classification of an image to an appropriate template (Cui et al. 2008), and limiting the search region by predicting future positions of the tumor (Cui et al. 2007a).

Similar approaches have been applied to rotational acquisitions, such as CBCT acquisition (Hugo et al. 2007). A radiotherapy linac gantry-mounted CBCT acquisition is simply a set of radiographs acquired over a range of gantry angular positions. 2D templates were generated over the span of gantry angular positions and registered to the measured 2D images to estimate the 2D position of lung tumors in the CBCT projection space. The method described in Section 2.2.2.2 was used to estimate 3D tumor position from the set of rotational 2D positions. The tumor must be detected over the range of projection angles, and occlusions from the treatment table and overlying anatomy may introduce error into the template registration. For this reason, robust similarity metrics based on robust statistics were evaluated by Hugo et al. (2007).

Rather than performing 2D template-based registration to monitor a single, rigid object in a set of CBCT projections during CBCT acquisition, a 3D patient model could be deformed, using 2D/3D registration methods, to match each CBCT projection. Two groups have demonstrated feasibility in reduced, well-conditioned data sets or phantoms (Zeng et al. 2005; Docef and Murphy 2008). This approach would generate a 4D patient model representing deformation as a function of time, enabling 4D monitoring of multiple structures during treatment. However, much development remains before this approach will be available for real-time monitoring applications.

2.2.2.3 3D and Near-3D Methods

The per-scan dose and acquisition time to generate a volumetric image limits the usefulness of fan-beam CT and CBCT for real-time monitoring. However, for long treatment sessions (such as in SBRT), CT is useful to assess slowly trending motion such as musculoskeletal motion. A more practical approach under development is digital tomosynthesis (DTS) using an onboard imaging device (Godfrey et al. 2006). DTS enables the reconstruction of a single tomographic slice perpendicular to the detector plane through the use of a limited-angle scan (usually a few tens of degrees of gantry rotation). DTS could be acquired

as the gantry rotates between static beam delivery angles or continuously during arc-based therapies for monitoring.

2.2.2.4 Electromagnetic

A system capable of continuous, real-time monitoring of internal anatomy without the necessity of ionizing radiation presents several advantages over optical and radiographic systems. As mentioned above, optical systems are generally limited to monitoring external anatomy and surrogates for internal motion. Radiographic systems must be used with care, particularly in regard to imaging technique and imaging duration due to the radiographic dose the patient will receive. For these reasons, several investigators developed implantable devices that could be tracked with electromagnetic signals. Watanabe and Anderson (1997) investigated a commercial system consisting of a transmitter coil and a wired receiver coil for the localization of intraoperative brachytherapy sources. Because the receiver coil was wired, implantation of the receiver into internal anatomy was limited. Furthermore, they reported interference of the system with external materials such as metal surgical tools and external electromagnetic fields such as those present near the linear accelerator. Seiler et al. (2000) developed an in-house system generating fluctuating magnetic fields that induced an alternating current in a small, wired implantable receiver.

The interest in electromagnetic tracking has increased with the availability of a commercial system utilizing wireless transponders as receivers. The system (Calypso Medical Technologies, Seattle, Washington) uses a flat array panel to generate an oscillating magnetic field near the patient. The field induces a resonant signal inside each transponder (which can be tracked individually), each of which is detected by the array. These signals are used to determine the pose and position of each transponder in relation to the array. The array itself is localized using an infrared optical system. The lack of wires and small size enable the implantation of the transponders internally, similarly to radiographic fiducial markers. This system enables the continuous monitoring with a measured latency of 303 ms (Santanam et al. 2009). As with fiducial markers, marker positions must be calibrated relative to the positions of internal anatomy using imaging capable of resolving soft-tissue structures, such as CT.

2.2.2.5 Magnetic Resonance Imaging/Ultrasound

Magnetic resonance imaging (MRI) and ultrasonography would be useful for real-time monitoring because neither uses ionizing radiation. However, both have some limitations and have not been commonly used for this purpose to date for several reasons. Ultrasonography lacks the soft-tissue contrast of MRI that would enable robust, automated feature extraction and position assessment. Furthermore, ultrasound probes must be in continuous contact with the patient, which requires a tool to keep the probe correctly positioned and out of the radiation beam to limit the attenuation by the probe. MRI suffers from the same problem as other volumetric approaches in that the acquisition time for a single 3D volume limits real-time application. However, the ability of MRI systems to acquire near real-time 2D images at arbitrary orientations through the patient may enable this limitation to be overcome. If the anticipated pattern of motion is known, image acquisition can be oriented along the direction of maximum motion to enable the capture of this motion component.

Technical hurdles involving radiofrequency and magnetic field interaction must be overcome to enable the simultaneous imaging with an MRI while a treatment beam is delivered. Several groups are developing combined MRI–linac systems that would enable real-time surveillance and position detection with the MRI unit. A commercial system (Viewray, Oakwood Village, Ohio) is being developed combining MRI with multiple gamma-ray sources in an attempt to overcome the technical difficulties of MRI–linac integration. However, currently, none of these systems are available for clinical use in real-time monitoring.

2.2.2.6 Correlation between and Combinations of Methods (e.g., Optical and Radiographic)

As is hopefully apparent to the reader, each system for online monitoring has advantages and disadvantages. A robust approach is to use multiple systems in combination to capitalize on the advantages of each individual method. A common combined system is optical/radiographic, consisting of an optical system to continuously monitor the external anatomy and a radiographic system for periodic assessment of the internal anatomy. This type of combined system works well for monitoring gross musculoskeletal motion, as in the commercial ExacTrac system (BrainLAB AG, Feldkirchen, Germany). Stereoscopic radiographs are acquired and used to adjust the patient position based on bony anatomy. At the same time, these radiographs are used to calibrate the initial position of a set of external markers on the patient surface, which are then monitored with the optical system. Respiratory motion can also be monitored with this type of combined system. The CyberKnife (Accuray, Sunnyvale, CA) system monitors fiducial markers placed in or near the tumor with stereoscopic radiographs. A correlation map between the internal fiducial markers and a set of optically tracked external markers is formed with a short-duration, high-frequency radiographic acquisition. Respiratory motion is then monitored for some time solely with the optical system. The correlation map is then periodically updated with a simultaneous optical/radiographic acquisition at a low frequency (every few seconds).

Other (potential) combined systems include the following:

- Electromagnetic/CBCT. CBCT is used to initially calibrate transponder position relative to anatomy position at the beginning of a treatment session, with electromagnetic tracking used for monitoring during treatment.
- Radiographic/CBCT. CBCT used for periodic 3D position assessment, radiographs, or DTS used to assess motion of markers or anatomy during gantry rotation.
- Electromagnetic/optical. Electromagnetic could be used for tracking tumor motion; optical could be used for monitoring gross musculoskeletal motion of the patient or organs at risk.

2.2.3 Prediction

Image acquisition, feature detection and extraction, and position estimation all take measurable time due to the large size of medical images and the complex image processing that must be performed. Real-time monitoring must enable a decision to be made on intervention at the time an object is residing at the detected position. If a decision is made to intervene and an intervention is applied when the object has already moved to a new position, errors will be introduced into the radiotherapy delivery process. For this reason, monitoring systems with long "latency" (the time from the start of image acquisition to when an object's position is available from this image or when an intervention takes place) requires prediction strategies to reduce the effect of the latency time. As the motion pattern becomes more highly varied (i.e., position changes more rapidly in a fixed time) and as the latency time increases, the requirements on accuracy and robustness of the prediction method increase. Finally (and, most importantly, for some intervention strategies), the type of intervention will contribute to the overall system latency. For example, if the intervention is to just turn the beam on or off based on the measured target position, the contribution to system latency may be small. However, if the patient or beam position must be dynamically adjusted, then the latency may be larger due to the additional mechanical and computational requirements involved in this type of intervention. We will address this contribution to latency in more detail in Section 3.2.

Prediction is generally based on the output from the monitoring system, which is the estimated 3D position of an object or objects. As mentioned above, some investigators have used prediction methods in early steps of the monitoring process, to limit the search region in an image for an object, for example. However, in this section, we will concentrate on prediction methods to estimate future 3D positions of objects based on current and historical measurements.

2.2.3.1 Algorithms

Consider a 3D position $x(t)$ that is sampled at a periodic frequency throughout treatment. The goal of a prediction algorithm is to estimate a future position $x(t + \delta t)$ based on a set of measured values of x up to time t. Many varied approaches have been investigated relying on different subsets of historic measurements of $x(t)$, varying from using all previous data (including previous treatment sessions or data from patient populations) to using only a few measurements acquired immediately before time t. Most methods discussed here use filters to process and predict the future position, whereas others use analytical or statistical motion models. For a review of work on prediction methods up to 2005, the reader is directed to the paper by Murphy and Dieterich (2006).

2.2.3.1.1 Model Based

Model-based methods assume some prior knowledge about the expected pattern of motion on a global scale. Analytical models have been investigated for motion such as respiratory motion, where the motion pattern has a well-defined global structure. A sinusoidal-based model (Lujan et al. 1999) or state-based model (Wu et al. 2004) captures a large portion of the semiperiodic respiratory motion. However, irregular breathing that diverges from a model is commonly observed. Some examples of irregular breathing include large variations in end inhalation or end exhalation position, changes in the shape of the motion pattern at points during treatment, variation in the period of the motion pattern, and slow drifting of the mean position during motion. These types of irregularity require auxiliary prediction methods for the accurate position prediction.

For random motion patterns, statistical models have been investigated. In such a model, the set of prior measurements of $x(t)$ is used to construct a probability density function. The goal of the prediction algorithm, in this case, is not to predict the exact position but rather to (1) determine if there is a general "trend" in the motion pattern (such as a drift or change in the mean position) and (2) determine if the latest measurements are likely to be sampled from a distribution similar to the original distribution (i.e., has the "spread" of motion changed). Similarly, trend-measuring algorithms have also been used for nonrandom motion such as respiration to measure and predict for drifting trends (changes over time in the mean position of motion, averaged over many respiratory cycles) (Trofimov et al. 2008).

2.2.3.1.2 Linear, Nonlinear, and Adaptive Filters

Linear methods of prediction were first evaluated to solve the problems of model-based algorithms in predicting respiratory motion. For irregular breathing or noisy measurements, a historical sample of measurements of $x(t)$ better predicts a future position than a fixed model (Vedam et al. 2004). Shirato et al. (2000) examined a linear extrapolation method, which predicts the future velocity based on the previous two measurements. Isaksson et al. (2005) and Sharp et al. (2004) extended this approach to evaluate weighted linear models on larger samples of historical data and investigated linear and nonlinear artificial neural networks.

These approaches are "data driven" in that they do not generally require a model of the expected motion. However, often model-based rules are heuristically built into filter-based approaches. For example, training data or analytical models are used to tune free parameters in artificial neural networks. One issue with such trained filters is the difficulty in adjusting to changes in the inherent signal, because it varies from that present in the training data. Adaptive filters, which enable adjustment of the free parameters (e.g., recomputing the linear weights) periodically or continually, have been shown to reduce error between the predicted and the actual positions, particularly for nonstationary and irregular signals. A common adaptive filter is the Kalman filter, which provides an optimal prediction given known uncertainty distributions in process and measurement noise. The Kalman filter optimally estimates the current "state" (e.g., position and velocity) of a signal given the previous state, the current measurement, and a model for how the previous state is used to generate the new state, termed the state-transition

model. The Kalman filter is not empirical, because it requires these process and noise models, although these have often been estimated from training data (Sharp et al. 2004). Recently, Putra et al. (2008) extended the Kalman filter approach using multiple state transition models for various types of respiration.

In the past several years, the field of target position prediction has become even more active. One approach is to decompose the respiration signal into periodic (or semiperiodic) and nonperiodic components (McCall and Jeraj 2007). The mean periodic cycle shape over a number of respiration cycles is calculated; then, autoregressive moving average is applied to the residual nonperiodic component for prediction. Ruan et al. (2007) developed a method termed "local regression." Essentially, a state is formed as a vector of the current position and the previous n positions (n is determined experimentally). A weighted regression is performed where measured states most "similar" to the current state are given more weight. Similarity is assessed by the local shape and position of the curve—thus states or positions in the same respiratory phase and amplitude are considered more similar—and by temporal proximity (i.e., similarity fades over time).

A comparison of the prediction accuracy between methods can be difficult due to the covariance of error with the irregularity of the respiration trace used for testing and the wide range of breathing patterns observed. Fortunately, some of the works discussed above directly compare methods on the same data sets. Adaptive methods have been found to reduce prediction error, particularly for irregular respiration. Murphy and Dieterich (2006) found that nonlinear neural networks achieved less than 2 mm root mean square error at a 0.2 s horizon in a set of test cases. Ruan et al. (2007) found that linear nonadaptive methods generally achieve a root mean square error of 3 to 6 mm, at a horizon of 0.2 to 1.0 s ahead of the current measurement, in a different set of test cases. A nonlinear neural network reduced the error range to 2 to 5 mm, and local regression achieves less than 3 mm error up to 1.0 s (Ruan et al. 2007). Generally, most investigators have found that as the prediction horizon increases, accuracy decreases. Thus, interventions with higher latency will either exhibit larger prediction error or require more complex and accurate prediction algorithms.

2.3 Intervention for Motion

Once the position of the target has been measured, estimated, or predicted at a point in time, the question is what to do next. This may seem a silly question, because the obvious answer is either "move the target back to the intended position" or "move the beam." However, efficiency and necessity must be considered. If the goal is to deliver the planned dose to the target while limiting unnecessary irradiation of all other tissue, will continually moving the beam or patient achieve this goal? The answer to this question hinges on a number of factors, such as the following:

- The uncertainty in knowledge of the target position and position of critical risk structures (due to measurement, position estimation, and prediction errors)

- The selected safety margins on the target structure (larger margins increase the likelihood of normal tissue irradiation but require less accuracy in target position)
- Available frequency of monitoring
- Available methods of intervention (i.e., if the user must turn off the beam and reposition the patient manually to correct for target position errors, one is less likely to correct for all but the largest errors)

All of these factors should go into selection of an appropriate "action level." The action level is a tolerance around the planned position the target should reside in after localization at the beginning of a treatment session. The action level and the planned position define a set of allowed positions for the target during the treatment session (this action level is for monitoring during treatment and is separate from any action level defined for localization). For ease in this section, this set of allowed positions will be termed the "treatment region." The goal of an intervention strategy is to determine if the current or predicted position is inside the treatment region, make a determination of whether to intervene, and then correct for targeting error to return the target to the treatment region. In the following sections, specific strategies for intervention are discussed.

2.3.1 Beam Gating

The simplest method to intervene is to discontinue the treatment beam if the target moves outside the treatment region. This method is known as "gating" and is generally implemented for two distinct types of motion:

- The beam is held off due to the target drifting outside of the treatment region, and the patient is repositioned. This type of gating is generally used for musculoskeletal or digestive motion where the target is not anticipated to return inside the treatment region without repositioning.
- The beam is held off temporarily, with the expectation that the target will return to the treatment region periodically. This type of gating is used for respiratory motion and is termed respiration gating.

Respiration gating was originally developed for particle beam therapy (Ohara et al. 1989) and adapted for external beam radiotherapy by Kubo and Hill (1996). The gating window is the portion of the breathing cycle during which beam delivery is allowed. This window can be set either on the target position or displacement of a surrogate signal (which is termed "amplitude gating") or on the phase of respiration (termed "phase gating"). Phase gating is generally chosen when the monitoring signal correlates with the phase of the respiratory-driven target position but not necessarily with the amplitude (or exact position). Amplitude gating is generally more useful when the target position can be measured or estimated directly (e.g., with fiducial marker detection). The duty cycle is the fraction (or percentage) of the total time during which the treatment beam is on. Larger gating windows (either in amplitude or phase) necessarily

produce a longer duty cycle. For a periodic, regular breathing cycle, the difference between phase and amplitude gating is minimal, with the key difference being in how the gating window is defined (Figure 2.4). For irregular respiration, phase gating can produce substantial residual error if there is substantial cycle-to-cycle variability or a slow, drifting trend in target position (Figure 2.5). In such a case, amplitude gating is recommended. Otherwise, coaching methods (audio, visual, or audiovisual) have been shown to be useful in regularizing breathing and making phase gating more amenable.

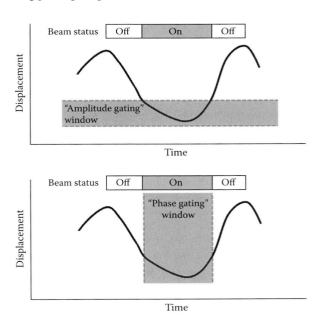

FIGURE 2.4 Amplitude and phase gating based on a 1D respiration signal. The gating window defines the signal magnitude or phase at which the beam is enabled. For regular, periodic respiration, phase and amplitude gating performance will be comparable.

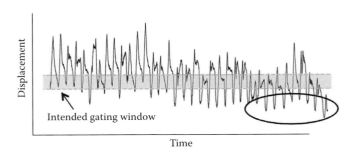

FIGURE 2.5 Beam gating for irregular or drifting respiration. If the intended gating window is defined at the beginning of the treatment session, drifts and irregular respiration may introduce unintended delivery errors with both amplitude and phase gating. For phase gating, the beam will be enabled outside the intended window (in the area highlighted by the ellipse) due to signal drift. For amplitude gating, the beam is enabled in the correct positional window; however, gating occurs at a different phase of respiration, which may introduce discrepancies between the planned and the delivered dose distribution.

Beam gating can be implemented with relatively technical ease (gating generally requires a simple interface for disabling/enabling the treatment beam, which many linear accelerator vendors provide). Latency is only a factor when the beam is enabled or disabled; during the middle of the gating cycle, the beam is stable. Modern accelerators are capable of disabling the treatment beam within a few hundreds of microseconds, but reestablishing stable beam parameters can take several hundreds of microseconds and result in some error in beam delivery. One detriment to gating is a duty cycle less than 100%; thus, treatments take longer than if the beam can be delivered continually.

One simple method to increase the duty cycle in gated irradiation is to have the patient hold their breath at a defined, reproducible point in the respiratory cycle. Breath hold is an extension of beam gating, where the gating window is simply placed about the breath hold location. Common points in the respiration cycle for breath hold have included end expiration and end inspiration, deep inspiration, and moderate deep inspiration (80% of maximum inspiration). Breath hold can be either self-controlled by the patient or computer controlled with a technique known as "active breathing control," where a spirometer measures air flow, which is converted to air volume inhaled and exhaled. A computer-controlled valve enables breath hold at a predefined lung volume based on the spirometry measurement. Because spirometry does not provide a direct measure of tumor position, investigators have combined spirometry-based systems with optical and radiographic monitoring.

Breath hold may increase the duty cycle compared with free breathing gating, although the rest time between successive breath holds must be considered, because it will prolong the overall treatment time. One of the commonly stated drawbacks of breath hold methods is the belief that patients with lung tumors and other lung disease may have trouble successfully completing the maneuver. Some reports suggest that the majority of patients cannot successfully perform the deep inspiration procedure (Keall et al. 2006). However, others do not observe as low a compliance rate for breath hold at the end of normal inspiration (Murphy et al. 2003). Recently, Glide-Hurst et al. (2008a) reported on a hybrid method combining breath hold with the aspects of free-breathing gating—breath hold at the end of normal inspiration, shorting the duration of breath hold to a few seconds and gating the accelerator directly with an active breathing control system—to improve the compliance rate.

2.3.2 Beam/Patient Synchronization

Beam gating relies on the assumption that the target position will return to the intended treatment region, either periodically (as in respiration gating) or by operator intervention (e.g., by shifting the patient to a more correct position). This type of system can be considered an "open-loop" feedback system. Another approach is to use a closed-loop feedback system, where the measurement of target position is used to adjust a control system, which actively synchronizes the target position with the treatment beam.

2.3.2.1 Robotic Synchronization

The first synchronization system was introduced on the CyberKnife, a compact linear accelerator mounted on a robotic arm. The entire accelerator can be repositioned in real time during treatment. Such a system is termed a "tracking" system because a monitoring system is used to determine target position, enabling the accelerator (and therefore the treatment beam) to follow, or track with, the target position. In such a tracking system, the monitoring system operates at a frequency of 20 Hz, and latency in repositioning of the treatment beam is expected to be longer. Thus, a robust prediction algorithm is necessary to ensure that the treatment beam is not "lagging behind" the target position. Such lag, if present, will introduce a systematic offset between the intended and the actually delivered dose locations.

2.3.2.2 Beam Aperture/Beam Line Synchronization

Rather than reposition the entire accelerator to reposition the treatment beam, the collimator can be adjusted in real time or the treatment beam itself can be deflected or scanned. Scanning beam methods are most commonly used in particle therapy, where magnets can control the beam location. For external beam photon therapy, where the beam is less easily steered, the preferred method is to adjust the collimator position to adjust the beam position. Most investigations with real-time tracking of the collimator have been performed in the venue of thoracic radiotherapy. In such a case, the collimator is made to "breathe" with the patient, oscillating in a manner that tracks the respiratory motion of the target. For this approach, each leaf in the multileaf collimator (MLC) is repositioned to shift the treatment beam in the plane normal to the beam axis. Note that, for a single fixed beam, target position changes along the beam axis cannot be corrected optimally. Generally, however, such motion along a beam axis causes minimal error in dose delivery to the target. Keall et al. (2001) developed methods to incorporate real-time position changes into dynamic MLC delivery of IMRT, given a known target position motion pattern and a rigid target. Several investigators have extended this approach to find optimal and practical leaf sequences to account for deforming targets, tissue density changes, and semiperiodic motion. However, to date, these algorithms remain experimental and have not been translated into clinical studies.

2.3.2.3 Couch

Tracking is the most commonly investigated closed-loop control system in radiotherapy, but another approach is to automatically adjust the patient position with the purpose of repositioning the target. Modern treatment couches have been shown to be capable of fast, accurate response to requested position changes. This approach appears feasible to adjust to low velocity, drifting changes in target position. However, one unanswered question is whether high-frequency, high-velocity position changes (such as those exhibited by prostate and lung tumors) could be accurately tracked due to couch adjustments possibly introducing secondary, inertial motion in the nonrigid patient. Also of concern, is whether patients would accept treatment on a couch undergoing such a motion pattern.

2.3.3 Issues Related to Rotational Therapy (Helical Tomotherapy and Volumetric Modulated Arc Therapy)

Due to the combination of modulated beam delivery with MLC and the simultaneous rotation of the gantry (and therefore the treatment beam) around the patient, intervention in rotational therapies is more complex than for delivery at static angles. Added complexity is due to the addition of rotation, but also to the added degrees of freedom in the delivery for some systems, including gantry rotational speed and dose rate modulation. To implement simple beam gating on a conventional gantry-based accelerator delivering rotational therapy, for example, one would need to gate not only beam delivery but also gantry rotation. Mechanically, this requirement presents challenges with current technology.

Most of the work in intervention for rotational therapy has been performed using helical tomotherapy. Briefly, helical tomotherapy is the delivery of modulated beams as the accelerator rotates around the patient while the patient is moved in the head-to-feet direction. This delivery produces a helical beam path in relation to the patient. Zhang et al. (2007b) evaluated a method to generate an optimal beam path given a predefined target motion pattern. This approach relies on breath coaching to maintain the predefined motion pattern. Lu (2008) developed a tracking technique for tomotherapy more flexible to irregular motion.

2.3.4 Intrafraction Dose Accumulation

As discussed in Section 2.1.3, many motion-related factors affect the delivered dose distribution, including the shape and magnitude of motion, the gradient and modulation of the dose distribution, and the position of structures and tissue density in relation to the dose distribution. A variety of methods to assess the cumulative dose delivered while considering moving anatomy are available, ranging from the very simple to the sophisticated and complex. The accuracy and necessity of a given method depend on the relative influence of these different parameters. In this section, we will first describe how dose mapping is used to generate the cumulative dose. Then, specific dose mapping algorithms will be described, paying particular attention to assumptions or limitations and expected accuracy.

Incidentally, similar algorithms are used to calculate the cumulative dose within a single fraction and over multiple fractions. The basic algorithms remain the same. However, the methods to accumulate dose over multiple treatment sessions must deal with the possibility of a change in patient mass (e.g., change in bladder or rectal filling, weight loss, and tumor regression due to treatment response) and other long-term pathologic changes (e.g., atelectasis forming or dissolving and radiation-induced fibrotic changes). Intrafraction dose accumulation algorithms

are generally simpler because one can assume that mass is conserved over the timescale of a single treatment session. Bladder, rectal, and gastric filling are exceptions to this mass conservation rule.

2.3.5 Basic Dose Mapping Theory

To compute the cumulative dose, the dose delivered on the anatomy during treatment delivery is summed. If the anatomy is stationary during treatment, one simply sums the dose delivered over time at each discrete spatial location in the anatomy, for all points of interest in the anatomy. If the anatomy is not stationary, the location of each point must be known at each instance of time. Note that, for a dense set of points (e.g., all voxels in a CT image or surface points on a delineated structure), it is implied that the correspondence between points at different time instances is known. Let us define the location of all anatomical points of interest at one instance of time an "anatomical instance." Generally, one selects a particular anatomical instance as a reference instance. This anatomical instance defines the anatomy, density, and spatial coordinate system upon which the cumulative dose will be evaluated. On a practical note, a high-quality image should be selected to represent the reference instance as artifacts present in a reference image may impact the evaluation of the cumulative dose to the anatomy.

The basic goal of dose mapping is to provide a means to map dose at one anatomical instance (termed the target instance) to the reference instance. Once dose has been mapped from the target instance space into the reference space, it is trivial to calculate the cumulative dose by simple summation at each point of interest. The key issues that differentiate the available dose mapping algorithms detailed below are the following:

- Spatial invariance of dose (does a change in anatomical position affect the shape of the dose distribution?)
- Complexity of mapping required to capture positional variability (can a rigid shift approximate the observed anatomical motion?)
- Applicability of motion models (can an analytical, empirical, or statistical model represent the observed motion?)

2.3.6 Dose Mapping Methods

2.3.6.1 Convolution

If the motion pattern can be considered "stationary," such that, on average, it does not vary in magnitude and shape and does not drift, and if the dose distribution is delivered by a static (non-IMRT) field, then it has been shown (Yan and Lockman 2001) that the motion-influenced dose distribution will be equal to the original, static dose distribution convolved with the probability density function of the motion pattern. Lujan et al. (1999) introduced such a convolution model for respiratory motion based on an analytical motion model. The analytical model was used to generate a statistical model of the motion by estimating the probability density function of the analytical model.

The cumulative dose is calculated by convolving the probability density function with the "static" dose distribution (the dose distribution calculated on the reference anatomy, without considering motion). A dose convolution method such as this one relies on several assumptions for high accuracy: each point of interest must move with exactly the same motion pattern (i.e., no deformation is present), motion does not change the shape of the dose distribution (dose is spatially invariant), and the motion pattern does not deviate statistically from the defined probability distribution.

In sites such as the thorax, the shape of the dose distribution will vary measurably with motion due to the large range of tissue densities present near the treatment site. For this reason, Chetty et al. (2003) and, separately, Beckham et al. (2002) introduced a fluence convolution method. In this method, rather than convolve the dose directly with a probability density function, the entrance fluence is convolved with the density function, and then the dose is calculated using this motion-blurred fluence. Fluence convolution has been used to model motion effects in the pelvis and thorax.

Fluence convolution methods, while reducing the effects of spatial dose variance, still rely on a rigid motion model, assuming that no deformation of the anatomy is present. Glide-Hurst et al. (2008b) developed a voxel-specific probability distribution model using dose convolution to correct for deformation effects. In this model, each point of interest is assigned a unique probability distribution based on a measured motion pattern. Dose convolution is then used to calculate the cumulative dose by convolving the probability distribution for each point of interest with the local dose distribution. To reduce spatial dose variance effects, a mean density representation was used, rather than the reference anatomy, for the dose calculation. In this approach, the average intensity at each spatial location due to motion is calculated to form the mean density for dose calculation.

Although corrections have been applied for spatial variance of the dose and nonrigid motion, convolution-based approaches still rely on an accurate statistical model of motion for accuracy. One approach is to generate the probability function directly from the measured motion pattern for each treatment session. This method eliminates the possible deviations from the motion pattern. If the motion pattern cannot be directly measured, another approach is to estimate likely deviations in the motion pattern from historical data. This method will generate a "worst-case" estimate of the cumulative dose but does not directly provide the cumulative dose for a single treatment session.

Statistical approaches essentially mimic continuous delivery of a static fluence under repetitious motion. The "interplay effect" (see Section 2.1.3) where time-varying fluence and time-varying motion interact to produce variations from the statistically expected cumulative dose, cannot be represented by simple convolution approaches. For fractionated therapy and semiperiodic motion, interplay effects have been shown to be small (Bortfeld et al. 2002). As the number of fractions decreases, for nonperiodic motion (such as digestive motion) and for tomotherapy delivery, interplay effects may be larger.

2.3.6.2 4D Imaging Methods

Statistical models decouple the delivery of radiation from the anatomical motion at the time of delivery. If coupling of anatomical location and delivery is required, one could image the patient periodically during a treatment session, calculate the dose on each instance of anatomy, and map and sum the dose to form the cumulative dose (Yan et al. 1999). We will term this method the "4D imaging method," because it relies on generating a high-quality "4D" representation of the 3D patient anatomy and 1D time variation of the 3D anatomy during treatment. This method would produce an exact representation of the cumulative dose assuming a high enough frequency of imaging and no error in the imaging, dose calculation, and mapping algorithms; however, it is generally limited by these exacting assumptions. The required imaging frequency and the effect of errors in dose calculation, imaging, and dose mapping (including image registration) are only beginning to be investigated, so the deliverable accuracy and precision of 4D imaging methods for dose accumulation are not well understood to date.

Generally, the approach has been to image in a manner to capture the major anatomical variability in a treatment session and to build in corrections for second-order variability. 4D imaging methods have mainly been used in the thorax and upper abdomen (e.g., for targets in the liver) as the motion is semiperiodic and can therefore be generally represented with a small number of images. Experience with 4D imaging methods for accumulating in sites affected by digestive motion is limited due to large number of images necessary to capture the unpredictable motion at sufficient time resolution and due to the generally accurate results of convolution methods in such sites.

Rosu et al. (2005) introduced a method to map dose from a target instance to a reference instance by interpolating the dose calculated on images from a 4D CT scan. Voxel correspondence between a target instance and reference instance is determined by deformable image registration between the target image and the reference image. If a dose calculation grid is associated with a particular target image (because the dose distribution for that target instance was calculated on the target image), then the registration can be used to interpolate the dose grid and generate a mapping back to the reference image. The accuracy of this method was investigated as a function of dose matrix resolution, number of 4D CT phase images required, and necessity of deformable image registration compared with rigid registration.

The interpolation method is analytically exact provided that the density of a voxel does not vary with respect to time in the 4D image (Siebers and Zhong 2008); however, lung expansion and contraction can introduce density variations with time. Heath and Seuntjens (2006) introduced a method to account directly for such density variations. Deformable image registration was used to directly deform the dose calculation grid from the rectilinear grid on the reference image to an arbitrary grid on the target image, where each grid voxel on the target image preserved the density of the corresponding grid voxel on the reference image. Monte Carlo methods were used to calculate dose on each arbitrary-shaped target grid from each target image of a 4D CT scan, and because the dose grid correspondences were known, cumulative dose was calculated by simply summing the dose at target grid locations to the corresponding reference grid locations.

Similar to the deformed dose grid method introduced by Heath and Seuntjens, the energy transfer method seeks to correct density-based errors in the interpolation method (Siebers and Zhong 2008). In this method, energy transferred to mass is calculated using Monte Carlo methods in a target instance-based rectilinear dose grid. Separately, the tissue mass within the grid is calculated using the target image. Energy transferred and mass are separately interpolated and mapped to a reference grid, where the quantities are separately accumulated. Dose is then calculated by taking the ratio of the accumulated energy transferred and the accumulated mass.

2.4 Summary and Conclusions

Because treatment sessions become longer due to the dramatic increase in hypofractionated treatment protocols, the assumption of the stationary patient becomes less valid. Precise and accurate monitoring of the target in real time has become necessary to ensure the precise and accurate delivery required to limit normal tissue toxicity due to the lack of protection offered by fractionation. These requirements have created the burgeoning field of patient monitoring and tracking. The field is beginning to see the introduction of optimal integrated systems that seek to provide safe, effective surveillance of the target and risk tissue in real time. Armed with an understanding of the available tools and their strengths and limitations, clinical users will be able to select and combine tools for their own needs. Finally, the user must always keep in mind the realistic requirements and performance expectations of surveillance and intervention for the clinical sites of interest. How to develop such requirements from knowledge of the expected motion, required tolerances, planned dose distributions, and uncorrected uncertainties is a key task, hopefully elucidated in this chapter.

References

Adamson, J. and Q. Wu. 2008. Prostate intrafraction motion evaluation using kV fluoroscopy during treatment delivery: A feasibility and accuracy study. *Medical Physics, 35*(5), 1793–1806.

Aubin, S., L. Beaulieu, S. Pouliot et al. 2003. Robustness and precision of an automatic marker detection algorithm for online prostate daily targeting using a standard V-EPID. *Medical Physics, 30*, 1825.

Balter, J. M., C. A. Pelizzari and G. T. Chen. 1992. Correlation of projection radiographs in radiation therapy using open curve segments and points. *Medical Physics, 19*(2), 329–334.

Baroni, G., G. Ferrigno, R. Orecchia and A. Pedotti. 2000. Real-time three-dimensional motion analysis for patient positioning verification. *Radiotherapy and Oncology, 54*(1), 21–27.

Beckham, W. A., P. J. Keall and J. V. Siebers. 2002. A fluence-convolution method to calculate radiation therapy dose distributions that incorporate random set-up error. *Physics in Medicine and Biology, 47*(19), 3465–3473.

Bentel, G. 1999. *Patient Positioning and Immobilization in Radiation Therapy.* McGraw-Hill, New York.

Berbeco, R. I., H. Mostafavi, G. C. Sharp and S. B. Jiang. 2005. Towards fluoroscopic respiratory gating for lung tumours without radiopaque markers. *Physics in Medicine and Biology, 50*(19), 4481–4490.

Bortfeld, T., K. Jokivarsi, M. Goitein, J. Kung and S. B. Jiang. 2002. Effects of intra-fraction motion on IMRT dose delivery: Statistical analysis and simulation. *Physics in Medicine and Biology, 47*(13), 2203–2220.

Buck, D., M. Alber and F. Nüsslin. 2003. Potential and limitations of the automatic detection of fiducial markers using an amorphous silicon flat-panel imager. *Physics in Medicine and Biology, 48*(6), 763–774.

Chetty, I. J., M. Rosu, N. Tyagi et al. 2003. A fluence convolution method to account for respiratory motion in three-dimensional dose calculations of the liver: A Monte Carlo study. *Medical Physics, 30*(7), 1776–1780.

Cho, B., P. R. Poulsen, A. Sloutsky, A. Sawant and P. J. Keall. 2009. First demonstration of combined kV/MV image-guided real-time dynamic multileaf-collimator target tracking. *International Journal of Radiation Oncology Biology Physics, 74*(3), 859–867.

Cho, Y., D. Moseley, J. Siewerdsen and D. Jaffray. 2005. Accurate technique for complete geometric calibration of cone-beam computed tomography systems. *Medical Physics, 32*, 968.

Court, L. E., M. Wagar, D. Ionascu, R. Berbeco and L. Chin. 2008. Management of the interplay effect when using dynamic MLC sequences to treat moving targets. *Medical Physics, 35*(5), 1926–1931.

Cui, Y., J. Dy, G. Sharp, B. Alexander and S. Jiang. 2007a. Robust fluoroscopic respiratory gating for lung cancer radiotherapy. *Physics in Medicine and Biology, 52*, 741–755.

Cui, Y., J. G. Dy, G. C. Sharp, B. Alexander and S. B. Jiang. 2007b. Multiple template-based fluoroscopic tracking of lung tumor mass without implanted fiducial markers. *Physics in Medicine and Biology, 52*(20), 6229–6242.

Cui, Y., J. G. Dy, B. Alexander and S. B. Jiang. 2008. Fluoroscopic gating without implanted fiducial markers for lung cancer radiotherapy based on support vector machines. *Physics in Medicine and Biology, 53*(16), N315–N327.

Docef, A. and M. Murphy. 2008. Reconstruction of 4D deformed CT for moving anatomy. *International Journal of Computer Assisted Radiology and Surgery, 3*(6), 591–598.

Gilhuijs, K. G. and M. van Herk. 1993. Automatic on-line inspection of patient setup in radiation therapy using digital portal images. *Medical Physics, 20*(3), 667–677.

Glide-Hurst, C. K., G. Hugo and A. Galerani. 2008a. Evaluation of intra- and interfraction reproducibility of a hybrid breath-hold gating technique throughout the course of radiation therapy. *International Journal of Radiation Oncology Biology Physics, 72*(1, Supplement 1), S627.

Glide-Hurst, C. K., G. D. Hugo, J. Liang and D. Yan. 2008b. A simplified method of four-dimensional dose accumulation using the mean patient density representation. *Medical Physics, 35*(12), 5269–5277.

Godfrey, D. J., F.-F. Yin, M. Oldham, S. Yoo and C. Willett. 2006. Digital tomosynthesis with an on-board kilovoltage imaging device. *International Journal of Radiation Oncology Biology Physics, 65*(1), 8–15.

Heath, E. and J. Seuntjens. 2006. A direct voxel tracking method for four-dimensional Monte Carlo dose calculations in deforming anatomy. *Medical Physics, 33*(2), 434–445.

Hugo, G., J. Liang and D. Yan. 2007. Direct detection of the tumor trajectory using raw cone beam CT projections. *Medical Physics, 34*(6), 2545.

Isaksson, M., J. Jalden and M. J. Murphy. 2005. On using an adaptive neural network to predict lung tumor motion during respiration for radiotherapy applications. *Medical Physics, 32*(12), 3801–3809.

Jähne, B. 2005. *Digital Image Processing: Concepts, Algorithms and Scientific Applications,* 6th ed., Springer, Berlin.

Keall, P. J., V. R. Kini, S. S. Vedam and R. Mohan. 2001. Motion adaptive X-ray therapy: A feasibility study. *Physics in Medicine and Biology, 46*(1), 1–10.

Keall, P. J., G. S. Mageras, J. M. Balter et al. 2006. The management of respiratory motion in radiation oncology report of AAPM Task Group 76. *Medical Physics, 33*(10), 3874–3900.

Keall, P. J., A. D. Todor, S. S. Vedam et al. 2004. On the use of EPID-based implanted marker tracking for 4D radiotherapy. *Medical Physics, 31*(12), 3492–3499.

Kubo, H. and B. Hill. 1996. Respiration gated radiotherapy treatment: A technical study. *Physics in Medicine and Biology, 41,* 83–91.

Lam, K. L., R. K. Ten Haken, D. L. McShan and A. F. Thornton. 1993. Automated determination of patient setup errors in radiation therapy using spherical radio-opaque markers. *Medical Physics, 20*(4), 1145–1152.

Langen, K. M., T. R. Willoughby, S. L. Meeks et al. 2008. Observations on real-time prostate gland motion using electromagnetic tracking. *International Journal of Radiation Oncology Biology Physics, 71*(4), 1084–1090.

Lindeberg, T. 1993. Detecting salient blob-like image structures and their scales with a scale-space primal sketch: A method for focus-of-attention. *International Journal of Computer Vision, 11*(3), 283–318.

Liu, H. H., P. Balter, T. Tutt et al. 2007. Assessing respiration-induced tumor motion and internal target volume using four-dimensional computed tomography for radiotherapy of lung cancer. *International Journal of Radiation Oncology Biology Physics, 68*(2), 531–540.

Low, D. A., P. J. Parikh, W. Lu et al. 2005. Novel breathing motion model for radiotherapy. *International Journal of Radiation Oncology Biology Physics, 63*(3), 921–929.

Lu, W. 2008. Real-time motion-adaptive delivery (MAD) using binary MLC: II. Rotational beam (tomotherapy) delivery. *Physics in Medicine and Biology, 53*(22), 6491–6511.

Lujan, A., E. Larsen, J. Balter and R. Ten Haken. 1999. A method for incorporating organ motion due to breathing into 3D dose calculations. *Medical Physics, 26,* 715.

Mao, W., N. Riaz, L. Lee, R. Wiersma and L. Xing. 2008a. A fiducial detection algorithm for real-time image guided IMRT based on simultaneous MV and kV imaging. *Medical Physics, 35*(8), 3554–3564.

Mao, W., R. D. Wiersma and L. Xing. 2008b. Fast internal marker tracking algorithm for onboard MV and kV imaging systems. *Medical Physics, 35*(5), 1942–1949.

McCall, K. C. and R. Jeraj. 2007. Dual-component model of respiratory motion based on the periodic autoregressive moving average (periodic ARMA) method. *Physics in Medicine and Biology, 52*(12), 3455–3466.

Murphy, M. J. 1997. An automatic six-degree-of-freedom image registration algorithm for image-guided frameless stereotaxic radiosurgery. *Medical Physics, 24*(6), 857–866.

Murphy, M. J., S. D. Chang, I. C. Gibbs et al. 2003. Patterns of patient movement during frameless image-guided radiosurgery. *International Journal of Radiation Oncology Biology Physics, 55*(5), 1400–1408.

Murphy, M. J. and S. Dieterich. 2006. Comparative performance of linear and nonlinear neural networks to predict irregular breathing. *Physics in Medicine and Biology, 51*(22), 5903–5914.

Nederveen, A., J. Lagendijk and P. Hofman. 2000. Detection of fiducial gold markers for automatic on-line megavoltage position verification using a marker extraction kernel (MEK). *International Journal of Radiation Oncology Biology Physics, 47*(5), 1435–1442.

Ohara, K., T. Okumura, M. Akisada et al. 1989. Irradiation synchronized with respiration gate. *International Journal of Radiation Oncology Biology Physics, 17*(4), 853–857.

Ozhasoglu, C. and M. J. Murphy. 2002. Issues in respiratory motion compensation during external-beam radiotherapy. *International Journal of Radiation Oncology Biology Physics, 52*(5), 1389–1399.

Poulsen, P. R., B. Cho, K. Langen, P. Kupelian and P. J. Keall. 2008. Three-dimensional prostate position estimation with a single X-ray imager utilizing the spatial probability density. *Physics in Medicine and Biology, 53*(16), 4331–4353.

Putra, D., O. C. L. Haas, J. A. Mills and K. J. Burnham. 2008. A multiple model approach to respiratory motion prediction for real-time IGRT. *Physics in Medicine and Biology, 53*(6), 1651–1663.

Rosu, M., I. J. Chetty, J. M. Balter et al. 2005. Dose reconstruction in deforming lung anatomy: Dose grid size effects and clinical implications. *Medical Physics, 32*(8), 2487–2495.

Ruan, D., J. A. Fessler and J. M. Balter. 2007. Real-time prediction of respiratory motion based on local regression methods. *Physics in Medicine and Biology, 52*(23), 7137–7152.

Sadowsky, O., J. D. Cohen and R. H. Taylor. 2006. Projected tetrahedra revisited: A barycentric formulation applied to digital radiograph reconstruction using higher-order attenuation functions. *IEEE Transactions on Visualization and Computer Graphics, 12*(4), 461–473.

Santanam, L., C. Noel, T. R. Willoughby et al. 2009. Quality assurance for clinical implementation of an electromagnetic tracking system. *Medical Physics, 36*(8), 3477–3486.

Seiler, P. G., H. Blattmann, S. Kirsch, R. K. Muench and C. Schilling. 2000. A novel tracking technique for the continuous precise measurement of tumour positions in conformal radiotherapy. *Physics in Medicine and Biology, 45*(9), N103–N110.

Sharp, G. C., S. B. Jiang, S. Shimizu and H. Shirato. 2004. Prediction of respiratory tumour motion for real-time image-guided radiotherapy. *Physics in Medicine and Biology, 49*(3), 425–440.

Shirato, H., S. Shimizu, T. Kunieda et al. 2000. Physical aspects of a real-time tumor-tracking system for gated radiotherapy. *International Journal of Radiation Oncology Biology Physics, 48*(4), 1187–1195.

Siebers, J. V. and H. Zhong. 2008. An energy transfer method for 4D Monte Carlo dose calculation. *Medical Physics, 35*(9), 4096–4105.

Trofimov, A., C. Vrancic, T. C. Y. Chan, G. C. Sharp and T. Bortfeld. 2008. Tumor trailing strategy for intensity-modulated radiation therapy of moving targets. *Medical Physics, 35*(5), 1718–1733.

Vedam, S. S., P. J. Keall, A. Docef et al. 2004. Predicting respiratory motion for four-dimensional radiotherapy. *Medical Physics, 31*(8), 2274–2283.

Vedam, S. S., V. R. Kini, P. J. Keall et al. 2003. Quantifying the predictability of diaphragm motion during respiration with a noninvasive external marker. *Medical Physics, 30*(4), 505–513.

Watanabe, Y. and L. L. Anderson. 1997. A system for nonradiographic source localization and real-time planning of intraoperative high dose rate brachytherapy. *Medical Physics, 24*(12), 2014–2023.

Wiersma, R. D., N. Riaz, S. Dieterich, Y. Suh and L. Xing. 2009. Use of MV and kV imager correlation for maintaining continuous real-time 3D internal marker tracking during beam interruptions. *Physics in Medicine and Biology, 54*(1), 89–103.

Wu, H., G. C. Sharp, B. Salzberg et al. 2004. A finite state model for respiratory motion analysis in image guided radiation therapy. *Physics in Medicine and Biology, 49*(23), 5357–5372.

Yan, D., D. A. Jaffray and J. W. Wong. 1999. A model to accumulate fractionated dose in a deforming organ. *International Journal of Radiation Oncology Biology Physics, 44*(3), 665–675.

Yan, D. and D. Lockman. 2001. Organ/patient geometric variation in external beam radiotherapy and its effects. *Medical Physics, 28*(4), 593–602.

Yu, C. X., D. A. Jaffray and J. W. Wong. 1998. The effects of intrafraction organ motion on the delivery of dynamic intensity modulation. *Physics in Medicine and Biology, 43*(1), 91–104.

Zeng, R., J. A. Fessler and J. M. Balter. 2005. Respiratory motion estimation from slowly rotating X-ray projections: Theory and simulation. *Medical Physics, 32*(4), 984–991.

Zhang, Q., A. Pevsner, A. Hertanto et al. 2007a. A patient-specific respiratory model of anatomical motion for radiation treatment planning. *Medical Physics, 34*, 4772.

Zhang, T., W. Lu, G. H. Olivera et al. 2007b. Breathing-synchronized delivery: A potential four-dimensional tomotherapy treatment technique. *International Journal of Radiation Oncology Biology Physics, 68*(5), 1572–1578.

3

Validation of Image Registration

Kristy K. Brock
University of Michigan

Michael Velec
University Health Network

Jenny H. M. Lee
University Health Network

3.1 Introduction

Validation of image registration is a challenging task. The vast majority of image registration algorithms rely on models of human tissue that often deviate substantially from the true biomechanics. Even when biomechanical models are employed, the algorithms must still rely on other approximations and models. The combination of these approximate models with the reliance of the registration algorithms on the images themselves (i.e., the accuracy of the registration algorithm is tied to the complexity, clarity, and uncertainties in the image) leads to a difficulty in validating them. One can simplify the relationship between the two images to be registered (e.g., image a physical phantom or generate a mathematical phantom) and therefore know the "correct" registration result, or one can estimate the "correct" registration and use realistic, clinical images. This complexity in validation differs greatly from other aspects of radiotherapy, such as validation of dose calculation algorithms. Given these circumstances, a multistep validation approach is likely the best method to ensure an accurate and robust registration algorithm.

The goal of image registration is to find the correspondence of each voxel between image A and image B. Unlike the goal of autosegmentation, where the algorithm only needs to find the boundary of the organ, image registration has the additional burden of needing to find the correspondence of voxels within the organs as well. This becomes particularly important for advanced applications of deformable registration, such as dose accumulation, adaptive radiotherapy, and response assessment.

3.2 Methods of Validation

There are several different methods to validate image registration. They can be broadly categorized into the following groups:

1. Propagation of region of interest (ROI) contours between registered images
2. Propagation of identified points between registered images
3. Registration of images of physical phantoms with known/identified offsets
4. Registration of mathematical phantoms with known transformations for each voxel

Categories 1 and 2, the propagation of contours and points between registered images, are typically performed on clinical images where the transformation or offsets are not known; however, they can also be used for images of phantoms, either physical or mathematical.

Several issues confound the ability to employ these validation techniques, including uncertainties in identifying the true boundary of ROI, uncertainties in finding an accurately identifying naturally occurring or implanted fiducials within the images, creating physical phantoms that represent the complexity of the human body, and generating mathematical phantoms that represent the complexity of both the human body and the uncertainties and noise contribution of imaging devices.

3.2.1 Propagation of ROI Contours

Every image contains some number of ROIs, including tumors, organs, or other identifiable substructures (e.g., vessels and stents). Once an expert user has contoured these ROIs, they can be used to validate the image registration algorithm. In an ideal situation, the expert user would perfectly contour the ROIs on each image (contours A on image A and contours B on image B) and the image registration algorithm would propagate the contours on image A onto image B (resulting in contours A′). The differences between contours A′ and contours B would be due to uncertainties in the image registration algorithm. The additional uncertainties arise due to the expert user's ability to contour ROIs on the image A and image B. If the expert user contoured the ROIs in image A multiple times, there would be variations in these contours. These uncertainties in contours A and contours B must be separated from the uncertainties in the registration of image A to image B.

Repeat contouring is needed to assess the uncertainty in the experts' ability to contour the ROIs. If different experts contoured structures in the corresponding images (i.e., expert 1 contoured image A and expert 2 contoured image B), then the uncertainty between experts must be assessed. Once the uncertainty between the repeat contouring is assessed, the difference between the propagated contours from image A (contours A′) and the actual contours on image B (contours B) can be compared with the uncertainty in the repeat contours. Statistical tests can then be used to determine if the differences between contours A′ and contours B can be distinguished from the uncertainties in the contours themselves.

There are several quantitative metrics that can be used to assess the differences between contours A′ and contours B and the uncertainty in repeat contouring by the observer. They can be divided into overlap-based metrics and surface distance-based metrics.

The two most common surface distance-based metrics are the mean distance to agreement and the maximum distance to agreement. The distance to agreement is computed by representing the contours as a series of points (typically as triangular meshes) and then finding the shortest Euclidean distance between each vertex on contours A′ and the surface of contours B and then averaging over all distances (or finding the maximum). When computing the distances (either maximum or mean), it is usually prudent to evaluate the distance from contours A′ to contours B and from contours B to contours A′. For smooth, regular contours, the results will be very similar (Figure 3.1, model I); however, for irregular contours (Figure 3.1, model II), the maximum difference can be significant and the mean may be effected if the deviation is significant.

The most common overlap metric is the Dice similarity coefficient (DSC), which is defined as two times the overlap of A′ and B, divided by the sum of the volume of A′ and B, as shown below:

$$DSC = 2(A′ \cap B)/(A′ + B)$$

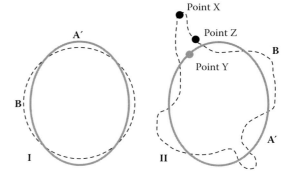

FIGURE 3.1 When comparing the distance between the dashed black (B) and solid gray (A′) contours in model I, the evaluation in each direction (i.e., A′ to B or B to A′) will result in the same average distance. However, for model II, the direction will impact the results. For example, the closest distance of point X on B will be point Y on A′; however, the closest distance to point Y on A′ will be point Z on B. Note that the point correspondence between surfaces is typically defined as the closest distance between surfaces and is not restricted following a trajectory that is perpendicular to the surface.

If A′ and B have no overlap, then the DSC is 0, and as the contours become identical, the DSC approaches a value of 1.

Other quantitative metrics include positive predictive value and sensitivity:

$$PPV = TP/(TP + FP) = (A′ \cap B)/[(A′ \cap B) + (A′ - (A′ \cap B))]$$

$$Sensitivity = TP/(TP + FN) = (A′ \cap B)/[(A′ \cap B) + (B - (A′ \cap B))]$$

where PPV is positive predictive value and TP, FP, and FN are the number of true positives, false positives, and false negatives, respectively.

There are several strengths in the use of contour propagation to evaluate the accuracy of image registration. Increasing the number of contours used will increase the comprehensiveness of the accuracy of the registration (e.g., if every structure in the image is contoured, the results will evaluate the accuracy over the entire image). Contouring is a common activity in radiotherapy, so expert users are easy to identify. It is a technique that can be used on clinical images, physical phantoms, and mathematical phantoms.

There are also several weaknesses to the use of contour propagation to evaluate the accuracy of image registration. The uncertainty in contouring some ROIs can be large for regions that do not have clear boundaries, resulting in a large inherent uncertainty in assessing the registration. The results only evaluate the registration at the boundary of an ROI, not in the volume throughout the ROI. Because the ROI boundary typically has a stronger gradient in the voxel intensity compared with the internal volume of the ROI, the registration algorithm (often driven by these gradients) will likely perform better at these boundaries. This may result in a higher estimate of the accuracy than that exists throughout the entire image volume.

3.2.2 Propagation of Identified Points

In addition to ROIs, corresponding points can also be identified on each image to be registered. The points can be either anatomical (e.g., vessel bifurcations and calcifications), as shown in Figure 3.2, or implanted fiducials (e.g., prostate markers and surgical clips). An expert user or a validated automated algorithm can identify the points on each image. The known transformation can then be calculated from the two known positions. This can be compared with the predicted transformation of the point from the deformable registration algorithm.

Target registration error (TRE) was introduced by Fitzpatrick et al. (1998) and is the distance between corresponding points after registration. The TRE will be affected by the error in localizing the fiducial points used in the analysis. TRE is defined as:

$$\text{TRE} = \frac{1}{n}\sum_{i=1}^{n} T(\mathbf{X}_i) - \mathbf{Y}_i,$$

where *n* is the number of corresponding fiducials (identified to represent the target to be register but not used to drive the registration) identified in the images, and *T* is the transformation applied to point $\mathbf{X}_i$ in image *X* to align it to image *Y*, where it should be aligned with $\mathbf{Y}_i$. $\mathbf{X}_i$ and $\mathbf{Y}_i$ are vectors; therefore, the TRE is the 3D vector magnitude.

In addition to the calculation of the TRE, the error in the propagation of identified points can also be represented as the average (signed) or average absolute error in each of the cardinal directions [left–right (LR), anterior–posterior (AP), and superior–inferior (SI)]. It is typically reported with the corresponding standard deviation.

3.2.3 Registration of Images of Physical Phantoms with Known/Identified Offset

Physical phantoms can also be used to validate both rigid and nonrigid image registration. For rigid registration, geometric or anthropomorphic phantoms can be imaged and then translated and/or rotated by a known amount and reimaged. These images then have a known transformation that has been applied and the registration results can be compared against these known transformations.

For physical phantoms that include deformation, the validation is more complex. Although the overall motion can be known, the exact motion of each point in the image obtained from imaging the phantom at different deformation states is not known. Typically developers of deformable phantoms have included markers in or on their phantoms. These markers are either nonradiopaque and on the surface of the phantom or radiopaque internal markers that are then masked out once the images are acquired. To mask out the points, the region of the marker in the image is identified and the intensities in these voxels are replaced with noise that is characteristic of the surrounding voxels. In addition to using the masked out points as known transformations in the image, the ROI methods (described in

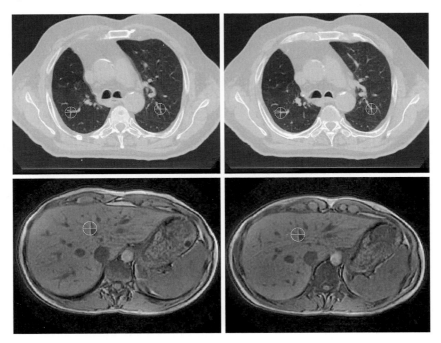

FIGURE 3.2 Selection of anatomical landmarks (in this case, vessel bifurcations) for registration validation. Landmarks (white crosshairs) are identified in the left and right lungs on the inhale 4D CT (top left) and the corresponding landmarks in the exhale 4D CT (top right). Similarly, landmarks are identified on MR for liver at inhale (bottom left) and exhale (bottom right). Note that anatomical landmarks could also be used for multimodality image registration.

Section 3.2.1) can also be used to validate the registration performed on deformable phantoms.

Numerous rigid physical phantoms have been developed, both commercially and in a research environment. Deformable phantoms have also been developed, although far fewer in number than rigid. Among the deformable phantoms are two phantoms to simulate lung motion (Kashani et al. 2007; Serban et al. 2008), both with internally placed radiopaque fiducials that could be masked for independent registration, and a deformable pelvic phantom by Kirby et al. (2011), which has nonradiopaque markers on the surface and whose motion is captured by a camera.

3.2.4 Registration of Mathematical Phantoms with Known Transformations for Each Voxel

One limitation of deformable physical phantoms is the inability to know the transformation of each voxel, reducing the ability to thoroughly validate the transformation generated from the deformable registration algorithm. An alternative to deformable physical phantoms is a deformable mathematical phantom. A deformable mathematical phantom is an image, either obtained from a phantom or a patient or generated through simulation, which then has a known transformation map applied to it. In this way, the transformation of each voxel in the image is known, because it was deliberately applied to each voxel. The limitation of mathematical phantoms is to ensure that the transformation is realistic and that the generated deformed image has the appropriate noise variation from the original image that would be expected for a repeat image of an actual patient. To ensure that the applied transformation represents the complexity of human anatomy, patient data are often studied and models are developed in attempt to reproduce this complexity. Care must be taken to avoid applying a deformation using the same model (e.g., B-splines) that the user employs to resolve the deformation, as this would create a deformation that was biased to the results expected from the deformable registration algorithm. Simulated noise patterns can also be applied to the generated images to produce noise variations between the two image sets that would be consistent with the noise variation seen in repeat images of a patient. In addition, mathematical phantoms can be developed that introduce deformation and simulate an image of a different modality [e.g., simulated images from a computed tomography (CT) and magnetic resonance (MR) with deformation introduced].

The benefit of mathematical phantoms is that the deformation of each voxel is precisely known. This allows the unique opportunity to validate the accuracy of the image registration throughout the entire image, which is not possible using the other techniques described. One can argue that the assessment of the deformable registration algorithm accuracy using a mathematical phantom represents the "best-case scenario" because the phantom does not likely include the complexity of the human anatomy and imaging systems. However, one may also argue that if the simulated deformation deviates largely from what is expected from the human anatomy and the deformable registration algorithm is optimized to model the actual human, these simulations may represent a possible worst-case scenario. Either way, mathematical phantoms represent a powerful tool that must be used with caution when validating deformable registration algorithms.

3.3 Multiinstitutional Studies of Registration Accuracy

There have been two radiation oncology-focused multiinstitutional studies to evaluate the registration accuracy of deformable registration algorithms: one by Kashani et al. (2008), which used a deformable physical phantom and evaluated 8 registration algorithms, and another by Brock and the Deformable Registration Accuracy Consortium (2010), which used clinical data and evaluated 21 registration algorithms.

3.3.1 Multiinstitution Physical Phantom Study

In the study by Kashani et al. (2007), a physical phantom was developed, which mimicked a breathing lung with a surrounding rib cage. Forty-eight small plastic markers were placed throughout the phantom, deliberately capturing the motion of the high-contrast areas as well as the areas of relatively uniform intensity. In addition, markers were placed at the areas of discontinuities in motion (i.e., at the lung–rib interface, where the lung tissue was sliding in the superior–inferior direction and the ribs had significantly less motion). CT images (resolution $0.78 \times 0.78 \times 1$ mm) were acquired of the phantom at two states of deformation. The phantom was constructed by modifying an anthropomorphic plastic chest wall with a skeleton to include a compressible section made of high-density foam with four spheres simulating tumors of different sizes. The "exhale" scan compressed the phantom by initiating 30 mm of motion of the compression plate, which mimicked the diaphragm motion, relative to the "inhale" state of no compression. The 48 markers were manually localized on each image and then digitally removed from the images, replacing the high-contrast voxels with intensities of the immediate surrounding region.

Eight algorithms were evaluated with the phantom: two used a thin-plate spline (TPS) model, three used a B-spline model, and the remaining three used a Demons algorithm, fluid flow, and a free form with calculus of variations. One of the TPS algorithms and one B-spline algorithm cropped the image to include only the "lung" anatomy and the other two B-spline algorithms masked the vertebrae. The image registration error was defined as the difference between the manually measured exhale marker position and the estimated position based on the deformation map from each registration technique. The difference was quantified in each direction (LR, AP, and SI) as well as the 3D vector distance. The maximum error in each component was reported. The mean and standard deviation of the 3D error were reported across all markers for each algorithm. To describe the distribution of the errors, a differential histogram was also presented to describe the distribution of the errors (in 2 mm increment bins) for each algorithm.

The average 3D vector magnitude of the error ranged from 1.7 to 3.9 mm and the standard deviation ranged from 1.1 to 3.0 mm. The maximum 3D vector error ranged from 5.1 to 15.4 mm. It is interesting to note that the algorithm with the smallest average error (1.7 mm) also had the smallest standard deviation (1.1 mm) and one of the smallest maximum errors (5.5 mm; smallest vector error was 5.1 mm), and the algorithm with the largest average error (3.9 mm) also had the largest standard deviation (3.0 mm) and the largest maximum error (15.4 mm). The maximum LR component errors ranged from 1.3 to 7.7 mm, AP ranged from 1.0 to 7.3 mm, and SI ranged from 4.1 to 15.2 mm. The maximum SI component error was the largest for all algorithms, except one where the maximum LR error was larger, and corresponded to the largest component of motion between the image sets.

Two interesting points were noted from this study. The first was that the distributions of errors from the three B-spline algorithms were considerably different. Although two algorithms had approximately 75% of the points registered with a 3D vector error of less than 2 mm, the third B-spline method had less than 50% of the points registered with an error of less than 2 mm. In addition, the algorithm that had reduced accuracy also had 3D errors exceeding 15 mm, where the others had maximum errors of less than 8 mm each. It was noted by the author that the B-spline algorithm that had the largest errors used a single-resolution knot spacing, whereas the other two B-spline algorithms used a multiresolution approach for the knot spacing. This result highlights that the implementation of algorithms can make a significant difference and that the reported results from one algorithm implementation cannot be applied to another implementation of the same type of algorithm.

The second note of interest from the results was the evaluation of the algorithms together with respect to the intensity variation at each point. For example, when evaluating the overall error of each point (across all algorithms), it was noted that one marker had a large mean error (across all algorithms) and a relatively small standard deviation, indicating that most algorithms failed to accurately register this point. It was noted that this marker was in a region with relatively low intensity distribution and was located next to a high-intensity region with a large change in intensity between inhale and exhale but without a large amount of deformation. Evaluation across all algorithms also revealed a slight increase in the average error as the magnitude of the motion of the marker increased; however, the magnitude of motion did not directly predict for error magnitude.

3.3.2 Multiinstitution Clinical Data Study

In the study led by Brock and the Deformable Registration Accuracy Consortium (2010), anonymized clinical data sets were used, including inhale and exhale reconstructed phases from a 4D CT scan of a lung patient, inhale and exhale reconstructed phases from a 4D CT scan of a liver patient as well as an MR

image at exhale, and repeat MR images of a prostate patient with a full and empty rectum. The lung images (resolution $0.98 \times 0.98 \times 2.5$ mm) had the lungs, tumor, and external surface delineated and were provided to the 22 participants. Seventeen bronchial bifurcations were identified in the left lung and an additional 17 in the right lung. Two calcifications were identified in the heart and two in the aorta. The SI motion of the lungs, based on the bifurcations, ranged from 0 to 15 mm and the calcification motion ranged from 0 to 5 mm in the SI direction.

The liver images (CT resolution $0.98 \times 0.98 \times 2.5$ mm and MR resolution $1.7 \times 1.7 \times 7.0$ mm) had the liver, kidneys, bowel, duodenum, esophagus, and tumor contoured on the exhale data set per clinical practice. The external surface and liver were also contoured in the inhale data set and the MR data set. Intravenous contrast was delivered before the CT scan, which enabled the identification of the tumor and the vessels within the liver. Twenty-five vessel bifurcations were identified in the liver, five vessel bifurcations were identified in the left kidney, and six vessel bifurcations were identified in the right kidney. The SI motion of the liver ranged from 7.5 to 15 mm and the SI motion of the kidneys ranged from 2.5 to 12.5 mm. Seven pairs of vessel bifurcations in the liver were identified between the exhale reconstruction of the 4D CT scan and the exhale MR scan.

The repeat prostate MR images (resolution $0.70 \times 0.70 \times 2$ mm) had the prostate, rectum, and bladder contoured on each image. In addition, three implanted gold markers (1×5 mm) were identified on each image. Significant AP deformation was introduced due to the rectal filling.

Each institution was sent all image sets and corresponding contours but was not sent the information on the identified fiducial points in each image. The participants were instructed to deform the exhale liver CT to the inhale liver CT, inhale lung CT to exhale lung CT, MR1 prostate to MR2 prostate, and liver MR to liver exhale CT. In return, the participants sent back a summary of their algorithm, their processing times, and the resulting deformation vector field for each deformation processed. The known displacement for each identified point was compared with the result of each participant's deformation vector field.

Twenty-two participants returned the data for the lung 4D CT, 17 for the liver 4D CT, 3 for the liver MR-CT, and 3 for the prostate MR-MR studies. For the lung 4D CT study, all algorithms had a mean absolute error in each direction of less than 2.5 mm. Twenty (out of 22) of the algorithms had a standard deviation in each direction of less than 2.5 mm. Nine algorithms had a maximum error in each direction of less than 5.0 mm and 17 had a maximum error in each direction of less than 7.5 mm. The largest maximum error was 1.2 cm.

For the liver 4D CT registration, seven (out of 17) algorithms had a mean absolute error in the liver of less than 2.5 mm in each direction and eight (out of 17) had a mean absolute error in the liver of less than 5.0 mm in each direction. Twelve algorithms had an absolute standard deviation of less than 2.5 mm in each direction. All algorithms had a maximum error of greater than 5.0 mm in at least one direction, and three algorithms had a maximum error of greater than 1 cm, occurring in the SI direction. Overall, the algorithms did a better job registering the kidneys, because more algorithms had mean absolute errors less than 2.5 mm in each direction (11 for the right kidney and 16 for the left kidney), and the maximum error in any direction was 5.6 mm.

Only three participants took part in the registration involving MR images. The mean absolute error in each direction for the liver MR-CT ranged from 1.1 to 5.0 mm. The absolute standard deviations were less than 2.5 mm for all algorithms and the maximum error was less than 7.0 mm. For the prostate, the mean absolute error ranged from 0.4 to 6.2 mm, the standard deviations ranged from 0.3 to 3.4 mm, and the maximum errors from 5.0 to 8.7 mm.

Although the study was limited to four data sets, several interesting observations could be made. First, all algorithms had a reasonable accuracy in the lung (absolute errors less than 2.5 mm), which is very promising. The potential for large maximum errors should be noted, however, and caution should be used when validating algorithms to ensure that there are not local areas of large misregistration. Overall, the high level of registration accuracy did not always translate from the lung application to the liver, because not all algorithms had an absolute error of less than 2.5 mm. This is especially noteworthy for clinical implementation, indicating that translating an algorithm from one clinical site to another will require an additional testing to ensure sufficiently accurate results. Very few algorithms reported registration results for the multimodality and MR-MR registration. This indicates a clear need for development as multimodality images are becoming more prevalent in radiation oncology clinical practice. In addition, similar to the physical phantom study described above, algorithm implementation was shown to have an effect on the registration results. Three algorithms reported using the Demons algorithm for liver 4D CT registration and the resulting accuracy ranged from 2.3 to 4.8 mm. The variation was even more extreme for algorithms using TPS, where the accuracy ranged from 2.1 to 7.8 mm for the liver 4D CT registration.

3.4 Uncertainty in the Measurement of the Error

There is always an inherent uncertainty when measuring the error of the deformable registration, whether that is in the selection of points, delineation of structures of interest, or creation of the mathematical phantom. Several authors have evaluated, through repeat identification, the uncertainty in picking points. The uncertainty in identifying corresponding bifurcations of the vessels and bronchial tree in the lung and bifurcations of the vessels in the liver is smaller than the image voxel size (Coselmon et al. 2004; Brock et al. 2005). In addition to the uncertainty in contour and point identification, several other uncertainties exist, two of which will be highlighted here: the effect of the number and distribution of the points and the effect of the slice thickness of the image.

3.4.1 Effect of Sample Points

A recent paper by Castillo et al. (2009) investigates the impact of the number of points used for evaluation of a deformable registration algorithm. An interface was developed to streamline and manage the selection of more than 1000 corresponding pulmonary landmark features from 4D CT data. For this study, the inhale and exhale reconstructions of the 4D CT from five patients were obtained. The images were cropped to include the rib cage and down-sampled to a resolution of 1.16×1.16 mm (from 0.97×0.97 mm) with a slice thickness of 2.5 mm.

APRIL, an in-house developed MATLAB®-based software interface, was used to facilitate the manual selection of pairs of features between image sets. To improve the efficiency and enable the selection of a large number of points, the software automatically localizes the corresponding feature on the second image using normalized crosscorrelation. Following the completion of the point selection, a summary file is generated with the Cartesian and spherical coordinates of all points and displacement magnitudes.

Landmark pairs were identified on five 4D CT sets. The total number of landmark pairs per image set ranged from 1166 to 1561, and these points were distributed quite uniformly between the left and right lungs for each patient. The average displacements ranged from 4.01 to 9.42 mm across patients, with a maximum displacement of 12.65 to 24.78 mm in each patient, indicating that the patients represented a fairly typical breathing motion magnitude. Randomly selected subsets of points were identified to evaluate the reproducibility of selection by having a second reader identify the corresponding point on the secondary image in addition to having the primary reader reselect the corresponding points. The repeated mean selection error ranged from 0.61 to 1.11 mm for the primary reader and from 0.74 to 1.14 mm for the secondary reader, indicating a high fidelity in the point correspondence.

To evaluate the utility of the large landmark set, the authors then performed image registration using an optical flow deformable image registration (DIR) algorithm and a landmark-based DIR. Through these evaluations and with numerical simulations, the authors demonstrated that "the statistical uncertainty of the DIR spatial error estimate is inversely proportional to the square root of the number of landmark point pairs and directly proportional to the standard deviation of the spatial error specific to the DIR." They subsequently recommended that a large set of validation landmarks (more than 1000) be used for thorough evaluation of spatial accuracy in the lung. In their conclusion, the authors recognize that this large number of landmark pairs is not feasible for routine quality assurance but may play a role in clinical acceptance.

3.4.2 Effect of Voxel Size and Image Orientation

Registration error vector, defined as $T(\mathbf{X}_i) - \mathbf{Y}_i$ from Section 3.2.2, needs to be interpreted in the content of voxel size, fiducial localization error (FLE) (Fitzpatrick and West 2001), and magnitude of the motion corrected for by the registration method. Voxel size quantifies the discretization error of the continuous space to the discrete image space. FLE is the error in localizing a point by an observer or algorithm, where the point could be an implanted fiducial marker or a naturally occurring landmark.

A few simple observations can be made.

- When there is translation, the voxel size in the direction of translation needs to be smaller than the typical scale of motion in that direction. Otherwise, the TRE would be due entirely to FLE and is meaningless in terms of the accuracy of registration.
- Voxel size is typically anisotropic in radiation therapy, such that FLE and discretization error are greater in the z-direction (i.e., slice thickness) than the x- and y-directions (i.e., the in-plane resolution). If the motion involves rotation about the x- or y-axis, the FLE and discretization error from the z-direction would be projected onto the y- and x-directions.
- For image registration involving MR images, the angle of acquisition must be taken into account. The imaging plane of an MR image can be taken at any angle to the imaging table to give the best anatomical view of an organ. If the image pairs were acquired with different acquisition angle, then there are additional uncertainties in the identification of landmark pairs and accuracy assessment.
- When there are subregions within the image that have local deformation that cannot be resolved with rigid registration, the TRE can only fairly assess the deformable registration method if the points are in those regions (i.e., the accuracy of local deformation must be assessed by landmarks identified in the local deformation region).

Because MR images are becoming more common in the radiotherapy environment, it is useful to evaluate the impact of obliqueness specifically. Assuming no FLE and perfect rigid registration, the following thought experiment demonstrates how the components of TRE depend on voxel size, rotation, and oblique angle of the image plane. FLE is not considered here for simplicity; that is, assume that the exact centroid of a marker is known (although it could be modeled as a random variable under a normal distribution).

Consider a set of images whose imaging plane is acquired at an angle α to an axis that is perpendicular to the imaging table, as shown in Figure 3.3. Further, consider a 2D problem from the $y - z$ perspective, where the pixel size is denoted as dy and dz, as shown in Figure 3.4a. Assuming no FLE, consider a point inside the pixel and denote its coordinate in the continuum space as $x \in \Re^2$. Because one can only detect its location with the discrete image space, any point inside the pixel will have the coordinates $\bar{x} \in \bar{\Re}^2$, where $\bar{\Re}^2$ is the discrete image space, and $\bar{x}$ is at the center of the pixel as per DICOM imaging standard.

The error between x and $\bar{x}$ is greatest if the point is located at one of four corners of the pixel. The four possible locations will be denoted as x_i, such that, for $i = 1, \ldots 4$, they would give the same discrete coordinate $\bar{x}$, $\bar{x} = x_i + e_i$.

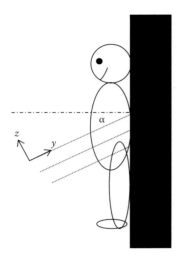

FIGURE 3.3 Depiction of an oblique acquisition of an MR image obtained at angle α.

From Figure 3.4b, e_i is the discretization error vector, defined through vectors a and b, where

$$e_1 = -a + b, \ e_2 = -a - b, \ e_3 = a - b, \ e_4 = a + b.$$

From Figure 3.4c, $\angle ox_4 n = \alpha$ by parallel lines. Because $\angle \bar{x}on + \angle x_4 on = 90°$ and $\alpha + \angle x_4 on = 90°$, then $\angle \bar{x}on = \alpha$. Thus, the vector a and b are defined as

$$a = \begin{bmatrix} \dfrac{dy}{2}\cos\alpha \\[2mm] \dfrac{dy}{2}\sin\alpha \end{bmatrix}, b = \begin{bmatrix} \dfrac{dz}{2}\sin\alpha \\[2mm] -\dfrac{dz}{2}\cos\alpha \end{bmatrix}.$$

Therefore, the discretization error vector depends on the voxel size and oblique angle, $e_i = e_i(dy, dz, \alpha)$.

Consider that the point is rotated and translated, and the transformed coordinate of x_i is $X_i \in \Re^2$, such that $i = 1, \ldots 4$,

$$X_i = Rx_i + T,$$

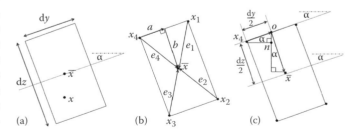

FIGURE 3.4 (a) Evaluation of the thought experiment in 2D focused on the $y - z$ plane of data acquired at an angle α. (b) Discretization error vectors defined through vector a and b. (c) Evaluation of error vectors as a function of angle.

where T is the translation vector, and $R = R(\theta)$ is the rotation matrix that depends on the angle of rotation θ.

Consider that a second image is acquired at the transformed point with an oblique angle β and pixel size dY and dZ. For each X_i, the starting image position of the second image could be different such that it could be inside any pixel. The discretization error would be the greatest when X_i is at a corner of a pixel. Thus, for each X_i, its discrete coordinate on the second image is denoted as $\bar{X}_{ij} \in \mathfrak{R}^2$, where

$$\bar{X}_{ij} = X_i + E_j,$$

for $j = 1, \ldots 4$. Similar as before, the discretization error vector $E_j = E_j(dY, dZ, \beta)$ depends on pixel size and oblique angle and is defined through

$$E_1 = -A + B, \, E_2 = -A - B, \, E_3 = A - B, \, E_4 = A + B,$$

where

$$A = \begin{bmatrix} \dfrac{dY}{2}\cos\beta \\[2mm] \dfrac{dY}{2}\sin\beta \end{bmatrix}, B = \begin{bmatrix} \dfrac{dZ}{2}\sin\beta \\[2mm] -\dfrac{dZ}{2}\cos\beta \end{bmatrix}.$$

In this thought experiment, where R and T are known exactly so that the rigid registration method is perfect, the registration maps the discrete coordinate $\bar{x}$ from the first image to second image via $R\bar{x} + T$. Then, the registration error vector between the mapped points and the discrete coordinate in the second image is

$$
\begin{aligned}
R\bar{x} + T - \bar{X}_{ij} &= R(x_i + e_i) + T - X_i - E_j \\
&= Rx_i + Re_i + T - Rx_i - T - E_j. \\
&= Re_i - E_j
\end{aligned}
$$

This shows that the registration error vector depends on the voxel size and the oblique angle of the two images through $e_i = e_i(dy, dz, \alpha)$ and $E_j = E_j(dY, dZ, \beta)$. Moreover, regardless of having a perfect registration, the registration error vector depends on the rotation, $R = R(\theta)$, which can project the larger discretization error, dz, from the z-direction onto the y-direction.

3.5 Examples in Literature

To provide some perspective into the implementation and accuracy assessment of deformable registration for radiotherapy applications, the following subsections will highlight publications evaluating accuracy in the head and neck, lung, liver, and prostate. A table summarizes recent publications for each site and highlights the DIR algorithm used, the accuracy assessment

performed, and the results. For comprehensive details on each study, the reader is referred to the authors' specific manuscript.

3.5.1 Head and Neck

All of the works referenced in Table 3.1 refer to the studies evaluating CT to CT or CT to in-room imaging, including kilovoltage cone-beam CT (CBCT) and megavoltage CT (MVCT). The algorithms used include linear elastic, TPS, B-splines, Demons, and various implementations of free-form, intensity-based methods. Many of the manuscripts highlighted preliminary evaluation or investigation of an algorithm in the region of the head and neck. As a result, the number of cases on which the algorithms are evaluated is often limited. Eight of the studies reported the results on five patients or phantoms or fewer. Only six investigators evaluated the registration results on more than five cases, and the largest cohort was 12. Given the complexity of motion and deformation in the head-and-neck region, these small cohorts clearly represent an area of additional investigation that is needed to evaluate this technology.

The majority of the evaluations use a form of contour or ROI overlap (i.e., DSC, volume overlap index, intersection, and overlap index), which provides a validation for contour propagation purposes and gross organ or ROI alignment; however, it does not address the accuracy of the deformable registration to correlate the internal volume of the structures. The majority of investigators reported DSC, intersection, union, or overlaps of 0.77 to 0.95 and surface or contour differences of submillimeter to 1 to 2 mm.

Several of the reports do evaluate the accuracy using TRE. Ireland, Nithiananthan, and Malsch identified primarily the points corresponding to bony anatomy. Vasquez Osorio identified the landmarks associated with soft tissue using the interface between structures, such as the middle point of the interface between the parotid glad and the mandible and styloid process. These identified landmarks are useful in evaluating the accuracy of dominant deformation, such as neck flexion; however, they do not address the need to evaluate the accuracy of alignment of the internal volume of soft tissue (i.e., the volumetric mapping of the entire parotid volume), which exhibit dramatically different deformation compared with the bony structure. This is clearly a challenging task because structures in the head and neck do not have prominent landmarks, such as vessel and bronchial bifurcations in the lung. However, research must continue to investigate methods to evaluate the accuracy in these regions because the applications of deformable registration move toward adaptive radiotherapy and dose accumulation. The interstudy comparisons on the accuracy reported are challenging due to the variation in cases evaluated, voxel size, and deformation that was present between the image sets. In the studies highlighted here, the majority of investigators reported a mean TRE of approximately 1.5 to 3.2 mm, with standard deviations approximately 1 mm. In addition, the number of identified landmarks was typically small, especially for the head and neck, where the region is large, the deformation is complex and varied, and the deformation field is very localized.

TABLE 3.1 Accuracy Assessment of DIR in Head and Neck

Author(s) Year	Regularization, Similarity Metric, Data, Imaging Modalities	Accuracy Evaluation	
Al-Mayah et al. 2010a	• Linear elastic/finite element, contour matching • 4 patients, CT to CBCT	DSC, AVG ± SD: GTV 0.86 ± 0.08 Lt parotid 0.84 ± 0.11 Rt parotid 0.89 ± 0.04	Δ COM, AVG ± SD: GTV 2.3 ± 1.0 mm Lt parotid 2.5 ± 0.8 mm Rt parotid 2.0 ± 0.9 mm
Chao et al. 2010	• TPS, scale-invariant feature transformation • 1 digital phantom, CT to CT • 4 patients, CT to CBCT (*n* = 1–3)	Δ phantom GTV contour: AVG (max): 1.3 (3.0) mm	Δ patient GTV contour: AVG range: 1.3–2.7 mm Max range 3.3–5.8 mm Patient GTV volume intersection/union: Range 89.8%–94.1%
Faggiano et al. 2011	• B-spline, MI • 10 patients, CT to MVCT	DSC, AVG (range): Parotids 0.773 (0.678–0.846)	
Hou et al. 2011	• Demons • 12 patients, CT to CBCT (*n* = 5–7)	TRE (9 landmarks), AVG ± SD: 2.6 ± 0.6 mm	VOI, AVG ± SD: 76.2% ± 4.6%
Ireland et al. 2007	• Pseudoelastic, SSD • 5 patients, CT to CT (*n* = 2)	RMS error (5 landmarks), AVG ± SD: Treatment position 2.8 ± 0.8 mm Diagnostic position 3.2 ± 1.2 mm	
Lu, Olivera et al. 2006b	• Free-form, calculus of variations • 5 patients, CT to MVCT (*n* = 20–40)	NCC, AVG range: 0.80–0.90[a]	
Malsch et al. 2006	• TPS, NCC • 2 patients, CT to CT	Δ (78, 79 landmarks), |AVG (max)|: LR 0.7, 0.6 (2.1, 2.0) mm AP 0.9, 1.1 (2.1, 2.9) mm SI 1.2, 0.8 (3.2, 3.0) mm	
Nithiananthan et al. 2009	• Demons • 1 cadaver, CBCT to CBCT • 10 patients, CBCT to CBCT	Cadaver NCC: 0.991 Cadaver TRE (7 landmarks), AVG ± SD (range): 0.8 ± 0.3 (0.4–2.0) mm	Patients NCC: Range 0.986–0.995 Patients TRE (8 landmarks, AVG ± SD (range): 1.6 ± 0.8 (0.4–4.4) mm
Olteanu et al. 2012	• Proprietary free-form, intensity-based • 12 patients, CT to CT	Jaccard index, AVG (range): GTV 0.4 (0–0.7) CTV 0.6 (0.3–0.8) Parotids 0.7 (0.6–0.8)	Overlap index, AVG (range): GTV 0.8 (0–1.0) CTV 0.8 (0.6–1.0) Parotids 0.8 (0.7–0.9)
Paquin et al. 2009	• B-spline/multiscale, landmark-intensity hybrid • 1 digital phantom, CT to CBCT	MI: 0.89	
Vasquez Osorio et al. 2008, 2009	• TPS, surface matching • 10 patients, CT to CT (2008) • TPS, surface matching • 2 patients, CT to CT (2009)	Δ surface vertices, AVG ± SD: 0.6 ± 0.5 mm Δ (12 landmarks), AVG ± SD: 1.5 ± 0.8 mm	Δ surface vertices after inverse transformation, AVG ± SD: 1.5 ± 0.7 mm
Wang et al. 2005b, 2008	• Demons • 1 digital phantom, CT to CT (2005) • Demons • 8 patients, CT to CT (*n* = 11–14) (2008)	Δ displacement, AVG ± SD: 0.2 ± 0.6 mm VOI, AVG ± SD: Lt parotid 99.4% ± 0.7% Rt parotid 98.7% ± 0.6% CTV 97.1% ± 1.5%	Δ |AVG| surface distance, AVG ± SD: Lt parotid 0.0 ± 0.0 mm Rt parotid 0.1 ± 0.1 mm CTV 0.4 ± 0.2 mm
Zhang et al. 2007	• Free-form, SSD • 7 patients, CT to CT (*n* = 3–6)	DSC, range[a]: Lt parotid 0.70–0.86 Rt parotid 0.71–0.84	Contour distance transformation, AVG ± SD: Lt parotid 1.1 ± 2.2 mm Rt parotid 0.5 ± 1.9 mm

Note: AP, anterior–posterior; AVG, average; CBCT, cone-beam CT; CC, correlation coefficient; COM, center of mass; CT, computed tomography; CTV, clinical target volume; DSC, dice similarity coefficient; GTV, gross tumor volume; LR, left–right; Lt, left; MI, mutual information; MR, magnetic resonance; MSD, mean squared difference; MVCT, megavoltage CT; NCC, normalized crosscorrelation; RMS, root mean square; Rt, right; SD, standard deviation; SI, superior–inferior; SSD, sum of squared difference; TPS, thin-plate spline; TRE, target registration error; VOI, volume overlap index.

[a] Values estimated from graphs.

3.5.2 Lung

An impressive amount of work has been performed by many investigators in the deformable registration of the lung (Table 3.2). The vast majority of studies have evaluated the accuracy of DIR using landmark-based metrics based on bronchial and vessel bifurcations in the lung. In addition, the majority of these evaluations were done on clinical data due to the ease of identifying landmarks. The challenge is to ensure that the accuracy throughout the entire thorax is evaluated, including the ribs and surrounding normal tissue, if they are included in the registration (e.g., several algorithms segment out the lung and only

TABLE 3.2 Accuracy Assessment of DIR in Lung

Author(s) Year	Regularization, Similarity Metric, Data, Imaging Modalities	Accuracy Evaluation	
Bender et al. 2012	• Demons • 4 patients, exhale/inhale 4D CT	TRE (300 landmarks), AVG ± SD: LR 0.4 ± 0.4 mm AP 0.5 ± 0.5 mm SI 0.9 ± 0.7 mm	
Bai and Brady 2009	• B-spline, MSD • 1 thorax phantom, 4D-PET	GTV position estimation error, AVG: LR 0.17% AP 0.2% SI 0.65%	GTV diameter estimation error, AVG: LR 5.7% AP 6.1% SI 13.0%
Brock et al. 2005, Al-Mayah et al. 2008, 2009, 2010b, 2011	• Linear elastic/finite element, contour matching • 5 volunteers, MR to MR (2005)	Δ (6–14 landmarks), |AVG| ± SD: LR 2.3 ± 1.4 mm AP 3.6 ± 1.7 mm SI 1.9 ± 1.4 mm	
	• Linear elastic/finite element, contour matching • 14 patients, exhale/inhale 4D CT (2011)	Δ (~48 landmarks), |AVG| ± SD: LR 0.9 ± 0.8 mm AP 1.3 ± 1.1 mm SI 1.9 ± 1.6 mm 3D 2.8 ± 1.7 mm	
	• Hyperelastic/finite element, contour matching • 1 patient, exhale/inhale 4D CT (2008)	Δ (90 lung landmarks), AVG: LR –0.7 mm AP 0.0 mm SI 0.9 (SD 2.1) mm	Δ (90 GTV landmarks), AVG: LR –1.4 mm AP –0.6 mm SI –0.8 (SD 1.7) mm
	• Hyperelastic/finite element, contour matching • 16 patients, exhale/inhale 4D CT (2009)	Δ (53–113 landmarks), |AVG| ± SD: LR 1.0 ± 0.7 mm AP 1.2 ± 1.0 mm SI 1.7 ± 1.4 mm	Δ (GTV landmarks, 7 patients), |AVG| ± SD: LR 1.0 ± 0.6 mm AP 0.9 ± 0.7 mm SI 1.4 ± 1.0 mm
	• Hyperelastic/finite element, contour matching • 10 patients, exhale/inhale 4D CT (2010)	Δ (~40 landmarks), |AVG| range (SD range): LR 0.5–1.3 (0.4–1.1) mm AP 0.7–2.0 (0.5–1.8) mm SI 0.9–2.3 (0.7–2.9) mm 3D 1.6–3.8 (0.8–2.9) mm	
Chao, Li et al. 2008	• Narrowband B-spline, MI • 4 digital phantoms, 4D CT • 4 patients, exhale/inhale 4D CT	Phantom contour Δ, AVG (max): 1.0 (1.5) mm	Bidirectional contour mapping, AVG: Phantoms 1.8 mm Patients <3 mm
Coselmon et al. 2004	• TPS, MI • 11 patients, CT to CT	Δ (6 landmarks), AVG ± SD: LR 0 ± 1.7 mm AP –0.5 ± 3.1 mm SI 0.4 ± 3.6 mm	
Du et al. 2012	• B-spline, sum of squared volume difference • 9 patients, exhale/inhale 4D CT • 3 animals, exhale/inhale 4D CT	TRE (100–140 landmarks): AVG range: 0.7–3.7 mm SD range: 0.4–1.9 mm 90th percentile <5 mm for 8/9 patients	
Eom et al. 2010	• Hyperelastic/finite element • 4 patients, 4D CT (all phases)	Δ (39–48 landmarks): AVG range: 2.0–4.5 mm SD range: 1.0–3.3 mm	
Gu et al. 2010	• Demons • 5 patients, exhale/inhale 4D CT	Δ (1166–1561 landmarks): AVG range: 1.5–1.8 mm SD range: 1.5–2.0 mm	

(continued)

TABLE 3.2 Accuracy Assessment of DIR in Lung (Continued)

Author(s) Year	Regularization, Similarity Metric, Data, Imaging Modalities	Accuracy Evaluation	
Guerrero et al. 2004; Castillo et al. 2010	• Optical flow • 2 digital phantoms, CT to CT (2004)	RMS error, AVG range: LR 0.01–0.04 mm AP 0.01–0.02 mm SI 0.04–0.20 mm	
	• Optical flow • 10 patients, 4D CT (exhale phases) (2010)	Δ (342–1561 total landmarks), AVG ± SD: LR 0.4 ± 0.7 mm AP 0.5 ± 0.7 mm SI 0.7 ± 1.3 mm 3D 1.3 ± 1.4 mm	
Heath et al. 2007	• Free-form/linear elastic, CC • 5 patients, exhale/inhale 4D CT	Δ (20–30 landmarks), AVG ± SD: 1.6 ± 0.6 mm Δ GTV COM, AVG ± SD: 2.0 ± 0.4 mm	NCC, AVG: 0.995 Surface DTA, AVG ± SD: 1.8 ± 0.4 mm
Kaus et al. 2007	• TPS, elastic body or Wendland function, surface matching • n = 5, MR to MR	Δ Rt lung (6–14 landmarks), AVG ± SD: LR 0.5 ± 2.7 mm AP –1.1 ± 5.2 mm SI –2.1 ± 3.8 mm	Δ Lt lung (6–14 landmarks), AVG ± SD: LR 1.1 ± 2.5 mm AP –2.1 ± 3.2 mm SI –1.3 ± 2.8 mm
Li et al. 2008a	• TPS, intensity and bronchial matching • 5 patients, exhale/inhale 4D CT • 1 animal, CT to CT	Δ patient data (landmarks), AVG: 0.4 mm	Δ animal data (13 fiducials), AVG (max): 1.9 (3.7) mm
Li et al. 2008b	• Fluid flow plus nonlinear/finite element, SSD • 5 patients, exhale/inhale 4D CT	Δ (16–39 landmarks), |AVG| range (SD range): LR 0.8–1.5 (0.6–0.9) mm AP 1.3–3.5 (1.1–1.9) mm SI 1.1–3.7 (0.7–2.3) mm 3D 2.9–4.4 (1.2–2.1) mm	
Liu et al. 2012	• Free-form, calculus of variations • 1 digital phantom, CT to CT (n = 5)	|AVG| (95th percentile): LR 0.5 (2.2) mm AP 0.8 (2.6) mm SI 3.6 (7.6) mm 3D 3.8 (2.0) mm	
Lu et al. 2006a,b	• Free-form, calculus of variations • 1 phantom, CT to CT (2006a) • 6 patients, 4D CT (all phases) (2006a)	Contour matching index measure, range: Phantom data 0.84–0.98 Patient data 0.83–0.98	
	• Free-form, calculus of variations • 2 patients, CT to MVCT (n = 20–40) (2006b)	NCC, range: 0.88–0.93[a]	
McClelland et al. 2006, 2011	• B-spline, SSD • 6 patients, CT to CT (2006)	TRE, AVG ± SD (max): 1.3 ± 0.7 (6.2) mm	
	• B-spline, MI • 6 patients, 4D CT (all phases) (2011)	TRE, AVG (99th percentile): 1.0 (3.2) mm	
Pekar et al. 2006	• Elastic body, SSD • 1 patient, CT to PET	CC: 0.964	
Pevsner et al. 2006	• Viscous fluid flow, SSD • 6 patients, exhale/inhale 4D CT	Δ (41 total landmarks), AVG (90% confidence interval): 2.9 (7.3) mm	AVG GTV surface Δ, AVG (range): 2.6 (1.1–5.1) mm
Rietzel and Chen 2006	• B-spline, SSD • 5 patients, 4D CT (all phases)	Δ (5 landmarks), |AVG| ± SD: LR 0.8 ± 0.8 mm AP 1.1 ± 1.0 mm SI 0.9 ± 1.5 mm 3D 2.1 ± 1.5 mm	
Sarrut et al. 2006, Boldea et al. 2008	• Demons, SSD • 4 patients, CT to CT (2006)	Δ (14–25 landmarks), AVG ± SD (max): 2.7 ± 1.1 (15) mm	Voxels with negative Jacobian, AVG range: 0.3%–2.4%
	• Demons, SSD • 5 patients, 4D CT (all phases) (2008)	TRE (60 anatomic landmarks): AVG range: 1.9–2.9 mm SD range: 1.2–2.2 mm	

(continued)

TABLE 3.2 Accuracy Assessment of DIR in Lung (Continued)

Author(s) Year	Regularization, Similarity Metric, Data, Imaging Modalities	Accuracy Evaluation	
Schreibmann and Xing 2006	• B-spline, NCC • 3 patients, exhale/inhale 4D CT	Difference image, max: Initial difference 875 HU Residual difference 250 HU (0.01% >20 HU)	
Shekhar et al. 2007	• B-spline, MSD • 5 patients, exhale/inhale 4D CT	Surface RMS error, range: Lungs 1.2–2.6 mm GTV 1.8–4.5 mm Hausdorff distance, range: Lungs 14.0–26.3 mm GTV 3.9–18.5 mm	Δ COM, range: Lungs 0.1–1.1 mm GTV 0.2–4.1 mm Volume ratio of intersection/union, range: Lungs 0.94–0.99 GTV 0.56–0.91
Shusharina and Sharp 2012	• B-spline and radial basis functions, point matching • 5 patients, exhale/inhale 4D CT	Δ (~300 landmarks), AVG: 3.6 mm	
Stancanello et al. 2005	• B-spline, MI • 7 patients, CT to CT	Δ target COM, range: LR –3.7 to 2.7 mm AP –0.8 to 4.7 mm SI 0.2–4.5 mm 3D 2.1–4.9 mm	Target volume overlap, range: 80%–95%
Staring et al. 2007	• B-spline, MI • 5 patients, CT to CT	Lungs DSC, AVG ± SD: 0.97 ± 0.02	
Sohn et al. 2008	• Featurelets with B-spline, CC or MI • 1 digital phantom, 4D CT • 4 patients, exhale/inhale 4D CT	Δ digital phantom, AVG ± SD: 1.1 ± 1.2 mm	Δ patient data (11–15 landmarks), AVG ± SD (max): LR –0.3 ± 0.8 mm AP 0.0 ± 0.9 mm SI 0.1 ± 1.5 mm 3D 1.6 ± 1.0 (4.6) mm
Vandemeulebroucke et al. 2011, 2012	• B-spline, MSD • 6 patients, 4D CT (all phases) (2011) • B-spline, MSD • 16 patients, exhale/inhale 4D CT (2012)	TRE (100 landmarks/phase), AVG ± SD: 1.4 ± 1.5 mm TRE (100–300 landmarks), AVG ± SD: Lungs 1.8 ± 1.5 mm Chest wall 2.6 ± 2.5 mm Diaphragm 1.7 ± 1.6 mm	DSC, AVG ± SD: Bones 92.3% ± 2.3% Trachea/bronchi 81.1% ± 4.2%
Wang et al. 2008	• Demons • 9 patients, 4D CT (all phases)	VOI, AVG ± SD: Lung 99.3% ± 0.7% GTV 98.3% ± 1.1%	
Werner et al. 2009	• Linear elastic/finite element, contour matching • 12 patients, exhale/inhale 4D CT	Δ (~45 landmarks), AVG ± SD (max): LR 0.0 ± 1.7 mm AP –0.1 ± 2.2 mm SI –0.6 ± 2.7 mm 3D 3.3 ± 2.1 mm	
Wijesooriya et al. 2008	• Diffeomorphic • 13 patients, 4D CT (all phases)	Δ GTV COM, AVG ± SD: 0.5 ± 1.5 mm GTV fractional volume, AVG ± SD: 0.2 ± 0.1	GTV surface congruence, AVG ± SD: 0.0 ± 1.1 mm
Wolthaus et al. 2008	• Optical flow • 1 patient, 4D CT (all phases)	Δ (40 landmarks), |AVG| ± SD: LR 0.1 ± 0.5 mm AP –0.1 ± 0.5 mm SI –0.3 ± 0.7 mm	
Wu et al. 2008	• B-spline or Demons, MSD • 4 patients, exhale/inhale 4D CT	Δ (17–56 landmarks/lung), |AVG|: B-spline 2.8 mm Demons 2.7 mm	Δ (22–63 landmarks/rib), |AVG|: B-spline 1.8 mm Demons 1.7 mm
Yang et al. 2008	• Optical flow, MSD • 1 digital phantom, CT to CT • 4 patients, exhale/inhale 4D CT	Δ digital phantom, AVG ± SD: 0.9 ± 0.7 mm	Δ patient data (landmarks), range: AVG 0.96–1.48 mm SD 0.47–1.34 mm Max 2.78–5.87 mm
Yim et al. 2010	• Demons, SSD • 8 patients, CT to CT	Δ (27 landmarks), AVG ± SD (max): 2.8 ± 1.5 (5.4) mm	DSC, AVG ± SD: 90.1% ± 3.7%

(continued)

TABLE 3.2 Accuracy Assessment of DIR in Lung (Continued)

Author(s) Year	Regularization, Similarity Metric, Data, Imaging Modalities	Accuracy Evaluation	
Yin et al. 2010	• B-spline, Demons, or level-set • 10 patients, CT to CT	TRE, AVG: B-spline 1.9 mm Demons 1.1 mm Level set 0.7 mm	Voxels with negative Jacobian: B-spline 0.26% Demons 37.9% Level set 24.7%
Yin et al. 2009, 2011	• B-spline, sum of squared volume difference • 6 patients, CT to CT (2009)	Δ (120–210 landmarks), AVG ± SD: Landmarks ≤20 mm apart 0.6 ± 0.0 mm Landmarks ≥60 mm apart 1.9 ± 0.4 mm	
	• B-spline, sum of squared volume difference and landmark matching • 6 patients, exhale/inhale 4D CT (2011)	Δ (100–240 landmarks): AVG range: 0.71–1.69 mm SD range: 0.39–1.45	
Xie et al. 2009	• TPS, scale-invariant feature transformation • 1 digital phantom, 4D CT • 3 patients, 4D CT (all phases)	Δ phantom (15 feature points), AVG ± SD: 0.5 ± 0.2 mm	Δ patient (3 landmarks for 1 patient), AVG ± SD: 2.3 ± 1.6 mm Δ diaphragm for 1 patient, AVG ± SD: 3.3 ± 1.9 mm GTV DSC for 1 patient: 91.3%
Zhong et al. 2010, 2012	• B-spline or demons • 1 digital phantom, 4D CT (2010)	Δ B-spline, AVG: Whole CT 1.5 mm Lung only 0.5 mm	Δ Demons, AVG: Whole CT 1.3 mm Lung only 0.8 mm
	• Demons (±linear elastic/finite element), feature matching • 1 digital phantom, 4D CT (2012)	Δ AVG (max): Demons 1.7 (12.0) mm Demons with finite element 1.1 (4.0) mm	

Note: AP, anterior–posterior; AVG, average; CBCT, cone-beam CT; CC, correlation coefficient; COM, center of mass; CT, computed tomography; CTV, clinical target volume; DSC, dice similarity coefficient; GTV, gross tumor volume; LR, left–right; Lt, left; MI, mutual information; MR, magnetic resonance; MSD, mean squared difference; MVCT, megavoltage CT; NCC, normalized crosscorrelation; RMS, root mean square; Rt, right; SD, standard deviation; SI, superior–inferior; SSD, sum of squared difference; TPS, thin-plate spline; TRE, target registration error; VOI, volume overlap index.

[a] Values estimated from graphs.

focus on the internal volume of the lung). It is important to note that a limitation in many of these studies is the narrow focus of the registration on the lung tissue. The complexity of the deformation at the lung–chest wall interface challenges many DIR algorithms, because the lung may be moving 1 to 2 cm in the SI direction; however, the ribs typically exhibit very small motion in this direction during breathing. This is clearly an area of importance for research investigations. Image registration algorithms studied include Demons, B-spline, linear elastic, hyperelastic, TPS, optical flow, free-form, fluid flow, elastic body, and viscous fluid flow.

The average accuracy ranges from submillimeter to 2 to 3 mm, with the majority of investigators able to optimize their registration algorithms to achieve an accuracy of less than 2 mm in each direction. The comparisons between reported metrics can be evaluated in several of the studies. For example, Sohn et al. (2008) evaluated a B-spline–based registration algorithm on a digital phantom (accuracy 1.1 ± 1.2 mm) as well as clinical data using internal landmarks (1.6 ± 1.0 mm), indicating that, for this case, the digital phantom closely predicted the clinical results. In this study, the digital phantom was constructed by using an independent DIR algorithm to register an exhale and inhale CT image. The deformation map was then applied to the inhale image to create a synthetic exhale image with a known deformation map. When implementing this technique, it is important to

use a DIR that is not the algorithm that is being tested for accuracy. For this study, the authors used a free-form DIR to generate the synthetic image and the validation was performed for a B-spline algorithm.

3.5.3 Liver

The majority of DIR evaluations in the liver have included quantitative point-based assessment using vessel bifurcations (Table 3.3). The DIR techniques include TPS, linear elastic, B-spline, free-form, viscous fluid, and Demons. It is important to note that, across the studies, the TRE-based registration accuracy has been reported in different formats, including average absolute error in each direction, average (signed) error and standard deviation in each direction, and average vector magnitude error and standard deviation. When reporting the average (signed) error and standard deviation, the average indicates if there is a bias in the registration (i.e., an underlying rigid error that was not removed). This number should be very close to zero. The standard deviation indicates the spread in the data. When the average absolute error is reported, this indicates the expected error that would be associated with the typical point and the standard deviation indicates the spread around that value. Therefore, for the signed average in each direction, one would expect the mean to be small and the standard deviation

TABLE 3.3 Accuracy Assessment of DIR in Liver

Author(s) Year	Regularization, Similarity Metric, Data, Imaging Modalities	Accuracy Evaluation	
Brock et al. 2003	• TPS, MI • 6 patients, CT to CT	Δ (6–18 landmarks), |AVG| ± SD: LR 1.3 ± 1.0 mm AP 1.5 ± 1.2 mm SI 1.5 ± 1.4 mm	
Brock et al. 2005, 2006; Voroney et al. 2006	• Linear elastic/finite element, contour matching • 5 volunteers, MR to MR (2005)	Δ (7–10 landmarks), |AVG| ± SD: LR 1.2 ± 0.7 mm AP 1.7 ± 1.4 mm SI 1.4 ± 1.0 mm	
	• Linear elastic/finite element, contour matching • 5 patients, CT to MR (2006a)	Δ (4–5 landmarks), AVG ± SD (range): 4.2 ± 1.4 (4.0–4.4) mm	
	• Linear elastic/finite element, contour matching • 17 patients, CT to MR (2006b)	Δ (5 landmarks), AVG ± SD: 4.2 ± 1.7 mm	
Kaus et al. 2007	• TPS, elastic body or Wendland function, surface matching • 5 volunteers, MR to MR • 5 patients, CT to MR	Δ MR data (7–10 landmarks), AVG ± SD: LR −0.7 ± 2.2 mm AP −0.7 ± 3.5 mm SI −0.8 ± 2.5 mm	Δ CT/MR data (7–10 landmarks), AVG ± SD: LR −0.6 ± 5.3 mm AP −0.9 ± 3.4 mm SI 0.3 ± 4.4 mm
Lee et al. 2011	• B-spline, vessel/surface gradients • 5 patients, US to CT	Distance measure, AVG ± SD: Vessel center lines 1.9 ± 0.3 mm Organ surfaces 1.7 ± 0.4 mm FRE (vessels), AVG ± SD: 2.4 ± 1.1 mm	Δ GTV COM, AVG (range): 2.8 (0.5–4.5) mm GTV overlap measure, AVG (range): 92 (86–100)%
Liu et al. 2012	• Free-form, calculus of variations • 1 digital phantom, CT to CT ($n = 5$) • 1 abdominal phantom, CT to CT • 7 patients, exhale/inhale 4D CT	Digital phantom, AVG (95th percentile): LR 1.0 (2.4) mm AP 0.8 (2.1) mm SI 3.3 (7.5) mm 3D 3.7 (1.8) mm Δ abdominal phantom (5 fiducials), AVG ± SD: 3.6 ± 2.8 mm	Δ patient data (21 total landmarks), AVG ± SD: LR 0.6 ± 0.4 mm AP 1.2 ± 2.0 mm SI 1.0 ± 1.2 mm
Piper et al. 2012	• Viscous fluid and elastic, MI • 25 patients, CT to CT ($n = 16$)	Δ (~19 surface landmarks), AVG ± SD (99th percentile): LR 0.9 ± 1.1 (5.9) mm AP 0.8 ± 1.2 (4.8) mm SI 1.1 ± 2.9 (19.0) mm 3D 3.7 mm	
Rohlfing et al. 2004	• B-spline, MI • 4 volunteers, MR to MR	Δ surface/center line, AVG range: Liver 2.5–5.1 mm Inferior vena cava 1.7–2.3 mm Hepatic veins 2.2–4.3 mm	
Shekhar et al. 2007	• B-spline, MSD • 4 patients, exhale/inhale 4D CT	Surface RMS error, range: Liver 2.5–3.4 mm GTV 2.7–3.6 mm Hausdorff distance, range: Liver 6.6–13.7 mm GTV 6.1–9.1 mm	Δ COM, range: Liver 0.5–2.9 mm GTV 1.2–3.8 mm Volume ratio of intersection/union, range: Liver 0.91–0.95 GTV 0.67–0.82
Stancanello et al. 2005	• B-spline, MI • 7 patients, CT to CT	Δ target COM, range: LR −0.4 to 5.5 mm AP −2.8 to 3.9 mm SI 2.5–4.9 mm 3D 2.5–7.4 mm	Target volume overlap, range: 77%–93%
Vasquez Osorio et al. 2012	• TPS, vessel-midline matching • 7 patients, MR to CT	Δ (10–15 vessel landmarks), AVG (range): 1.6 (1.3–1.9) mm	Δ (4–6 other landmarks), AVG (range): 1.5 (1.1–2.3) mm

(continued)

TABLE 3.3   Accuracy Assessment of DIR in Liver (Continued)

Author(s) Year	Regularization, Similarity Metric, Data, Imaging Modalities	Accuracy Evaluation	
Xie et al. 2011	• TPS, scale-invariant feature transformation • 6 patients, exhale/inhale 4D CT	Δ (50 landmarks), range: AVG 1.1–1.8 mm SD 0.9–1.6 mm 3D AVG 2.1–2.8 mm	Liver contours, range: |AVG| 2.1–3.1 mm |SD| 1.3–2.1 mm |Max| 8.3–16.8 mm GTV DSC (*n* = 1): 93.6%
Zhang et al. 2012	• B-spline or Demons, MSD • 3 patients, 4D CT (all phases)	Δ (3 fiducials), max range: B-spline 3–6 mm Demons 3–7 mm	

Note: AP, anterior–posterior; AVG, average; CBCT, cone-beam CT; CC, correlation coefficient; COM, center of mass; CT, computed tomography; CTV, clinical target volume; DSC, dice similarity coefficient; GTV, gross tumor volume; LR, left–right; Lt, left; MI, mutual information; MR, magnetic resonance; MSD, mean squared difference; MVCT, megavoltage CT; NCC, normalized crosscorrelation; RMS, root mean square; Rt, right; SD, standard deviation; SI, superior–inferior; SSD, sum of squared difference; TPS, thin-plate spline; TRE, target registration error; VOI, volume overlap index.

to be approximately 2 mm for a well-performing algorithm. For the absolute average in each direction, one would expect the mean to be 1 to 2 mm for a well-performing algorithm and the standard deviation to be less than 1 mm to indicate a robust algorithm across the anatomy. The vector magnitude simplifies the reporting of the accuracy, however, and limits the readers' ability to assess the distribution of the accuracy in terms of direction. It is inherently an absolute value, so the average is the expected error and the standard deviation represents the consistency of this expected error across the landmarks identified.

For algorithms reporting the average absolute error in each direction, the accuracy is approximately 1.5 mm, with standard deviations approximately 1 to 1.5 mm. For algorithms reporting the average (signed) error in each direction, the mean is typically less than 1 mm, with standard deviations ranging from 2.2 to 5.3 mm. For algorithms reporting the vector magnitude, the mean error ranges from 1.6 to 4.2 mm. In addition, several studies evaluated the liver surface agreement or liver and tumor center of mass (COM) agreement. The COM is typically smaller (from 0.5 to 1.5 mm) because it is less sensitive to local variation in accuracy and the average surface agreement ranges from 2 to 5 mm.

3.5.4 Prostate

A large number of investigations have evaluated accuracy in the prostate (Table 3.4). A wide range of DIR algorithms have been investigated, including linear elastic, B-spline, Demons, viscous fluid flow, TPS, free-form deformation, optical flow, and elastic body. Because MR imaging is often used for prostate cancer detection, several of the studies are evaluated on MR images; however, only one study evaluates multimodality registration, performing alignment between an MR and a CT.

DIR accuracy was assessed using volume overlap measurements, that is, DSC, surface distances, and TRE, often of implanted fiducials. In addition, several investigators used digital phantoms to evaluate the registration accuracy. ROI overlap measures (i.e., overlap ratio, DSC, and volume intersection) ranged from 0.77 to 0.95, which is likely on the order of the contour uncertainty in the identification of the prostate boundary.

Overall, the accuracy in the prostate is on the order of 1 to 2 mm. Sixteen publications reported quantitative accuracy measurements based on landmarks. The accuracy ranged from submillimeter to 3.5 mm in vector magnitude. When comparing results of image registration algorithms across studies, it is important to bear in mind the impact on the registration results due to voxel size, deformation complexity, image noise/distortion, number of landmarks identified, and number of patients evaluated.

Seven publications evaluated registration accuracy using both digital phantoms and patient data. This allows a comparison between the two assessment techniques and to evaluate how the phantom results translate into clinical results. In Chao et al.'s work, the digital phantom results were significantly better than the clinical results for the rectum, with the phantom having an average contour difference of less than 1.3 mm and the clinical data ranging between 2.0 and 8.0 mm. For Chen and Schreibmann, the phantom data and clinical data results were much closer. The results from Wang et al. allow the comparison of phantom point-based accuracy and corresponding intensity similarity metric [correlation coefficient (CC)] with patient data. Here, a point-based measurement with accuracy of less than 1 mm translates into a CC measurement of 0.8 to 0.9. This CC measurement in a phantom translates into a CC measurement following registration of 0.944. It is important to keep in mind the challenges in ensuring that phantom data have the complexity and noise distribution that the clinical data will have. Clinical data may have more texture to it, which will provide more information for intensity-based registration and also more variation in the calculation of the CC used for accuracy assessment (e.g., a homogeneous phantom CC measurement may be high even if the registration is not correct just due to lack of variation across the phantom).

TABLE 3.4 Accuracy Assessment of DIR in Prostate

Author(s) Year	Regularization, Similarity Metric, Data, Imaging Modalities	Accuracy Evaluation	
Alterovitz et al. 2006	• Linear elastic/finite element (2D), contour matching • 10 patients, MR to MR	DSC, AVG ± SD: Balloon probe ($n = 5$) 97.5 ± 0.7 Rigid probe ($n = 5$) 98.1 ± 0.4	Δ anatomic/surface landmarks, AVG ± SD: Balloon probe ($n = 5$) 2.0 ± 0.2 mm Rigid probe ($n = 5$) 1.0 ± 0.5 mm
Bharatha et al. 2001	• Linear elastic/finite element, surface matching • 10 patients, MR to MR	DSC, AVG (95% CI): Total prostate 0.94 (0.89, 0.99) Central zone 0.86 (0.77, 0.95) Peripheral zone 0.76 (0.62, 0.91)	Δ (2 landmarks), AVG ± SD (max): First point 1.0 ± 0.6 (2.3) mm Second point 0.7 ± 0.4 (1.6) mm
Chao et al. 2008	• Narrowband B-spline, NCC • 1 digital phantom, CT to CT • 5 patients, CT to CBCT	Δ contour: Phantom rectum, AVG <1.3 mm Patients prostate, AVG range: 2.0–2.5 mm Patients rectum, AVG range: 2.0–8.0 mm	
Chen et al. 2010	• Demons-based • 1 digital phantom, CT to CT • 15 patients, CT to CBCT	Volume similarity measure, range: Prostate 87.4%–94.5% Fiducial markers 91.0%–95.0% Seminal vesicles 87.6%–92.9% Δ (3–4 fiducials), AVG range: 0.11–0.23 mm	Δ prostate surface: AVG range: 0.9–1.9 mm SD range: 0.8–1.6 mm Phantom RMS error: 0.26 mm
Foskey et al. 2005	• Viscous fluid flow, SSD • 5 patients, CT to CT ($n = 13$)	Δ COM: AVG range: 2.1–3.7 mm SD range: 1.3–2.2 mm	DSC: AVG range: 0.78–0.84 SD range: 0.6–0.8
Godley et al. 2009	• Demons • 5 patients, CT to CT ($n = 8$)	DSC, AVG: Prostate 77.6 Rectum 93.2 Bladder 98.1	
Hensel et al. 2007; Brock et al. 2008	• Linear elastic/finite element, contour matching • 19 patients, MR to MR (2007)	Δ (3 fiducials), AVG ± SD (max): LR 0.1 ± 0.9 mm AP 0.3 ± 1.6 mm SI −0.3 ± 1.6 mm 3D 2.2 ± 0.9 (4.2) mm	Δ surface, AVG ± SD LR 0.0 ± 0.6 mm AP 0.1 ± 0.7 mm SI −0.5 ± 0.7 mm 3D 1.5 ± 0.6 mm
	• Linear elastic/finite element, contour matching • 21 patients, MR to MR (2008)	Δ (3 fiducials), AVG ± SD (|AVG|): LR 0.3 ± 0.6 (0.8) mm AP −0.4 ± 1.3 (1.3) mm SI −0.2 ± 1.1 (1.1) mm	
Karnik et al. 2010	• TPS or B-spline, surface matching or MI • 16 patients, US to US	TRE (2–6 landmarks), AVG ± SD: TPS 2.1 ± 0.8 mm B-spline 1.5 ± 0.8 mm	
Kaus et al. 2007	• TPS, elastic body or Wendland function, surface matching • 10 patients, MR to MR	Δ (3 fiducials), AVG ± SD: LR 0.1 ± 1.6 mm AP −0.4 ± 2.3 mm SI −0.5 ± 2.5 mm	
Lian et al. 2004	• TPS, surface matching • 4 phantoms, 2D/2D radiographs • 3 patients, CT to MR	Phantom data Δ (10–15 landmarks): AVG range: 0.5–0.6 mm SD range: 0.4–0.5 mm Max range: 1.0–1.1 mm	Patient data, AVG ± SD: Δ COM 0.56 ± 0.09 mm Coincidence index 93.1% ± 5.0%
Lu et al. 2004, 2006b	• Free-form, calculus of variations • 1 pelvic phantom, CT to CT ($n = 5$) (2004) • 3 patients, CT to CT ($n = 15–18$) (2004)	Phantom data, Δ (320 fiducials): All <1 mm	Patient data, AVG range: MI 1.25–1.6 CC 0.98–0.99
	• 2 patients, CT to MVCT ($n = 20–40$) (2006)	NCC, AVG range: 0.91–0.92[a]	
Malsch et al. 2006	• TPS, NCC • 1 patient, CT to CT	Δ (67 landmarks), |AVG (max)|: LR 0.7 (1.9) mm AP 0.8 (2.2) mm SI 1.2 (3.1) mm	

(continued)

TABLE 3.4 Accuracy Assessment of DIR in Prostate (Continued)

Author(s) Year	Regularization, Similarity Metric, Data, Imaging Modalities	Accuracy Evaluation	
Paquin et al. 2009	• B-spline/multiscale, landmark-intensity hybrid • 1 digital phantom for prostate, CT to CBCT • 1 digital phantom for rectum/pelvis, CT to simulated CBCT (*n* = 50)	Rectum phantom sum of mean absolute difference measure: AVG 0.0257 Median 0.0251 Max 0.0415	Prostate phantom, MI: 0.81
Pekar et al. 2006	• Elastic body, SSD • 1 patient, CT to CT	CC, range: 0.975–0.982	Δ landmarks: AVG range: 3.6–3.8 mm
Rodriguez-Vila et al. 2010	• Optical flow • 4 patients, CT to CT	Maximum surface error, range: 4–9 mm	
Schaly et al. 2005	• TPS, contour matching • 10 patients, CT to CT (*n* = 4–7)	TRE, AVG ± SD: 3.0 ± 1.9 mm	
Schreibmann and Xing 2005	• Narrowband B-spline, NCC • 1 digital phantom, MR to MR • 2 patients, MR to CT	Visual inspection of landmarks: all <2 mm	
Vasquez Osorio et al. 2009	• TPS, surface matching • 2 patients, CT to CT	Δ (3 fiducials, 3 features), range: Prostate 1.3–3.0 mm Vesicles 2.0–3.7 mm	
Venugopal et al. 2005	• TPS, point matching • 1 patient, MR to MR (*n* = 2)	Volume intersection: Prostate AVG 97% Intraprostatic nodules range 63%–93%	
Wang et al. 2005a,b 2008	• Demons • 1 pelvic phantom, CT to CT (2005a) • 1 digital phantom, CT to CT (2005a) • Demons • 1 pelvic phantom, CT to CT (2005b) • 1 digital phantom, CT to CT (2005b) • 1 patient, CT to CT (2005b)	Δ displacement, AVG ± SD: Pelvic phantom 0.8 ± 0.5 mm Digital phantom 0.5 ± 1.5 mm Δ digital phantom (control points), |AVG| ± SD (max): LR 0.2 ± 0.6 (5.0) mm AP 0.3 ± 1.1 (14.0) mm SI 0.3 ± 1.2 (13.0) mm 3D 0.5 ± 1.5 (N/A) mm	 Δ pelvic phantom (23 fiducials), AVG ± SD (max): 0.8 ± 0.5 (2.7) mm CC: Digital phantom 0.991 Pelvic phantom 0.816 Patient data 0.944
Yang et al. 2009	• Optical flow • 3 patients, CT to MVCT	Δ (3 fiducials), |AVG| ± SD: 2.5 ± 1.2 mm	DSC, AVG: Prostate 0.92 Bladder 0.95 Rectum 0.81
Zhong et al. 2010	• B-spline or Demons • 1 digital phantom, CT to CT	Δ landmarks, AVG: B-spline 1.6 mm Demons 2.0 mm	
Zhou et al. 2010	• Superquadric/finite element, surface matching • 5 patients, CT to CBCT (*n* = 1–2)	Δ landmarks, range: AVG 0.4–2.2 mm RMS error 0.5–2.4 mm	Overlap ratio, range: 85.2%–95.0%

Note: AP, anterior–posterior; AVG, average; CBCT, cone-beam CT; CC, correlation coefficient; COM, center of mass; CT, computed tomography; CTV, clinical target volume; DSC, dice similarity coefficient; GTV, gross tumor volume; LR, left–right; Lt, left; MI, mutual information; MR, magnetic resonance; MSD, mean squared difference; MVCT, megavoltage CT; NCC, normalized crosscorrelation; RMS, root mean square; Rt, right; SD, standard deviation; SI, superior–inferior; SSD, sum of squared difference; TPS, thin-plate spline; TRE, target registration error; VOI, volume overlap index.

[a] Values estimated from graphs.

3.6 Summary

In summary, the validation of image registration and its use for contour propagation and volumetric analysis is challenging. Unlike other applications in radiation therapy, such as dose calculation where the physical interaction of the photon with the patient is modeled, image registration algorithms often rely on models that do not have a basis in the physics underlying the problem (i.e., the use of splines to model breathing motion). This adds an additional complexity to validating the algorithms because the simplification of the problem (i.e., the use of phantoms) can often significantly change the problem and what is being tested. However, an accurate and comprehensive validation of deformable registration is critical during algorithm development and before its use in radiation therapy. Because all models will have an associated uncertainty, it is also important to develop methods to understand these uncertainties and their impact on the clinical use of the resulting registration. When the algorithms are used for autosegmentation, the uncertainties in the results can be addressed by modifying the resulting contours. However, when deformable registration is used for dose accumulation and adaptive radiotherapy, manual adjustments are discouraged and often not possible. The uncertainties in the algorithm are also often masked in the simple visual verification techniques provided. It is therefore critical that medical physicists ensure the validation of these algorithms before their clinical use. Multiple methods of assessment, including phantoms and clinical data, can ensure an understanding of the algorithms as well as an assessment of their performance.

References

Al-Mayah, A., J. Moseley and K. K. Brock (2008). "Contact surface and material nonlinearity modeling of human lungs." *Phys Med Biol* 53(1): 305–317.

Al-Mayah, A., J. Moseley, S. Hunter, M. Velec, L. Chau, S. Breen and K. Brock (2010a). "Biomechanical-based image registration for head and neck radiation treatment." *Phys Med Biol* 55(21): 6491–6500.

Al-Mayah, A., J. Moseley, M. Velec and K. Brock (2011). "Toward efficient biomechanical-based deformable image registration of lungs for image-guided radiotherapy." *Phys Med Biol* 56(15): 4701–4713.

Al-Mayah, A., J. Moseley, M. Velec and K. K. Brock (2009). "Sliding characteristic and material compressibility of human lung: parametric study and verification." *Med Phys* 36(10): 4625–4633.

Al-Mayah, A., J. Moseley, M. Velec, S. Hunter and K. Brock (2010b). "Deformable image registration of heterogeneous human lung incorporating the bronchial tree." *Med Phys* 37(9): 4560–4571.

Alterovitz, R., K. Goldberg, J. Pouliot, I. C. Hsu, Y. Kim, S. M. Noworolski and J. Kurhanewicz (2006). "Registration of MR prostate images with biomechanical modeling and nonlinear parameter estimation." *Med Phys* 33(2): 446–454.

Bai, W. and M. Brady (2009). "Regularized B-spline deformable registration for respiratory motion correction in PET images." *Phys Med Biol* 54(9): 2719–2736.

Bender, E. T., N. Hardcastle and W. A. Tome (2012). "On the dosimetric effect and reduction of inverse consistency and transitivity errors in deformable image registration for dose accumulation." *Med Phys* 39(1): 272–280.

Bharatha, A., M. Hirose, N. Hata, S. K. Warfield, M. Ferrant, K. H. Zou, E. Suarez-Santana et al. (2001). "Evaluation of three-dimensional finite element-based deformable registration of pre- and intraoperative prostate imaging." *Med Phys* 28(12): 2551–2560.

Boldea, V., G. C. Sharp, S. B. Jiang and D. Sarrut (2008). "4D-CT lung motion estimation with deformable registration: Quantification of motion nonlinearity and hysteresis." *Med Phys* 35(3): 1008–1018.

Brock, K. K., L. A. Dawson, M. B. Sharpe, D. J. Moseley and D. A. Jaffray (2006). "Feasibility of a novel deformable image registration technique to facilitate classification, targeting, and monitoring of tumor and normal tissue." *Int J Radiat Oncol Biol Phys* 64(4): 1245–1254.

Brock, K. K. and C. Deformable Registration Accuracy (2010). "Results of a multiinstitution deformable registration accuracy study (MIDRAS)." *Int J Radiat Oncol Biol Phys* 76(2): 583–596.

Brock, K. K., A. M. Nichol, C. Menard, J. L. Moseley, P. R. Warde, C. N. Catton and D. A. Jaffray (2008). "Accuracy and sensitivity of finite element model-based deformable registration of the prostate." *Med Phys* 35(9): 4019–4025.

Brock, K. K., M. B. Sharpe, L. A. Dawson, S. M. Kim and D. A. Jaffray (2005). "Accuracy of finite element model-based multiorgan deformable image registration." *Med Phys* 32(6): 1647–1659.

Brock, K. M., J. M. Balter, L. A. Dawson, M. L. Kessler and C. R. Meyer (2003). "Automated generation of a four-dimensional model of the liver using warping and mutual information." *Med Phys* 30(6): 1128–1133.

Castillo, R., E. Castillo, R. Guerra, V. E. Johnson, T. McPhail, A. K. Garg and T. Guerrero (2009). "A framework for evaluation of deformable image registration spatial accuracy using large landmark point sets." *Phys Med Biol* 54(7): 1849–1870.

Castillo, R., E. Castillo, J. Martinez and T. Guerrero (2010). "Ventilation from four-dimensional computed tomography: Density versus Jacobian methods." *Phys Med Biol* 55(16): 4661–4685.

Chao, M., T. Li, E. Schreibmann, A. Koong and L. Xing (2008). "Automated contour mapping with a regional deformable model." *Int J Radiat Oncol Biol Phys* 70(2): 599–608.

Chao, M., Y. Xie, E. G. Moros, Q. T. Le and L. Xing (2010). "Image-based modeling of tumor shrinkage in head and neck radiation therapy." *Med Phys* 37(5): 2351–2358.

Chao, M., Y. Xie and L. Xing (2008). "Auto-propagation of contours for adaptive prostate radiation therapy." *Phys Med Biol* 53(17): 4533–4542.

Chen, T., S. Kim, S. Goyal, S. Jabbour, J. Zhou, G. Rajagopal, B. Haffty and N. Yue (2010). "Object-constrained meshless deformable algorithm for high speed 3D nonrigid registration between CT and CBCT." *Med Phys* 37(1): 197–210.

Coselmon, M. M., J. M. Balter, D. L. McShan and M. L. Kessler (2004). "Mutual information based CT registration of the lung at exhale and inhale breathing states using thin-plate splines." *Med Phys* 31(11): 2942–2948.

Du, K., J. E. Bayouth, K. Cao, G. E. Christensen, K. Ding and J. M. Reinhardt (2012). "Reproducibility of registration-based measures of lung tissue expansion." *Med Phys* 39(3): 1595–1608.

Eom, J., X. G. Xu, S. De and C. Shi (2010). "Predictive modeling of lung motion over the entire respiratory cycle using measured pressure-volume data, 4DCT images, and finite-element analysis." *Med Phys* 37(8): 4389–4400.

Faggiano, E., C. Fiorino, E. Scalco, S. Broggi, M. Cattaneo, E. Maggiulli, I. Dell'Oca, N. Di Muzio, R. Calandrino and G. Rizzo (2011). "An automatic contour propagation method to follow parotid gland deformation during head-and-neck cancer tomotherapy." *Phys Med Biol* 56(3): 775–791.

Fitzpatrick, J. M. and J. B. West (2001). "The distribution of target registration error in rigid-body point-based registration." 20(9): 917–927.

Fitzpatrick, J. M., J. B. West and C. R. Maurer, Jr. (1998). "Predicting error in rigid-body point-based registration." IEEE Trans Med Imaging 17(5): 694–702.

Foskey, M., B. Davis, L. Goyal, S. Chang, E. Chaney, N. Strehl, S. Tomei, J. Rosenman and S. Joshi (2005). "Large deformation three-dimensional image registration in image-guided radiation therapy." *Phys Med Biol* 50(24): 5869–5892.

Godley, A., E. Ahunbay, C. Peng and X. A. Li (2009). "Automated registration of large deformations for adaptive radiation therapy of prostate cancer." *Med Phys* 36(4): 1433–1441.

Gu, X., H. Pan, Y. Liang, R. Castillo, D. Yang, D. Choi, E. Castillo, A. Majumdar, T. Guerrero and S. B. Jiang (2010). "Implementation and evaluation of various demons deformable image registration algorithms on a GPU." *Phys Med Biol* 55(1): 207–219.

Guerrero, T., G. Zhang, T. C. Huang and K. P. Lin (2004). "Intrathoracic tumour motion estimation from CT imaging using the 3D optical flow method." *Phys Med Biol* 49(17): 4147–4161.

Heath, E., D. L. Collins, P. J. Keall, L. Dong and J. Seuntjens (2007). "Quantification of accuracy of the automated nonlinear image matching and anatomical labeling (ANIMAL) nonlinear registration algorithm for 4D CT images of lung." *Med Phys* 34(11): 4409–4421.

Hensel, J. M., C. Menard, P. W. Chung, M. F. Milosevic, A. Kirilova, J. L. Moseley, M. A. Haider and K. K. Brock (2007). "Development of multiorgan finite element-based prostate deformation model enabling registration of endorectal coil magnetic resonance imaging for radiotherapy planning." *Int J Radiat Oncol Biol Phys* 68(5): 1522–1528.

Hou, J., M. Guerrero, W. Chen and W. D. D'Souza (2011). "Deformable planning CT to cone-beam CT image registration in head-and-neck cancer." *Med Phys* 38(4): 2088–2094.

Ireland, R. H., K. E. Dyker, D. C. Barber, S. M. Wood, M. B. Hanney, W. B. Tindale, N. Woodhouse, N. Hoggard, J. Conway and M. H. Robinson (2007). "Nonrigid image registration for head and neck cancer radiotherapy treatment planning with PET/CT." *Int J Radiat Oncol Biol Phys* 68(3): 952–957.

Karnik, V. V., A. Fenster, J. Bax, D. W. Cool, L. Gardi, I. Gyacskov, C. Romagnoli and A. D. Ward (2010). "Assessment of image registration accuracy in three-dimensional transrectal ultrasound guided prostate biopsy." *Med Phys* 37(2): 802–813.

Kashani, R., M. Hub, J. M. Balter, M. L. Kessler, L. Dong, L. Zhang, L. Xing et al. (2008). "Objective assessment of deformable image registration in radiotherapy: a multiinstitution study." *Med Phys* 35(12): 5944–5953.

Kashani, R., M. Hub, M. L. Kessler and J. M. Balter (2007). "Technical note: A physical phantom for assessment of

accuracy of deformable alignment algorithms." *Med Phys* 34(7): 2785–2788.

Kaus, M. R., K. K. Brock, V. Pekar, L. A. Dawson, A. M. Nichol and D. A. Jaffray (2007). "Assessment of a model-based deformable image registration approach for radiation therapy planning." *Int J Radiat Oncol Biol Phys* 68(2): 572–580.

Kirby, N., C. Chuang and J. Pouliot (2011). "A two-dimensional deformable phantom for quantitatively verifying deformation algorithms." *Med Phys* 38(8): 4583–4586.

Lee, D., W. H. Nam, J. Y. Lee and J. B. Ra (2011). "Non-rigid registration between 3D ultrasound and CT images of the liver based on intensity and gradient information." *Phys Med Biol* 56(1): 117–137.

Li, B., G. E. Christensen, E. A. Hoffman, G. McLennan and J. M. Reinhardt (2008). "Pulmonary CT image registration and warping for tracking tissue deformation during the respiratory cycle through 3D consistent image registration." *Med Phys* 35(12): 5575–5583.

Li, P., U. Malsch and R. Bendl (2008). "Combination of intensity-based image registration with 3D simulation in radiation therapy." *Phys Med Biol* 53(17): 4621–4637.

Lian, J., L. Xing, S. Hunjan, C. Dumoulin, J. Levin, A. Lo, R. Watkins et al. (2004). "Mapping of the prostate in endorectal coil-based MRI/MRSI and CT: A deformable registration and validation study." *Med Phys* 31(11): 3087–3094.

Liu, F., Y. Hu, Q. Zhang, R. Kincaid, K. A. Goodman and G. S. Mageras (2012). "Evaluation of deformable image registration and a motion model in CT images with limited features." *Phys Med Biol* 57(9): 2539–2554.

Lu, W., M. L. Chen, G. H. Olivera, K. J. Ruchala and T. R. Mackie (2004). "Fast free-form deformable registration via calculus of variations." *Phys Med Biol* 49(14): 3067–3087.

Lu, W., G. H. Olivera, Q. Chen, M. L. Chen and K. J. Ruchala (2006a). "Automatic re-contouring in 4D radiotherapy." *Phys Med Biol* 51(5): 1077–1099.

Lu, W., G. H. Olivera, Q. Chen, K. J. Ruchala, J. Haimerl, S. L. Meeks, K. M. Langen and P. A. Kupelian (2006b). "Deformable registration of the planning image (kVCT) and the daily images (MVCT) for adaptive radiation therapy." *Phys Med Biol* 51(17): 4357–4374.

Malsch, U., C. Thieke, P. E. Huber and R. Bendl (2006). "An enhanced block matching algorithm for fast elastic registration in adaptive radiotherapy." *Phys Med Biol* 51(19): 4789–4806.

McClelland, J. R., J. M. Blackall, S. Tarte, A. C. Chandler, S. Hughes, S. Ahmad, D. B. Landau and D. J. Hawkes (2006). "A continuous 4D motion model from multiple respiratory cycles for use in lung radiotherapy." *Med Phys* 33(9): 3348–3358.

McClelland, J. R., S. Hughes, M. Modat, A. Qureshi, S. Ahmad, D. B. Landau, S. Ourselin and D. J. Hawkes (2011). "Inter-fraction variations in respiratory motion models." *Phys Med Biol* 56(1): 251–272.

Nithiananthan, S., K. K. Brock, M. J. Daly, H. Chan, J. C. Irish and J. H. Siewerdsen (2009). "Demons deformable registration for

CBCT-guided procedures in the head and neck: Convergence and accuracy." *Med Phys* 36(10): 4755–4764.

Olteanu, L. A., I. Madani, W. De Neve, T. Vercauteren and W. De Gersem (2012). "Evaluation of deformable image coregistration in adaptive dose painting by numbers for head-and-neck cancer." *Int J Radiat Oncol Biol Phys* 83(2): 696–703.

Paquin, D., D. Levy and L. Xing (2009). "Multiscale registration of planning CT and daily cone beam CT images for adaptive radiation therapy." *Med Phys* 36(1): 4–11.

Pekar, V., E. Gladilin and K. Rohr (2006). "An adaptive irregular grid approach for 3D deformable image registration." *Phys Med Biol* 51(2): 361–377.

Pevsner, A., B. Davis, S. Joshi, A. Hertanto, J. Mechalakos, E. Yorke, K. Rosenzweig et al. (2006). "Evaluation of an automated deformable image matching method for quantifying lung motion in respiration-correlated CT images." *Med Phys* 33(2): 369–376.

Piper, J., Y. Ikeda, Y. Fujisawa, Y. Ohno, T. Yoshikawa, A. O'Neil and I. Poole (2012). "Objective evaluation of the correction by non-rigid registration of abdominal organ motion in low-dose 4D dynamic contrast-enhanced CT." *Phys Med Biol* 57(6): 1701–1715.

Rietzel, E. and G. T. Chen (2006). "Deformable registration of 4D computed tomography data." *Med Phys* 33(11): 4423–4430.

Rodriguez-Vila, B., F. Gaya, F. Garcia-Vicente and E. J. Gomez (2010). "Three-dimensional quantitative evaluation method of nonrigid registration algorithms for adaptive radiotherapy." *Med Phys* 37(3): 1137–1145.

Rohlfing, T., C. R. Maurer, Jr., W. G. O'Dell and J. Zhong (2004). "Modeling liver motion and deformation during the respiratory cycle using intensity-based nonrigid registration of gated MR images." *Med Phys* 31(3): 427–432.

Sarrut, D., V. Boldea, S. Miguet and C. Ginestet (2006). "Simulation of four-dimensional CT images from deformable registration between inhale and exhale breath-hold CT scans." *Med Phys* 33(3): 605–617.

Schaly, B., G. S. Bauman, J. J. Battista and J. Van Dyk (2005). "Validation of contour-driven thin-plate splines for tracking fraction-to-fraction changes in anatomy and radiation therapy dose mapping." *Phys Med Biol* 50(3): 459–475.

Schreibmann, E. and L. Xing (2005). "Narrow band deformable registration of prostate magnetic resonance imaging, magnetic resonance spectroscopic imaging, and computed tomography studies." *Int J Radiat Oncol Biol Phys* 62(2): 595–605.

Schreibmann, E. and L. Xing (2006). "Image registration with auto-mapped control volumes." *Med Phys* 33(4): 1165–1179.

Serban, M., E. Heath, G. Stroian, D. L. Collins and J. Seuntjens (2008). "A deformable phantom for 4D radiotherapy verification: Design and image registration evaluation." *Med Phys* 35(3): 1094–1102.

Shekhar, R., P. Lei, C. R. Castro-Pareja, W. L. Plishker and W. D. D'Souza (2007). "Automatic segmentation of phase-correlated CT scans through nonrigid image registration using geometrically regularized free-form deformation." *Med Phys* 34(7): 3054–3066.

Shusharina, N. and G. Sharp (2012). "Analytic regularization for landmark-based image registration." *Phys Med Biol* 57(6): 1477–1498.

Sohn, M., M. Birkner, Y. Chi, J. Wang, Y. Di, B. Berger and M. Alber (2008). "Model-independent, multimodality deformable image registration by local matching of anatomical features and minimization of elastic energy." *Med Phys* 35(3): 866–878.

Stancanello, J., E. Berna, C. Cavedon, P. Francescon, D. Loeckx, P. Cerveri, G. Ferrigno and G. Baselli (2005). "Preliminary study on the use of nonrigid registration for thoraco-abdominal radiosurgery." *Med Phys* 32(12): 3777–3785.

Staring, M., S. Klein and J. P. Pluim (2007). "Nonrigid registration with tissue-dependent filtering of the deformation field." *Phys Med Biol* 52(23): 6879–6892.

Vandemeulebroucke, J., O. Bernard, S. Rit, J. Kybic, P. Clarysse and D. Sarrut (2012). "Automated segmentation of a motion mask to preserve sliding motion in deformable registration of thoracic CT." *Med Phys* 39(2): 1006–1015.

Vandemeulebroucke, J., S. Rit, J. Kybic, P. Clarysse and D. Sarrut (2011). "Spatiotemporal motion estimation for respiratory-correlated imaging of the lungs." *Med Phys* 38(1): 166–178.

Vasquez Osorio, E. M., M. S. Hoogeman, A. Al-Mamgani, D. N. Teguh, P. C. Levendag and B. J. Heijmen (2008). "Local anatomic changes in parotid and submandibular glands during radiotherapy for oropharynx cancer and correlation with dose, studied in detail with nonrigid registration." *Int J Radiat Oncol Biol Phys* 70(3): 875–882.

Vasquez Osorio, E. M., M. S. Hoogeman, L. Bondar, P. C. Levendag and B. J. Heijmen (2009). "A novel flexible framework with automatic feature correspondence optimization for nonrigid registration in radiotherapy." *Med Phys* 36(7): 2848–2859.

Vasquez Osorio, E. M., M. S. Hoogeman, A. Mendez Romero, P. Wielopolski, A. Zolnay and B. J. Heijmen (2012). "Accurate CTMR vessel-guided nonrigid registration of largely deformed livers." *Med Phys* 39(5): 2463–2477.

Venugopal, N., B. McCurdy, A. Hnatov and A. Dubey (2005). "A feasibility study to investigate the use of thin-plate splines to account for prostate deformation." *Phys Med Biol* 50(12): 2871–2885.

Voroney, J. P., K. K. Brock, C. Eccles, M. Haider and L. A. Dawson (2006). "Prospective comparison of computed tomography and magnetic resonance imaging for liver cancer delineation using deformable image registration." *Int J Radiat Oncol Biol Phys* 66(3): 780–791.

Wang, H., L. Dong, M. F. Lii, A. L. Lee, R. de Crevoisier, R. Mohan, J. D. Cox, D. A. Kuban and R. Cheung (2005). "Implementation and validation of a three-dimensional deformable registration algorithm for targeted prostate cancer radiotherapy." *Int J Radiat Oncol Biol Phys* 61(3): 725–735.

Wang, H., L. Dong, J. O'Daniel, R. Mohan, A. S. Garden, K. K. Ang, D. A. Kuban, M. Bonnen, J. Y. Chang and R. Cheung

(2005). "Validation of an accelerated 'demons' algorithm for deformable image registration in radiation therapy." *Phys Med Biol* 50(12): 2887–2905.

Wang, H., A. S. Garden, L. Zhang, X. Wei, A. Ahamad, D. A. Kuban, R. Komaki et al. (2008). "Performance evaluation of automatic anatomy segmentation algorithm on repeat or four-dimensional computed tomography images using deformable image registration method." *Int J Radiat Oncol Biol Phys* 72(1): 210–219.

Werner, R., J. Ehrhardt, R. Schmidt and H. Handels (2009). "Patient-specific finite element modeling of respiratory lung motion using 4D CT image data." *Med Phys* 36(5): 1500–1511.

Wijesooriya, K., E. Weiss, V. Dill, L. Dong, R. Mohan, S. Joshi and P. J. Keall (2008). "Quantifying the accuracy of automated structure segmentation in 4D CT images using a deformable image registration algorithm." *Med Phys* 35(4): 1251–1260.

Wolthaus, J. W., J. J. Sonke, M. van Herk and E. M. Damen (2008). "Reconstruction of a time-averaged midposition CT scan for radiotherapy planning of lung cancer patients using deformable registration." *Med Phys* 35(9): 3998–4011.

Wu, Z., E. Rietzel, V. Boldea, D. Sarrut and G. C. Sharp (2008). "Evaluation of deformable registration of patient lung 4DCT with subanatomical region segmentations." *Med Phys* 35(2): 775–781.

Xie, Y., M. Chao and L. Xing (2009). "Tissue feature-based and segmented deformable image registration for improved modeling of shear movement of lungs." *Int J Radiat Oncol Biol Phys* 74(4): 1256–1265.

Xie, Y., M. Chao and G. Xiong (2011). "Deformable image registration of liver with consideration of lung sliding motion." *Med Phys* 38(10): 5351–5361.

Yang, D., S. R. Chaudhari, S. M. Goddu, D. Pratt, D. Khullar, J. O. Deasy and I. El Naqa (2009). "Deformable registration of abdominal kilovoltage treatment planning CT and tomo-therapy daily megavoltage CT for treatment adaptation." *Med Phys* 36(2): 329–338.

Yang, D., W. Lu, D. A. Low, J. O. Deasy, A. J. Hope and I. El Naqa (2008). "4D-CT motion estimation using deformable image registration and 5D respiratory motion modeling." *Med Phys* 35(10): 4577–4590.

Yim, Y., H. Hong and Y. G. Shin (2010). "Deformable lung registration between exhale and inhale CT scans using active cells in a combined gradient force approach." *Med Phys* 37(8): 4307–4317.

Yin, L. S., L. Tang, G. Hamarneh, B. Gill, A. Celler, S. Shcherbinin, T. F. Fua et al. (2010). "Complexity and accuracy of image registration methods in SPECT-guided radiation therapy." *Phys Med Biol* 55(1): 237–246.

Yin, Y., E. A. Hoffman, K. Ding, J. M. Reinhardt and C. L. Lin (2011). "A cubic B-spline-based hybrid registration of lung CT images for a dynamic airway geometric model with large deformation." *Phys Med Biol* 56(1): 203–218.

Yin, Y., E. A. Hoffman and C. L. Lin (2009). "Mass preserving non-rigid registration of CT lung images using cubic B-spline." *Med Phys* 36(9): 4213–4222.

Zhang, T., Y. Chi, E. Meldolesi and D. Yan (2007). "Automatic delineation of on-line head-and-neck computed tomography images: Toward on-line adaptive radiotherapy." *Int J Radiat Oncol Biol Phys* 68(2): 522–530.

Zhang, Y., D. Boye, C. Tanner, A. J. Lomax and A. Knopf (2012). "Respiratory liver motion estimation and its effect on scanned proton beam therapy." *Phys Med Biol* 57(7): 1779–1795.

Zhong, H., J. Kim and I. J. Chetty (2010). "Analysis of deformable image registration accuracy using computational modeling." *Med Phys* 37(3): 970–979.

Zhong, H., J. Kim, H. Li, T. Nurushev, B. Movsas and I. J. Chetty (2012). "A finite element method to correct deformable image registration errors in low-contrast regions." *Phys Med Biol* 57(11): 3499–3515.

Zhou, J., S. Kim, S. Jabbour, S. Goyal, B. Haffty, T. Chen, L. Levinson, D. Metaxas and N. J. Yue (2010). "A 3D global-to-local deformable mesh model based registration and anatomy-constrained segmentation method for image guided prostate radiotherapy." *Med Phys* 37(3): 1298–1308.

II

Registration

4

Similarity Metrics

He Wang
*The University of Texas M.D.
Anderson Cancer Center*

4.1 Introduction

As highlighted in the previous chapters, the goal of image registration is to find a geometric transformation (rigid or nonrigid) that aligns two given images. When two images are deemed "registered," this indicates that they are best matched or most similar compared with the initial relationship between the images. With this concept in mind, similarity metrics are arguably the most critical element of a registration problem. The metric defines what the goal of the registration process is, and it measures how well one image is matched to another after the transformation has been applied. The aim of this chapter is to describe some of the basic similarity measures that are used in the process of image registration.

A large number of image similarity metrics have been proposed in medical imaging and computer vision community. Broadly, they can be categorized into two classes: feature-based metrics and intensity-based metrics. Feature-based metrics use corresponding points (also called landmarks or "fiducial markers") or corresponding surfaces in the two images. These features can be extracted from both images manually or automatically. The advantages of feature-based similarity metrics include the independence of anatomy, the fast computation due to the limited number of points involved, and the exact match at the feature points for some of the algorithms. However, feature-based methods are sensitive to the accuracy of feature extraction algorithms, and user interactions are often required. In contrast, intensity-based metrics use the intensities in the two images alone without the requirement to manually segment or delineate the corresponding points or structures. These methods use all or a large proportion of the data in each image and thus tend to average out any errors caused by the noise or random fluctuation of image intensity. The disadvantage of intensity-based similarity metrics is that the computation may be time-consuming.

A "generic" image similarity metric does not exist, but there are a set of metrics that are appropriate for particular applications. The selection of the metric should depend on the types of objects to be registered and the expected precision of the resultant alignment. Some metrics work well for a larger capture region with lower precision, whereas other metrics can provide a higher precision for the registration of subtle or detailed structures but are usually required to be initialized quite close to the optimal value. In some cases, it could be advantageous to use a particular metric to get an initial approximation of the transformation and then switch to another more sensitive metric to achieve a better precision in the final result.

The choice of an image similarity metric also depends on the modality of the images to be registered. Although computed tomography (CT) is the most common modality used in radiotherapy treatment planning to calculate dose distributions, more and more clinics are using magnetic resonance (MR) imaging or positron emission tomography imaging for a better delineation of the tumor target and a more focused radiation treatment. In addition, the advancement of volumetric imaging in the treatment room, using in-room CT, kilovoltage cone-beam CT, megavoltage cone-beam CT, and megavoltage CT imaging, has provided the imaging data needed to perform adaptive radiation therapy. The fusion of information from different imaging modalities requires appropriate similarity metrics for each application.

The applications of similarity metrics in radiation therapy include the definition of cost functions for image registration and fusion, the evaluation of the performance of different registration algorithms before they can be released to the clinic, and the verification of the daily position of the patient before radiation treatment. Similarity metrics alone are mainly based on local image correspondence. For model-based (elastic, viscous fluid, and biomechanical) image registration, similarity metrics are often used in collaboration with other energy functions, which constrains the image deformation according to certain physical characteristics of the human body (i.e., soft tissue and bone), to achieve a more reasonable registration result. This will be described in later chapters.

In the following sections, the terms similarity, distance, or disparity metric will be used interchangeably. The similarity metric is the opposite of distance and disparity metric. While similarity will be maximized, distance and disparity function are normally minimized during the registration process.

4.2 Intensity-Based Metrics

A variety of intensity-based metrics have been studied and are currently used in medical imaging (Hill and Hawkes 1994; Penney et al. 1998; Holden et al. 2000; Skerl et al. 2006; Wu et al. 2009). The most popular metrics include the sum of squared intensity difference, correlation coefficient, and mutual information (MI). The intensity difference-based metrics are computationally attractive, but they require that the two objects have intensity values in the same range; the correlation coefficient requires that the intensities of the two images are related by a linear transformation, whereas MI is the metric of choice when images from different modalities need to be registered. Many other similarity metrics are also used in medical imaging, including pattern intensity, gradient correlation, and gradient difference. The evaluation of these similarity metrics in different applications have been published (Penney et al. 1998; Skerl et al. 2006; Wu et al. 2009).

4.2.1 Intensity Differences

Here, let image A be the reference static image and image B be the moving image to be transformed to match with image A. $A(\vec{x})$ and $B(\vec{x})$ are the intensity values at the locations with spatial coordinates $\vec{x}$ in images A and B, respectively. Images A and B can be 2D or 3D images. The symbol T will be used to represent a registration transformation, for example, $T(B(\vec{x}))$ means the intensity value at location $\vec{x}$ of image B after the transformation T is applied to it.

The simplest pixel similarity metrics are based on the difference in image intensities at corresponding points between two images. The most commonly used metric is the mean squared difference MSD(A,B), which is minimized during registration (Friston et al. 1995; Hajnal et al. 1995; Woods et al. 1998; Wolberg and Zokai 2000; Wijesooriya et al. 2008):

$$\text{MSD}(A,B) = \frac{1}{N}\sum_{\vec{x}}[A(\vec{x}) - T(B(\vec{x}))]^2, \quad (4.1)$$

where N is the number of pixels in the overlapping region in the two images that contribute to the above calculations. These measures are normalized so that they are not affected by the number N. The MSD measures make the implicit assumption that, when two images are aligned, they only differ by Gaussian noise. Therefore, they can never be used for intermodality registration.

The MSD measure is very sensitive to a small number of pixels that have very large intensity differences between images A and B, for example, when contrast material is used for one of the scans. The effect of such "outlier" pixels can be reduced by using the mean absolute difference MAD(A,B) rather than MSD. MAD is defined as

$$\text{MAD}(A,B) = \frac{1}{N}\sum_{\vec{x}}\left|A(\vec{x}) - T(B(\vec{x}))\right|.$$

4.2.2 Correlation-Based Method

Cross-correlation is a basic statistical criterion to measure similarity in signal and image processing. It is usually used for template matching, pattern recognition, and image registration. For two digital images A and B, the cross-correlation C is defined as

$$C = \sum_{\vec{x}} A(\vec{x})T(B(\vec{x})). \quad (4.2)$$

Image B can be an overlap region or a small feature template.

The cross-correlation metric has the disadvantage of being sensitive to changes in image amplitude of A and B. For example, doubling all intensity values of A doubles the value of C. Also, the range of C is dependent on the size of the overlapping region or the size of template B. An approach frequently used to overcome this difficulty is to perform matching via the correlation coefficient (CC) metric, or the normalized cross-correlation (Gonzalez and Woods 1992), which is defined as

$$\text{CC} = \frac{\sum_{\vec{x}}(A(\vec{x}) - \bar{A})(T(B(\vec{x})) - \bar{B})}{\sqrt{\sum_{\vec{x}}(A(\vec{x}) - \bar{A})^2 \sum_{\vec{x}}(T(B(\vec{x})) - \bar{B})^2}}, \quad (4.3)$$

where $\bar{A}$ is the mean pixel value in image A in the overlapping region and $\bar{B}$ is the mean pixel value of image B in the overlapping region.

The correlation coefficient method assumes a linear relationship between the intensity values in the images; therefore, it can deal with differences in image contrast and brightness. The correlation coefficient–based registration works well to align images with translations (Moseley and Munro 1994; Penney et al. 1998; Clippe et al. 2003), but it has some difficulties when it is applied to images with rotation and scaling in spatial domain. Fortunately, the correlation can also be carried out in the frequency domain, and this special property converts the image rotation to a simple shift of the angular coordinate in the frequency domain (Kassam and Wood 1996); thus, the rotation detection can be solved. Several generalized versions of CC have also been used to handle more complicated geometric deformations, such as affine transformation (Berthilsson 1998), deformation caused by perspective projection, and image distortion due to imperfect lens (Simper 1996).

The main drawbacks of the correlation-based methods are the flatness of the similarity measure maxima, which is due to the self-similarity of the images. The maximum can be sharpened by the preprocessing of the image (Pratt 1974). Several authors employed edge-based correlation methods (van Herk and Kooy 1994; Penney et al. 1998; Sawada et al. 2005; Wu et al. 2009), which are less sensitive to intensity differences between two images. These methods computed the correlation on the edges extracted from the images rather than on the original images.

Usually, CC is not suited for multimodality image registration because a global linear transformation function of the gray values cannot be presumed. However, some studies have shown that the correlation coefficient can be applied to subregions of the images, where the assumption of a linear relationship is valid in a small neighborhood, enabling rigid and deformable registration to be performed for multimodality images (Weese et al. 1999).

4.2.3 Information Theory Metrics

Information theory is based on probability theory and statistics. The most important quantities of information are entropy, the information in a random variable, and MI, the amount of information in common between two random variables.

4.2.3.1 Entropy Metric

The most commonly used metric of information in signal and image processing is the Shannon–Wiener entropy metric, which was originally developed as part of communication theory in the 1940s (Shannon 1948). Entropy gives a measure of the average information provided by a set of symbols. The entropy for a random variable is defined by

$$H = -\sum_i p_i \log p_i, \qquad (4.4)$$

where p_i is the probability of the ith symbol of the variable. In the application of imaging, the symbols are intensity values occurring in the image of interest.

In information theory, entropy is a measure of randomness, uncertainty, variability, or complexity of a discrete random variable X with probability function p_i. The more random a variable is, the more entropy it will have. Entropy is always positive, that is, $H \geq 0$, because $0 \leq p_i \leq 1$ for all i. Entropy will have a maximum value if all symbols have equal probability of occurring

and have a minimum value of zero if the variable is deterministic, that is, the probability of one symbol occurring is 1 and the probabilities of all the other occurring are 0. Any change in the data that tends to equalize the probabilities of the symbols (i.e., that makes the histogram more uniform) increases the entropy. Blurring the data reduces noise and so sharpens the histogram and results in reduced entropy.

4.2.3.2 Joint Entropy

Given a pair of images A and B, the joint entropy $H(A,B)$ of this pair of discrete random variables is defined as

$$H(A,B) = -\sum_a \sum_b p_{AB}(a,b) \log p_{AB}(a,b), \qquad (4.5)$$

where $p_{AB}(a,b)$ is the joint probability of pairs of image values (value a in image A and value b in image B) occurring together. The number of elements in calculating the probability distribution can either be determined by the range of intensity values in the two images or from a reduced number of intensity "bins." For example, MR and CT images usually use 12 bits to store intensity values. This gives a very sparse probability distribution with 4096×4096 elements in the joint histogram. The original intensity values can be rescaled or selected to generate smaller number of bins. In practice, it is more common to use 32 to 256 bins. In the above equation, a and b represent either the original image intensities or the selected intensity bins. The reduced bins directly improve the estimation of joint distribution. However, it also lowers the intensity sensitivity, for example, different features that are represented by similar intensity values can be grouped into the same bin. Further work to improve the estimation of joint probability distribution includes Parzen window density estimate (Wells et al. 1996; Thevenaz and Unser 1998) and the discrete histogram estimates (Studholme et al. 1999).

The joint entropy measures the amount of information available in the combined images. Some characteristics of joint entropy are summarized as follows:

- Like other entropies, $H(A,B) \geq 0$.
- If A and B are statistically independent, then the joint entropy will be the sum of the entropies of the individual images.
- The more similar (i.e., less independent) the images, the lower the joint entropy compared with the sum of the individual entropies.
- Two systems, considered together, can never have more entropy than the sum of the entropy in each of them.

$$H(A,B) \leq H(A) + H(B)$$

- On the contrary, the joint entropy is always at least equal to the entropies of the original system; adding a new system can never reduce the available uncertainty.

$$H(A,B) \geq H(A)$$

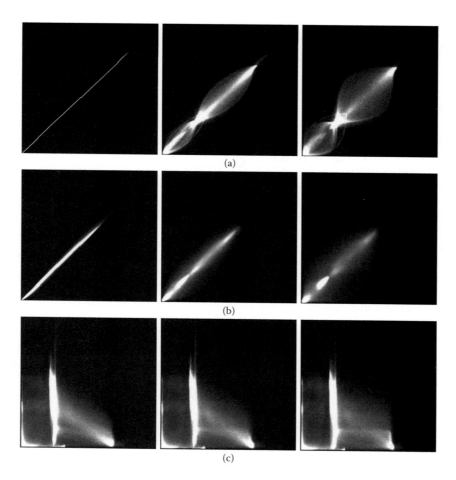

FIGURE 4.1 Examples of joint probability distribution for (a) identical two diagnostic CT images, (b) MR image and its Gaussian smoothed image, and (c) CT (horizontal axis) and MR (vertical axis) images. Left: Images are aligned. Middle: Images are misaligned by a 2 pixel translation laterally. Right: Images are misaligned by a 4 pixel translation.

The concept of joint entropy can be visualized using a joint probability distribution calculated from images A and B. Examples are shown in Figure 4.1. Notice in Figure 4.1a that there exists a linear relationship in the joint histogram for aligned images; however, during misalignment, it becomes dispersed. Figure 4.1b and c also shows that the misalignment results in dispersion or blurring of the histogram and thus increases the joint entropy. Table 4.1 shows the joint entropies for the joint probability distributions in Figure 4.1.

Joint entropy is minimized during image registration. As seen in Figure 4.1 and Table 4.1, with constant marginal entropies of the two images of interest, joint entropy is minimized when the two images are aligned. However, there are two disadvantages when joint entropy is used for image registration. One is due to the fact that the marginal entropies for the moving image B may not be constant during the image registration process. Image transformation and interpolation involved in the registration process will alter both the marginal and joint probability distributions. Assuming that the marginal entropy of image A, $H(A)$, is fixed, the joint entropy $H(A,B)$ and the marginal entropy $H(B)$

will vary because the transformation estimate and image overlap vary. Therefore, the minimized joint entropy may not be a solution of an optimal alignment. The second disadvantage of joint entropy is that, by minimizing joint entropy, a transformation estimate may be selected that simply finds the overlap that contains the least information, instead of finding the image overlap that contains the most corresponding information, which is what is expected and desired. In this situation, the transformation may produce a small number of histogram elements with high probabilities, which will minimize the marginal entropy $H(B)$. Registration can therefore be thought of as trying to find the transformation that maximizes the "sharpness"

TABLE 4.1 Joint Entropies Calculated for Joint Probability Distributions in Figure 4.1

	Aligned	2 Pixel Shifted	4 Pixel Shifted
CT-CT	3.885	5.876	6.332
MR-MR	6.579	8.029	8.559
CT-MR	7.572	7.996	8.071

of the histogram, thereby minimizing the joint entropy. What is desired is to relate the changes in the value of the joint entropy $H(A,B)$ to the marginal entropies of the two images $H(A)$ and $H(B)$ derived from their region of overlap.

4.2.3.3 Mutual Information

A solution to the overlap problem that the joint entropy method suffers from is to use MI or relative entropy between the two images. MI was initially introduced by Shannon (1948) as the "rate of transmission of information." It was proposed independently and simultaneously by researchers in Leuven, Belgium (Collignon et al. 1995; Maes et al. 1997), and at the Massachusetts Institute of Technology in the United States (Viola 1995; Wells et al. 1996) for intermodality medical image registration.

The MI $I(A,B)$, when formulated using the Kullback–Leibler measure (Vajda 1989), is defined as

$$I(A,B) = \sum\sum p_{AB}(a,b)\log\frac{p_{AB}(a,b)}{p_A(a)p_B(b)} . \tag{4.6}$$

MI is related to marginal and joint entropies by the equation

$$I(A,B) = H(A) + H(B) - H(A,B).$$

A useful way of visualizing the relationship between these entropies is provided by the Venn diagram as shown in Figure 4.2. Here, the size of each of the circles represents the value of the particular entropy. The overlapping area represents the MI. MI can qualitatively be thought of as a measure of how well one image explains the other. During image registration, the MI is maximized, and solutions can be sought to have low joint entropy together with high marginal entropies.

To better understand Figure 4.2, it is useful to think more about probabilities. The conditional probability $p(a|b)$ is the probability that A will take the value a given that B has the value b. The conditional entropy of A given B is defined as

$$H(A|B) = -\sum_a\sum_b p_{AB}(a,b)\log p_{A|B}(a|b).$$

The conditional entropy of A conditional on B refers to the average entropy of A conditional on the value of B, averaged over all possible values of B. The relationship between joint entropy and conditional entropy is

$$H(A,B) = H(B) + H(A|B).$$

TABLE 4.2	Properties of MI
Nonnegativity	$I(A,B) \geq 0$
Independence	$I(A,B) = 0 \Leftrightarrow p_{AB}(a,b) = p_A(a) \bullet p_B(b)$
Invariance	$I(A,T(A)) = I(A,A)$
	(T is a one-to-one mapping)
Symmetry	$I(A,B) = I(B,A)$
Self-information	$I(A,A) = H(A)$
Boundedness	$I(A,B) \leq \min(H(A),H(B))$
	$\leq (H(A)+H(B))/2$
	$\leq \max(H(A),H(B))$
	$\leq H(A<B)$
	$\leq H(A)+H(B)$
Data processing inequality	$I(A,B) \geq I(A,T(B))$
	(T is a processing operation)

This states that the total uncertainty about the value of A and B is equal to the uncertainty about B plus the (average) uncertainty about A once you know B. Similarly,

$$H(A,B) = H(A) + H(B|A).$$

The previous expression for MI using conditional entropy can be rewritten as

$$I(A,B) = H(A) - H(A|B) = H(B) - H(B|A).$$

This relationship can be seen in Figure 4.2. The maximization of MI, therefore, involves minimizing the conditional entropy with respect to the entropy of one of the images.

Table 4.2 shows some properties of MI. The last property in this table, data processing inequality, states that if a random variable B tells us something about another random variable A, further information about A cannot be extracted by performing additional processing operations, random or deterministic, on B. That is, the MI between A and B is greater or equal to the MI between A and any function of B alone.

4.2.3.4 Normalized MI

MI does not entirely solve the overlapping problem described in Section 4.2.3.2. In particular, changes in the overlap of very low intensity regions of the image (especially noise around the patient) can disproportionately contribute to the MI. Alternative normalizations of the joint entropy have been proposed to overcome this problem.

Three normalization schemes have thus far been proposed in the literature to address this problem. Maes et al. (1997) proposed two normalized MI in the discussion section of his paper:

$$\tilde{I}_1(A,B) = \frac{2I(A,B)}{H(A)+H(B)} \tag{4.7}$$

$$\tilde{I}_2(A,B) = H(A,B) - I(A,B) \tag{4.8}$$

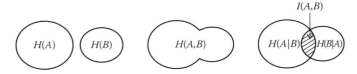

$I(A,B)$

$H(A)$ $H(B)$ $H(A,B)$ $H(A|B)$ $H(B|A)$

FIGURE 4.2 Set theory representation of the entropies and MI.

Studholme et al. (1999) have proposed an alternative normalization devised to overcome the sensitivity of MI to change in image overlap. This metric involves normalized MI with respect to the joint entropy of the overlapped volume.

$$\tilde{I}_3(A,B) = \frac{H(A) + H(B)}{H(A,B)}. \tag{4.9}$$

The third version of normalized MI has been shown to be considerably more robust than standard MI for intermodality registration in which the overlap volume changes substantially (Studholme et al. 1999). However, another study showed that the performance of MI and of normalized MI are equivalent for serial MR registration, when two images have virtually identical fields of view (Holden et al. 2000).

It can be shown that the versions of the normalized MI in Equations 4.7 and 4.9 are closely related:

$$\tilde{I}_3(A,B) = \frac{H(A) + H(B)}{H(A,B)} = \frac{I(A,B)}{H(A,B)} + 1 = \frac{1}{\tilde{I}_1(A,B) - 2}.$$

4.2.3.5 Other Variations of MI

In theory, the MI of two images does not take spatial information into account. Sometimes, it is useful to incorporate spatial information in registration procedure to achieve much better results. The MI of two random variables conditioned on an extra variable such as spatial location can be found as

$$I(A,B|C) = H(A|C) + H(B|C) - H(A,B|C).$$

Conditioning on a third random variable may either increase or decrease the MI, but it is always true that

$$I(A,B|C) \geq 0$$

for discrete, jointly distributed random variables *A*, *B*, and *C*.

Spatial information can be introduced into MI by different ways, such as labeling, local intensity, and local structure. A good review for these applications can be found in the tutorial at the International Conference on Medical Image Computing and Computer Assisted Intervention 2009 (Wells et al. 2009).

4.2.4 Probabilistic Framework

Assuming that the intensity values in two registered images are probabilistically related, the search for the optimal transformation from one image to another can be cast into parameter estimation problem. In this framework, the appropriate similarity measure to formulate registration can be represented by the likelihood function that is associated with the image intensity values as a function of spatial transformations. The optimal transformation parameters can be subsequently estimated by maximizing the likelihood function over the transformation space.

Given images A and B and model parameters θ, the maximum likelihood estimates are found by

$$\arg\max_{\theta} p(A = \{a_1, \ldots, a_n\}|\theta) = \arg\max_{\theta} \prod_{i=1}^{n} p(a_i|\theta),$$

where *p*(*A*) is the probability density function of *A*. In general, the maximum likelihood methods have good convergence properties, are relatively simple to implement, and are asymptotically unbiased and efficient as the sample size increases.

It is often more convenient to use log-likelihood function to reduce the burden of multiplication:

$$\log L(A|\theta) = \sum_{i=1}^{n} \log p(a_i|\theta).$$

Maximum likelihood approaches for image registration can be found in many literatures (Viola and Wells 1997; Roche et al. 2000; Olsen 2002; Zhu and Cochoff 2002; Munbodh et al. 2009; Wells et al. 2009; Chen et al. 2011; Risholm et al. 2011). Roche et al. (2000) derived similarity measures corresponding to different specific modeling assumptions from general maximum likelihood and retrieved some well-known intensity-based measures (correlation coefficient, correlation ratio, and MI), whereas Viola and Wells (1997) also showed that, under certain conditions, the conditional log likelihood of the image is a multiple of the conditional entropy of the image. Incorporating information about the probability density function of the image intensity into similarity measure allows the development of more powerful methodology for intensity-based image registration.

4.3 Feature-Based Metrics

Whereas intensity-based methods compare intensity patterns in images via correlation metrics, feature-based registration algorithms use sparse representations of the image features to make the algorithms fast once the segmentation has been undertaken. Features can be landmark points, contours, or surfaces. The feature-based method establishes a correspondence between a number of points in images. By knowing this correspondence, a transformation between two images can be determined.

In the application of radiation therapy, the contours and surfaces are mostly represented by ordered points. An overall image similarity *S* can be expressed for a transformation *T* between the two images as the sum of individual similarity measurement *S*(.) at *N* different landmark points:

$$S(T) = \sum_{i=1}^{N} S(A(\vec{x}_i), B(T(\vec{x}_i))). \tag{4.10}$$

The similarity measurement *S*(.) can take many forms, such as absolute distance, squared distance, and weighted distance, and they are also called disparity metrics because they are all spatial distance related.

4.3.1 Landmark-Based Metrics

Landmark-based similarity metrics require the identification of the corresponding 2D or 3D points in the two images to be aligned. The corresponding points are sometimes called homologous landmarks to emphasize that they should represent the same feature in the different images. Assume that we have two sets of N selected points, that is, $X = \{\vec{x}_i\}$ in image A and $Y = \{\vec{y}_i\}$ in image B, where $i = 1, 2, \ldots, N$ and $\vec{x}_i$ and $\vec{y}_i$ are the spatial position of the ith points in images A and B, respectively. An example of the similarity measure between these two point sets can be expressed as the sum of the squared distance of each point pair:

$$SM = \sum_{i=1}^{N} \left\| \vec{x}_i - \vec{y}_i \right\|^2 . \tag{4.11}$$

Landmark-based similarity metrics can be used in the cost function to be minimized for the registration of two images, which have affine transformation relative to each other. Assume a transformation T applied on image B as well as its corresponding landmark set Y; the cost function is normally expressed as one of the following:

$$D(T) = \sum_{i=1}^{N} \left\| \vec{x}_i - T(\vec{y}_i) \right\|^2 \tag{4.12}$$

$$D(T) = \frac{1}{N} \sum_{i=1}^{N} \left\| \vec{x}_i - T(\vec{y}_i) \right\|^2 . \tag{4.13}$$

Ideally, after registration, $\vec{x} = T(\vec{y})$. The task of finding the transformation T for this expression is known mathematically as the orthogonal Procrustes problem (Hill and Batchelor 2001).

Landmark-based similarity metrics can also be used as an evaluation metrics for alignment. To establish the "ground truth" of alignment, radiopaque markers can be customized in test phantoms (rigid or deformable) (Wang et al. 2005; Brock et al. 2008; Serban et al. 2008) or feature points be selected from mathematical phantoms (Lu et al. 2004; Wang et al. 2005) or clinical images (Brock et al. 2005; Castillo et al. 2009). These are called "fiducial markers." The performance of the registration technique is then evaluated to compare the registration results $Y = \{T(\vec{y}_i)\}$ with the corresponding "ground truth" $X = \{\vec{x}_i\}$. The most used quantified performance index is the mean absolute distance:

$$D = \frac{1}{N} \sum_{i} \left\| \vec{x}_i - \vec{y}_i \right\| . \tag{4.14}$$

Besides the distance, the standard deviation of the spatial error can also provide a basic understanding of the behavior of each image registration technique.

In another application, landmark-based similarity metrics can help with the patient setup procedure in radiation therapy (West et al. 1997; Shirato et al. 2004; Kashani et al. 2008; Nelson et al. 2008). Before the radiation treatment, the setup can be verified by comparing marker positions on daily portal images to pretreatment digitally reconstructed radiographs using the similarity metric such as Equation 4.14. Markers such as these can be either detected automatically in the images or identified by hand. Figure 4.3 shows an example of daily tumor alignment based on fiducial landmarks during radiation treatment.

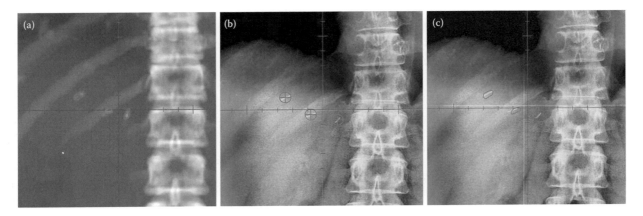

FIGURE 4.3 **(See color insert.)** Fiducials as landmarks for liver tumor alignment for daily radiation treatment. (a) Reference digitally reconstructed radiography from planning CT with fiducials contoured. (b) Daily portal image with fiducials marked manually. (c) Alignment of tumor based on fiducials using sum of squared distance.

4.3.2 Contour/Surface-Based Metrics

The contour and surface of an anatomic structure can be used for medical image registration. Compared with landmarks, contours are normally represented as ordered points, and surfaces are normally represented as a set of contours or surface meshes to characterize the geometric features. Therefore, a surface can still be considered as a point set. Similar to the techniques described above, the approach for solving the contour- or surface-based registration problem is to search for the transformation that minimizes some disparity measure between the two point sets X and Y. The disparity measure is generally a distance. The distance expressed in Equation 4.12 was successfully used for curve-based registration of two radiographs (Balter et al. 1992; Cai et al. 1998).

The disparity metric used for contour- or surface-based image registration is normally an averaged, and optionally weighted, distance between points on one contour or surface and corresponding points on the other. Assuming a set of feature points X on the surface in image A and a set of points $Y = \{\vec{y}_i, i = 1, 2, ..., N\}$ on the surface in image B, the distance is normally defined as

$$D(X, T(Y)) = \sum_{i=1}^{N} w_i d^2(X, T(\vec{y}_i)) = \sum_{i=1}^{N} w_i \left\| \vec{x}_i' - T(\vec{y}_i) \right\|^2, \quad (4.15)$$

where

$$\vec{x}_i' = C(X, T(\vec{y}_i))$$

is a point on the surface X "corresponding" to the point $T(\vec{y}_i)$. C can be called "correspondence" function, and w_i is the weight associated with point $\vec{y}_i$. Note that Equation 4.15 is similar to Equation 4.12 of landmark point-based measure. However, for the contour- or surface-based method, the total number of points in X may not be the same as the number of points in Y, and the two point sets X and Y may not be homological. The corresponding point for $\vec{y}_i$ in Y will be determined by the correspondence function from X and may not be any one of the points in X. One example of the correspondence function is the closest point operator. Another can be a projection function, which projects point $\vec{y}_i$ in Y to surface X with the flexibility in the projection to maintain the relationship among the points on the surface. The projection is primarily perpendicular to the surface. The availability of point correspondence information is the principal difference between point-based registration and surface-based registration. The lack of exact point correspondence information causes surface-based registration algorithms to be based on iterative search. Equation 4.15 merely provides approximate point correspondence information for a particular transformation T during an iterative search.

The first surface-based registration to a medical problem is called the "head and hat" algorithm (Pelizzari et al. 1989). In this algorithm, the static surface is called head. It can be represented as a stack of CT or MR discs, an unconnected point set, a connected point set describing a mesh, or a surface described by some parameters such as a B-spline surface. The moving surface is referred to as hat and is mostly represented as a list of unconnected 3D points. In the registration process, for each point on the hat surface, a corresponding nearest point on the head is identified. The measure of closeness of fit used is the mean squared distance. The registration transformation is determined by iteratively transforming the rigid hat surface with respect to the head surface until the closest fit of the hat onto the head was found. This algorithm can be improved by preprocessing the head images using distance transform (Pelizzari et al. 1989; Jacobs et al. 1999; Lee et al. 2005). The most commonly used surfaces are the skin and organ boundaries.

Besl and McKay (1992) presented an iterative closest point (ICP) algorithm for general-purpose, representation-independent, and shape-based matching that can be used with a variety of geometric primitives including point sets, line segment sets, triangle sets (faceted surfaces), and implicit and parametric curves and surfaces. One shape is assigned to be the "data" shape and the other shape to be the "model" shape. For surface-based registration, the shapes are surfaces. The data shape is first decomposed into a point set (if it is not already in point set form). Then, the data shape is registered to the model shape by iteratively searching the closest model points to the data points by minimizing the disparity measure of Equation 4.12 and applying the resulting transformation to the data points. This algorithm has been widely used in medical image application (Li et al. 2006; Clements et al. 2008) and was also enhanced to solve nonrigid image registration (Declerck et al. 1997; Chui and Rangarajan 2003).

4.4 Summary

In this chapter, basic similarity metrics used in medical imaging were described. Feature-based similarity measures are normally used when localized accuracy is important and image segmentation is already undertaken. They can normally be used for both single and multimodality images, because they are independent of imaging modality. The limitation of the feature-based method lies in that it is highly dependent on the accuracy of the feature extraction. The intensity-based methods are attractive because they do not require the detection of geometric features, such as points, contours, or surfaces, from the images. MI and normalized MI are the most popular image similarity measures for intermodality image registration, whereas they are also used for intramodality image registration. Cross-correlation and the sum of squared intensity differences are commonly used for intramodality image registration.

It is important to understand that the similarity metric is only an approximation to the homological relationship of two images. Because it is only an approximation, a successful registration that maximizes the similarity metric may not represent the best mapping for the homology function. The selection of similarity metrics is normally based on the application and the accuracy requirement of the application. Sometimes, two or more similarity metrics can be used in a single image registration process simultaneously or sequentially to achieve a better approximation to the homological relationship. Recent studies show that hybrid methods that combine intensity- and feature-based similarity measures can also be used to achieve global accuracy as well as localized accuracy (Russakoff et al. 2004; Gan et al. 2008; Wells et al. 2009).

References

Balter, J. M., Pelizzari C.A. et al. (1992). "Correlation of projection radiographs in radiation therapy using open curve segments and points." *Med. Phys.* 19(2): 329–334.

Berthilsson, R. (1998). Affine correlation. Proceedings of the International Conference on Pattern Recognition ICPR'98, Brisbane, Australia, 1458–1460.

Besl, P. J. and McKay, H. D. (1992). "A method for registration of 3-D shapes." 14(2): 239–256.

Brock, K. K., Nichol, A. M. et al. (2008). "Accuracy and sensitivity of finite element model-based deformable registration of the prostate." *Med. Phys.* 35(9): 4019–4025.

Brock, K. K., Sharpe, M. B. et al. (2005). "Accuracy of finite element model-based multi-organ deformable image registration." *Med. Phys.* 32(6): 1647–1659.

Cai, J., J. Chu, C. H. et al. (1998). "A simple algorithm for planar image registration in radiation therapy." *Med. Phys.* 25(6): 824–829.

Castillo, R., Castillo, E. et al. (2009). "A framework for evaluation of deformable image registration spatial accuracy using large landmark point sets." *Phys. Med. Biol.* 54(7): 1849–1870.

Chen, S., Guo, Q. et al. (2011). "A maximum likelihood approach to joint image registration and fusion." *IEEE Trans. Med. Imaging* 20(5): 1363–1371.

Chui, H. and Rangarajan, A. (2003). "A new point matching algorithm for non-rigid registration." *Comput. Vis. Image Understand.* 89(2–3): 114–141.

Clements, L. W., Chapman, W. C. et al. (2008). "Robust surface registration using salient anatomical features for image-guided liver surgery: Algorithm and validation." *Med. Phys.* 35(6): 2528–2540.

Clippe, S., Sarrut, D. et al. (2003). "Patient setup error measurement using 3D intensity-based image registration techniques." *Int. J. Radiat. Oncol. Biol. Phys.* 56(1): 259–265.

Collignon, A., Maes, F. et al. (1995). Automated multi-modality image registration based on information theory, in *Information Processing in Medical Imaging*, Y. Bizais, C. Barillot, and R. Di Paola, eds., Kluwer Academic, Dordrecht, The Netherlands, 263–274.

Declerck, J., Feldmar, J. et al. (1997). "Automatic registration and alignment on a template of cardiac stress and rest reoriented SPECT images." *IEEE Trans. Med. Imag.* 16(6): 727–737.

Friston, K. J., Ashburner, J. et al. (1995). "Spatial registration and normalization of images. *Human Brain Map.*, 3: 165–189.

Gan, R., Chung, A. C. S., Liao, S. (2008). "Maximum distance-gradient for robust image registration." *Med. Image Anal.* 12: 452–468.

Gonzalez, R. C. and Woods, R. E. (1992). *Digital Image Processing*, Addison Wesley.

Hajnal, J. V., Saeed, N. et al. (1995). "A registration and interpolation procedure for subvoxel matching of serially acquired MR images." *J. Comput. Assist. Tomogr.*, 19(2), 289–296.

Hill, D. L. G. and Batchelor, P. (2001). Ragistration methodology: Concepts and algorithms, in *Medical Image Registration*, J. V. Hajnal, D. L. G. Hill, D. J. Hawkes eds., CRC Press, Boca Raton, FL, 39–70.

Hill, D. L. G. and Hawkes, D. J. (1994). "Voxel similarity measures for automated image registration." *Proc. SPIE-Visualization Biomed. Computing* 2359: 205–216.

Holden, M., Hill, D. L. G. et al. (2000). "Voxel similarity measures for 3D serial MR image registration." *IEEE Trans. Med. Imaging* 19: 94–102.

Jacobs, M. A., Windham, J. P. et al. (1999). "Registration and warping of magnetic resonance images to histological sections." *Med. Phys.* 26(8): 1568–1578.

Kashani, R., Hub, M. et al. (2008). "Objective assessment of deformable image registration in radiotherapy: A multi-institution study." *Med. Phys.* 35(12): 5944–5953.

Kassam, A. and Wood, M. L. (1996). "Fourier registration of three-dimensional brain MR images: Exploiting the axis of rotation." *J. Magn. Reson. Imaging* 6: 894–902.

Lee, C. W.-C., Tublin, M. E. et al. (2005). "Registration of MR and CT images of the liver: Comparison of voxel similarity and surface based registration algorithms." *Comput. Methods Prog. Biomed.* 78(2): 101–114.

Li, S., Liu, D. et al. (2006). "Real-time 3D-surface-guided head refixation useful for fractionated stereotactic radiotherapy." *Med. Phys.* 33(2): 492–503.

Lu, W., Chen, M.-L. et al. (2004). "Fast free-form deformable registration via calculus of variations." *Phys. Med. Biol.* 49(14): 3067–3087.

Maes, F., Collignon, A. et al. (1997). "Multimodality image registration by maximization of mutual information." *IEEE Trans. Med. Imaging* 16: 187–198.

Moseley, J. and Munro, P. (1994). "A semiautomatic method for registration of portal images." *Med. Phys.* 21(4): 551–558.

Munbodh, R., Tagare, H. D. et al. (2009). "2D-3D registration for prostate radiation therapy based on a statistic model of transmission images." *Med. Phys.* 36(10): 4555–4568.

Nelson, C., Balter, P. et al. (2008). "A technique for reducing patient setup uncertainties by aligning and verifying daily positioning of a moving tumor using implanted fiducials." *Med. Phys.* 9(4): 110–122.

Olsen, C. F. (2002). "Maximum-likelihood image matching." Pattern analysis and machine intelligence. *IEEE Trans.* 24(6): 853–857.

Pelizzari, C., Chen, G. et al. (1989). "Accurate three-dimensional registration of CT, PET, and/or MR images of the brain." *J. Comp. Assisted Tomogr.* 13(1): 20–26.

Penney, G. P., Weese, J. et al. (1998). "A comparison of similarity measures for use in 2-D-3-D medical image registration." *IEEE Trans. Med. Imaging* 17(4): 586–595.

Pratt, W. K. (1974). "Correlation techniques of image registration." *IEEE Trans. Aerospace Electron. Syst.* 10: 353–358.

Risholm P., Fedorov, A. et al. (2011). "Probabilistic non-rigid registration of prostate images: Modeling and quantifying uncertainty." *IEEE Int. Symposium Biomed. Imaging* 553–556.

Roche, A., Malandain, G., Ayache, N. "Unifying maximum likelihood approaches in medical image registration." *Int. J. Imaging Syst. Technol. Spec. Issue 3D Imaging, 11*: 71–80.

Russakoff, D. B., Rohlfing, T. et al. (2004). "Intensity-based 2D-3D spine image registration incorporating a single fiducial marker." *Acad. Radiol. 12*: 37–50.

Sawada, A., Yoda, K. et al. (2005). "Patient positioning method based on binary image correlation between two edge images for proton-beam radiation therapy." *Med. Phys. 32*(10): 3106–3111.

Serban, M., Heath, E. et al. (2008). "A deformable phantom for 4D radiotherapy verification: Design and image registration evaluation." *Med. Phys.* 35(3): 1094–1102.

Shannon, C. E. (1948). "The mathematical theory of communication." *Bell Syst. Tech. J.* 27: 379–423, 623–656.

Shirato, H., Oita, M. et al. (2004). "Three-dimensional conformal setup (3D-CSU) of patients using the coordinates system provided by three internal fiducial markers and two orthogonal diagnostic X-ray systems in the treatment room." *Int. J. Radiat. Oncol. Biol. Phys.* 60(2): 607–612.

Simper, A. (1996). Correcting general band-to-band misregistrations, in *Proceedings of the IEEE International Conference on Image Processing ICIP'96*, Lausanne, Switzerland, 597–600.

Skerl, D., Tomazevic, D. et al. (2006). "Evaluation of similarity measures for reconstruction-based registration in image-guided radiotherapy and surgery." *Int. J. Radiat. Oncol. Biol. Phys.* 65(3): 943–953.

Studholme, C., Hill, D. L. G. et al. (1999). "An overlap invariant entropy measure of 3D medical image alignment." *Pattern Recognit.* 32: 71–86.

Thevenaz, P. and Unser, M. (1998). An efficient mutual information optimizer for multiresolution image registration, in *Proceedings of the IEEE International Conference on Image Processing*, Chicago, Illinois, 833–837.

Vajda, I. (1989). *Theory of Statistical Inference and Information*, Kluwer, Dordrecht, The Netherlands.

van Herk, M. and Kooy, H. M. (1994). "Automated three-dimensional correlation of CT-CT, CT-MRI and CT-SPECT using chamfer matching." *Med. Phys.* 16: 443–448.

Viola, P. (1995). Alignment by maximization of mutual information, Massachusetts Institute of Technology, Ph.D. thesis.

Viola, P. and Wells, W. M. (1997). "Alignment by Maximization of Mutual Information", *Int. J. Computer Vision* 24(2): 137–154.

Wang, H., Dong, L. et al. (2005). "Validation of an accelerated 'demons' algorithm for deformable image registration in radiation therapy." *Phys. Med. Biol.* 50(2): 2887–2905.

Weese, J., Roesch, P. et al. (1999). Gray-value based registration of CT and MR images by maximation of local correlation, in MICCAI'99, C. Taylor and A. Colchester, eds., Lecture Notes in Computer Science, 656–663.

Wells, W. M. I., Voila, P. et al. (1996). "Multi-modal volume registration by maximization of mutual information." *Med. Image Anal.* 1: 35–51.

Wells, W. M., Maes, F., Pluim, J. (2009). Tutorial at MICCAI 2009: Information theoretic similarity measures for image registration and segmentation.

West, J., Fitzpatrick, J. et al. (1997). "Comparison and evaluation of retrospective intermodality brain image registration techniques." *J. Comp. Assisted Tomogr.* 21(4): 554–568.

Wijesooriya, K., Weiss, E. et al. (2008). "Quantifying the accuracy of automated structure segmentation in 4D CT images using a deformable image registration algorithm." *Med. Phys.* 35(4): 1251–1260.

Wolberg, G. and Zokai, A. (2000). Image registration for perspective deformation recovery, in *SPIE Conference on Automatic Target Recognition X*, Orlando, Florida, USA, 12.

Woods, R. P., Grafton, S. T. et al. (1998). "Automated image registration: 1. General methods and intrasubject, intramodality validation." *J. Comp. Assisted Tomogr.* 22(1): 139–152.

Wu, J., Kim, M. et al. (2009). "Evaluation of similarity measures for use in the intensity-based rigid 2D-3D registration for patient positioning in radiotherapy." *Med. Phys.* 36(12): 5391–5403.

Zhu, Y. M. and Cochoff, S. M. (2002). "Likelihood maximization approach to image registration." *IEEE Trans. Med. Imag.* 11(12): 1417–1426.

5

Parametric Image Registration

Raj Shekhar
Children's National Medical Center

William Plishker
University of Maryland

5.1 Introduction

Rigid and affine image registration, although adequate for many registration tasks mostly involving the brain, is insufficient for the registration problems that pertain to the extracranial anatomy. Soft-tissue organs of the abdomen and the thorax, for example, often deform leading to variations of shape and size from factors such as respiration, beating of the heart, and gravity. The registration of images of such organs or organ systems requires nonrigid image registration.

As in rigid and affine image registration, the nonrigid image registration problem is also set up as the maximization or minimization of an image similarity measure (i.e., cost function) between the two images. What is different is the transformation model, which must now include additional degrees of freedom to capture and describe the nonrigid deformation of the underlying anatomy. Whereas the transformation model of rigid and affine registration is linear, the nonrigid transformation model is nonlinear, which, in physical terms, means that each image sample of the source image undergoes a unique transformation in the course of nonrigid registration.

There are many approaches to formulate the nonlinear transformation model (Maintz and Viergever 1998; Rueckert 2001; Crum et al. 2004; Goshtasby 2005; Zagorchev and Goshtasby 2006; Holden 2008). Many are physically inspired and attempt to model the deformation of the anatomy in terms of well-studied models of fluid flow or the deformation of a viscoelastic material. These so-called nonparametric transformation models are described in detail in the following chapters. The topic of this chapter is the parametric transformation models that are often described by a set of basis functions. In parametric nonrigid medical image registration, the basis function is a spline. Although the origins of many of the splines can be traced back

to some physical constructs, their correlation with the way human anatomy deforms remains weak at best. These models are used in nonrigid image registration more for providing the mathematical framework to interpolate or approximate the nonrigid image registration solutions available at a subset of image locations, usually referred to as control points or landmarks, over the entire image space.

Control points are a necessary element of parametric image registration and often correspond to easily distinguishable anatomic landmarks in a pair of images to be registered. When registering images of the abdomen, the dome of the liver, for example, could serve as a landmark. Manual identification, indeed, is one way to select the landmarks. Several semi-automated as well as fully automated approaches have been suggested as well. The focus here is not on the specific steps involved in landmark selection but rather on the process of coordinate transformation that follows after landmarks have been identified.

Assuming that a set of corresponding landmarks exists, the process of parametric registration can be viewed as (1) applying the known displacements to the landmarks in the source image, such that their shifted locations correspond with their known locations in the target image, and (2) performing the necessary interpolation to define displacements at all remaining (nonlandmark) locations in the source image to obtain a complete displacement field, which is the desired solution of nonrigid image registration. Mathematically, let p_i and q_i with $i = 1, 2, \ldots N$, represent the 3D coordinates of N corresponding landmarks in the source and the target images, respectively. Parametric registration determines a coordinate transformation such that, at location p_i, the displacement vector is $d_i = (q_i - p_i)$ for $i = 1, 2, \ldots N$. At all other locations, the displacement is an interpolated value of the displacements at the N landmarks.

Parametric image registration can be categorized by the spatial arrangement of landmarks or control points, which, in turn, determines the type of interpolant functions or splines that fit the problem. When anatomic landmarks make up the control points, these, not unexpectedly, sample the image space nonuniformly. Two common parametric transformation models that fit this scenario are based on thin-plate spline (TPS) and elastic body spline (EBS). A third common parametric transformation model based on B-spline does not rely on anatomic landmarks. It instead uses a grid of automatically placed control points within the image space. The mathematical foundations of TPS, EBS, and B-spline–based parametric registration approaches are described in considerable detail below.

To give the reader an appreciation for the relative merits of these algorithms, the performance (accuracy and speed) of these approaches is compared on the same test images. All test images are 3D; therefore, all registration examples included here pertain to 3D images. For the same reason, the mathematics behind various parametric image registration approaches are discussed assuming the nonrigid registration of 3D images.

5.2 Thin-Plate Spline

The TPS transformation belongs to the family of transformations based on radial basis functions (RBFs), which are globally supported, radially symmetric functions of distance between a point of interest and any arbitrary 3D point in space. In the context of image registration, the points of interest are the control points, and the RBFs are centered at these control points. A linear summation of shifted RBFs provides the necessary coordinate transformation for nonrigid image registration. In its most general form, the RBF-based transformation is given as

$$t(p) = ap + b + \sum_{i=1}^{N} c_i R(p - p_i), \qquad (5.1)$$

which describes the spatial mapping of a source image point p to its transformed location $t(p)$ in the target image. The first part of the equation, involving coefficients a and b, is the affine transformation part, which helps recover any global misregistration present between the source and the target images. The second part, a weighted summation of RBFs with c_i weights, describes the nonaffine part and helps recover localized nonrigid misregistration.

The RBFs can take on many forms from Gaussian to multiquadrics to inverse multiquadrics, etc., and the use of each in image registration has been reported (Goshtasby 2005; Holden 2008). Because of its favorable properties, TPS has enjoyed more success in image registration than the other RBF-based transformations and is therefore a primary focus of this chapter. TPS was originally associated with describing the bending of thin pieces of metal, subject to loads, used in aircraft and ship building. Mathematically, the RBF for the TPS is derived from the square Laplacian, $\nabla^4 u(p) = c\delta(0, 0, 0)$, where $u(p)$ is the displacement

at point p and c is a constant, and is given in 3D as $R_{TPS}(r) = r$, where $r = (p - p_i)$ (Holden 2008). The corresponding RBF in two dimensions is the familiar logarithmic function $R_{TPS}(r) = r^2 \log r$.

The TPS transformation represents three separate 3D hypersurfaces, one for each of the three dimensions. Each hypersurface yields an interpolated x, y, or z location (in target image) for a given point in the source image. The physical interpretation of TPS interpolation is easier to understand in two dimensions. This will be described and the 3D extension will be left for the reader to perform. In two dimensions, the TPS interpolation surface can be viewed as a flat surface that then undergoes undulations because of the weights located at control point locations. The weights can be either positive or negative. A positive weight would cause the surface to dip, whereas a negative weight will cause it to rise. The deflection caused by each weight is circularly symmetric and extends to infinity, that is, TPS is symmetric and has global support. In fact, most RBFs are such that the deflection asymptotically approaches zero (no displacement) away from the location of the load. Moreover, the loads balance one another and the surface does not see a net translation or rotation. This will be referred to as the stability criterion.

When expanded, Equation 5.1 produces a system of $3N$ linear equations with $(3N + 12)$ unknowns, which are affine and nonaffine coefficients. An additional 12 equations are needed for a fully determined system. These equations come from the stability criterion in that the summation of all loads as well as the summation of first moments must be zero (Goshtasby 2005). The solution of TPS, therefore, means solving the following system of equations:

$$\begin{pmatrix} K & P \\ P^T & 0 \end{pmatrix} \begin{pmatrix} C \\ A \end{pmatrix} = \begin{pmatrix} Q \\ 0 \end{pmatrix},$$

where A is a 4×3 matrix of affine coefficients, C is a $3N \times 3$ matrix of nonaffine coefficients, K has elements $K_{ij} = R_{TPS}(p_i - p_j)$, and P and Q are vectors containing known landmarks in source and target images, respectively. Solving this system of equation using standard technique yields the TPS coefficients.

The performance of TPS, and that of most other parametric registration methods, is affected by the number of control points, the distribution of control points, and the accuracy with which control points are identified. The effect of control point selection on TPS and other parametric image registration algorithms will be described later in the chapter. Because TPS are interpolating, that is, the resulting solution is constrained to pass through the identified landmarks, any positional error in the identification of the landmarks degrades the quality of resulting registration. Approximating TPS relaxes this constraint (i.e., the resulting solution does not need to pass through the landmarks) and thus is more suitable when there is some uncertainty with landmark identification (Rohr et al. 2001). An alternative strategy is to refine the correspondence of identified landmarks using some measure of image similarity in the vicinity of the landmarks (Meyer et al. 1997).

5.3 Elastic Body Spline

As with TPS, the EBS algorithm achieves nonrigid image registration by interpolating a set of known displacements at control points. The difference is in the interpolation function, which, although a 3D spline, is derived from solving the Navier partial differential equations (PDEs). The EBS was first proposed and developed for nonrigid image registration by Davis et al. (1997), and the reader is referred to their original source for additional details.

The Navier PDEs describe the deformation of an elastic body made of homogeneous isotropic material in response to forces applied to it. The elastic nature of human tissues offers some physical basis for modeling deformation using Navier PDEs, but, on the whole, the application of EBS to images of the human body with a mix of tissue types and material properties represents an approximation. For small deformations, however, this approximation works acceptably well. It should be noted that the Navier PDEs are employed not to model human tissues as isotropic homogeneous elastic bodies exclusively but rather to propagate a known displacement at a given location throughout the image space, assuming the tissue deforms like an elastic body. This should be contrasted with deformation of thin metal pieces in TPS. The EBS resulting from the solution of the Navier PDEs in response to a single force, represented by a priori displacement at a control point, provides a means to propagate that displacement throughout the space. The complete nonrigid image registration solution then is a weighted sum of EBS solved for all forces (i.e., known displacements) at control points.

The Navier PDEs are given as

$$\mu \nabla u(p) + (\mu + \lambda)\nabla[\nabla \cdot u(p)] = f(p),$$

where $u(p)$ is the displacement of a point from its original location upon application of the force, ∇^2 and ∇ denote the Laplacian and gradient operators, respectively, and μ and λ are the Lamé coefficients pertaining to the material's elastic properties. $f(p)$ is the force field applied to the elastic body.

A simple version of the force field would have nonzero point forces located at the control points and proportional in magnitude to the known displacements at those locations. At all other locations, the force would be zero. Such a force field leads to the singularities in the solution of the Navier PDEs. Davis et al. (1997) suggested using a force field of the form $f(p) = cr(p)^{2k+1}$, where c is a constant, k is an integer, and $r(p)$ is the radial distance from the control point. The resulting force field is then smooth and radially symmetric with a gradual variation in the force amplitude away from the control points. These investigators considered two values of k (0 and –1), resulting in force fields of $cr(p)$ for $k = 0$ and $cr(p)^{-1}$ for $k = -1$. The solutions to Navier PDEs for either force field have the same general form with a slight variation as given below:

$$u(p) = cR_{\text{EBS}}(p),$$

where, for $f(p) = cr(p)$, $R_{\text{EBS}}(p) = [\alpha r(p)^2 I - 3pp^T](p)$ and, for $f(p) = cr(p)^{-1}$, $R_{\text{EBS}}(p) = [\beta r(p)I - pp^T/r(p)]$.

In this solution, $\alpha = 12(1 - v) - 1$ and $\beta = 8(1 - v) - 1$, where $v = \lambda/[2(\lambda + \mu)]$ is the Poisson's ratio. Furthermore, I is a 3×3 identity matrix and pp^T is an outer product.

The EBS itself, much like TPS, is a linear combination of the translated versions of the RBFs resulting from the solution of the Navier PDEs, or

$$t(p) = ap + b + \sum_{i=1}^{N} c_i R_{\text{EBS}}(p - p_i), \tag{5.2}$$

where, as before, $(ap + b)$ is the affine part of the EBS and the remainder is the nonaffine part. As in Equation 5.1, Equation 5.2 describes the transformation of a source image voxel to its corresponding location in the target image.

When expanded, Equation 5.2 leads to $3N$ equations with $(3N + 12)$ unknowns, which are the force and affine transformation coefficients. As in the case of TPS, additional 12 equations come from the stability condition to ensure that there is no net rotation of the elastic body and the net displacement at infinity (i.e., sufficiently away from the control points) is described by the affine coefficients only. The $(3N + 12)$ EBS coefficients are found by solving the following system of equations, which has a similar form as that for TPS:

$$\begin{pmatrix} K & P \\ P^T & 0 \end{pmatrix} \begin{pmatrix} C \\ A \end{pmatrix} = \begin{pmatrix} Q \\ 0 \end{pmatrix},$$

where A is a 4×3 matrix of affine coefficients, C is a $3N \times 3$ matrix of force coefficients, K has elements $K_{ij} = R_{\text{EBS}}(p_i - p_j)$, and P and Q are vectors containing known landmarks in source and target images, respectively.

5.4 B-Spline

In nonrigid registration problems, B-splines can be the basis functions of a free-form deformation (FFD) that defines the underlying nonrigid coordinate transformation (Rueckert et al. 1999). The points of FFD are not landmarks to be explicitly aligned but instead define the deformation field as described by B-splines placed over a uniform grid of automatically selected control points. The desired deformation is achieved by adjusting the 3D locations of the control points of the underlying grid.

Unlike TPS and EBS, the identification of corresponding landmarks is not a prerequisite for this algorithm. Another defining feature of the algorithm is that it offers local support as opposed to global. The RBFs in TPS and EBS algorithms have infinite support; therefore, moving a control point influences the displacement field throughout the image. Comparatively, the control points in this algorithm have local support, meaning that moving a control point affects the displacement field only locally. Furthermore, the denser the control point grid, the smaller the region of support of a control point. As demonstrated in the experimental results, the B-spline algorithm benefits from this property in recovering localized deformations.

The B-spline algorithm also separates net deformation into affine and nonaffine transformations:

$$T(x, y, z) = T_{\text{affine}}(x, y, z) + T_{\text{nonrigid}}(x, y, z).$$

The affine transformation models the global motion and is fixed throughout the image space. The nonrigid transformation models the local deformations representing localized tissue compressions or expansions. The algorithm starts with an explicit affine registration step, which determines 12 parameters of $T_{\text{affine}}(x, y, z)$ that represent the combined effect of translation, rotation, scale, and shear. The FFD coefficients are determined next.

The modeling of FFD for nonrigid transformation begins by placing a uniform grid of control points over the image. Let $\phi_{i,j,k}$ define such a control point grid of size $n_i \times n_j \times n_k$ with a grid spacing of $\delta_x(t) \times \delta_y(t) \times \delta_z(t)$. The FFD then is a tensor product of 1D cubic B-splines as follows:

$$T_{\text{nonrigid}}(x,y,z) = \sum_{l=-1}^{2}\sum_{m=-1}^{2}\sum_{n=-1}^{2} B_l(u)B_m(v)B_n(w)\phi_{i+l,j+m,k+n}$$

where $i = [x/\delta_x(t)]$, $j = [y/\delta_y(t)]$, $k = [z/\delta_z(t)]$, $u = x/\delta_x(t) - i$, $v = y/\delta_y(t) - j$, $w = z/\delta_z(t) - k$, and

$$B_{-1}(r) = (1-r)^3/6$$

$$B_0(r) = (3r^3 - 6s^2 + 4)/6$$

$$B_1(r) = (-3r^3 + 3r^2 + 3r + 1)/6$$

$$B_2(r) = r^3/6$$

where r is either u, v, or w.

The control point grid directly affects the accuracy of image registration. Given that a control point can be moved in any of the three directions in the 3D space, the algorithm offers $3 \times$ (# of control points) degrees of freedom. The B-spline algorithm lends itself well to hierarchical registration (Lester and Arridge 1999) in which the density of the control point grid can be increased as the algorithm progresses. In some implementations, including the one reported here, the image resolution is also varied. This leads to a coarse-to-fine approach to modeling localized deformations with increasing conformity afforded by increasing degrees of freedom. Given an initial control point grid size of $n_i \times n_j \times n_k$ and grid spacing of $\delta_x \times \delta_y \times \delta_z$, a common approach is to double the grid size (e.g., $2n_i \times 2n_j \times 2n_k$) while simultaneously halving the grid spacing (e.g., $\delta_x/2 \times \delta_y/2 \times \delta_z/2$) as one moves through resolution levels. In the experiments conducted here on $256 \times 256 \times 256$ images, a two-level grid resolution was used.

Because no corresponding control points with known displacements are needed, the FFD algorithm does not lead to a closed-form solution obtained from solving a system of linear equations as in TPS and EBS. Instead, an optimization algorithm that either maximizes or minimizes a cost function must be used. Rueckert et al. (1999) used normalized mutual information as a measure of image similarity in their demonstration registration application that involved precontrast and postcontrast breast magnetic resonance images. As reported in the earlier chapters, mutual information and its variant normalized mutual information are widely regarded as the most reliable and accurate image similarity measures currently known for both single-modality and multimodality image registration (Pluim et al. 2003). The FFD approach, however, does not restrict the choice of image similarity measure, and alternative measures, such as cross-correlation and mean squared difference, are equally applicable if these are appropriate for registering the image pair in consideration.

The general form of the cost function to minimize in this algorithm is

$$C_{\text{similarity}}(\text{SI, TI}, T(x, y, z)) + \lambda C_{\text{smooth}}(T(x, y, z)),$$

where the $C_{\text{similarity}}(\cdot)$ term is defined such that minimizing it improves registration between the target image (TI) and the source image (SI) subject to the latest coordinate transformation. The second term, $C_{\text{smooth}}(T(x, y, z))$, is a regularization term as that maintains smoothness of the control point grid and the resulting FFD. λ controls the balance of the two terms. Although alternative forms are possible, Rueckert et al. (1999) define the regularization term based on the second-order derivatives of the FFD that represent bending energy of a 3D object and thus minimizing this term means discouraging excessive deformation of the FFD. Mathematically,

$$C_{\text{smooth}}(T(x,y,z)) = \frac{1}{V}\int_0^X\int_0^Y\int_0^Z \left[\left(\frac{\partial^2 T}{\partial x^2}\right)^2 + \left(\frac{\partial^2 T}{\partial y^2}\right)^2 + \left(\frac{\partial^2 T}{\partial z^2}\right)^2 + \left(\frac{\partial^2 T}{\partial xy}\right)^2 \right.$$
$$\left. + \left(\frac{\partial^2 T}{\partial yz}\right)^2 + \left(\frac{\partial^2 T}{\partial zx}\right)^2\right]\mathrm{d}x\mathrm{d}y\mathrm{d}z,$$

where V is the volume of the image space.

B-spline-based FFD is inherently smooth, but, at high resolutions, image-derived forces may lead to unrealistic deformations, an example of which is grid folding, resulting from crisscrossing of neighboring control points. The regularization terms penalize such physiologically unattainable deformations. A λ of 0.05 has been empirically found to provide the right balance between the two parts for most applications. Other forms of regularization are possible. Some notable examples include the Jacobian of the FFD that penalizes excessive local volume expansion and compression (Rohlfing et al. 2003), the checking for the consistency between forward and backward deformation fields (Christensen and Johnson 2001), and the use of geometric constraints that explicitly bar grid folding (Shekhar et al. 2007).

The solution of the B-spline algorithm is arrived at through iterative minimization of the cost function until the changes in the cost function reduce to a preset threshold. The registration needs to be carried out in two steps: affine registration followed by nonrigid registration. In the first step, the values of 12 affine transformation parameters are searched through the minimization of the aforementioned cost function albeit without the FFD regularization component. In the second step, the coefficients of the nonrigid transformation, corresponding with the degrees of freedom, are searched through the minimization of the same cost function, which now includes the regularization component. The number of parameters to be optimized is $3N$, where N is the number of control points (or grid points). This number can be in thousands at the highest grid resolution. Almost any standard minimization algorithm capable of dealing with a high number of search parameters can be used (Press et al. 2007). Rueckert et al. (1999) used the gradient descent algorithm in their implementation. The application of this algorithm means computing a gradient vector composed of partial derivatives of the cost function with respect to all $3N$ search parameters and updating the locations of the control point based on it. The process continues, as before, until the change in cost function is smaller than the threshold for the current control point grid and image resolution. The cycle repeats for all resolutions with a

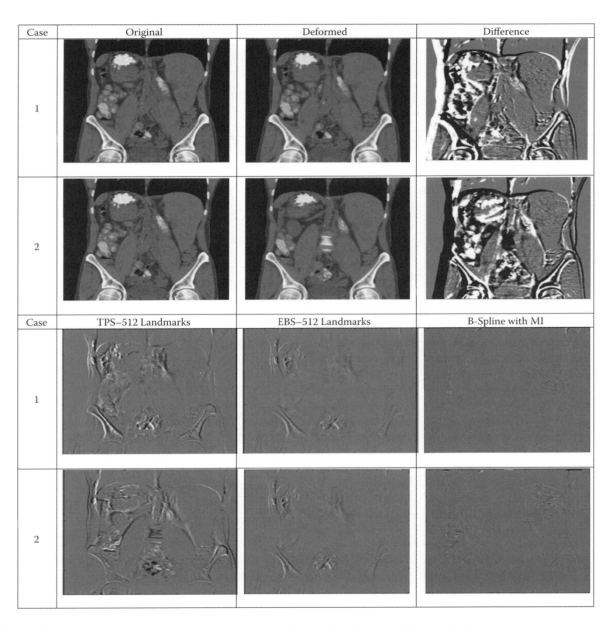

FIGURE 5.1 Summary of the registration setup and final results with two of the five cases. MI, mutual information.

decreasing threshold for change in the value of the cost function. The algorithm terminates when the minimization converges for the finest resolution level.

5.5 Empirical Evaluation

The theoretical foundations of parametric image registration algorithms provide distinctly different ways of solving the registration problem. However, with the proper similarity measure, the speed and the accuracy of each algorithm can be compared directly for specific problems. To this end, an evaluation scenario has been devised to try specific instances of these registration approaches. These results provide a glimpse into how these algorithms compare under real-world registration scenarios.

For an empirical evaluation of the results, five computed tomography (CT)-to-CT nonrigid image registration cases of the abdomen have been selected. To construct the registration pairs, a known deformation field was applied to an actual CT image based on a composite of typical registration results from previous CT-to-CT abdomen registration cases. Therefore, whereas the deformed image was created through artificial means, the

Number of landmarks	TPS	EBS
8	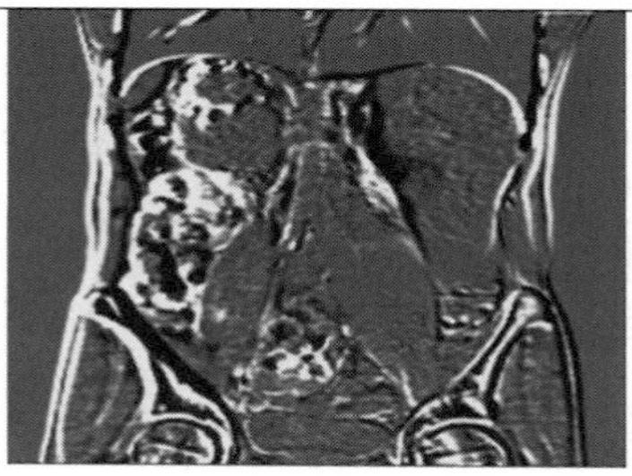	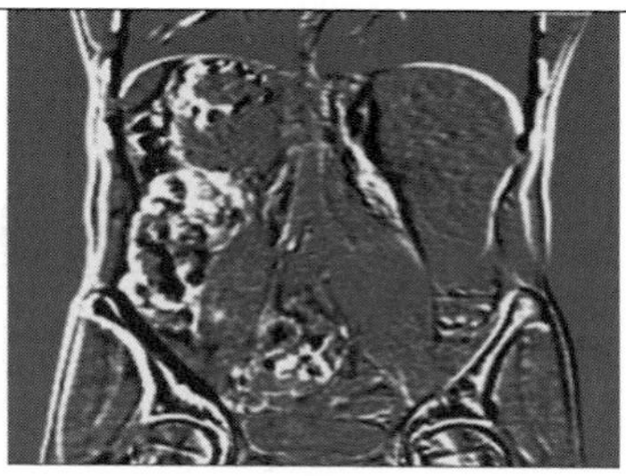
32	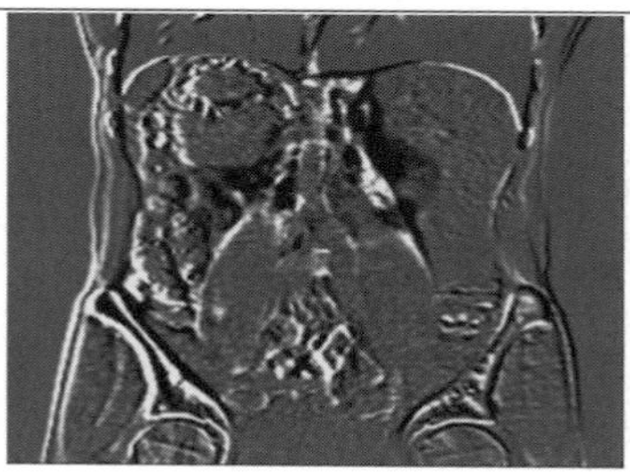	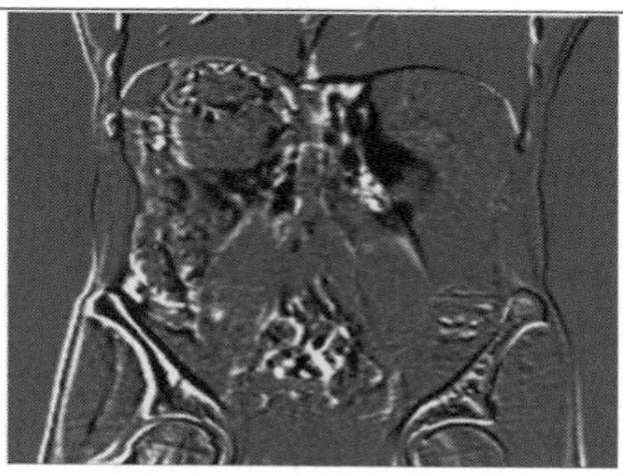
128	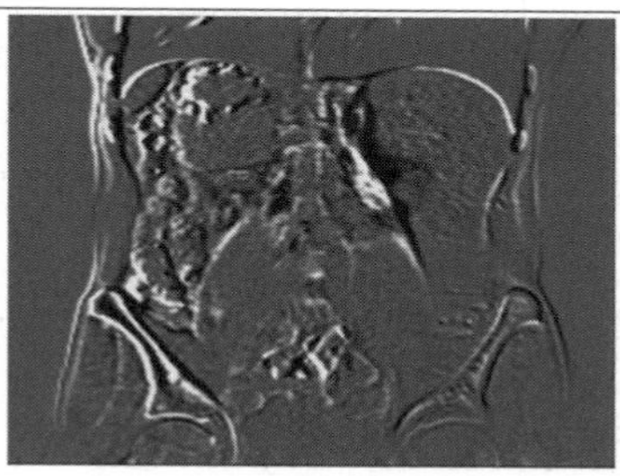	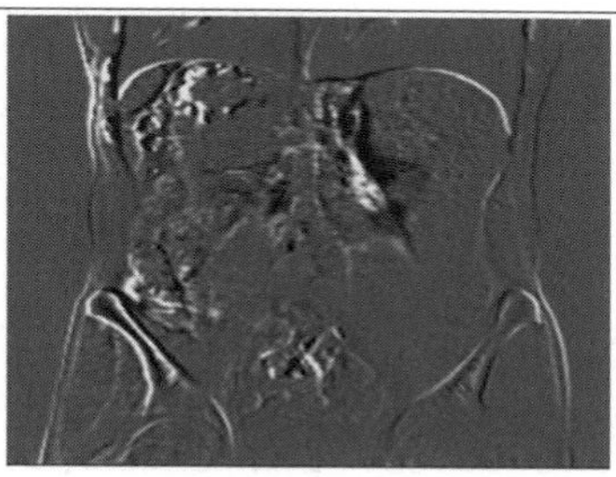
512	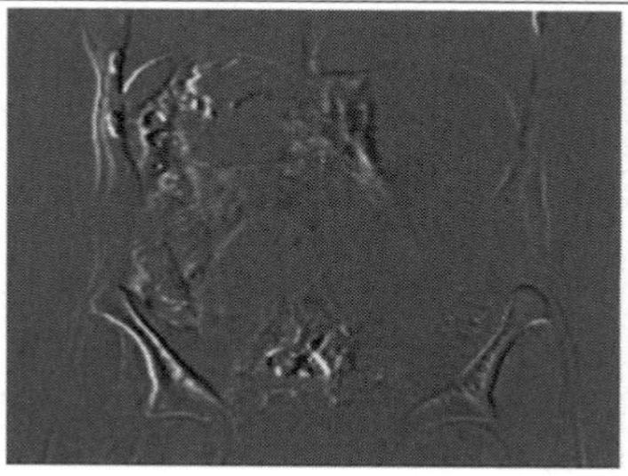	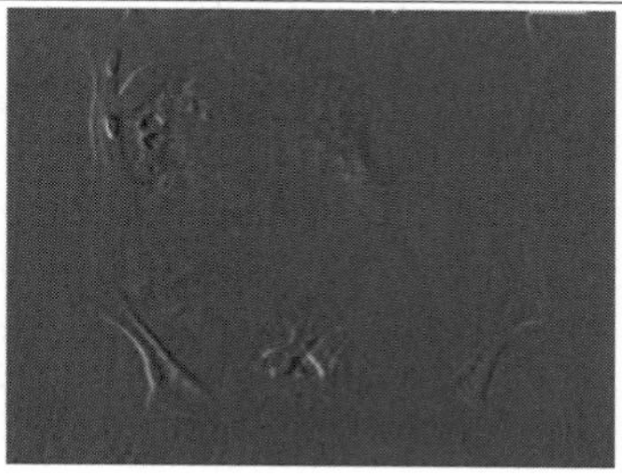

FIGURE 5.2 Case 1's difference images of TPS and EBS with a different number of landmarks.

applied deformation field was based on an actual physiologically realistic deformation. With these five pairs of images, the three different algorithms described above attempted to recover the deformation. The first two methods were TPS and EBS. The third method was the B-spline method that optimized mutual information. Well-tested implementations of these algorithms provided in the Insight Segmentation and Registration Toolkit (ITK), an open-source system providing a rich set of image analysis algorithms, were used (Ibanez et al. 2005). Furthermore, ITK version 3.20 was used on a computer with dual-core Intel Xeon 3 GHz CPU with 3 GB of main memory.

A TPS implementation and an EBS implementation were built based on an existing example within the registration examples provided. Each implementation accepted a set of nonuniformly spaced landmarks that were matched in both images and returned a deformation field constructed from these landmarks. In both cases, random points were chosen from the space of the target image and their displacement vectors looked up in the deformation field. In other words, landmarks were derived from the ground-truth deformation field, providing each of these approaches a set of exact matching points from which to recover the rest of the deformation field. To evaluate the execution time

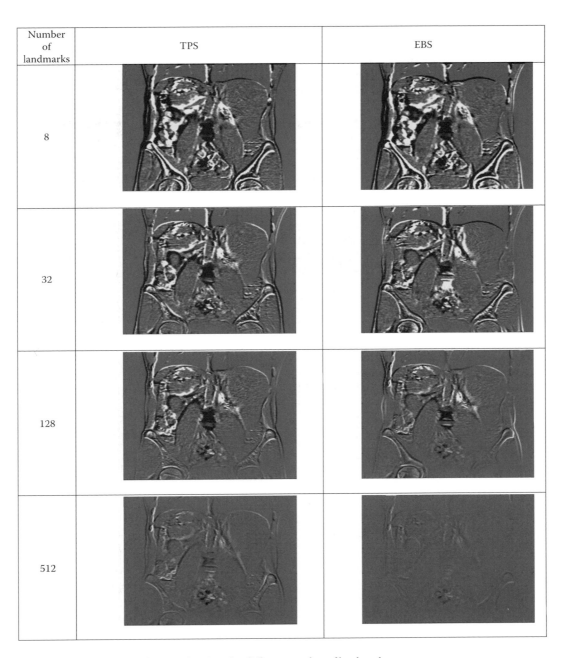

FIGURE 5.3 Case 2's difference images of TPS and EBS with a different number of landmarks.

and accuracy, the number of landmarks was varied by powers of two, and the time required for creating the deformation field, the target registration error defined as an average of the difference in the distance between points in the derived and given deformation field, and the mean squared difference between the two images were tracked. When increasing the number of landmarks, the previously derived landmarks were retained and new landmarks were added to them. For example, with the eight-landmark instance, the same four landmarks were used from the four-landmark instance plus four new pairs of landmarks.

A mutual information implementation was constructed of the same problem using ITK for the B-spline algorithm. An affine registration was performed, followed by a nonrigid registration. The transform resulting from the affine registration was used as the bulk transform of a B-spline–based deformable registration that uses a gradient descent optimization approach to arrive at a solution. This implementation leveraged a multiresolution, grid-based deformation, correcting for coarse deformations with a coarse mesh ($5 \times 5 \times 5$) and a significantly subsampled image and fine deformations with a fine mesh ($20 \times 20 \times 20$) and little subsampling of the images.

A summary of the end results for two of the five image pairs is in Figure 5.1. In the first rows, a coronal slice of the original and deformed images and their differences are shown, revealing significant nonrigid misalignment between the two instances. The last two rows show the difference images of the same slices after registration with the different techniques. Both TPS and EBS have 512 of the original points as anchors (i.e., control points), so much of the image is well aligned; but some features between these anchors are not interpolated accurately, leaving some features still visible in the difference image. Figures 5.2 and 5.3 show qualitatively the effect of changing the number of control points (8, 32, 128, and 512) in the same two cases as the summary.

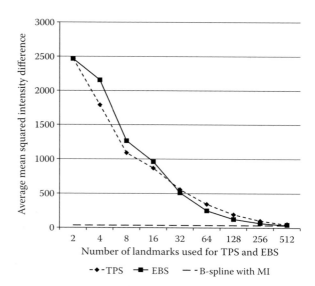

FIGURE 5.5 Average mean squared difference for each approach, averaged over five cases.

Figures 5.4 through 5.6 show the average performance of each algorithm in terms of average target registration error, average mean squared difference, and registration time, averaged across all five cases. TPS and EBS perform comparably with similar numbers of control points, with the quality of registration consistently improving with the number of landmarks. The additional quality comes at the expense of more computation and slower registration times, which, for EBS at least, surpasses those of the B-spline method when the number of control points grows. Compared with TPS, EBS shows better image similarity with more landmarks. This may be attributed to the superior

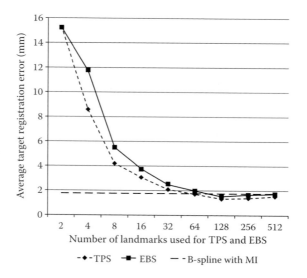

FIGURE 5.4 Average target registration error for each approach, averaged over five cases.

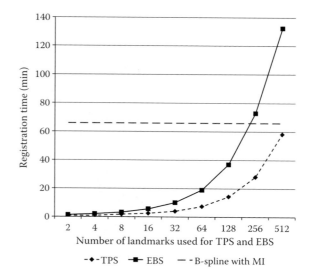

FIGURE 5.6 Registration time for each approach, averaged over five cases.

interpolation capability of EBS once points are close enough to better model tissue deformation. The mutual information-based B-spline algorithm performs well on the image set too, providing accuracy comparable with TPS or EBS with hundreds of landmarks with a similar registration time required. This algorithm uses a fixed automatic scheme of landmark selection; thus, its quality and execution time (approximately 1 h) do not change. Comparatively, the execution time of TPS and EBS, ranging from a couple of minutes to approximately an hour or more, scales with the control points. Indeed, a major difference between the B-spline algorithm and the TPS or EBS is the fact that no landmarks were required to achieve a quality registration with the former.

5.6 Summary

Interpolating point correspondences to perform nonrigid image registration is the essence of parametric image registration. Interpolating functions are basis functions whose characteristics can differ considerably depending on the exact approach. In TPS and EBS, which share many common traits, the basis functions are radially symmetric with infinite support. Although this feature is helpful in nonrigid registration even with sparsely and nonuniformly distributed point correspondences, the infinite support leads to changes in the solution, the displacement field, when a landmark is moved. The identification of landmarks is also a necessary first step for these methods. Because anatomic landmarks are limited in number and their identification is laborious if done interactively or semi-automatically, these methods are often executed with a small number of control points (fewer than a hundred in most cases). Although a solution is guaranteed regardless of the number of control points, the quality of registration generally correlates with the number and, in addition, happens to be higher near the control points and degrades away from those. The physical distance is the measure of similarity in these methods and is minimized. A great advantage of these methods is that these are computationally efficient in most practical settings and lead to a closed-form solution.

The B-spline-based approach represents another class of parametric image registration. This method is also based on control points, but control points are automatically identified, are a part of a uniform grid overlaid on the image, and do not necessary correspond with a clearly identifiable anatomic feature. The cubic B-splines, defined over four neighboring control points in each axis, provide local control. The approach requires a measure of image similarity, subject to some regularization, which guides the registration process in an iterative fashion. During this process, the correspondence of automatically selected control point pairs is improved using the image similarity measure of choice. The algorithm conceptually converges when each control point correspondence is optimized, and the final solution is a displacement field that interpolates these correspondences using B-splines. The advantage of this method is that it obviates the need for landmark selection, but this comes at an expense of an iterative, search-based paradigm that accompanies high computational costs and lengthy execution times. High-speed implementations and algorithmic enhancement have been proposed to lower the runtime. Overall, parametric image registration is conceptually straightforward, relatively computationally efficient, and currently the most successful and widely used approach to nonrigid image registration.

References

Christensen, G. E. and H. J. Johnson 2001. Consistent image registration. *IEEE Trans Med Imaging* 20(7): 568–582.

Crum, W. R., T. Hartkens, and D. L. Hill 2004. Non-rigid image registration: Theory and practice. *Br J Radiol* 77(Spec No 2): S140–S153.

Davis, M. H., A. Khotanzad, D. P. Flamig, and S. E. Harms 1997. A physics-based coordinate transformation for 3-D image matching. *IEEE Trans Med Imaging* 16(3): 317–328.

Goshtasby, A. A. 2005. *2-D and 3-D Image Registration for Medical, Remote Sensing, and Industrial Applications.* Hoboken: Wiley-Interscience.

Holden, M. 2008. A review of geometric transformations for nonrigid body registration. *IEEE Trans Med Imaging* 27(1): 111–128.

Ibanez, L., W. Schroeder, L. Ng, and J. Cates 2005. *The ITK Software Guide.* Clifton Park, NY: Kitware, Inc.

Lester, H. and S. R. Arridge 1999. A survey of hierarchical nonlinear medical image registration. *Pattern Recognit* 32(1): 129–149.

Maintz, J. B. and M. A. Viergever 1998. A survey of medical image registration. *Med Image Anal* 2(1): 1–36.

Meyer, C. R., J. L. Boes, B. Kim et al. 1997. Demonstration of accuracy and clinical versatility of mutual information for automatic multimodality image fusion using affine and thin-plate spline warped geometric deformations. *Med Image Anal* 1(3): 195–206.

Pluim, J. P., J. B. Maintz, and M. A. Viergever 2003. Mutual-information-based registration of medical images: a survey. *IEEE Trans Med Imaging* 22(8): 986–1004.

Press, W. H., S. A. Teukolsky, W. T. Vetterling, and B. P. Flannery. 2007. *Numerical Recipes in C++: The Art of Scientific Computing.* Cambridge, UK: Cambridge University Press.

Rohlfing, T., C. R. Maurer Jr, D. A. Bluemke, and M. A. Jacobs 2003. Volume-preserving nonrigid registration of MR breast images using free-form deformation with an incompressibility constraint. *IEEE Trans Med Imaging* 22(6): 730–741.

Rohr, K., H. S. Stiehl, R. Sprengel, T. M. Buzug, J. Weese, and M. H. Kuhn 2001. Landmark-based elastic registration using approximating thin-plate splines. *IEEE Trans Med Imaging* 20(6): 526–534.

Rueckert, D. 2001. Nonrigid registration: Concepts, algorithms, and applications. In *Medical Image Registration*, eds. J. V. Hajnal, D. L. G. Hill, and D. J. Hawkes, 281–301. Boca Raton: CRC Press.

Rueckert, D., L. I. Sonoda, C. Hayes, D. L. Hill, M. O. Leach, and D. J. Hawkes 1999. Nonrigid registration using free-form deformations: application to breast MR images. *IEEE Trans Med Imaging* 18(8): 712–721.

Shekhar, R., P. Lei, C. R. Castro-Pareja, W. L. Plishker, and W. D. D'Souza 2007. Automatic segmentation of phase-correlated CT scans through nonrigid image registration using geometrically regularized free-form deformation. *Med Phys* 34(7): 3054–3066.

Zagorchev, L. and A. Goshtasby 2006. A comparative study of transformation functions for nonrigid image registration. *IEEE Trans Image Process* 15(3): 529–538.

6

Toward Realistic Biomechanical-Based Modeling for Image-Guided Radiation

Adil Al-Mayah
University of Waterloo

Kristy K. Brock
University of Michigan

6.1 Introduction

Soft tissues are nonhomogeneous and anisotropic materials that most likely experience large deformations. Their complex structures, geometry, and materials make the finding of closed-form solution "almost impossible to obtain" (Holzapfel 2004) and hence intensify the need for computational modeling (Humphrey 2003). Computational biomechanical model has the potential to overcome these challenges while addressing the needs for understanding the behavior of soft tissues under different loading conditions. Clinically, a biomechanical model can improve diagnostics, treatment, planning, and interventions in addition to its academic and industrial contributions (Holzapfel 2004).

A biomechanical model is a physical-based model that incorporates the mechanical properties of the soft tissue in addition to its geometry and boundary conditions. Including material properties provides a unique method to address, more realistically, the anatomical features and the variation of the organ under consideration. Biomechanical models have been applied for deformable image registration for image-guided surgery and radiation therapy.

Biomechanical-based image registration includes a number of techniques, namely, the spring-mass, finite difference, boundary element, and finite element modeling (FEM). In the spring mass technique, the anatomy is modeled as a set of springs with mass attached at the end, which can be interpreted as a set of particles. The spring properties can be linear (Hookean) or nonlinear. This representation of the anatomy is a fast-processed model due to its simplicity but at the expense of the accuracy (Maciel et al.

2003). The lack of accuracy of modeling a realistic deformation map across the tissue is mostly related to the unrealistic characterization of the material properties where a hypothetical material parameter, such as modulus of elasticity, is used to describe an object, not the material itself (Van Gelder 1998). In addition, the method is unable to model the heterogeneous nature of the soft tissues and thin structures. A more realistic modeling of soft tissues can be achieved by a continuum representation of the region using boundary element models, finite difference models, and FEMs. Because the boundary element model requires the surface mesh only, it is an efficient way to reduce the complexity of mesh generation. However, it is not capable to model heterogeneous material unless the region is divided into small areas. Similarly with finite difference, a square grid network is required to model the region under consideration. Because a model with fine square grid is required to encompass the complex geometry of organs, the limitation is obvious. On the contrary, the most popular modeling technique that has been used in the biomechanical modeling is the FEM, which has the potential to provide a realistic model of the soft tissue while addressing the tissues' large deformation, response, variation, and complications due to radiation therapy. This chapter will concentrate on the FEM of the soft tissue.

6.2 Elements of FEMs

6.2.1 Geometry

A realistic biomechanical model requires an accurate geometry of anatomy. Patient-specific FEMs are built by modeling the exact anatomy of the patients using segmentation. This segmentation can be manual and/or automatic. Manual segmentation is a time-consuming process that requires expertise in the anatomy under consideration. A number of automatic segmentation methods have been proposed, including thresholding, region-based methods, edge-based techniques, deformable contour models, fuzzy connectivity, and morphologic image processing. Research is under way to improve the automatic segmentation (Barber and Hose 2005). A detailed chapter has been assigned for image segmentation in this book.

6.2.2 Biomechanical Material Properties

The primary methods for applying biomechanical material properties for modeling soft tissues can be divided into three categories depending on the parameters involved in the simulation. In the case of no time involvement of the material properties, the stress–strain relationship can be linear or nonlinear elastic. The linear elastic material properties are represented by Hooke's law, where two material constants are used, namely, modulus of elasticity (Young's modulus) and compressibility parameter (Poisson's ratio). Roy (1880) stated that the Hooke's law "does not hold good" to characterize soft tissues as nonlinear behavior was observed. Therefore, nonlinear elastic material properties are applied using different strain energy functions. However, if very low strain is applied to the tissue, it does not make significant difference if it is modeled as linear elastic materials (Chabanas et al. 2004). Furthermore, because soft tissues have large water content, their

behavior is a combination of fluid-like and solid-like materials (Humphrey 2003). This coins the term "visco (fluid-like)-elastic (solid-like) material" to address the time-dependent properties.

6.2.2.1 Linear Elastic Model

Elastic material returns to its original state upon the removal of the loads (i.e., force/stress or displacement). In a linear elastic model, the material is characterized by two parameters: elastic modulus (E) and Poisson's ratio (v). The elastic modulus represents the slope of the stress–strain plot. Poisson's ratio is the ratio of the transverse strain caused by an applied strain in the perpendicular direction, as shown in Figure 6.1a through c. When the load is applied parallel to the surface, shear stress is generated causing a change in the shape represented by an angular deformation (Figure 6.1d). The ratio of shear stress (τ) to the angular deformation ($\Delta l/l$) is the shear modulus (G). In other loading conditions, an equal triaxial pressure is applied to the volume causing a volumetric change, unlike the shape change caused by the shear loading above (Figure 6.1e). The ratio of the load (p) to the volumetric change ($\Delta V/V$) is known as the bulk modulus (K). The relationships between G, K, E, and v are illustrated in Equation 6.1:

$$
\begin{aligned}
G &= \frac{E}{2(1+v)} \\
K &= \frac{E}{3(1-2v)}
\end{aligned}
\qquad (6.1)
$$

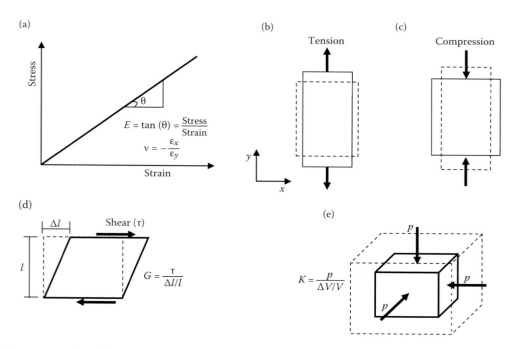

FIGURE 6.1 (a) Stress–strain plot of elastic material under (b) tension and (c) compression loads, where the tangent of the plot represents the modulus of elasticity and the negative ratio of the lateral to longitudinal strains is the Poisson's ratio, (d) angular deformation caused by shear, and (e) volumetric deformation under triaxial loading.

The relationship between stress and strain in the elastic materials is described by the generalized Hooke's law:

$$\sigma_{ij} = C_{ijkl}\varepsilon_{kl}, \tag{6.2}$$

where σ and ε are the stress and the strain, respectively. C_{ijkl} is referred to by a variety of names, including elastic moduli, elastic constants, and stiffness coefficients. This matrix has 81 constants; however, there is no material with that number of constants as a result of stress and strain tensors symmetry. Therefore, the anisotropic material has 36 constants. The number is reduced to only 12 coefficients in the isotropic materials with only two material constants required, namely, modulus of elasticity (E) and Poisson's ratio (v), as shown in the matrix below of 3D isotropic materials:

$$C_{ij} = \frac{E}{(1+v)(1-2v)} \begin{bmatrix} 1-v & v & v & 0 & 0 & 0 \\ v & 1-v & v & 0 & 0 & 0 \\ v & v & 1-v & 0 & 0 & 0 \\ 0 & 0 & 0 & \frac{(1-2v)}{2} & 0 & 0 \\ 0 & 0 & 0 & 0 & \frac{(1-2v)}{2} & 0 \\ 0 & 0 & 0 & 0 & 0 & \frac{(1-2v)}{2} \end{bmatrix} \tag{6.3}$$

6.2.2.2 Hyperelastic Model (Green Elastic Material Named after Green in 1839)

An experimental investigation of rubber materials shows a nonlinear elastic stress–strain relationship that requires a method of analysis that deviates from linear elastic behavior. Soft tissues have a similar characteristic of that of the rubber material (Fung 1993). Therefore, a "rubber-like material" term is used to describe soft tissues. This nonlinear elastic behavior is called hyperelastic (Figure 6.2).

Hyperelastic material can be defined as an elastic material that has a strain–energy function (W), which refers to the energy stored in the material caused by deformation. This is different from Cauchy elastic materials (named after Cauchy in 1789–1857, who formulated the stress–strain relationship for isotropic linear elastic material in 1822), which have a nonconservative structure and therefore the stress cannot be derived from a scalar potential function (strain-energy function).

Characterizing the mechanical behavior of an isotropic hyperelastic material is based on the relationship between the strain–energy density and the material stretches. These stretches are characterized by three invariants (I_1, I_2, and I_3). They are

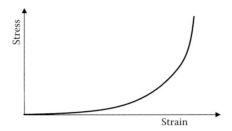

FIGURE 6.2 Typical nonlinear stress–strain relationship of hyperelastic material.

called invariants because they are the same regardless of the coordinate system:

$$\begin{aligned} I_1 &= \lambda_1^2 + \lambda_2^2 + \lambda_3^2 \\ I_2 &= \lambda_1^2\lambda_2^2 + \lambda_2^2\lambda_3^2 + \lambda_3^2\lambda_1^2, \\ I_3 &= \lambda_1^2\lambda_2^2\lambda_3^2 \end{aligned} \tag{6.4}$$

where λ_1, λ_2, and λ_3 are the principal stretches. For incompressible materials (Poisson's ratio = 0.5), I_3 is 1.

There are a number of hyperelastic models available in the literature. The efficiency of each model is based on its ability to satisfactorily describe the behavior of the material under specific load with a minimum number of material parameters. In other words, if the proposed model results fit the experimental data of the material, the model is considered efficient to model the material under this loading condition. Reviews on these models are presented in Boyce and Arruda (2000), Marckmann and Verron (2006), and Martins et al. (2006). Based on these reviews, a brief description is provided here for a number of frequently used models for the biomechanical modeling.

The Mooney–Rivlin model has been widely used to model elastomer material (Boyce and Arruda 2000). This is related to its high accuracy in addition to its history as a first model to capture the nonlinear elastic behavior of rubber-like materials (Martins et al. 2006). It is suitable for a median strain range of 200% to 250% modeling:

$$W = C_{10}(I_1 - 3) + C_{01}(I_2 - 3). \tag{6.5}$$

An extension of this model is proposed by Rivlin who developed a general strain–energy model in the form of polynomial series:

$$W = \sum_{i,j=0}^{\infty} C_{ij}(I_1 - 3)^i (I_2 - 3)^j, \tag{6.6}$$

where C_{ij} is a material parameter.

Considering the first term only of the Rivlin model, the neo-Hookean model is derived:

$$W = C_{10}(I_1 - 3). \tag{6.7}$$

This model is recommended for modeling small strain (150%) because of its ability to predict the material behavior under different loading conditions in addition to its simplicity (Marckmann and Verron 2006).

A high-order term of I_1 or I_2 may be applied to account for a wider spectrum of deformation. Yeoh added a higher order of I_1, which was shown to accurately capture the large deformation. The Yeoh model is

$$W_{\text{Yeoh}} = C_{10}(I_1 - 3) + C_{20}(I_1 - 3)^2 + C_{30}(I_1 - 3)^3. \qquad (6.8)$$

Unlike other models that use the stretch invariants, Ogden (1984) used the principal stretches to derive the strain–energy function. The model has been widely used to for large strain problems:

$$W_O = \sum_{i=1}^{n} \frac{\mu_i}{\alpha_i} \left(\lambda_1^{\alpha_i} + \lambda_2^{\alpha_i} + \lambda_3^{\alpha_i} - 3 \right), \qquad (6.9)$$

where μ_i and α_i are real numbers representing material parameters, whereas n is a positive integer.

6.2.2.3 Viscoelastic Model

Viscoelasticity is the time-dependent mechanical properties of materials. This can be classified into different material responses to the applied load or deformation when time is considered as a parameter in calculating the response.

If the material is suddenly deformed under sudden increase of strain (ε) from zero to a value of ε_0 at time $t = 0$ and held fixed afterward, the resulting stress decreases with time. Monitoring this stress pattern that is required to maintain the constant strain history reveals different stages. At time $t = 0$, an instantaneous increase in stress is applied. This is followed by a gradual nonconstant decrease in stress that may not reach zero. This is known as a relaxation as illustrated in Figure 6.3a, where two samples are subjected to two levels of uniaxial tensile loading with a displacement control machine.

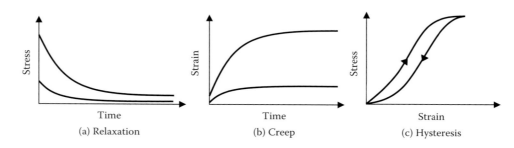

(a) Relaxation (b) Creep (c) Hysteresis

FIGURE 6.3 (a) Stress–time behavior of a viscoelastic material under two constant strain levels, (b) strain–time behavior of a viscoelastic material subjected to two constant stress levels, and (c) hysteresis of stress–strain plot of a viscoelastic material under cyclic loading.

TABLE 6.1 Linear Viscoelastic Models

Model	Configuration	Creep Function	Relaxation Function
Maxwell			
Kelvin–Voigt			
Standard linear solid (Maxwell form)			

When the material is subjected to a sudden increase in stress and the stress is maintained constant over time, the strain increases instantaneously and it continues strained with time as shown in Figure 6.3b. This is known as creep. The two samples are loaded with two different strain levels using a load control testing machine.

A third feature of the time-dependent properties of soft tissue is hysteresis, as shown in Figure 6.3c. This appears when the tissue is subjected to a cyclic loading where the stress–strain in loading path is different from that of the unloading. It is related to the energy dissipation during the cycle of loading.

Viscoelasticity models can be classified as linear, quasi-linear, and nonlinear. An overview of the nonlinear viscoelasticity is provided by Wineman (2009). The linear viscoelasticity is explained here for the sake of simplicity.

In the linear viscoelastic model, the material is divided into two parts: elastic and visco as indicated by the name. The elastic part is normally modeled as a linear spring, whereas the visco section is modeled as a dashpot. The most commonly used linear viscoelastic models are Maxwell, Kelvin–Voigt, and standard linear solid. An overview of the viscoelastic characteristic of materials, the creep, and relaxation functions of the models is shown in Table 6.1.

6.2.3 Loading and Boundary Conditions

The boundary conditions applied in the model have a significant role in the accuracy of the model. In fact, it has been stated that the boundary conditions and loading are more significant than material properties (Carter et al. 2005). This is an important feature in the biomechanical model because there is always uncertainty in material properties of soft tissues, as stated by Miller et al. (2010). Similarly, Tanner et al. (2006) found that the boundary conditions in addition to the Poisson's ratio have a significant role on the accuracy of breast biomechanical models.

There are different ways to apply the loading on the biomechanical models including force, pressure, or displacement. Applying displacement on the surface of the organs to find the field displacement distribution is preferable due to difficulties associated with measuring applied forces (Wittek et al. 2007).

Biomechanical modeling of soft tissue for radiotherapy application can be either single-organ or multiorgan models. The organs in the multiorgan studies have been modeled as attached or sliding organs. Although some regions are anatomically attached to each other, most of the organs slide relative to each other, where sliding is an important feature for their normal functioning, as in the case of lungs (D'Angelo et al. 2004; Loring et al. 2005; Widmaier et al. 2006). Modeling breathing motion of the lungs with sliding shows a significant improvement on the accuracy of registration (Al-Mayah et al. 2008). Similar observation has been reported in the case of brain modeling where a gap between the skull and the brain is required for the brain to move inside the cranial cavity (Wittek and Omori 2003; Hu et al. 2007).

6.3 Applications

Biomechanical-based deformable image registration has been used in different anatomical sites, including the lungs, liver, breast, and prostate. The schematic procedures of developing and analyzing the biomechanical-based deformable image registration using an in-house developed algorithm, MORFEUS (Brock et al. 2005), are illustrated in the Figure 6.4. The lung model is used as an example where a number of biomechanical modeling issues are addressed, including sliding contact and heterogeneity by including the bronchial tree and tumor (Al-Mayah et al. 2009b). Although the deformation of the anatomical site can be caused by tumor shrinkage, bowel movement, gas, bladed filling, and weight loss, the breathing motion is applied in this example. The modeling process starts by acquiring inhale and exhale images. Surface meshes are constructed using the segmented lungs, bronchial trees, tumor, and body. The differences between inhale and exhale positions of the lungs and external body are found using surface projection. A volumetric mesh is applied for all regions, except the bronchial tree that is modeled as a set of hollow tubes made of shell elements. The lungs are allowed to slide inside the chest cavity.

An overview of the biomechanical investigations of the human lungs, liver, breast, and prostate is presented here. This includes the material properties and FEM.

6.3.1 Lungs

The biomechanics of the lungs was pioneered by Mead et al. (1970) who modeled the lungs as a set of springs paving the way for numerous research projects in the solid mechanics of the lungs (Fung 1974; Fung et al. 1978; Lee 1978; Liu and Lee 1978; Vawter et al. 1979; De Wilde et al. 1981; Maksym and Bates 1997; Denny and Schroter 2006).

Different studies have investigated the mechanical properties of the lungs. Poisson's ratio and shear modulus of lungs are found to be age-related properties with values increasing as age increases (Lai-Fook and Hyatt 2000). Zeng et al. (1987) experimentally investigated the stress–strain relationship of human lungs. Nonlinear stress–strain relationship was found where the stiffness increased with the level of applied loading. Similar findings were reported by Lai-Fook and Hyatt (2000), where the shear modulus increased with the increase in applied loading.

The first FEM of the lung investigated the human and dog lungs under its own weight (Matthews and West 1972). An approximate geometrical representation of the lung was modeled. Although the material properties of human lungs are different from dog lungs (Zeng et al. 1987), the material properties of dog lungs were applied for both human and dog models. Using the same material properties, Sundaram and Feng (1977) developed a 3D FEM of a half of human thorax, including the lungs and heart.

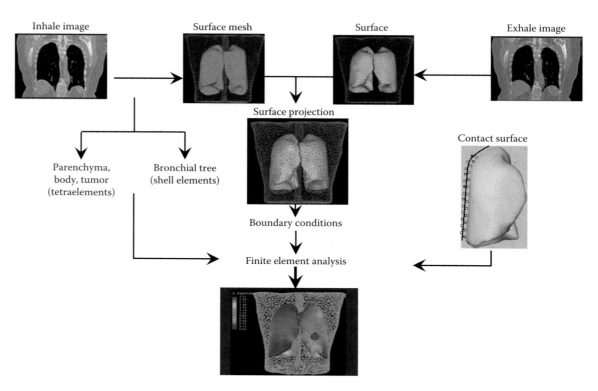

FIGURE 6.4 Model development starts by acquiring CT images for both exhale and inhale phases. 3D surface meshes are created and used for the projection of the inhale and exhale surfaces to find the boundary conditions. The inhale surface meshes of all components, except the bronchial tree, are tetrameshed leaving the bronchial trees as shell structures. After applying contact surface on the lungs, finite element analysis is conducted to find the deformation map. (Republished from Al-Mayah, A. et al., *Medical Physics, 37*(9), 4560–4571, 2010. With permission.)

A large variation in the mechanical properties can be found in the finite element studies of lungs. This is clearly demonstrated by the values of Poisson's ratio ranging from 0.2 to 0.499 (Matthews and West 1972; Sundaram and Feng 1977; Zhang et al. 2004; Brock et al. 2005; Villard et al. 2005; Al-Mayah et al. 2009). Similar variation can be observed in the modulus of elasticity ranging from 0.25 kPa (Matthews and West 1972), 0.73 kPa (De Wilde et al. 1981), 0.82 kPa (Villard et al. 2005), 4.0 kPa (Zhang et al. 2004), to 5.0 kPa (Brock et al. 2005).

Parametric studies have been conducted to investigate the effect of the Poisson's ratio and modulus of elasticity. The minimal effect of Poisson's ratio (between 0.25 and 0.45) and modulus of elasticity (between 0.1 and 10 kPa) on the models outcomes was reported (Werner et al. 2009). Villard et al. (2005) found that increasing Poisson's ratio resulted in an increase in displacement difference between the surface of the lungs and the surrounding body nodes. Therefore, by allowing the lungs to slide relative to the chest cavity, Al-Mayah et al. (2009) found that the effect of changing the Poisson's ratio between 0.35 and 0.499 is minor. However, this effect is more pronounced in models without sliding.

Using the experimental data from Zeng et al. (1987), Al-Mayah et al. (2008) reduced the registration error in the SI direction from 0.68 cm in the linear elastic model to 0.33 cm by applying hyperelastic material properties. This registration error was reduced further to 0.09 cm by modeling the sliding between the lungs and

chest cavities. This is mainly attributed to the pleural membranes surrounding the lungs: one covering the lungs and the other attached to the chest. This anatomical feature makes sliding model as a realistic technique that simulates breathing motion of the lung (Zhang et al. 2004; Al-Mayah et al. 2008; Werner et al. 2009). Furthermore, the thin layer of pleural liquid that fills the gap between the two pleural layers in each lung acts as a lubricant facilitating the sliding of the lungs inside the chest cavities. Investigating the effect of pleural lubrication using coefficients of friction of 0.0, 0.1, and 0.2, Al-Mayah et al. (2009) found that the frictionless surface is an accurate assumption, although the model is not very sensitive to the coefficient of friction.

6.3.2 Liver

The liver is described as "the most movable organ in both normal and standardized breathing" (Suramo et al. 1984). Most biomechanical studies of the liver have concentrated on its mechanical properties. Yeh et al. (2002) reported that the liver and tumor modulus of elasticity is strain dependent, with values of modulus of elasticity ranging between 0.64 and 2.0 kPa for 5% and 15% preloading, respectively. The tumor behaves similarly, but higher than normal tissues, with modulus of elasticity of 3.0 and 12 kPa for 5% and 15% preloading, respectively. In addition, the fibrousness of the liver alters its stiffness. The stiffness of the liver increased from 7.2 to 18.2 kPa as the fibrosis degree increased from 2 to 4 in the liver of patients with hepatitis B (Marcellin et al. 2009). Although Carter et al. (2001) reported that the stress–strain relationship is nonlinear, an average value of the modulus of elasticity of 270 kPa was calculated. Lower values have been reported in the literature with a value of approximately 6.0 kPa (Lim et al. 2009; Muller et al. 2009). Higher values of modulus of elasticity of 20 and 60 kPa for long and instantaneous linear elastic modulus were reported by Nava et al. (2008). The large difference was attributed to the inclusion of the liver capsule in the analysis (Nava et al. 2008).

Ahn et al. (2008) investigated the material properties of human liver cancer cells using inverse FEM. The hyperelastic model of the tissues has been characterized, where the C_{10} and C_{01} parameters in neo-Hookean and Mooney–Rivlin were found to be 0.91 and 0.33 Pa, respectively. The modulus of elasticity is also calculated for the cancer cell with a reported value of 7.44 kPa. The viscoelastic properties of the human liver were also investigated. The standard linear viscoelastic properties were 1.16 ± 0.28 kPa, 1.97 ± 0.30 kPa, and 7.3 ± 2.3 Pa s for E_1, E_2, and η, respectively (Asbach et al. 2008). The quasi-linear viscoelastic properties were reported by Nava et al. (2008) using an in vivo measurement.

Finite element-based deformable image registration has been proven to be effective in integrating multimodality images in the radiation treatment of liver cancer patients (Brock et al. 2006). Deformable image registration was conducted based on pretreatment computed tomography (CT), magnetic resonance imaging (MRI), during treatment cone-beam CT, and a follow-up CT of five liver cancer patients. The multiorgan finite element-based program (MORFEUS) was applied. The program was able to provide accurate geometrical and spatial information of the target for the treatment purpose. Archip et al. (2007) used FEM for deformable image registration of the liver using planning MRI and intraprocedural CT images for radiofrequency tumor ablation. The average registration errors of 2.35, 3.04, and 1.64 mm were reported using B-spline, Demons, and FEM, respectively, based on the edge anatomical landmarks of the liver. Brock et al. (2008b) evaluated the accuracy of the online tumor location on the deformable image registration of the liver by registering the daily kilovolt (kV) cone-beam CT and planning CT images. The study showed that although the deformable image registration may not be required for some patients, it improved the target localization in 33% of patients.

6.3.3 Breast

In the area of biomechanical material properties, Samani et al. (2007) measured the elastic modulus of normal and pathologic breast tissues using elastography under small deformation. These tissues included the fat, fibroglandular, and different types of benign and malignant breast tumors. It was found that the fat and fibroglandular tissues have a similar modulus of elasticity of 3.25 ± 0.91 and 3.24 ± 0.61 kPa, respectively. This value increases by threefold to sixfold for pathologic tissues. Sinkus et al. (2005) characterized the viscoelastic properties

of different breast tissues, including cancerous and surrounding breast tissues, using magnetic resonance elastography. The shear modulus and shear viscosity were measured. The shear modulus of the cancer and the surrounding tissues are 2.9 ± 0.3 and 0.87 ± 0.15 kPa, respectively, whereas the shear viscosity values of 2.4 ± 1.7 and 0.55 ± 0.12 Pa s are reported for the cancer and surrounding tissues, respectively. A review of the in vivo mechanical properties of the breast tissues using magnetic resonance elastography is presented in Van Houton et al. (2003).

FEM of the breast has been widely investigated to address different related issues, such as simulating the mammographic compression (Samani et al. 2001), guiding the biopsy needle (Azar et al. 2002), predicting the shape of the breast under its own weight (Samani and Plewes 2002; Del Palomar et al. 2008), assessing and improving other different registration techniques (Schnabel et al. 2003), modeling gravity and compression (Rajagopal et al. 2008), and finding breast density distribution (Yaffe et al. 2009). A review of the biomechanical modeling, including FEM of the breast, was presented by Rajagopal et al. (2009).

The effect of linear and nonlinear material properties, in addition to the boundary conditions on the accuracy of the model, was investigated by Tanner et al. (2006) using MRI images of precompression and postcompression of 20% of the breast. It was reported that the boundary conditions and Poisson's ratio have more influence on the accuracy of the model than the other material properties.

FEM is used for intramodality and intermodality image registration. Intramodality image registration used for the breast mainly concentrates on x-ray mammogram, where the craniocaudal (CC) and a mediolateral oblique (MLO) images are registered (Zhang et al. 2007). The intermodality image registration studies include magnetic resonance/x-ray mammogram (Ruiter et al. 2004) and MRI/positron emission tomography (Krol et al. 2006).

6.3.4 Prostate

The investigation of the prostate biomechanical properties has attracted attention since the pioneering research conducted by Parker et al. (1990), who reported that the value of the modulus of elasticity of the excised normal human prostate ranges between 1 and 6 kPa. The different value of modulus of elasticity was found in benign prostate hyperplasia human prostate with values ranging between 0.95 and 7.0 kPa. The difficulty in measuring the transverse strain resulted in the assumption of 0.495 for the Poisson's ratio. Nonlinearity and hysteresis were observed in a stress–strain plot; therefore, the measurements of the modulus of elasticity are considered for the portion of the plot near the origin. The nonlinearity of the material is clearly demonstrated later by Parker et al. (1993), who reported an initial modulus of elasticity of 2.15 ± 0.81 kPa and a higher value of 17.3 ± 18.0 kPa by considering a higher strain level between 0% and 10%. However, Krouskop et al. (1998) reported that a little difference in material properties was found under different strain ranges and loading frequency, which led to the conclusion that the tissue can be modeled as an elastic material.

The different moduli of elasticity of 55 ± 14, 62 ± 19, 38 ± 8, and 96 ± 19 kPa were reported for the normal anterior, posterior, benign prostate hyperplasia, and cancer, respectively (Krouskop et al. 1998), where the prostate was tested under compression using 0.1 Hz frequency. An in vitro investigation of normal and cancerous human prostate shows a good correlation ($R^2 = 0.97$) between the relaxation test and the viscoelastic Kelvin–Voigt model, with mean values of the complex modulus of elasticity of 15.9 ± 5.9 kPa and 40.4 ± 15.7 kPa for the normal and cancerous prostate at 150 Hz, respectively (Zhang et al. 2008).

FEM has been used in different applications, such as the estimation of material properties of the prostate (Alterovitz et al. 2006), the assessment of other deformable image registration (Zhang et al. 2007), and the evaluation of the effect of contour variation and model parameters on image registration (Brock et al. 2008a). In most cases, the displacement boundary conditions are applied directly to the prostate; however, in other studies, the deformation of the prostate is a result of indirect applied displacement on the surrounding tissue either with contact surface modeling (Boubaker et al. 2009) or without contact surface modeling (Hensel et al. 2007).

A number of research projects have used FEM-based registration for the prostate radiation treatment. Yan et al. (1999) implemented the registration to calculate the accumulated dose distribution in deformed prostate using daily CT images. Similarly, Wu et al. (2006) used the FEM image registration between treatment and planning CT after a rigid registration and setup correction are conducted at each fraction. The accumulative dose was compensated either once or weekly if it was significantly different from the prescribed dose. The weekly compensation was proven to be more effective.

In brachytherapy treatment, Bharatha et al. (2001) used 3D FEM for image registration of preoperative and intraoperative images of the prostate. The registration was conducted using the preoperative image while the patients are in supine position with endorectal probe and the intraoperative images while the patients are in the lethotomy position with a rectal obturator in place. The change in position and the rectal filling were the main source of deformation in this case. Alignment of the center of mass was conducted and followed by deformable registration. The linear material properties were applied. The similarity Dice index in the total gland increased from 0.81 to 0.94 using rigid and deformable registration, respectively.

6.4 Conclusion

The principles of biomechanical modeling for deformable image registration have been introduced with a focus on the FEM technique. The biomechanical material properties, geometry, and boundary conditions are discussed. Although hyperelastic and viscoelastic properties are most suitable to describe soft-tissue properties, their modeling applications are limited compared with the linear properties. Biomechanical studies have been conducted in different anatomical regions, including the lungs, liver, breast, and prostate. The biomechanical material properties of these sites are reviewed. The FEM and its applications for deformable image registration are then presented for each site. A more detailed model of the lungs has been developed recently, which addresses the effect of including the bronchial tree on the deformable image registration of the lungs. This work has been under review for a potential publication in the near future.

References

Ahn, B., Kim, Y., and Kim, J. 2008. Biomechanical characterization with inverse FE model parameter estimation: Macro and micro applications, in *International Conference on Control, Automation and Systems*, IEEE, Piscataway, NJ, 1769–1772.

Al-Mayah, A., Moseley, J., and Brock, K.K. 2008. Contact surface and material nonlinearity modeling of human lungs. *Physics in Medicine and Biology* 53: 305–317.

Al-Mayah, A., Moseley, J., Velec, M., and Brock, K.K. 2009a. Sliding characteristic and material compressibility of human lung: Parametric and verification. *Medical Physics, 36*: 4625–4633.

Al-Mayah, A., Moseley, J., Velec, M., and Brock, K. 2009b. Effect of heterogeneous material of the lung of deformable image registration. *Proceedings of SPIE, 7261*: 1–8.

Alterovitz, R., Goldberg, K., Pouliot, J. et al. 2006. Registration of MR prostate images with biomechanical modeling and nonlinear parameter estimation. *Medical Physics, 33*: 446–454.

Archip, N., Tatli, S., Morrison, P., Jolesz, F., Warfield, S.K., and Silverman, S. 2007. Non-rigid registration of pre-procedural MR images with intra-procedural unenhanced CT images for improved targeting of tumors during liver radiofrequency ablations. *Medical Image Computing and Computer Assisted Intervention, 10*(Part 2): 969–977.

Asbach, P., Klatt, D., Hamhaber, U., Braun, J. et al. 2008. Assessment of liver viscoelasticity using multifrequency MR elastography. *Magnetic Resonance in Medicine, 60*: 373–379.

Azar, F. S., Metaxas, D.N., and Schnall, M.D. 2002. Methods for modeling and predicting mechanical deformations of the breast under external perturbations. *Medical Image Analysis* 6: 1–27.

Barber, D.C. and Hose, D.R. 2005. Automatic segmentation of medical images using image registration: Diagnostic and simulation applications. *Journal of Medical Engineering and Technology, 29*: 53–63.

Bharatha, A., Hirose, M., Hata, N. et al. 2001. Evaluation of three-dimensional finite element-based deformable registration of pre- and intraoperative prostate imaging. *Medical Physics, 28*: 2551–2560.

Boubaker, M.B., Haboussi, M, Ganghoffer, J., and Aletti, P. 2009. Finite element simulation of interactions between pelvic organs: Predictive model of the prostate motion in the context of radiotherapy. *Journal of Biomechanics, 42*: 1862–1868.

Boyce, M.C. and Arruda, E.M. 2000. Constitutive models of rubber elasticity: A review. *Rubber Chemistry and Technology, 73*: 504–523.

Brock, K.K., Nichol, A.M, Ménard, C. et al. 2008a. Accuracy and sensitivity of finite element model-based deformable registration of the prostate. *Medical Physics, 35*: 4019–4025.

Brock, K.K., Hawkins, M., Eccles, C. et al. 2008b. Improving image-guided target localization through deformable registration. *Acta Oncologica, 47*: 1279–1285.

Brock, K.K., Sharpe, M.B., Dawson, L.A., Kim, S.M., and Jaffray D.A. 2005. Accuracy of finite element model-based multi-organ deformable image registration. *Medical Physics 32*: 1647–1659.

Brock, K.K., Dawson, L.A., Sharpe, M.B., Moseley, D.J., and Jaffray, D.A. 2006. Feasibility of a novel deformable image registration technique to facilitate classification, targeting, and monitoring of tumor and normal tissue. *International Journal of Radiation Oncology Biology Physics, 64*: 1245–1254.

Carter, F.J., Frank, T.G., Davies, P.J., McLean, D., and Cuschieri, A. 2001. Measurement and modelling of the compliance of human and porcine organs. *Medical Image Analysis, 5*: 231–236.

Carter, T.J., Sermesant, M., Cash, D.M., Barratt, D.C., Tanner, C., and Hawkes, D.J. 2005. Application of soft tissue modelling to image-guided surgery. *Medical Engineering and Physics, 27*: 893–909.

Chabanas, M., Payan, Y., Marécaux, C., Swider, P., and Boutault, F. 2004. Comparison of linear and non-linear soft tissue models with post-operative CT scan in maxillofacial surgery. *Medical Simulation, Lecture Notes in Computer Science, 3078*: 19–27.

D'Angelo, E., Loring, S.H., Gioia, M.E., Pecchiari, M., and Moscheni, C. 2004. Friction and lubrication of pleural tissues. *Respiratory Physiology and Neurobiology, 142*: 55–68.

De Wilde, R., Clement, J., Hellemans, J.M. et al. 1981. Model of elasticity of the human lung. *Journal of Applied Physiology, 51*: 254–261.

Del Palomar, A.P., Calvo, B., Herrero, J., Lopez, J., and Doblare, M. 2008. A finite element model to accurately predict real deformations of the breast. *Medical Engineering and Physics, 30*: 1089–1097.

Denny, E. and Schroter, R.C. 2006. A model of non-uniform lung parenchyma distortion. *Journal of Biomechanics, 39*: 652–663.

Fung, Y.C. 1974. A theory of elasticity of the lung. *Journal of Applied Mechanics, 41*: 8–14.

Fung, Y.C. 1993. *Biomechanics; Mechanical Properties of Living Tissues*, Springer, Berlin.

Fung, Y.C., Tong, P., and Patitucci, P. 1978. Stress and strain in the lung. *Journal of the Engineering Mechanics Division, 104*: 201–223.

Hensel, J.M., Ménard, C., and Chung, P.W.M. 2007. The development of a multi-organ finite element based prostate deformation model enabling the registration of endorectal coil magnetic resonance images (ERC-MRI) for radiotherapy planning. *International Journal of Radiation Oncology Biology Physics, 68*: 1522–1528.

Holzapfel, G.A. 2004. Computational biomechanics of soft biological tissue, in *Encyclopedia of Computational Mechanics. Solids and Structures Vol. 2*, Wiley, Chichester, 605–635, Chapter 18.

Hu, J., Jin, X., Lee, J.B. et al. 2007. Intraoperative brain shift prediction using a 3D inhomogeneous patient-specific finite element model. *Journal of Neurosurgery, 106*: 164–169.

Humphrey, J.D. 2003. Continuum biomechanics of soft biological tissues. *Proceedings of the Royal Society of London Series A, 459*: 3–46.

Krol, A., Unlu, M.Z., and Baum, K.G. 2006. MRI/PET nonrigid breast-image registration using skin fiducial markers. *Physica Medica, 21*: 39–43.

Krouskop, T.A., Wheeler, T.M., Kallel, F., Garra, B.S., and Hall, T. 1998. Elastic moduli of breast and prostate tissues under compression. *Ultrason Imaging, 20*: 260–274.

Lai-Fook, S.J. and Hyatt R.E. 2000. Effect of age on elastic moduli of human lungs. *Journal of Applied Physiology, 89*: 163–168.

Lee, G.C. 1978. Solid mechanics of lungs. *Journal of the Engineering Mechanics Division, 104*: 177–199.

Lim, Y.J. Deo, D., Singh, T.P., Jones, D.B., and De, S. 2009. In situ measurement and modeling of biomechanical response of human cadaveric soft tissues for physics-based surgical simulation. *Surgical Endoscopy, 23*: 1298–1307.

Liu, J.T. and Lee, G.C. 1978. Static finite deformation analysis of the lung. *Journal of the Engineering Mechanics Division, 104*: 225–239.

Loring, S.E., Brown, R.E., Gouldstone, A., and Butler, J.P. 2005. Lubrication regimes in mesothelial sliding. *Journal of Biomechanics, 38*: 2390–2396.

Maciel, A., Boulic, R., and Thalmann, D. et al. 2003. Deformable tissue parameterized by properties of real biological tissue, in *International Symposium on Surgery Simulation and Soft Tissue Modeling*, 74–87.

Maksym, G.N. and Bates, J.H. 1997. A distributed nonlinear model of lung elasticity. *Journal of Applied Physiology, 82*: 32–41.

Marcellin, P., Zoil, M., Bedossa, P. et al. 2009. Non-invasive assessment of liver fibrosis by stiffness measurement in patients with chronic hepatitis B. *Liver International, 29*: 242–247.

Marckmann, G. and Verron, E. 2006. Comparison of hyperelastic models for rubber-like materials. *Rubber Chemistry and Technology, 79*: 835–858.

Martins, P.A.L.S., Natal Jorge, R.M., and Ferreira A.J.M. 2006. A comparative study of several material models for prediction of hyperelastic properties: Application to silicone-rubber and soft tissues. *Strain, 42*: 135–147.

Matthews, F.L. and West, J.B. 1972. Finite element displacement analysis of a lung. *Journal of Biomechanics, 5*: 591–600.

Mead, J., Takishima, T., and Leith, D. 1970. Stress distribution in lungs: A model of pulmonary elasticity. *Journal of Applied Physiology, 28*: 596–608.

Miller, K., Wittek, A., Joldes, G. et al. 2010. Modelling brain deformations for computer-integrated neurosurgery. *International Journal for Numerical Methods in Biomedical Engineering, 26*: 117–138.

Muller, M., Gennisson, J.L., Deffieux, T., Tanter, M., and Fink, M. 2009. Quantitative viscoelasticity mapping of human liver using supersonic shear imaging: Preliminary in vivo feasibility study. *Ultrasound in Medicine and Biology, 35*: 219–229.

Nava, A., Mazza, E., Furrer, M. et al. 2008. In vivo mechanical characterization of human liver. *Medical Image Analysis, 12*: 203–216.

Ogden, R.W. 1984. *Non-linear Elastic Deformations*, Dover Publications, Inc., New York.

Parker, K.J., Huang, S.R., Musulin, R.A., and Lerner, R.M. 1990. Tissue response to mechanical vibrations for sonoelasticity imaging. *Ultrasound in Medicine and Biology, 16*: 241–246.

Parker, K.J., Huang, S.R., Lerner, R.M., Lee Jr., F., Rubens, D., and Roach, D. 1993. Elastic and ultrasonic properties of the prostate. *Proceedings of the IEEE Ultrasound Symposium, 2*: 1035–1038.

Rajagopal, V., Lee, A., Chung, J. et al. 2008. Creating individual-specific biomechanical models of the breast for medical image analysis. *Academic Radiology, 15*: 1425–1436.

Rajagopal, V., Nielsen, P.M.F., and Nash, M. P. 2009. Modeling breast biomechanics for multi-modal image analysis—Successes and challenges. *Wiley Interdisciplinary Reviews: Systems Biology and Medicine, 2*(3): 293–304.

Roy, C.S. 1880. The elastic properties of the arterial wall. *Philosophical Transactions of the Royal Society of London Series B, 99*: 1–31.

Ruiter, N.V., Stotzka, R., Muller, T.O., and Gemmeke, H. 2004. Model-based registration of X-ray mammograms and MR images of the female breast. *IEEE Nuclear Science Symposium Conference Record, 5*: 3290–3294.

Samani, A., Bishop, J., and Plewes, D.B. 2001. Biomechanical 3-D finite element modeling of the human breast using MRI data. *IEEE Transactions on Medical Imaging 20*: 271–279.

Samani, A. and Plewes, D.B. 2002. Finite element model of MRI image updating for breast surgery, in *OCITS Workshop: Computational and Numerical Modeling for Image-Guided Therapy and Surgery*, Toronto, Canada.

Samani, A., Zubovits, J., and Plewes, D.B. 2007. Elastic moduli of normal and pathological human breast tissues: An inversion-technique-based investigation of 169 samples. *Physics in Medicine and Biology, 52*: 1565–1576.

Schnabel, J.A., Tanner, C., Castellano-Smith, A.D. et al. 2003. Validation of nonrigid image registration using finite-element methods: Application to breast MR images. *IEEE Transactions on Medical Imaging, 22*: 238–247.

Sinkus, R., Tanter, M., Xydeas, T., Catheline, S., Bercoff, J., and Fink. M. 2005. Viscoelastic shear properties of in vivo breast lesions measured by MR elastography. *Magnetic Resonance Imaging, 23*: 159–165.

Sundaram, S.H. and Feng, C.C. 1977. Finite element analysis of the human thorax. *Journal of Biomechanics, 10*: 505–516.

Suramo, I., Paivansalo, M., and Myllyla, V. 1984. Cranio-caudal movements of the liver, pancreas and kidneys in respiration. *Acta Radiologica: Diagnosis, 25*: 129–131.

Tanner, C., Schnabel, J.A., Hill, D.L.G., Hawkes, D.J., Leach M.O., and Hose, D.R. 2006. Factors influencing the accuracy of bio-mechanical breast models. *Medical Physics, 33*: 1758–1769.

Van Gelder, A. 1998. Approximate simulation of elastic membranes by triangulated spring meshes. *Journal of Graphics Tools, 3*: 21–41.

Van Houten, E.E., Doyley, M.M., Kennedy, F.E. et al. 2003. Initial in vivo experience with steady-state subzone-based MR elastography of the human breast. *Journal of Magnetic Resonance Imaging, 17*: 72–85.

Vawter, D.L., Fung, Y.C., and West, J.B. 1979. Constitutive equation of lung tissue elasticity. *Journal of Biomechanical Engineering, 101*: 38–45.

Villard, P., Beuve, M., Shariat, B., Baudet, V., and Jaillet, F. 2005. Simulation of lung behaviour with finite elements: Influence of bio-mechanical parameters, in *Proceedings of the Third International Conference on Medical Information Visualisation—BioMedical Visualisation*, MediVi, V2005, IEEE Computer Society, London, 9–14.

Werner, R., Ehrhardt, J., Schmidt, R., and Handels, H. 2009. Patient-specific finite element modeling of respiratory lung motion using 4D CT image data. *Medical Physics, 36*: 1500–1511.

Widmaier, E.P., Raff, H., and Strang, K. T. 2006. *Vander's Human Physiology: The Mechanisms of Human Body Function*, 10th edn., McGraw-Hill, New York.

Wineman, A. 2009. Nonlinear viscoelastic solids: A review. *Mathematics and Mechanics of Solids 14*: 300–366.

Wittek, A. Miller, K., Kikinis, R., and Warfiel S.K. 2007. Patient-specific model of brain deformation: Application to medical image registration. *Journal of Biomechanics, 40*: 919–929.

Wittek, A. and Omori, K. 2003. Parametric study of effects of brain-skull boundary conditions and brain material properties on responses of simplified finite element brain model under angular acceleration in sagittal plane. *JSME International Journal 46*: 1388–1398.

Wu, Q., Liang, J. and Yan, D. 2006. Application of dose compensation in image-guided radiotherapy of prostate cancer. *Physics in Medicine and Biology, 51*: 1405–1419.

Yaffe, M., Boone, J. Packard, N. et al. 2009. The myth of the 50-50 breast. *Medical Physics, 36*: 5437–5443.

Yan, D., Jaffray, D.A., and Wang, J.W. 1999. A model to accumulate fractionated dose in a deforming organ. *International Journal of Radiation Oncology Biology Physics, 44*: 665–675.

Yeh, W.C., Li, P.C., Jeng, Y.M. et al. 2002. Elastic modulus measurements of human liver and correlation with pathology. *Ultrasound in Medicine and Biology, 28*: 467–474.

Zeng, Y.J., Yager, D., and Fung, Y.C. 1987. Measurement of the mechanical properties of the human lung tissue. *Journal of Biomechanical Engineering, 109*: 169–174.

Zhang, T., Orton, N.P., Mackie, T.R., and Paliwal, B.R. 2004. Technical note: A novel boundary condition using contact elements for finite element based deformable image registration. *Medical Physics 31*: 2412–2415.

Zhang, M., Nigwekar, P., Castaneda, B. et al. 2008. Quantitative characterization of viscoelastic properties of human prostate correlated with histology. *Ultrasound in Medicine and Biology, 34*: 1033–1042.

Zhang Y., Qiu, Y., Goldgof, D.B., Sarkar, S., and Li, L. 2007. 3D finite element modeling of nonrigid breast deformation for feature registration in X-ray and MR images source, in *IEEE Workshop on Applications of Computer Vision*, p. 6.

Nonparametric: Demons, Diffusion, and Viscous-Fluid Approaches

David Sarrut
Université de Lyon

Jef Vandemeulebroucke
Vrije Universiteit Brussell

7.1 Demons Approach and Variations

7.1.1 Principles of the Demons Algorithm

Probably one of the most popular methods for deformable image registration (DIR) is the so-called "Demons" method, initially proposed by Thirion (1996, 1998). DIR between images I_1 and I_2 can be considered as the minimization of an energy function, representing a tradeoff between image similarity and deformation regularity. In the Demons algorithm, similarity and regularity are optimized in consecutive but independent operations. The algorithm consists of an iterative procedure composed of two steps. The first step aims at defining an active force at each voxel. This force is directed in the opposite direction of the image gradient, with a magnitude proportional to the gray-level difference between the two images. The second step is a Gaussian smoothing of the resulting vector field. This two-step procedure is iterated until convergence.

We define $\mathbf{u}(\mathbf{x})$ as the displacement of a point at location $\mathbf{x}$, and $\varphi(\mathbf{x}) = \mathbf{x} + \mathbf{u}(\mathbf{x})$ as the related transformation. In the Demons method, dissimilarity (D) between images is measured using the sum of squared differences (SSD), $D_{\text{SSD}}(I_1, I_2, \phi) = \sum_{\mathbf{x} \in \Omega} (I_1(\mathbf{x}) - I_2(\phi(\mathbf{x})))^2$, computed over the overlapping image domain Ω. The minimization is typically performed by steepest

gradient descent, and the local gradient (i.e., gradient in each point) of the SSD has to be computed.

Pennec et al. (1999) proposed an expression, denoted by ∇D_{SSD}, which allows limiting the local displacement at each iteration using an additional parameter α (see Equation 7.1). $\mathbf{u}(\mathbf{x})$ denotes the displacement at point $\mathbf{x}$ and $\nabla I_1(\mathbf{x})$ denotes the gradient of image I_1 at point $\mathbf{x}$. This criterion was shown to be an approximation of a second-order gradient descent of the SSD (Pennec et al. 1999; Cachier and Ayache 2004). For small displacements, as is the case at each iteration, it is equivalent to expressing ∇D_{SSD} according to the gradient of image I_1 or of the transformed image I_2 (by inverting the transformation). It is, however, simpler and faster to use ∇I_1 because it does not require the computation of ∇I_2 at each iteration (Lu et al. 2004):

$$\nabla D_{\text{SSD}}(\mathbf{x}, \mathbf{u}) = \frac{I_1(\mathbf{x}) - I_2(\mathbf{x} + \mathbf{u}(\mathbf{x}))}{\left\| \nabla I_1(\mathbf{x}) \right\|^2 + \alpha^2 (I_1(\mathbf{x}) - I_2(\mathbf{x} + \mathbf{u}(\mathbf{x})))^2} \nabla I_1(\mathbf{x}). \quad (7.1)$$

In the originally proposed algorithm (Thirion 1998), a Gaussian convolution is used as a regularization filter: for one given voxel at location $\mathbf{x}$, the local iterative update schemes as proposed in Cachier and Ayache (2004) is expressed in Equation 7.2. It consists of the application of a 3D Gaussian filter to the three components of the vector field, resulting in a smoother field:

$$\mathbf{u}_{i+1}(\mathbf{x}) = G_\sigma(\nabla D_{SSD}(\mathbf{x}, \mathbf{u}_i) \circ \mathbf{u}_i(\mathbf{x})). \qquad (7.2)$$

$$G_\sigma(\mathbf{x}) = \frac{1}{\sqrt{2\pi}\sigma} e^{-\frac{\mathbf{x}^2}{2\sigma^2}}. \qquad (7.3)$$

$\mathbf{u}_i$ denotes the displacement field at iteration i, $G_\sigma(.)$ denotes a Gaussian kernel of variance $\sigma > 0$ (large σ values result in a smoother vector field), and $\circ$ is the composition operator. The Gaussian operator has the advantage of being both isotropic and separable. Hence, it can be applied independently in each dimension, for example, using the fast recursive filter as proposed by Deriche (1993). Thirion (1998) related the Gaussian filtering to the diffusion of heat in homogeneous material by analogy with the Maxwell's demons. Bro-Nielsen and Gramkow (1996) showed that such Gaussian filtering may be considered as an approximation of the linear elastic filter used in the viscous-fluid modeling.

Vercauteren (2008) noted that, in several implementations, including the one in the popular ITK (http://www.itk.org) framework, the composition operator in Equation 7.2 is replaced with an addition, $\mathbf{u}_{i+1}(\mathbf{x}) = G_\sigma(\mathbf{u}_i(\mathbf{x}) + \nabla D_{SSD}(\mathbf{x}, \mathbf{u}_i))$. Such a scheme can be considered as a rough first-order approximation of the composition scheme and could lead to slower convergence. However, using the full composition rule requires an additional warping step to compose the current field $\mathbf{u}_i$ with the update field.

In Equation 7.1, displacements are limited to $1/(2\alpha)$ for each iteration, but each iteration starts from the previous result and it can thus lead to large estimated displacements. In practice, α is typically set to 1.0 (voxel), the standard deviation of the Gaussian regularization operator G is commonly set to $\sigma = 1.0$ (Sarrut et al. 2006a), and these parameter values are not very critical. The stopping criteria of this iterative process are not easy to define. The maximal number of iterations depends on the application and can go up to 1000 iterations for large deformations. Other stopping criteria can be defined, such as stopping when the displacement field does not evolve more than a given threshold, but they should be examined for each application separately.

7.1.1.1 Some Practical Considerations

In practice, some preprocessing is applied to the initial images I_1 and I_2. Because the computation time is proportional to the number of voxels considered, the images are generally cropped to discard nonrelevant voxels (such as the air surrounding patient), or the processing is limited to a mask identifying the region for which motion estimation should be performed. Moreover, although not explicitly required for the procedure, images are often subsampled to isotropic voxels of larger size, the new size representing a compromise between computation time and accuracy. Voxel sizes of up to $2 \times 2 \times 2$ mm^3 are not uncommon. In addition, a multiresolution minimization strategy can be employed during which the estimation is progressively refined using smaller voxel sizes. This can provide shorter execution times, improve robustness, and result in a smoother

deformation vector field. It should be noted that the Demons criterion from Equation 7.1 involves evaluating the image intensities at nongrid positions [i.e., $I_2(\mathbf{x} + \mathbf{u}(\mathbf{x}))$], thus requiring interpolation. Linear interpolation is generally considered as a good compromise between speed and precision, whereas fast nearest neighbor interpolation may be sufficient in some cases. Cubic B-spline interpolation will result in a considerably slower algorithm but tends to lead to higher accuracies.

7.1.1.2 Implementation and Graphics Processing Units Acceleration

The implementation of a DIR method is an important step, and it has been reported that different implementations of the same method can lead to strongly differing results and performance. This was highlighted in a multiinstitution study (Brock and Deformable Registration Accuracy Consortium 2010) evaluating various DIR algorithms and included multiple implementations of the same method. An important well-known resource of open-source implementations is the Insight Toolkit ITK (http://www.itk.org), in which numerous types of DIR are grouped. Such open-source repositories have shown to be very important in the scientific community because it allows researchers to easily share their proposition and compare it with other approaches. DIR remains a computationally intensive process, and several efforts have been made to reduce the computation time through adapted hardware. In particular, graphics processing units (GPU), which are dedicated processors for graphics rendering, have been shown to be very efficient in some image processing computations. Several groups (Sharp et al. 2007; Noe et al. 2008; Gu et al. 2010) proposed GPU versions of DIR algorithms that are 20 to 70 times faster in comparison with CPU implementations executed on a single core. Depending on the image sizes, the number of iterations, or, more generally, the considered application, this can lead to entire registration procedures that complete within a few minutes or even seconds.

7.1.2 Variants of the Demons Algorithm

7.1.2.1 Specifics for Thoracic Computed Tomography Images

The deformation forces used in the Demons algorithm are closely related to the SSD similarity measure, limiting the method to images of the same modality. However, even with monomodal images, the intensity conservation assumption may not always be valid. In particular, for computed tomography (CT) images of the thorax, when registering inspiration images with expiration images, the assumption is only globally valid for volume elements outside the lungs. The assumption is violated inside the lung where the quantity of inspired air leads to a decrease in lung density and effectively modifies the Hounsfield unit (HU) values. The density decrease is known to be distributed in the whole lung volume (Milic-Emili et al. 1996), although it tends to be more important in the lower parts of the lungs than in the upper parts (Monfraix et al. 2005). Indeed, regionally specific thoracopulmonary compliance was found to increase with the lung distance from the lung apex.

Several authors attempted to take this phenomenon into account. In Sarrut et al. (2006a), the authors proposed a simple "a priori" lung density modification consisting of modulating the lung densities in one image according to the densities in the other to make them comparable. Castillo et al. (2009a) introduced a compressible combined local global (CCLG) method for registration of lung tissue that accounts for the compressible nature of media in an optical flow framework. Similarly, although applied to parametric registration using cubic B-splines, Yin et al. (2009) proposed a new similarity criterion, the sum of squared tissue volume difference (SSTVD), which takes into account the effect of dilation or contraction. Other multimodal similarity measures, such as mutual information or correlation ratio, can also be used in such a way.

7.1.2.2 Symmetric Forces Demons

In the original version, the Demons algorithm uses gradient information from a single image to compute the force field at each iteration. Wang et al. (2005c) proposed to add a "symmetric force" by introducing a force field $\mathbf{u}^S$ symmetric to the one in Equation 7.1 but using the image gradient of the moving image I_2 instead of the one of the reference image I_1 (Equation 7.4).

$$\nabla D_{SSD}^{sym}(\mathbf{x}, \mathbf{u}^S) = -\frac{I_2(\mathbf{x}) - I_1(\mathbf{x} + \mathbf{u}^S(\mathbf{x}))}{\left\|\nabla I_2(\mathbf{x})\right\|^2 + \alpha^2 (I_2(\mathbf{x}) - I_1(\mathbf{x} + \mathbf{u}^S(\mathbf{x})))^2} \nabla I_2(\mathbf{x}).$$

(7.4)

The two force fields $\mathbf{u}$ and $\mathbf{u}^S$ are combined in a single field. The method requires that the active term $\mathbf{u}^S$ is updated at each iteration, but the algorithm converges faster, requiring less iterations in total. An improvement of 40% speed is achieved together with an improved matching in the presence of large deformations. An open-source implementation is available in the ITK framework.

7.1.2.3 Inverse Consistent Demons

Yang et al. (2008) proposed another modification enforcing inverse consistency in the Demons method. In this version, the two images are symmetrically deformed toward each other until both deformed images are matched. This principle is called "consistent" because it implicitly ensures that the inverse deformation field exists. The computation time is typically higher than for the conventional Demons but lower than the symmetric forces Demons. Convergence speed and accuracy seem to be improved by this version.

7.1.2.4 Diffeomorphic Demons

Vercauteren et al. (2009) extended the Demons method to constrain the deformation to a diffeomorphism, which is a one-to-one, smooth, and continuous mapping with derivatives that are invertible. Such a deformation maintains the topology and

guarantees that connected regions of an image remain connected (Christensen and Johnson 2001). This approach leads to similar results in terms of accuracy as the ones given by the initial approach, but with a smoother transformation. Again, an open-source implementation is given in the ITK framework (Dru and Vercauteren 2009; Zhao and Johnson 2009 for using it in combination with a mask). More details on the Demons method can be found in Vercauteren et al. (2007).

7.2 Diffusion-Like Approaches

As stated before, DIR can be considered as a search for a spatial transformation $\varphi(\mathbf{x}) = \mathbf{x} + \mathbf{u}(\mathbf{x})$ that minimizes a given criteria or energy composed of two terms: a dissimilarity measure quantifying differences between the two images given the current transformation, and a regularization term penalizing spatial transformations that are unlikely to correspond to the underlying sought physical deformation. Such a minimization is expressed in Equations 7.5 and 7.6, where $D(\varphi)$ is the dissimilarity function, $R(\varphi)$ is the regularization function, and H is the space of admissible transformation functions:

$$\hat{\varphi} = \arg\min_{\varphi \in H} F(I_1, I_2, \varphi)$$

(7.5)

$$F(I_1, I_2, \varphi) = D(I_1, I_2, \varphi) + R(\varphi).$$

(7.6)

D corresponds to (the opposite of) a similarity measure that quantifies the match of the two images given the current φ. Well-known measures are described in Chapter 4. Generally, R is defined as a measure of the spatial variation of φ. Deformations that are too varying from one voxel to the other, or that exhibit other undesired properties, are penalized.

To solve the minimization problem, Euler–Lagrange differential equations can be used. This is analogous to finding the function φ for which the gradient of F is zero, $\nabla F = 0$. Generally, this cannot be solved directly and an iterative gradient descent strategy can be used. Given an initial estimate φ_0, at each iteration i, a new transformation φ_i is updated according to the function gradient $\varphi_{i+1} = \varphi_i - \varepsilon \nabla F(I_1, I_2, \varphi_i) = \varphi_i - \varepsilon (\nabla D(I_1, I_2, \varphi_i) + \nabla R(\varphi_i))$, with ε a parameter controlling the descent speed. One of the main steps consists of obtaining computationally tractable expressions for ∇R and ∇D.

R, which is often intended to penalize a nonsmooth transformation, generally involves first and/or second derivatives of the deformation field φ (Cachier and Ayache 2004; Hermosillo-Valadez 2002), aggregated over the whole range of voxels. The gradient ∇R thus involves local derivatives of φ.

7.2.1 Linear Elastic Regularization

Even if it is not, strictly speaking, a variant of the Demons algorithm, the same two-step iterative approach (alternate force field estimation and regularization) can be used with the same kind of D function (D_{SSD}) and other kinds of regularization, leading to a diffusion-like approach. Hence, whereas Gaussian filtering operates independently on each coordinate of the displacement field, a linear elastic filtering can be used instead, allowing "cross-effects." Such effects appear when deforming an object, for example, in horizontal direction, leads to a vertical stretching. Using such a linear elastic filter in this scheme comes down to using the update scheme given in Equation 7.7 instead of Equation 7.2:

$$\mathbf{u}_{i+1}(\mathbf{x}) = \mathbf{u}_i(\mathbf{x}) + \varepsilon(\gamma \nabla D_{SSD}(\mathbf{x}, \mathbf{u}_i) + (1 - \gamma)\nabla R_{LE}(\mathbf{x}, \mathbf{u}_i)), \quad (7.7)$$

$$\nabla R_{LE}(\mathbf{x}, \mathbf{u}) = (\lambda + \mu)\nabla(\nabla \mathbf{u}(\mathbf{x})) + \mu\Delta\mathbf{u}(\mathbf{x}). \quad (7.8)$$

$R_{LE}(.)$ denotes the linear elastic regularization operator, γ denotes the tradeoff between image similarity and regularization, and $\varepsilon > 0$ denotes the gradient descent step. Large γ values increase the weight of image similarity, whereas low values increase the weight of regularization ($\gamma \in [0:1]$). Large ε values could decrease the number of iterations required to converge, but also increases the possibility to get trapped in local minima. The gradient of linear elastic regularization is expressed in Equation 7.8; λ and μ are the Lamé parameters (μ is sometimes named the "shear modulus"), ∇A is the gradient of A (the matrix of first-order derivative), $\nabla.A$ or div A is the divergence of A (trace of the gradient), and ΔA is the Laplacian of A (sum of all unmixed partial derivatives). The λ and μ parameters can be expressed by the Young's modulus E and the Poisson's ratio v:

$$\lambda = \frac{vE}{(1+v)(1-2v)}, \quad \mu = \frac{E}{2(1+v)} \quad (7.9)$$

Chefd'Hotel et al. (2001) proposed to use a single parameter ξ, with $\nabla R_{LE}(\mathbf{x},\mathbf{u}) = (1 - \xi)\nabla(\nabla.\mathbf{u}(\mathbf{x})) + \xi\Delta\mathbf{u}(\mathbf{x})$, with $\frac{1}{2}<\xi\leq 1$ denoting the tradeoff between the Laplacian and the gradient of divergence. Low values of ξ are related to the lateral contraction due to longitudinal extension.

In practice, differential operators of the linear elastic regularization model can be approximated with a first-order Taylor expression (Hermosillo et al. 2002) and thus computed by finite differences. The resulting operators are listed explicitly in Appendix 7.A.

7.2.2 Anisotropic and Adaptive Diffusion

Alvarez et al. (2000) and Nagel and Enkelmann (1986) proposed to preserve motion discontinuities by using nonisotropic smoothing in an optical flow method. Areas in the image with high gradient magnitudes are considered more susceptible to present deformation discontinuities and are thus less smoothed. This is achieved by the operator given in Equations 7.10 and 7.11 (for 3D images), where div is the divergence and I_x, I_y, and I_z are the components of the gradient of I. The corresponding finite difference operators (Alvarez et al. 2000; Hermosillo et al. 2002) are given in Appendix 7.A.

$$\nabla R_{NE}(\mathbf{x},\mathbf{u}) = \begin{pmatrix} \mathrm{div}\left(T_I^\gamma \nabla \mathbf{u}_{i,1}\right) \\ \mathrm{div}\left(T_I^\gamma \nabla \mathbf{u}_{i,2}\right) \\ \mathrm{div}\left(T_I^\gamma \nabla \mathbf{u}_{i,3}\right) \end{pmatrix} \quad (7.10)$$

$$T_I^\gamma = \frac{1}{2\left(I_x^2+I_y^2+I_z^2\right)+3\gamma} \begin{pmatrix} I_y^2+I_z^2+\gamma & -I_xI_y & -I_xI_z \\ -I_xI_y & I_x^2+I_z^2+\gamma & -I_yI_z \\ -I_xI_z & -I_yI_z & I_x^2+I_y^2+\gamma \end{pmatrix} \quad (7.11)$$

The parameter γ acts as a threshold: in homogeneous regions, when the image gradient is low, $|\nabla I|^2 \ll \gamma$, the filter acts as an isotropic filter, whereas in areas with a high gradient, the deformation field is less regularized in the gradient direction.

Following the same principle, Cahill et al. (2009) proposed an image-driven, locally adaptive curvature regularizer that can be adapted to the Demons algorithm. Schmidt-Richberg et al. (2009) included an anisotropic regularizer to allow sliding along a precomputed segmentation. The regularizer is designed to allow deformation discontinuities in the tangential direction of the object boundaries (defined by the segmentation), while maintaining smoothing in the normal direction, thus avoiding gaps in the deformation field. More regularization filters can be used instead of the original Gaussian filtering proposed in the demons algorithm, an overview of which can be found in Cachier and Ayache (2004).

7.3 Viscous Fluid Modeling

One of the main issues with DIR is to define suitable sets of admissible transformation models, allowing one to represent the desired physiologic deformation. We previously showed that deformation fields could be a constraint with models based on continuum mechanics, particularly by considering the deforming material as linear elastic (Bajscy and Kovacic 1989). Strictly speaking, however, the equations derived are only valid for small deformations. An approach that can potentially circumvent this limitation is to consider material acting like a viscous fluid (Christensen et al. 1994; Bro-Nielsen and Gramkow 1996). In this approach, the transformation itself is not regularized, but its "evolution" is constrained to be smooth. Viscous fluid DIR is based on using the linear elastic constraint on the velocity field rather than on the deformation field. Hence, the partial differential equation (PDE) of the regularization function R for a viscous fluid model is similar to the linear elastic model but acts on the velocity field $\mathbf{v}$ (see Equation 7.14), where R_{LE} is the elastic energy of the deformation and E is the Green-St. Venant strain tensor (Equation (7.13)):

$$R_{LE}(\mathbf{u}) = \frac{1}{\lambda}(\mathrm{tr}E)^2 + \mu\,\mathrm{tr}E^2 \quad (7.12)$$

$$E = \frac{1}{2}(\nabla \mathbf{u} + \nabla \mathbf{u}^T) \qquad (7.13)$$

$$\nabla R_{\text{fluid}}(\mathbf{u}) = \nabla R_{\text{LE}}(\mathbf{v}) = (\lambda + \mu) \, \nabla(\nabla \mathbf{v}(\mathbf{x})) + \mu \Delta \mathbf{v}(\mathbf{x}) \qquad (7.14)$$

Using finite differences of the time derivative, we get the following rule for updating the deformation field from the velocity field at time (iteration) i: $u_{i+1}(\mathbf{x}) = \mathbf{u}_i(\mathbf{x}) + \nabla \varphi_i(\mathbf{x}) \mathbf{v}_i(\mathbf{x})$. To solve the PDE, Bro-Nielsen and Gramkow (1996) performed Euler integration over time using finite differences and derived a convolution filter allowing faster performance than previous approaches (Christensen et al. 1994). The Demons method has been shown to be an approximation of such a viscous fluid modeling.

7.4 Some Applications in Radiation Therapy

DIR is potentially useful for many applications related to radiation therapy, and we will list some examples of the use of nonparametric approaches. It should be noted that nonparametric approaches represent only a fraction of all different DIR types of algorithms proposed in this field. In particular, we do not describe parametric methods such as the popular B-spline-based approaches nor feature-based or hybrid methods relying on landmarks or segmentations.

One of the main interests of DIR in radiation therapy is related to the fraction-to-fraction variations in patient anatomy and setup. Such variations lead to dosimetric uncertainties, potentially leading to underdosage of the tumor and/or overdosage of healthy tissue. Due to this organ motion and to several other reasons, the delivered dose might not be identical to the predicted dose. Adaptive radiation therapy (ART) (Yan et al. 1997, 2005; Langen and Jones 2001) was developed to reduce these uncertainties. It relies on information obtained frequently over the course of treatment to make mid-course adjustments in the treatment plan that remains to be delivered. Reducing uncertainties would allow the reduction of margins, allowing the possibility of safe dose escalation and hopefully improvement in the treatment outcome. However, determining an ART strategy is a complex and time-consuming process. DIR can be used to (semi)automatically quantify image-to-image variations and represents a key enabling tool in image-guided ART.

7.4.1 Examples for the Prostate

Wang et al. (2005b) used a symmetric Demons approach to register prostate motion between two CT images acquired on different days to aid dose tracking, with an accuracy approximately 1 mm. Lu et al. (2004) proposed a nonparametric DIR method and applied it on daily CT images of prostate cancer patients. They also used an SSD criterion as dissimilarity measure and a regularization scheme based on the Frobenius's norm of the Jacobian, leading to an update scheme corresponding to the Laplacian operator (which can be viewed as related to the linear elastic operator R_{LE} with $\xi = 1$, without the cross-effect member).

It should be noted that several studies have shown that the deformation of prostate and seminal vesicles during the course of radiotherapy is small relative to organ motion (Deurloo et al. 2005; Kupelian et al. 2005). Hence, some authors suggest that "local" (on a user-defined region of interest) rigid registration can be sufficient (Smitsmans et al. 2004) even if significant deformation can occur in some cases (Kupelian et al. 2005).

7.4.2 Examples for the Lungs

Although the previous examples deal with interfraction motion, numerous studies applied DIR to deal also with intrafraction motion. In particular, registration of thoracic CT images to account for respiratory-induced motion has been studied extensively. Most authors (Fan et al. 2001; Li et al. 2003; Weruaga et al. 2003; Guerrero et al. 2004; Kaus et al. 2004; Lu et al. 2004) used SSD as dissimilarity measure and neglected the lung density variations due to breathing or proposed ad hoc approaches (Sarrut et al. 2006a; Yin et al. 2009). Sundaram and Gee (2005) used normalized cross-correlation on 2D magnetic resonance imaging (MRI) slices. Coselmon et al. (2004) used mutual information on lung images. Weruaga et al. (2003) computed a similarity measure that was a combination of cross-correlation and SSD. Keall et al. (2005) used a viscous fluid intensity-based approach (Christensen and Johnson 2001; Christensen et al. 2001) to map the transformation between a reference CT image (peak-inhale) to any other CT image in the 4D data set. The resulting deformation vector field allows the contours defined on one image to be automatically transferred to other phases images (eight images in total), thus allowing one to draw 4D contours. The previously described method from Lu et al. (2004) was also applied with thoracic images. Castillo et al. (2009a) tested their compressible optical flow method with thoracic images. Guerrero et al. (2005) described a method of quantifying regional lung ventilation to develop functional images for treatment planning and optimization. A diffusion-based DIR method (Guerrero et al. 2004) was used to obtain voxel correspondences, and local volume change due to inspiration was computed from the change in HU between corresponding voxels. Inversely, Sarrut et al. (2006a) generated intermediate voxel densities by taking into account the air volume change, for simulating a 4D image from two exhale–inhale breath-hold CT images.

7.4.3 Other Examples

Even with positron emission tomography (PET)–CT devices, functional PET images can be difficult to register with CT when deformation occurs (due to respiration, change in the position of the arms, etc.), leading to incorrect lesion volume estimation (Nehmeh et al. 2002). Shekhar et al. (2005) registered a whole-body functional PET with an anatomic CT to differentiate viable tumors from benign masses by using an elastic intensity-based approach with normalized mutual information. In this case, the transformation was derived from a combination of multiple local rigid body transformations. Another example deals with interpatient DIR, mainly for making anatomical atlases for automated organs segmentation. Li et al. (2003) built a normative atlas of the human lung from interpatient thoracic images with the method from Christensen and Johnson (2001). Bondiau et al. (2005) used a fluid approach to register brain MRI to another segmented MRI, allowing automatic delineation of brain structures.

7.5 Detailed Example of DIR for Radiation Therapy Planning in Lung Cancer

The goal of this section is to describe the use of the DIR method for a specific radiation therapy application related to lung cancer treatment.

Lung cancer. Lung cancer is the most common cause of death related to cancer for men and women worldwide (see Mathers and Loncar 2006). It is also the second most prevalent cancer after prostate cancer in men and breast cancer in women.

Lung cancer recently surpassed heart disease as the leading cause of smoking-related mortality. It is considered responsible for approximately 1.3 million deaths annually (in 2007). Considering all stages, the 5-year patient survival rates remain poor, approximately 8% in Europe and 14% in the United States. A distinction is made between small cell lung cancer (SCLC) and nonsmall cell lung cancer (NSCLC), the latter representing approximately 80% of lung cancer, with better survival rates for early stages. Whenever possible, in practice, for less than 25% of the cases, surgery is preferred as the primary treatment modality. In the remaining cases or in complement of surgery, chemotherapy and radiation therapy can be used.

DIR for thoracic CT images. In radiation therapy, DIR of CT images of the thorax has been used extensively for a variety of tasks (Kessler 2006; Sarrut et al. 2006b), including automatic contour propagation (Keall et al. 2005; Rietzel et al. 2005a; Boldea et al. 2006; Lu et al. 2006), 4D treatment planning (Keall 2004), dose deformation (as in Zhang et al. 2004), or combination of dose distributions (Brock et al. 2003; Rietzel et al. 2005) for lung or liver, quantification of residual motion in breath-hold

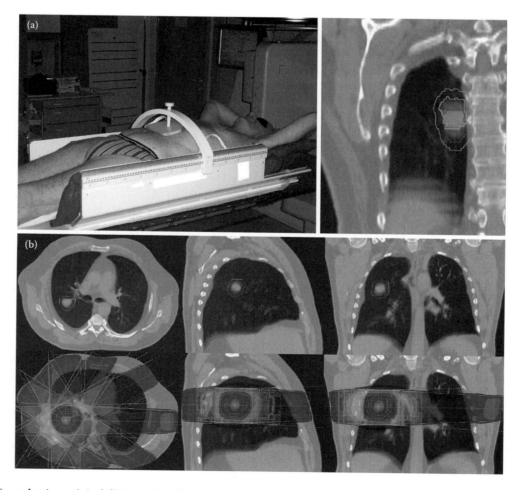

FIGURE 7.1 **(See color insert.)** (a, left) Example of lung cancer treatment with stereotactic body radiation therapy contention system and abdominal compression aiming at reducing motion amplitude. (a, right) Coronal slice showing contoured structures representing the tumor motion (exhale, inhale positions, and ITV). (b) Dose distribution obtained from treatment planning system with a 12-field plan.

CT scans (Sarrut et al. 2005), 4D dose estimation (Guerrero et al. 2005), dynamic ventilation imaging (Guerrero et al. 2006), construction of a mid-position reference planning image (Wolthaus et al. 2008), quantification of motion nonlinearity and hysteresis (Boldea et al. 2008), building respiratory motion models (McClelland et al. 2006), Monte Carlo simulations (Keall et al. 2004; Paganetti 2004; Wang et al. 2005a), and motion-compensated cone-beam reconstruction (Rit et al. 2009).

Scope of the example. This example deals with the Demons method in stereotactic body radiation therapy (SBRT). The goal was to allow clinicians to define personalized margins according to the specific tumor motion of each patient (Figure 7.1). To adapt tumor margins according to motion is a difficult task and is out of the scope of this section. Here, we only describe the first important task aiming at "quantitatively" evaluating the motion within the 4D CT image. DIR is intended to be as automated as possible; however, it remains necessary that experts perform careful verification. The procedure described in the following has also been used to evaluate the utility of diaphragmatic compression (Bouilhol et al. 2012).

7.5.1 Step 1: 4D CT Images of the Thorax

Several studies (e.g., Chen et al. 2004) show the impact of the breathing motion on CT image quality: artifacts can be observed when interference occurs between the organ motion and the scanner motion. Such artifacts compromise the reliable identification of tumor position, shape, and volume. To tackle this problem, several groups have proposed methods to acquire respiratory-correlated 4D CT scans (Low et al. 2003; Vedam et al. 2003; Keall 2004; Nehmeh et al. 2004; Pan et al. 2004; Rietzel et al. 2005b). Without going into details, the 4D CT acquisition process essentially consists of the acquisition of CT data throughout the respiratory cycle (Figure 7.2). 4D reconstruction relies on the simultaneous acquisition of a respiratory-correlated surrogate signal, based on which the acquired data are binned into consecutive respiratory frames. The resulting 4D image is composed of eight to 10 3D CT images representing different stages of the respiratory cycle. Here, we consider a 4D CT image of the thorax and the goal is to quantitatively and (mainly) automatically estimate motion within this data set (Figure 7.2).

7.5.2 Step 2: Image Preprocessing

Sliding motion issue. DIR can be used to estimate the motion of each frame of the 4D CT data set with respect to a reference frame. In the case of respiratory motion, the "sliding" of the lung along the chest wall leads to a discontinuity in the deformation field. As mentioned, DIR methods typically include a regularization mechanism that favors spatially smooth solutions, which renders it difficult to

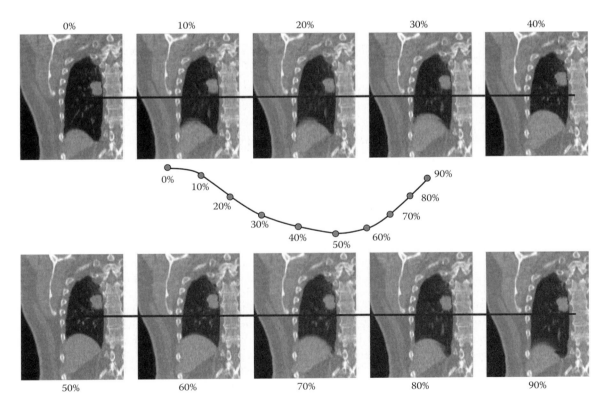

FIGURE 7.2   (See color insert.) Example of the 10 phases that compose a 4D CT image. (Reprinted from *Cancer Radiothérapie, 15*(2), Ayadi, M., Bouilhol, G., Imbert, L., Ginestet, C., and Sarrut, D., Scan acquisition parameter optimization for the treatment of moving tumors in radiotherapy, 115–122, Copyright 2010, with permission from Elsevier.)

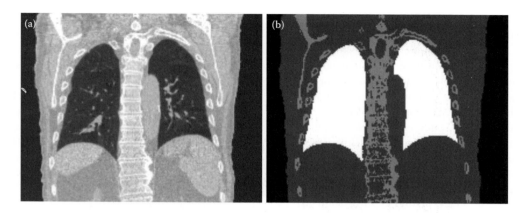

FIGURE 7.3 Example of an input CT image of the thorax (a) and the corresponding extracted features (b): the lungs (white), bony anatomy (light gray), and patient body (dark gray).

retrieve such discontinuous deformation fields (Wu et al. 2008; Schmidt-Richberg et al. 2009; Delmon et al. 2013). Note that similar sliding can also be observed for other organs (e.g., liver).

The issue of sliding motion in DIR has been addressed in a number of ways, including specifically designed regularization schemes (Ruan et al. 2008; Chun et al. 2009), tissue-dependent filtering (Wolthaus et al. 2008), finite element modeling (Al-Mayah et al. 2009; Werner et al. 2009b), and surface-based methods (Berg et al. 2007; Klinder et al. 2008). Another class of approaches consists of performing DIR separately on different anatomical regions that are assumed to move similarly (Rietzel and Chen 2006; Siebenthal et al. 2007; Wu et al. 2008; Werner et al. 2009a; Vandemeulebroucke et al. 2012). It thus requires the prior segmentation of the input image. As an example, we describe here a practical method (Vandemeulebroucke et al. 2012) for automatically dividing the thorax into regions with homologous, respiratory-induced motion. The main objective is to obtain an accurate interface where strong sliding motion occurs and to facilitate subsequent deformable registration.

Motion mask extraction. We divide the thorax into moving and less moving areas following the division proposed in Rietzel and Chen (2006). Note that this division relies on geometric and physiologic considerations rather than on organ boundaries. The core of the method is based on the level set framework of Osher and Sethian (1988), which allows incorporating geometric regularization in the segmentation procedure. The evolution of the level set is governed by a PDE in which two terms appear: the first corresponds to a propagation force, favoring an expansion or a contraction of the evolving contour. The other term corresponds to a local interface smoothing force. The motion mask is defined with respect to strong anatomical features. These features are incorporated in the algorithm as binary velocity maps defining two regions within the image: one in which the contour can evolve freely and one in which the contour is confined to its current position.

The velocity maps are obtained from the input image by extracting features through consecutive thresholding, region growing, and mathematical morphology. Three areas are segmented: the outer patient body contour, the bony anatomy, and the lungs (Figure 7.3). For the latter, the lung region is segmented as described in Hu et al. (2001) and Rikxoort et al. (2009). In brief, the lungs and airways are identified using region growing with a threshold obtained automatically by maximizing the separability between the considered regions (Otsu 1979). The trachea and large airways are extracted using explosion-controlled region growing (Mori et al. 1996) and removed from the result.

An initial small sphere is placed automatically within the patient's body at the upper abdomen. The level-set method is then used to progressively "inflate" the contour to fill the thoracic cavity, its evolution being guided by the previously extracted velocity maps. The obtained segmentation covers the patient's abdomen up until the anterior patient-to-air interface, the lungs, and the mediastinum. The final mask is illustrated in Figure 7.4.

7.5.3 Step 3: DIR

In this case, DIR of the 4D data set is performed as a series of consecutive 3D registrations from one reference frame to the remaining frames. Note that direct 4D approaches also have been proposed (Ledesma-Carbayo et al. 2005; Castillo et al. 2010; Vandemeulebroucke et al. 2011), which have the advantage of allowing temporal regularization of the deformation field. The Demons method (see Section 7.1.1) can be applied between each pair of images using the motion masks to modify the input images before the registration.

Figure 7.5 depicts two slices of the initial thoracic CT images of the exhale and inhale frames, with the deformation field for the lungs superimposed on the CT image (exhale frame). The deformation was computed with the Demons method and a motion mask between the end-exhale and end-inhale states. It is

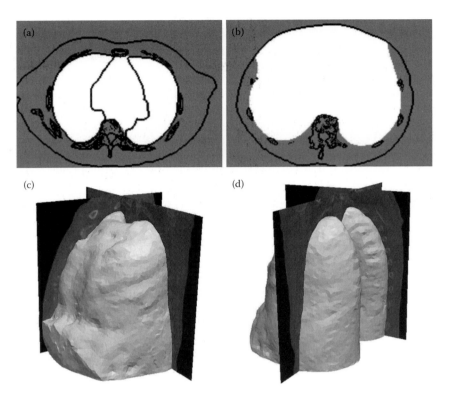

FIGURE 7.4 Two axial views of the mask: the first one (a) taken halfway the lungs and the second one (b) taken from the most inferior plane of the image. (c) and (d): anterior and posterior views of a 3D surface rendering of the motion mask. (From Vandemeulebroucke, J. et al., *Medical Physics*, 39(2), 1006–1015, 2012. With permission of the American Association of Physicists in Medicine.)

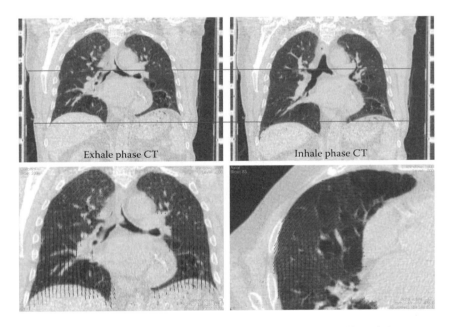

FIGURE 7.5 **(See color insert.)** Top: Initial exhale and inhale CT images to be registered. Red lines help to compare the two coronal slices. Bottom: Coronal and axial slices with deformation field superimposed. The vector field is only displayed in the lung region.

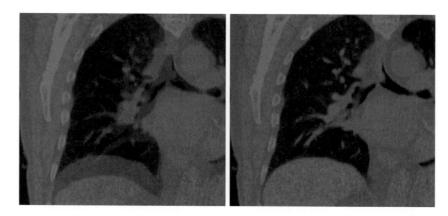

FIGURE 7.6 (**See color insert.**) Green–purple differences before and after registration.

used by medical physicists to design treatment margins personalized to the patient. By moving the mouse pointer on the image, the operator can instantly obtain the 3D displacement of any point in the lungs. The field is also used to automatically obtain contours on each phase or to derive nonisotropic margins to target volume. The vector field is only displayed on the lung region.

Figure 7.6 shows the differences between exhale and inhale images by overlay of the two images. Such overlays merge the two images as a function of the HU differences: if there are no differences, the initial gray level is displayed; if HU is greater in the first image, color tends to green; if HU is lower in the first image, the color tends to purple. On the left are the differences before the registration and on the right are the differences after the registration, once the inhale image has been warped with the deformation field using cubic B-spline interpolation.

7.5.4 Step 4: Validation

Once the deformation field has been computed, the last step consists of the validation of the results. Validation of DIR is known to be a challenging task because there is no standardized means of evaluating the results of a DIR method. Validation can be performed with synthetic simulated data, by checking consistency of the deformation field (Jannin et al. 2002a,b; Boldea et al. 2005), or by using phantom data (Wang et al. 2005c). With real images, the commonly used approach requires experts to manually define homologous landmarks in all images to be registered and to compare positions defined by experts to positions obtained with DIR (Sarrut et al. 2007; Brock and Deformable Registration Accuracy Consortium 2010). Selecting landmarks is a tedious task and specific graphical tools have been proposed (Murphy et al. 2008; Castillo et al. 2009b) to help experts in such a task. These semi-automatic methods help the observers to locate and identify corresponding anatomical features in each of the images.

The previously mentioned approaches do not allow validation on an individual patient basis in clinical routine. In practice, the application of DIR should therefore be thoroughly validated each time when applied to a new anatomical site, a different image modality, etc., during a prior pilot study. Afterward, for each additional patient processed, partial validation of the DIR result should be performed to reduce the risk of misregistration. Medical physicists can check the result of the DIR for the most critical regions by visually inspecting the deformation field using an adapted graphical user interface. Motion-compensated images, obtained by warping the 4D data set toward the reference using the found deformation fields, are a fast and simple way of verifying a result. It is a necessary condition that, for successful registrations, dynamic visualization of the motion-compensated sequence should still appear, with the exception of the acquisition noise present in the images. This procedure, however, does not give certainty on the quality of the registration, because the compensated images may seem visually acceptable, although the deformation field does not correspond to a physiologically admissible deformation.

7.6 Conclusion

Nonparametric DIR has been successfully applied to a variety of applications in the field of radiation therapy. The introduction of these methods into clinical routine should be done with caution. Custom-designed algorithms should be derived for the application at hand and be thoroughly validated for each situation separately. DIR is a fundamental image analysis tool for radiotherapy and will probably be included into all treatment planning systems in the near future.

Appendix 7.A Linear Elastic Operators through Finite Differences

For a 3D deformation field, the linear elastic regularization given in Equation 7.8 can be developed as in the following equation:

$$
\nabla R_{LE}(\mathbf{x},\mathbf{u}) = (\lambda+\mu)
\begin{pmatrix}
\dfrac{\partial^2 u_1}{\partial x_1^2}+\dfrac{\partial^2 u_2}{\partial x_1 \partial x_2}+\dfrac{\partial^2 u_3}{\partial x_1 \partial x_3} \\[2mm]
\dfrac{\partial^2 u_1}{\partial x_2 \partial x_1}+\dfrac{\partial^2 u_2}{\partial x_2^2}+\dfrac{\partial^2 u_3}{\partial x_2 \partial x_3} \\[2mm]
\dfrac{\partial^2 u_1}{\partial x_3 \partial x_1}+\dfrac{\partial^2 u_2}{\partial x_3 \partial x_2}+\dfrac{\partial^2 u_3}{\partial x_3^2}
\end{pmatrix}
+ \mu
\begin{pmatrix}
\dfrac{\partial^2 u_1}{\partial x_1^2}+\dfrac{\partial^2 u_1}{\partial x_2^2}+\dfrac{\partial^2 u_1}{\partial x_3^3} \\[2mm]
\dfrac{\partial^2 u_2}{\partial x_1^2}+\dfrac{\partial^2 u_2}{\partial x_2^2}+\dfrac{\partial^2 u_2}{\partial x_3^3} \\[2mm]
\dfrac{\partial^2 u_3}{\partial x_1^2}+\dfrac{\partial^2 u_3}{\partial x_2^2}+\dfrac{\partial^2 u_3}{\partial x_3^3}
\end{pmatrix}
\tag{7.A.1}
$$

It thus leads to the kernels given in Equation 7.A.2, which must be applied to the three deformation field components $\mathbf{u} = \{u_1, u_2, u_3\}$ (Hermosillo et al. 2002):

$$u_{i+1,1}(\mathbf{x}) = z:$$

Term 1 — $u_1(\mathbf{x})\mu$:

$$
z+1:\begin{bmatrix}0&0&0\\0&1&0\\0&0&0\end{bmatrix}\quad
z:\begin{bmatrix}0&1&0\\1&-6&1\\0&1&0\end{bmatrix}\quad
z-1:\begin{bmatrix}0&0&0\\0&1&0\\0&0&0\end{bmatrix}
$$

$+$

Term 2 — $u_1(\mathbf{x})(\lambda+\mu)$:

$$
z+1:\begin{bmatrix}0&0&0\\0&0&0\\0&0&0\end{bmatrix}\quad
z:\begin{bmatrix}0&0&0\\1&-2&1\\0&0&0\end{bmatrix}\quad
z-1:\begin{bmatrix}0&0&0\\0&0&0\\0&0&0\end{bmatrix}
$$

$+$

Term 3 — $u_2(\mathbf{x})\dfrac{1}{4}(\lambda+\mu)$:

$$
z+1:\begin{bmatrix}0&0&0\\0&0&0\\0&0&0\end{bmatrix}\quad
z:\begin{bmatrix}1&0&-1\\0&0&0\\-1&0&1\end{bmatrix}\quad
z-1:\begin{bmatrix}0&0&0\\0&0&0\\0&0&0\end{bmatrix}
$$

$+$

Term 4 — $u_3(\mathbf{x})\dfrac{1}{4}(\lambda+\mu)$:

$$
z+1:\begin{bmatrix}0&0&0\\-1&0&1\\0&0&0\end{bmatrix}\quad
z:\begin{bmatrix}0&0&0\\0&0&0\\0&0&0\end{bmatrix}\quad
z-1:\begin{bmatrix}0&0&0\\1&0&-1\\0&0&0\end{bmatrix}
$$

$$\tag{7.A.2}$$

$$u_{i+1,2}(\mathbf{x}) = z:$$

Term 1 — $u_2(\mathbf{x})\mu$:

$$
z+1:\begin{bmatrix}0&0&0\\0&1&0\\0&0&0\end{bmatrix}\quad
z:\begin{bmatrix}0&1&0\\1&-6&1\\0&1&0\end{bmatrix}\quad
z-1:\begin{bmatrix}0&0&0\\0&1&0\\0&0&0\end{bmatrix}
$$

$+$

Term 2 — $u_1(\mathbf{x})\dfrac{1}{4}(\lambda+\mu)$:

$$
z+1:\begin{bmatrix}0&0&0\\0&0&0\\0&0&0\end{bmatrix}\quad
z:\begin{bmatrix}1&0&-1\\0&0&1\\-1&0&0\end{bmatrix}\quad
z-1:\begin{bmatrix}0&0&0\\0&0&0\\0&0&0\end{bmatrix}
$$

$+$

Term 3 — $u_2(\mathbf{x})(\lambda+\mu)$:

$$
z+1:\begin{bmatrix}0&0&0\\0&0&0\\0&0&0\end{bmatrix}\quad
z:\begin{bmatrix}0&1&0\\0&-2&0\\0&1&0\end{bmatrix}\quad
z-1:\begin{bmatrix}0&0&0\\0&0&0\\0&0&0\end{bmatrix}
$$

$+$

Term 4 — $u_3(\mathbf{x})\dfrac{1}{4}(\lambda+\mu)$:

$$
z+1:\begin{bmatrix}0&-1&0\\0&0&1\\0&1&0\end{bmatrix}\quad
z:\begin{bmatrix}0&0&0\\0&0&0\\0&0&0\end{bmatrix}\quad
z-1:\begin{bmatrix}0&1&0\\0&0&0\\0&-1&0\end{bmatrix}
$$

$$\tag{7.A.3}$$

$$u_{i+1,1}(\mathbf{x}) = z :$$

$$
\begin{aligned}
z+1: \quad & \begin{bmatrix} 0 & 0 & 0 \\ 0 & 1 & 0 \\ 0 & 0 & 0 \end{bmatrix} && \begin{bmatrix} 0 & 0 & 0 \\ -1 & 0 & 1 \\ 0 & 0 & 0 \end{bmatrix} && \begin{bmatrix} 0 & -1 & 0 \\ 0 & 0 & 0 \\ 0 & 1 & 0 \end{bmatrix} && \begin{bmatrix} 0 & 0 & 0 \\ 0 & 1 & 0 \\ 0 & 0 & 0 \end{bmatrix} \\
z: \quad & \begin{bmatrix} 0 & 1 & 0 \\ 1 & -6 & 1 \\ 0 & 1 & 0 \end{bmatrix} + & & \begin{bmatrix} 0 & 0 & 0 \\ 0 & 0 & 0 \\ 0 & 0 & 0 \end{bmatrix} + & & \begin{bmatrix} 0 & 0 & 0 \\ 0 & 0 & 0 \\ 0 & 0 & 0 \end{bmatrix} + & & \begin{bmatrix} 0 & 0 & 0 \\ 0 & -2 & 0 \\ 0 & 0 & 0 \end{bmatrix} \\
z-1: \quad & \begin{bmatrix} 0 & 0 & 0 \\ 0 & 1 & 0 \\ 0 & 0 & 0 \end{bmatrix} && \begin{bmatrix} 0 & 0 & 0 \\ 1 & 0 & -1 \\ 0 & 0 & 0 \end{bmatrix} && \begin{bmatrix} 0 & 1 & 0 \\ 0 & 0 & 0 \\ 0 & -1 & 0 \end{bmatrix} && \begin{bmatrix} 0 & 0 & 0 \\ 0 & 1 & 0 \\ 0 & 0 & 0 \end{bmatrix} \\
& u_3(\mathbf{x})\mu && u_1(\mathbf{x})\tfrac{1}{4}(\lambda+\mu) && u_2(\mathbf{x})\tfrac{1}{4}(\lambda+\mu) && u_3(\mathbf{x})(\lambda+\mu)
\end{aligned}
\tag{7.A.4}
$$

References

Al-Mayah, A., Moseley, J., Velec, M., and Brock, K. K. (2009). Sliding characteristic and material compressibility of human lung: Parametric study and verification. *Medical Physics, 36*(10), 4625–4633.

Alvarez, L., Weickert, J., and Sánchez, J. (2000). Reliable estimation of dense optical flow fields with large displacements. *International Journal of Computer Vision, 39*(1), 41–56.

Ayadi, M., Bouilhol, G., Imbert, L., Ginestet, C., and Sarrut, D. (2010). Scan acquisition parameter optimization for the treatment of moving tumors in radiotherapy. *Cancer Radiothérapie, 15*(2), 115–122.

Bajscy, R. and Kovacic, S. (1989). Multiresolution elastic matching. *Computer Vision, Graphics, and Image Processing, 46*, 1–21.

Berg, J. von, Barschdorf, H., Blaffert, T., Kabus, S., and Lorenz, C. (2007). Surface based cardiac and respiratory motion extraction for pulmonary structures from multi-phase CT. *Proceedings of the SPIE, Medical Imaging 2007: Physiology, Function, and Structure from Medical Images,* eds. A. Manduca and X. P. Hu, 6511, 65110Y.

Boldea, V., Sarrut, D., and Carrie, C. (2005). Comparison of 3D dense deformable registration methods for breath-hold reproducibility study in radiotherapy. *SPIE Medical Imaging: Visualization, Image-Guided Procedures, and Display, 5747*, 222–230.

Boldea, V., Sharp, G., Jiang, S., Choi, N., Ginestet, C., Carrie, C., and Sarrut, D. (2006). Implementation and evaluation of automatic contour propagation in 4DCT of lung. *Medical Physics, 33*(6). In 48th American Association of Physicists in Medicine (AAPM) Annual Meeting, Orlando, FL, USA, 2019–2020.

Boldea, V., Sharp, G., Jiang, S. B., and Sarrut, D. (2008). 4D-CT lung motion estimation with deformable registration: Quantification of motion nonlinearity and hysteresis. *Medical Physics, 35*(3), 1008–1018.

Bondiau, P., Malandain, G., Chanalet, S., Marcy, P., Habrand, J., Fauchon, F., Paquis, P., Courdi, A. et al. (2005). Atlas-based automatic segmentation of MR images: Validation study on the brainstem in radiotherapy context. *International Journal of Radiation Oncology Biology Physics, 61*(1), 289–298.

Bouilhol, G., Ayadi, M., Rit, S., Thengumpallil, S., Schaerer, J., Vandemeulebroucke, J., Claude, L., and Sarrut, D. (2012). Is abdominal compression useful in lung stereotactic body radiation therapy? A 4DCT and dosimetric lobe-dependent study. *Physica Medica* (http://www.ncbi.nlm.nih.gov/pubmed/22617761).

Bro-Nielsen, M. and Gramkow, C. (1996). Fast fluid registration of medical images, in SPIE Visualization in Biomedical Computing, eds. Hone, K. and Kikinis, R., 1996, 1131, 267–276.

Brock, K. K. and Consortium, D. R. A. (2010). Results of a multi-institution deformable registration accuracy study (MIDRAS). *International Journal of Radiation Oncology Biology Physics, 76*(2), 583–596.

Brock, K., McShan, D., Ten Haken, R., Hollister, S., Dawson, L., and Balter, J. (2003). Inclusion of organ deformation in dose calculations. *Medical Physics, 30*(3), 290–295.

Cachier, P. and Ayache, N. (2004). Isotropic energies, filters and splines for vectorial regularization. *Journal of Mathematical Imaging and Vision, 20*(3), 251–265.

Cahill, N. D., Noble, J. A., and Hawkes, D. J. (2009). A Demons algorithm for image registration with locally adaptive regularization, in Medical Image Computing and Computer-Assisted Intervention, eds. G.-Z. Yang, et al., Lecture Notes in Computer Science, Vol. 5761, Springer, Heidelberg, 2009, 574–581.

Castillo, E., Castillo, R., Zhang, Y., and Guerrero, T. (2009a). Compressible image registration for thoracic computed tomography images. *Journal of Medical and Biological Engineering, 29*(5), 222–233.

Castillo, R., Castillo, E., Guerra, R., Johnson, V. E., McPhail, T., Garg, A. K., Guerrero, T. et al. (2009b). A framework for evaluation of deformable image registration spatial accuracy using large landmark point sets. *Physics in Medicine and Biology, 54*(7), 1849–1870.

Castillo, E., Castillo, R., Martinez, J., Shenoy, M., and Guerrero, T. (2010). Four-dimensional deformable image registration using trajectory modeling. *Physics in Medicine and Biology, 55*(1), 305–327.

Chefd'Hotel, C., Hermosillo, G., and Faugeras, O. (2001). A variational approach to multi-modal image matching, in *Proceedings of the IEEE Workshop on Variational and Level Set Methods,* IEEE Computer Society, Washington, DC, 21–28.

Chen, G., Kung, J., and Beaudette, K. (2004). Artifacts in computed tomography scanning of moving objects. *Seminars in Radiation Oncology, 14*(1), 19–26.

Christensen, G., Carlson, B., Chao, K., Yin, P., Grigsby, P., Nguyen, K., Dempsey, J. et al. (2001). Image-based dose planning of intracavitary brachytherapy: Registration of serial-imaging studies using deformable anatomic templates. *International Journal of Radiation Oncology Biology Physics, 51*(1), 227–243.

Christensen, G. and Johnson, H. (2001). Consistent image registration. *IEEE Transactions on Medical Imaging, 20*(7), 568–582.

Christensen, G. E., Rabbitt, R. D., and Miller, M. I. (1994). 3D brain mapping using a deformable neuroanatomy. *Physics in Medicine and Biology, 39*(3), 609–618.

Chun, S. Y., Fessler, J. A., and Kessler., M. L. (2009). A simple penalty that encourages local invertibility and considers sliding effects for respiratory motion. *Proceedings of the SPIE, Medical Imaging 2009: Image Processing, 7259,* 72592U.

Coselmon, M., Balter, J., McShan, D., and Kessler, M. (2004). Mutual information based CT registration of the lung at exhale and inhale breathing states using thin-plate splines. *Medical Physics, 31*(11), 2942–2948.

Delmon, V., Rit, S., Pinho, R., and Sarrut, D. (2013). Registration of sliding objects using direction dependent B-splines decomposition. *Physics in Medicine and Biology, 58*(5), 1303–1314.

Deriche, R. (1993). *Recursively Implementing the Gaussian and Its Derivatives.* Tech. Rep. 1893. Available from: http://www.inria.fr/rrrt/rr-1893.html.

Deurloo, K., Steenbakkers, R., Zijp, L., Bois J.A., de Nowak, P., Rasch, C., and van Herk, M. (2005). Quantification of shape variation of prostate and seminal vesicles during external beam radiotherapy. *International Journal of Radiation Oncology Biology Physics, 61*(1), 228–238.

Dru, F. and Vercauteren, T. (2009). An ITK implementation of the symmetric log-domain diffeomorphic Demons algorithm. *MIDAS Journal,* 1–10.

Fan, L., Chen, C., Reinhardt, J., and Ho man, E. (2001). Evaluation and application of 3D lung warping and registration model using HRCT images, in SPIE Medical Imaging, Vol. 4321, San Diego, CA, 2001, 234–243.

Gu, X., Pan, H., Liang, Y., Castillo, R., Yang, D., Choi, D., Castillo, E., Majumdar, A., Guerrero, T., and Jiang, S. B. (2010). Implementation and evaluation of various demons deformable image registration algorithms on a GPU. *Physics in Medicine and Biology, 55*(1), 207–219.

Guerrero, T., Sanders, K., Castillo, E., Zhang, Y., Bidaut, L., Pan, T., and Komaki, R. (2006). Dynamic ventilation imaging from four-dimensional computed tomography. *Physics in Medicine and Biology, 51*(4), 777–791.

Guerrero, T., Sanders, K., Noyola-Martinez, J., Castillo, E., Zhang, Y., Tapia, R., Guerra, R., Borghero, Y., and Komaki, R. (2005). Quantification of regional ventilation from treatment planning CT. *International Journal of Radiation Oncology Biology Physics, 62*(3), 630–634.

Guerrero, T., Zhang, G., Huang, T.-C., and Lin, K.-P. (2004). Intrathoracic tumour motion estimation from CT imaging using the 3D optical flow method. *Physics in Medicine and Biology 49*(17), 4147–4161.

Hermosillo, G., Chefd'hotel, C., and Faugeras, O. (2002). Variational methods for multimodal image matching. *International Journal of Computer Vision, 50*(3), 329–343.

Hermosillo-Valadez, G. (2002). Variational methods for multi-modal image matching. Doctoral dissertation. Universite de Nice-Sophia Antipolis.

Hu, S., Ho man, E., and Reinhardt, J. (2001). Automatic lung segmentation for accurate quantitation of volumetric X-ray CT images. *IEEE Transactions on Medical Imaging, 20*(6), 490–498.

Jannin, P., Fitzpatrick, M., Hawkes, D., Pennec, X., Shahidi, R., and Vannier, M. (2002a). Editorial: validation of medical image processing in image-guided therapy. *IEEE Transactions on Medical Imaging, 21*(11), 1445–1449.

Jannin, P., Fitzpatrick, J., Hawkes, D., Pennec, X., Shahidi, R., and Vannier, M.W. (2002b). Validation of medical image processing in image-guided therapy. *IEEE Trans. Med. Imaging, 21*(12), 1445–1449.

Kaus, M., Netsch, T., Kabus, S., Pekar, V., McNutt, T., and Fischer, B. (2004). Estimation of organ motion from 4D CT for 4D radiation therapy planning of lung cancer. In *Medical Image Computing and Computer-Assisted Intervention, Lecture Notes in Computer Science,* Vol. 3217, Springer-Verlag, Berlin, 1017–1024.

Keall, P. (2004). 4-Dimensional computed tomography imaging and treatment planning. *Seminars in Radiation Oncology, 14*(1), 81–90.

Keall, P., Joshi, S., Vedam, S., Siebers, J., Kini, V., and Mohan, R. (2005). Four-dimensional radiotherapy planning for DMLC-based respiratory motion tracking. *Medical Physics, 32*(4), 942–951.

Keall, P., Siebers, J., Joshi, S., and Mohan, R. (2004). Monte Carlo as a four-dimensional radiotherapy treatment-planning tool to account for respiratory motion. *Physics in Medicine and Biology, 49*(16), 3639–3648.

Kessler, M. L. (2006). Image registration and data fusion in radiation therapy. *British Journal of Radiology, 79*(Special No. 1), S99–S108.

Klinder, T., Lorenz, C., and Ostermann, J. (2008). Respiratory motion modeling and estimation, in First International Workshop on Pulmonary Image Analysis, eds. M. Brown, et al., New York, 2008, 53–62.

Kupelian, P., Willoughby, T., Meeks, S., Forbes, A., Wagner, T., Maach, M., and Langen, K. (2005). Intraprostatic fiducials for localization of the prostate gland: Monitoring inter-marker distances during radiation therapy to test for marker stability. *International Journal of Radiation Oncology Biology Physics, 62*(5), 1291–1296.

Langen, K. and Jones, D. (2001). Organ motion and its management. *International Journal of Radiation Oncology Biology Physics, 50*(1), 265–278.

Ledesma-Carbayo, M. J., Kybic, J., Desco, M., Santos, A., Sihling, M., Hunziker, P., and Unser, M. (2005). Spatio-temporal nonrigid registration for ultrasound cardiac motion estimation. *IEEE Transactions on Medical Imaging, 24*(9), 1113–1126.

Li, B., Christensen, G., Ho man, E., McLennan, G., and Reinhardt, J. (2003). Establishing a normative atlas of the human lung: Intersubject warping and registration of volumetric CT images. *Academic Radiology, 10*(3), 255–265.

Low, D. A., Nystrom, M., Kalinin, E., Parikh, P., Dempsey, J. F., Bradley, J. D., Mutic, S. et al. (2003). A method for the reconstruction of four-dimensional synchronized CT scans acquired during free breathing. *Medical Physics, 30*(6), 1254–1263.

Lu, W., Chen, M.-L., Olivera, G. H., Ruchala, K. J., and Mackie, T. R. (2004). Fast free-form deformable registration via calculus of variations. *Physics in Medicine and Biology, 49*(14), 3067–3087.

Lu, W., Olivera, G. H., Chen, Q., Chen, M.-L., and Ruchala, K. J. (2006). Automatic re-contouring in 4D radiotherapy. *Physics in Medicine and Biology, 51*(5), 1077–1099.

Mathers, C. D. and Loncar, D. (2006). Projections of global mortality and burden of disease from 2002 to 2030. *PLoS Med, 3*(11), e442.

McClelland, J. R., Blackall, J. M., Tarte, S., Chandler, A. C., Hughes, S., Ahmad, S., Landau, D. B., and Hawkes, D. J. (2006). A continuous 4D motion model from multiple respiratory cycles for use in lung radiotherapy. *Medical Physics, 33*(9), 3348–3358.

Milic-Emili, J., Henderson, J., Dolovich, M., Trop, D., and Kaneko, K. (1966). Regional distribution of inspired gas in the lung. *Journal of Applied Physiology, 21*(3), 749–759.

Monfraix, S., Bayat, S., Porra, L., Berruyer, G., Nemoz, C., Thomlinson, W., Suortti, P., and Sovijrvi, A. (2005). Quantitative measurement of regional lung gas volume by synchrotron radiation computed tomography. *Physics in Medicine and Biology, 50*, 1–11.

Mori, K., Hasegawa, J., Toriwaki, J., Anno, H., and Katada, K. (1996). Recognition of bronchus in three-dimensional X-ray CT images with application to virtualized bronchoscopy system. *Proceedings of the International Conference on Pattern Recognition, 3*, 528532.

Murphy, K., Ginneken, B. van, Pluim, J. P. W., Klein, S., and Staring, M. (2008). Semi-automatic reference standard construction for quantitative evaluation of lung CT registration. *Medical Image Computing and Computer-Assisted Intervention International Conference, 11*(Part 2), 1006–1013.

Nagel, H.-H. and Enkelmann, W. (1986). An investigation of smoothness constraints for the estimation of displacement vector fields from image sequences. *IEEE Transactions on Pattern Analysis and Machine Intelligence, 8*(5), 565–593.

Nehmeh, S., Erdi, Y., Ling, C., Rosenzweig, K., Squire, O., Braban, L., Ford, E. et al. (2002). Effect of respiratory gating on reducing lung motion artifacts in PET imaging of lung cancer. *Medical Physics, 29*(3), 366–371.

Nehmeh, S., Erdi, Y., Pan, T., Pevsner, A., Rosenzweig, K., Yorke, E., Mageras, G. et al. (2004). Four-dimensional (4D) PET/CT imaging of the thorax. *Medical Physics, 31*(12), 3179–3186.

Noe, K. O., Tanderup, K., Lindegaard, J. C., Grau, C., and Sorensen, T. S. (2008). GPU accelerated viscous-fluid deformable registration for radiotherapy. *Studies in Health Technology and Informatics, 132*, 327–332.

Osher, S. and Sethian, J. (1988). Fronts propagating with curvature dependent speed: Algorithms based on Hamilton-Jacobi formulations. *Journal of Computational Physics, 79*, 12–49.

Otsu, N. (1979). A threshold selection method from gray-level histograms. *IEEE Transactions on Systems, Man and Cybernetics, 9*(1), 62–66.

Paganetti, H. (2004). Four-dimensional Monte Carlo simulation of time-dependent geometries. *Physics in Medicine and Biology, 49*(6), N75–97.

Pan, T., Lee, T., Rietzel, E., and Chen, G. (2004). 4D-CT imaging of a volume influenced by respiratory motion on multi-slice CT. *Medical Physics, 31*(2), 333–340.

Pennec, X., Cachier, P., and Ayache, N. (1999). Understanding the Demon's algorithm: 3D non rigid registration by gradient descent, in Medical Image Computing and Computer-Assisted Intervention, eds. Taylor, C. and Colschester, A., Lecture Notes in Computer Science, Vol. 1679, Springer-Verlag, Cambridge, UK, 1999, 597–605.

Rietzel, E. and Chen, G. T. Y. (2006). Deformable registration of 4D computed tomography data. *Medical Physics, 33*(11), 4423–4430.

Rietzel, E., Chen, G., Choi, N., and Willet, C. (2005a). Four-dimensional image-based treatment planning: Target volume segmentation and dose calculation in the presence of respiratory motion. *International Journal of Radiation Oncology Biology Physics, 61*(5), 1535–1550.

Rietzel, E., Pan, T., and Chen, G. T. Y. (2005b). Four-dimensional computed tomography: Image formation and clinical protocol. *Medical Physics, 32*(4), 874–889.

Rikxoort, E. van, Hoop, B. de, Viergever, M., Prokop, M., and van Ginneken, B. (2009). Automatic lung segmentation from thoracic computed tomography scans using a hybrid approach with error detection. *Medical Physics, 36*(7), 2934–2947.

Rit, S., Sarrut, D., and Desbat, L. (2009). Comparison of analytic and algebraic methods for motion-compensated cone-beam CT reconstruction of the thorax. *IEEE Transactions on Medical Imaging, 28*, 1513–1525.

Ruan, D., Fessler, J. A., Balter, J. M., Berbeco, R. I., Nishioka, S., and Shirato, H. (2008). Inference of hysteretic respiratory tumor motion from external surrogates: A state augmentation approach. *Physics in Medicine and Biology, 53*(11), 2923–2936.

Sarrut, D., Boldea, V., Ayadi, M., Badel, J., Ginestet, C., Clippe, S., and Carrie, C. (2005). Nonrigid registration method to assess reproducibility of breath-holding with ABC in lung cancer. *International Journal of Radiation Oncology Biology Physics, 61*(2), 594–607.

Sarrut, D., Boldea, V., Miguet, S., and Ginestet, C. (2006a). Simulation of 4D CT images from deformable registration between inhale and exhale breath-hold CT scans. *Medical Physics, 33*(3), 605–617.

Sarrut, D., Boldea, V., Miguet, S., and Ginestet, C. (2006b). Simulation of four-dimensional CT images from deformable registration between inhale and exhale breath-hold CT scans. *Medical Physics, 33*(3), 605–617.

Sarrut, D., Delhay, B., Villard, P., Boldea, V., Beuve, M., and Clarysse, P. (2007). A comparison framework for breathing motion estimation methods from 4D imaging. *IEEE Transactions on Medical Imaging, 26*(12), 1636–1648.

Schmidt-Richberg, A., Ehrhardt, J., Werner, R., and Handels, H. (2009). Slipping objects in image registration: Improved motion field estimation with direction-dependent regularization. In *Medical Image Computing and Computer-Assisted Intervention*, eds. Yang, G.-Z. et al., Lecture Notes in Computer Science, Vol. 5761, Springer, Heidelberg, 755–762.

Sharp, G. C., Kandasamy, N., Singh, H., and Folkert, M. (2007). GPU-based streaming architectures for fast cone-beam CT image reconstruction and demons deformable registration. *Physics in Medicine and Biology, 52*(19), 5771–5783.

Shekhar, R., Walimbe, V., Raja, S., Zagrodsky, V., Kanvinde, M., Wu, G., and Bybel, B. (2005). Automated 3-dimensional elastic registration of whole-body PET and CT from separate or combined scanners. *Journal of Nuclear Medicine, 46*(9), 1488–1496.

Siebenthal, M. von, Székely, G., Gamper, U., Boesiger, P., Lomax, A., and Cattin, P. (2007). 4D MR imaging of respiratory organ motion and its variability. *Physics in Medicine and Biology, 52*(6), 1547–1564.

Smitsmans, M. H. P., Wolthaus, J. W. H., Artignan, X., Bois, J. de, Ja ray, D. A., Lebesque, J. V., and van Herk, M. (2004). Automatic localization of the prostate for on-line or off-line image-guided radiotherapy. *International Journal of Radiation Oncology Biology Physics, 60*(2), 623–635.

Sundaram, T. A. and Gee, J. C. (2005). Towards a model of lung biomechanics: Pulmonary kinematics via registration of serial lung images. *Medical Image Analysis, 9*(6), 524–537.

Thirion, J. (1996). Non-rigid matching using demons, in IEEE Computer Vision and Pattern Recognition, San Francisco, CA.

Thirion, J. (1998). Image matching as a diffusion process: An analogy with Maxwell's demons. *Medical Image Analysis, 2*(3), 243–260.

Vandemeulebroucke, J., Bernard, O., Kybic, J., Clarysse, P., and Sarrut, D. (2012). Automated segmentation of a motion mask to preserve sliding motion in deformable registration of thoracic CT. *Medical Physics, 39*(2), 1006–1015.

Vandemeulebroucke, J., Rit, S., Kybic, J., Clarysse, P., and Sarrut, D. (2011). Spatiotemporal motion estimation for respiratory-correlated imaging of the lungs. *Medical Physics, 38*(1), 166–178.

Vedam, S. S., Keall, P. J., Kini, V. R., Mostafavi, H., Shukla, H. P., and Mohan, R. (2003). Acquiring a four-dimensional computed tomography dataset using an external respiratory signal. *Physics in Medicine and Biology, 48*(1), 45–62.

Vercauteren, T. (2008). Image registration and mosaicing for dynamic in vivo fibered confocal microscopy. Doctoral dissertation. Ecole des Mines de Paris/INRIA Sophia-Antipolis.

Vercauteren, T., Pennec, X., Malis, E., Perchant, A., and Ayache, N. (2007). Insight into efficient image registration techniques and the demons algorithm. *Information Processing in Medical Imaging, 20*, 495–506.

Vercauteren, T., Pennec, X., Perchant, A., and Ayache, N. (2009). Diffeomorphic demons: Efficient non-parametric image registration. *Neuroimage, 45*(1 Suppl), S61–S72.

Wang, B., Goldstein, M., Xu, X., and Sahoo, N. (2005a). Adjoint Monte Carlo method for prostate external photon beam treatment planning: An application to 3D patient anatomy. *Physics in Medicine and Biology, 50*(5), 923–935.

Wang, H., Dong, L., Lii, M., Lee, A., de Crevoisier, R., Mohan, R., Cox, J., Kuban, D., and Cheung, R. (2005b). Implementation and validation of a three-dimensional deformable registration algorithm for targeted prostate cancer radiotherapy. *International Journal of Radiation Oncology Biology Physics, 61*(3), 725–735.

Wang, H., Dong, L., O'Daniel, J., Mohan, R., Garden, A., Ang, K., Kuban, D., Bonnen, M., Chang, J., and Cheung, R. (2005c). Validation of an accelerated "Demons" algorithm for deformable image registration in radiation therapy. *Physics in Medicine and Biology, 50*(12), 2887–2905.

Werner, R., Ehrhardt, J., Schmidt-Richberg, A., and Handels, H. (2009a). Validation and comparison of a biophysical modeling approach and nonlinear registration for estimation of lung motion fields in thoracic 4D CT data. *Proceedings of SPIE-The International Society for Optical Engineering.* Vol. 7259.

Werner, R., Ehrhardt, J., Schmidt, R., and Handels, H. (2009b). Patient-specific finite element modeling of respiratory lung motion using 4D CT image data. *Medical Physics, 36*(5), 1500–1511.

Weruaga, L., Morales, J., Nunez, L., and Verdu, R. (2003). Estimating volumetric motion in thorax with parametric matching constraints. *IEEE Transactions on Medical Imaging, 22*(6), 766–772.

Wolthaus, J. W. H., Sonke, J. J., Herk, M. van, and Damen, E. M. F. (2008). Reconstruction of a time-averaged midposition CT scan for radiotherapy planning of lung cancer patients using deformable registration. *Medical Physics, 35*(9), 3998–4011.

Wu, Z., Rietzel, E., Boldea, V., Sarrut, D., and Sharp, G. C. (2008). Evaluation of deformable registration of patient lung 4DCT with subanatomical region segmentations. *Medical Physics, 35*(2), 775–781.

Yan, D., Vicini, F., Wong, J., and Martinez, A. (1997). Adaptive radiation therapy. *Physics in Medicine and Biology, 42*(1), 12–32.

Yan, D., Lockman, D., Martinez, A., Wong, J., Brabbins, D., Vicini, F., Liang, J., and L., K. (2005). Computed tomography guided management of interfractional patient variation. *Seminars in Radiation Oncology, 15*(3), 168–179.

Yang, D., Li, H., Low, D. A., Deasy, J. O., and Naqa, I. E. (2008). A fast inverse consistent deformable image registration method based on symmetric optical flow computation. *Physics in Medicine and Biology, 53*(21), 6143–6165.

Yin, Y., Ho man, E. A., and Lin, C.-L. (2009). Mass preserving nonrigid registration of CT lung images using cubic B-spline. *Medical Physics, 36*(9), 4213–4222.

Zhang, T., Jeraj, R., Keller, H., Lu, W., Olivera, G., McNutt, T., Mackie, T., and Paliwal, B. (2004). Treatment plan optimization incorporating respiratory motion. *Medical Physics, 31*(6), 1576–1586.

Zhao, Y. and Johnson, H. (2009). Diffeomorphic Demons registration with mask filter. *MIDAS Journal*. Available from: http://hdl.handle.net/10380/3105.

Optimization in Image Registration

Sanjiv S. Samant
University of Florida

Jian Wu
University of Florida

Junyi Xia
*University of Iowa
Hospitals and Clinics*

Arun Gopal
New York Presbyterian Hospital

8.1 Role Image Registration in Radiation Therapy

Image registration is used to determine the spatial relationship between two image data sets. Images acquired by the same (i.e., intramodal) or different (i.e., intermodal) imaging modalities can be used for (i) the sole purpose of determining the spatial transformation between two imaging data sets usually differing temporally or (ii) combining information from two different image data sets differing temporally or by modality. In radiation therapy, image registration is used in treatment planning, treatment assessment, and patient-position verification. In treatment planning, image registration is used to align multiple image data sets (both intramodal and intermodal) to assist the physician in contouring anatomical structures and planning target volumes. In treatment assessment, intermodal imaging [e.g., computed tomography (CT) and cone-beam computed tomography (CBCT)] can be used to assess dose delivery in terms of intended dose (i.e., treatment planning) and delivered dose (i.e., treatment delivery). In patient-position verification, image registration is used to verify patient positioning before or during treatment. Rigid registration involves a single global displacement vector and/or rotational transformation applied to an entire image data set. It is used in treatment planning, treatment assessment, and patient-position verification. Deformable registration involves local transformations applied to an image data set in cases where

the patient may have undergone physical changes that render a rigid registration anatomically inadequate. It is used in treatment planning and treatment assessment. Whether rigid or deformable, image registration is the process of combining different imaging data sets onto a common coordinate system where there is a mapping between the voxels (i.e., pixels) of the two data sets. This process involves a fixed or reference data set (i.e., image) and a moving or target data set. The target is aligned to the reference data set. (It should be noted that in literature that one sometimes finds the terminology of source and target sets used, with the source indicating the moving image and the target indicating the fixed or reference image. We will not be using this terminology here.) Rigid registration involves a single displacement vector, consisting of three orthogonal translations, and three angular rotations, which are globally applied to the target image so that the same transformation is applied to each voxel in the target image. Deformable registration involves aligning the target image to the reference image, whereby each voxel is assigned a displacement vector; this displacement vector can be different for the neighboring voxels so that the shape and volume can be altered.

8.2 Image Registration

As indicated in Figure 8.1, image registration is based on up to four components: similarity measures, transformation models, regularization, and optimization.

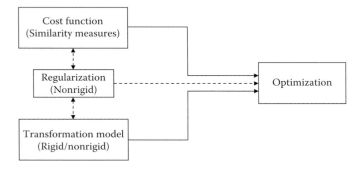

FIGURE 8.1 Image registration. Component diagram of image registration. Solid lines represent required input/output, whereas broken lines represent optional input/output depending on the registration problem.

8.2.1 Similarity Measures

Similarity measures compute the correspondence between two images. Image similarity is defined as mathematical or statistical measurements of the intensity distribution of the reference and target images that provide a quantitative representation of the similarity between two images. Similarity measures are algebraically incorporated within a scalar cost function, which is then used to evaluate the goodness of the registration. The cost function is used to determine the transformation between the target image and the reference image. The value of the cost function, for which the target and reference images are inputs, is used to guide the registration. Typically, the registration process is iterative, and a user-defined threshold for the cost function can serve as a termination point for the process.

Similarity measures, which are discussed in more detail in Chapter 4, are categorized into three groups here.

(1) *Feature-based measures, including points, curves, and surfaces.* Incorporating features is more likely to yield a registration that is anatomically valid and consistent with the underlying anatomical content of the imaging and physiologic constraints. The challenge for registrations using feature-based measures is to ensure minimal user intervention in feature selection and to ensure robustness of the registration result (i.e., small differences in feature selection do not lead to large differences in the registration).

(2) *Intensity-based measures.* These measures use the intensity assigned to each voxel. Imaging from the same modality (i.e., intramodal) will yield similar intensity values for voxels associated within the same anatomical structure. Imaging from different modalities (i.e., intermodal) will yield different intensity values for voxels within the same anatomical structure because the respective intensities represent different physiologic information. Intensity-based registrations can be used for intramodal and intermodal image registration as long as the two imaging data sets share common anatomical information. This anatomical information can have different voxel intensities

in each data set but should have similar intensity distribution based on the presence of different anatomical structures. Because fully automatic registrations usually include intensity-based measures, human intervention is reduced. The challenge for intensity-based measures is to ensure sufficient accuracy because, generally, there are no physiologically driven constraints to supervise the registration process.

(3) *Hybrid measures.* Such measures combine both the feature-based measures and the intensity-based measures. Registrations using hybrid models cannot be readily made fully automatic but can be used for situations where registration accuracy is more important than computational efficiency.

8.2.2 Transformation Models

Transformation models are required to transform or map the target image to match the reference image (i.e., how to move or deform the image content to improve the similarity between the two images). For rigid registration, the transformation model consists of up to six parameters: three translations and three rotations, which are globally applied to the voxels in the target image. In the local transformations encountered in deformable image registration, each voxel can move with a different displacement vector than that of its neighbors. The transformation model can be used to associate a displacement vector with each imaging voxel or with a subset of the original set of imaging voxels for increased computational efficiency and/or robustness of the registration. Examples of transformation models in deformable registration include (i) B-spline (Bookstein 1991; Xie and Farin 2004), which involves selecting a subset of voxels (i.e., nodes) to compute the registration and using interpolation to compute the displacements of the neighboring voxels, and (ii) dense-field models for which the displacement of every voxel in the target image is computed within the registration.

8.2.3 Regularization

Because the transformation model in rigid registration uses global parameters, the rigidity of the target image is maintained. However, in deformable image registration, this rigidity is lost. Hence, it leads to a problem peculiar to deformable registration: the transformation parameters that result in good registration based on computation of the similarity metric may not be physically meaningful (Crum et al. 2004). A common example of this problem is the change of deformation-field topology through tearing or folding to achieve a registration; however, these actions may result in transformation parameters that enhance similarities between two images (i.e., lower cost function value), although the resulting transformation is not physiologically valid. Therefore, relying on the cost function alone might not be a sufficient indication of good transformation parameters. Regularization can be used to ensure that the resulting transformation parameters are physically valid. The "regularizer" allows for computation of a solution to an ill-posed or ill-conditioned registration problem. The ill-posed aspect of image registration results in instability in the computation of the transformation model parameters, which is the result of having a highly nonconvex cost function. The regularizer seeks to restrict the transformations allowed, yielding a physiologically meaningful solution.

Image registration can be treated as a transformation or mapping, whereby, in the mapping $f:T \rightarrow R$, the target image (T) is made similar to the reference image (R). The energy or cost function in the optimization has two components: a similarity metric and a regularizer. The similarity metric, which drives the registration or the mapping f, describing the source and target image correspondence, ensures that the image $f(T)$ will be "similar" or equivalent to R. The regularization component determines the transformation function and is required to constrain the deformation field so that it is physically meaningful; thus, shearing and folding transformations are not used for anatomical structures that are topologically invariant in the reference and target imaging. The goal is to find the optimal deformation field u to minimize the energy function $E(u)$:

$$E(u) = E_{\text{sim}}(u) + \lambda E_{\text{reg}}(u)$$
$$u = \arg\min E(u) \tag{8.1}$$

where λ is the weighting factor for weighting the similarity and regularization terms. In this particular formulation, a smaller value of the similarity term indicates better matching. For certain similarity measures, such as cross-correlation, where larger values indicate better matching, the similarity term in Equation 8.1 can be represented by the negative of the similarity measure. This equation can be explained in a Bayesian framework; the similarity measure term acts like a likelihood term that expresses the probability of the source and target images matching. The regularization term serves to constrain the value of u based on *a priori* physiologic knowledge. Another way to understand Equation 8.1 is to consider the similarity term as a driving force that maximizes the correspondence between the target and reference images and the regularization term as a penalty function to constrain the transformation. In practice, this usually leads to preferential selection of small deformations over large deformations for approximately equivalent values of E_{sim}.

Alternatively, regularization can also be achieved by smoothing the deformation field u on an iterative basis, as in the case of optical flow models (Horn and Schunck 1981). The smoothing ensures minimization of variations in the displacement fields of neighboring voxels (i.e., the gradient across the deformation field is minimized):

$$E_{\text{reg}}(u) = \|\nabla u\|^2. \tag{8.2}$$

Regularizations can also be achieved using physical models. Linear elastic regularization (Bajcsy and Kovačič 1989; Christensen and Johnson 2001) assumes that the deformation is governed by the Navier equation from elasticity theory and can be written as

$$E_{\text{reg}}(u) = \mu\nabla^2 u + (\lambda + \mu) \nabla (\nabla u), \tag{8.3}$$

where ∇^2 is the Laplacian operator, ∇ is the gradient operator, ∇u is the divergence of u, and λ and μ are the Lamé parameters (i.e., characterize linear elasticity of a material). The elastic regularization assumes a small deformation. To handle large deformations, viscous fluid regularization (Bookstein 1991; Bro-Nielsen and Gramkow 1996; Freeborough and Fox 1998) has been considered. Viscous fluid regularization constrains the deformation analogous to the flow of a viscous fluid. It can be represented as follows:

$$\begin{aligned} &\mu\nabla^2 v + (\lambda + \mu)\nabla(\nabla v) \\ &v\frac{\partial u}{\partial t} + v\nabla u \end{aligned}, \tag{8.4}$$

where v is the velocity of the deformation. Because viscous fluid regularization allows large deformations, the deformation field must be checked to eliminate folding. Other regularizations include biharmonic (Bookstein 1991) and membrane (Terzopoulos 1986) models. Regularization techniques are described in detail in Chapters 6 and 7.

8.2.4 Optimization

After constructing a cost function and selecting the transformation model (which may require regularization) appropriate for the image registration problem, the final step is to obtain the transformation parameters that yield the best or optimal registration. Mathematically, this problem can be stated as follows: given a cost function f and unknown transformation parameters, find the optimal set of parameters that maximize (or minimize) the cost function. Generally, the choice of whether to maximize or minimize the cost function is trivial because maximizing f is equivalent to minimizing $-f$. Typically, except for the simplest test cases, registrations cannot be computed analytically; thus, a computational method is used. Optimization strategies, which seek to determine the parameters of the transformation model to maximize or minimize the cost function, are selected based on accuracy, computational efficiency, and robustness. A termination condition is often used in conjunction with a computational method because, in practice, the extremum is not precisely known and must be estimated. Due to the empirical nature of the termination condition, a simple optimization method, which yields a similar but not necessarily the best parameters compared with a more complex method, is usually preferred in radiation therapy applications due to the practical limits of registration accuracy in clinical implementation.

In radiation therapy, three cases of rigid image registration can be used for patient positioning: fiducial marker registration, registration based on volumetric imaging, and feature-based (i.e., curves) registration. Deformable registration, which is further discussed in Section 8.5, can also arise and affect patient position. However, it cannot be corrected through couch movement alone. In current clinical practice, deformable registration is used to estimate the corresponding effect on the dose distribution for the target and the normal/critical structures. In the case of fiducial marker registration, the cost function is the Euclidean distance between the markers, which typically number ≥3 but are still much fewer than the number of features in an image. In 3D, the set of transformations consists of three independent (orthogonal in the simplest case) translations and three independent rotations. For the case where an exact solution exists, Gaussian elimination can be used to determine the three translations and three angular rotations. In the case where an exact solution does not exist (i.e., due to fiducial marker migration or the lack of one-to-one identification between T and R, it may not be possible to align the markers from T onto R with zero residual error), a computational method such as singular value decomposition can be used to obtain a solution. Because of imperfect fiducial marker localization or marker migration, the Gaussian elimination method is not the optimal choice. Instead, given a corresponding point set, the transformation parameters can be calculated in a least-squares sense. Technically, this is equivalent to the orthogonal Procrustes optimal fitting problem, which can be solved using singular value decomposition (Schonemann 1966; Umeyama 1991) or general optimization techniques.

8.3 Review of Optimization Strategies for Rigid Image Registration

Solutions for many registration algorithms can be derived using existing optimization strategies, such as orthogonal Procrustes, downhill simplex method, (steepest) gradient descent method, conjugate gradient methods, quasi-Newton methods, and least-squares method. Press et al. (1992) provide an excellent review of wide variety of optimization methods. Here, we will review a few optimization methods commonly used in rigid registration in radiation therapy.

8.3.1 Orthogonal Procrustes Algorithm

An analytical solution to the orthogonal Procrustes problem was given by Schonemann (1966) and later applied to image registration. In the language of image registration, the algorithm can be described as follows.

Given two coordinate sets, A and B, find the orthogonal transformation matrix T so that the sums of squares of the residual matrix $E = AT − B$ [i.e., trace(EE^t)] is a minimum.

(i)   Find the centroids A_c and B_c of A and B, respectively.
(ii) Center the two coordinate systems $A' = A − A_c$ and $B' = B − B_c$.
       This step defines the translation $(B_c − A_c)$ between the two coordinate systems.
(iii) Calculate the singular value decomposition of $L = A'^t B'$.

(iv) The transformation and the root-mean-square (RMS) error are then given by

$$\begin{aligned} T &= U^t V, \\ E &= B' − A'T. \end{aligned} \tag{8.5}$$

This algorithm is fast and returns the RMS distance between the matching coordinate pairs. To determine the six independent Euler transformation parameters in the 3D space, at least six independent equations are required; thus, at least three fiducial marker pairs must be present.

Alternatively, instead of using features or points in the rigid registration, as in above case of the Procrustes algorithm, one can also carry out a rigid registration using the entire volume or suitably defined region-of-interest (ROI). In this case, the cost function can be the cross-correlation function (typically where T and R are from the same imaging modality) or mutual information (for the case where the imaging modalities may be different). Here, all voxels in the ROI are used to determine the transformation parameters, and the registration problem can involve 10^5 to 10^7 voxels. This approach is increasingly popular in radiation therapy due to the simplicity of requiring limited user input.

8.3.2 Downhill Simplex Method

The downhill simplex method, or the Nelder and Mead (1965) method, is a commonly used nonlinear optimization technique that is relatively easy to implement and does not require the gradient of the cost function to be calculated. A simplex is defined as a special polytope of $N + 1$ vertices in the N-dimensional parametric space. The algorithm starts with the initialization of a simplex with heuristically chosen vertices. Reflection, expansion, or contraction steps are then used to iteratively deform this simplex to move its vertices toward the minimum of the cost function. Convergence is declared when the fractional difference between the lowest and the highest function values evaluated at the vertices of the simplex is smaller than a given threshold.

The downhill simplex method usually requires more iterations compared with many other gradient-based nonlinear optimization methods. However, in practice, this method could be comparable or even better than the other methods because it does not require the calculation of the gradient, which is usually computationally expensive (Maes et al. 1999).

8.3.3 Gradient Descent Method

The gradient descent or steepest descent method is the most straightforward way to incorporate the gradient information into the optimization process. For a given point in the parametric space, the optimizer takes a step in the direction of the negative gradient, which leads to the fastest local decrease of the cost function value. For a real N-variable function $f(x)$, the optimization starts at point x_0. As many times as needed, the iteration moves from x_k to x_{k+1} by minimizing along the line from x_k

in the direction of the local downhill gradient $-\nabla f(x_k)$, which is given by

$$x_{k+1} = x_k - \alpha_k \nabla f(x_k), \qquad (8.6)$$

where α_k is the step size to minimize $f(x)$ along the searching line.

The gradient descent method is generally not an efficient searching algorithm. The line-search direction along the gradient does not necessarily point to the optimum even when the objective function is quadratic. Because consecutive steps are perpendicular to each other, many small steps are required to reach the optimum for a function that has a long and narrow valley.

8.3.4 Conjugate Gradient Methods

The conjugate gradient method tries to overcome the problems associated with the gradient descent method by proceeding along a number of conjugate directions instead of gradient directions. If one performs successive line minimizations of a function along a conjugate set of directions, it is not necessary to redo any of those directions. Thus, the optimization converges quadratically to the minimum.

As an iteration method, the conjugate gradient method initializes the search direction d_0 to the gradient vector g_0 at point x_0. In each iteration k, a line minimization is performed in the direction d_k, starting at point x_k, and leading to point x_{k+1}. Thereafter, a new direction d_{k+1} is constructed based on the current direction and the gradient information, which is

$$d_{k+1} = g_{k+1} + \gamma_k d_k, \qquad (8.7)$$

where g_{k+1} is the gradient evaluated at point x_{k+1}. A number of updating schemes to calculate γ_k have been proposed. Two of the best known formulas are the Fletcher and Reeves (1964) (FR) scheme:

$$\gamma_k = \frac{g_{k+1}^T g_{k+1}}{g_k^T g_k} \qquad (8.8a)$$

and the Polak–Ribiere (PR) scheme (Polak 1971):

$$\gamma_k = \frac{(g_{k+1} - g_k)^T g_{k+1}}{g_k^T g_k}. \qquad (8.8b)$$

For quadratic objective functions, these two formulas are equivalent. For nonquadratic functions, the choice of the formula may be determined heuristically.

8.3.5 Quasi-Newton Methods

The quasi-Newton method is based on Newton's method for determining the stationary point of a function where the gradient is zero. In Newton's method, the function value at point $(x_k + p)$

of an N-variable function $f(x)$ can be approximated by its Taylor series:

$$f(x_k + \Delta x) \approx f(x_k) + \nabla f(x_k)^T \Delta x + \frac{1}{2} \Delta x^T B \Delta x, \qquad (8.9)$$

where $\nabla f(x_k)$ is the gradient of $f(x)$ evaluated at x_k, B is the Hessian matrix, and T is the matrix transpose operator. The Hessian B matrix is a matrix of second-order partial derivatives of the function $f(x)$. Hence, Newton's method requires explicit computation of second-order derivatives, which can be computationally expensive. This equation is a quadratic model of the function $f(x)$ whose minimum can be calculated explicitly by computing the gradient of $f(x_k + \Delta x)$ with respect to Δx and setting the gradient to zero. Thus, the exact location x^*, for which $f(x)$ is the minimum, is given by

$$x^* = x_k - B^{-1} \nabla f(x_k). \qquad (8.10)$$

In the quasi-Newton method, the minimizer x^* is used as the search direction at each iteration. The parameter vector is updated as

$$x_{k+1} = x_k - \alpha_k B_k^{-1} \nabla f(x_k), \qquad (8.11)$$

where the step length α_k is chosen to satisfy the Wolfe conditions (Nocedal and Wright 1999). Instead of calculating the true Hessian, as in Newton's method, the approximate of the Hessian matrix B_k or the inverse $H_k = B_k^{-1}$ is updated at each iteration. Initially, B_0 is set to the identity matrix. Then, at each iteration, B_k and H_k are updated using the point-shift vector $\Delta x_k = x_{k+1} - x_k$ and the gradient difference $y_k = \nabla f(x_{k+1}) - \nabla f(x_k)$. Two popular updating strategies were proposed by Davidon–Fletcher–Powell (DFP) and by Broyden–Fletcher–Goldfarb–Shanno (BFGS) (Nocedal and Wright 1999). The DFP formula is

$$B_{k+1} = \left(I - \frac{y_k \Delta x_k^T}{y_k^T \Delta x_k} \right) B_k \left(I - \frac{\Delta x_k y_k^T}{y_k^T \Delta x_k} \right) + \frac{y_k y_k^T}{y_k^T \Delta x_k} \qquad (8.12)$$

$$H_{k+1} = H_k + \frac{\Delta x_k \Delta x_k^T}{y_k^T \Delta x_k} - \frac{H_k y_k y_k^T H_k^T}{y_k^T H_k y_k}. \qquad (8.13)$$

The BFGS formula is

$$B_{k+1} = B_k + \frac{y_k y_k^T}{y_k^T \Delta x_k} - \frac{B_k \Delta x_k (B_k \Delta x_k)^T}{\Delta x_k^T B_k \Delta x_k} \qquad (8.14)$$

$$H_{k+1} = \left(I - \frac{y_k \Delta x_k^T}{y_k^T \Delta x_k} \right) H_k \left(I - \frac{y_k \Delta x_k^T}{y_k^T \Delta x_k} \right) + \frac{\Delta x_k \Delta x_k^T}{y_k^T \Delta x_k}. \qquad (8.15)$$

Thus, in the quasi-Newton method, the Hessian is approximated using an iterative derivation that does not involve computationally expensive second-order derivatives, as in the case of Newton's method. For quadratic functions, both Newton and quasi-Newton methods converge to the exact Hessian B or its inverse H in N iterations, thereby requiring N gradient evaluations in all to arrive at the exact minimum of $f(x)$. Usually, Newton's method is preferred for quadratic cost functions, because it converges much faster, making up for the extra computation time of computing second-order derivatives. However, medical imaging typically involves cost functions with nonquadratic (i.e., third-order and higher) cost functions. For nonquadratic cost functions, for which the exact minimum is difficult to determine and typically stopping criteria are used, the quasi-Newton method is usually more efficient in practice than Newton's method because second derivatives are not required. Both DFP and BFGS schemes for an iterative computation of the Hessian can be used, and preference is application dependent.

Quasi-Newton methods and conjugate gradient methods represent two major families of algorithms for multidimensional optimization with calculation of first derivatives. Both families require a 1D line suboptimization, which can itself either use or not use the derivative information. Quasi-Newton methods require more storage (order of N^2) for their parameters compared with conjugate gradient methods (order of N). This factor may need to be considered when an optimization of a large number of parameters is required, as in deformable registrations. As neither family is dominant in all optimization applications, an empirical comparison may be the best way to determine which algorithm to choose.

8.3.6 Least-Squares Methods

Image registration can be considered a problem of determining an optimum N-dimensional transformation parameter vector β, such that the sum-of-squares error between the values of M corresponding voxels in two images is minimized. This can be formulated as

$$S(\beta) = \sum_{i=1}^{M} [y_i - f(x_i, \beta)]^2, \tag{8.16}$$

where y_i and $f(x_i, \beta)$ are the values of the M corresponding voxels in the fixed and the moving images, respectively. Note that $f(x_i, \beta)$ depends on the transformation parameter vector β.

A popular numerical solution to this problem was proposed by Levenberg and Marquardt (Nocedal and Wright 1999). The Levenberg–Marquardt algorithm iteratively updates the parameter vector β to minimize the sum-of-squares error. The iteration starts from an initial guess β_0. In each iteration step, β is replaced by a new estimate $\beta + \delta$. Here, δ can be determined by the first-order Taylor approximation of the functions $f(x_i, \beta + \delta)$:

$$f(x_i, \beta + \delta) \approx f(x_i, \beta) + J_i \delta, \tag{8.17}$$

where $J_i = \dfrac{\partial f(x_i + \beta)}{\partial \beta}$ is a row vector that represents the gradient of f with respect to β.

Substituting Equation 8.17 into Equation 8.16, the result is

$$S(\beta + \delta) \approx \sum_{i=1}^{M} [y_i - f(x_i, \beta) - J_i \delta]^2 \tag{8.18}$$

If we use the vector notation, the above equation can be written to yield the following error or difference function:

$$S(\beta + \delta) \approx \|y - f(\beta) - J\delta\|^2 \tag{8.19}$$

where J is the $M \times N$ Jacobian matrix, whose ith row equals J_i, and where y and $f(\beta)$ are column vectors with ith component y_i and $f(x_i, \beta)$, respectively. To minimize the above error function, we set the first derivative of the function to zero, which gives

$$(J^T J) \delta = J^T [y - f(\beta)]. \tag{8.20}$$

This is a system of linear equations that can be solved for δ. Levenberg proposed using a "damped version" of this equation, which is given by

$$(J^T J + \lambda I) \delta = J^T [y - f(\beta)], \tag{8.21}$$

where I is the identity matrix and λ is the damping factor. When λ approaches zero, Equation 8.21 reduces to the inverse Hessian method, giving us an exact solution if $S(\beta)$ is a quadratic function. When λ is sufficiently large, $J^T J + \lambda I$ is diagonally dominated and Equation 8.21 reduces to the gradient descent method. Levenberg's algorithm has the disadvantage that, if the value of damping factor λ is large, the term $J^T J + \lambda I$ has no effect on the solution. To address this problem, Marquardt proposed scaling each component of the gradient according to the curvature, resulting in larger movement along the directions where the gradient is smaller. This avoids slow convergence in the direction of the small gradient. Therefore, Marquardt replaced the identity matrix I with the diagonal of $J^T J$, resulting in the Levenberg–Marquardt algorithm:

$$[J^T J + \lambda \operatorname{diag}(J^T J)] \delta = J^T [y - f(\beta)]. \tag{8.22}$$

Unlike the other gradient-based methods, the Levenberg–Marquardt method does not require line minimization at each iteration, which significantly reduces the number of function evaluations.

8.4 Challenges to Optimization

8.4.1 Local Minima Trapping

A major challenge for an optimization strategy is local minima trapping, whereby the optimizer converges to the local minimum and becomes trapped based on the threshold criterion, yet the optimized parameters do not reflect the global minima. For example, in Figure 8.2, we consider optimizing a parameter x.

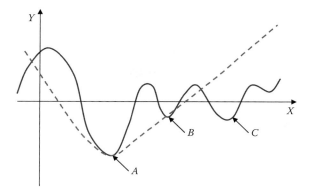

FIGURE 8.2 1D parameter space illustrating local minima. *X* axis is the parameter space, and *Y* axis is the value of the cost function; solid curve represents cost function. Points *B* and *C* on the solid curve are local minima and point *A* is global minima. The goal of the optimization algorithm is to find the parameters corresponding to the global minima and avoid local minima. Dotted curve represents the ideal convex parameter space and can be considered a simplification of the original cost function (solid curve). For the ideal convex case, the optimization method will always converge to global minima.

Point *A* is the ideal global minimum the optimization algorithm should find; however, it is very common for optimization algorithms to converge in the local minima (i.e., point *B* or point *C*). One of the approaches for avoiding the local minima is to construct an alternative convex cost function, such as the red curve in Figure 8.2. Unfortunately, the parameter space of cost function cannot always be easily represented as a convex function, especially for deformation image registrations that may contain thousands of parameters. Multiresolution is another widely used approach to reduce the possibility of trapping in the local minima; however, its convergence to the global minima is not guaranteed. Other strategies for minimizing local minima trapping include using simulated annealing optimization (Press et al. 1992) or variable step size rather than a uniformly diminishing step size.

8.4.2 Evaluation of Optimal Transformation Parameters

The similarity measure used for the cost function affects the performance of the registration in terms of accuracy, robustness, and local minima trapping. A number of methods have been proposed to objectively evaluate this performance. For sake of brevity, we mention here two specific methods that include a broad range of performance metrics for medical imaging registration evaluation. Skerl et al. (2006) proposed evaluating a similarity measure based on five properties: accuracy, distinctiveness of optimum (DO), capture range (CR), number of minima (NOM), and risk of nonconvergence (RON). Wu et al. (2009) proposed a modification of Skerl et al.'s method by using a modified or mean-capture range that better estimates robustness. These studies confirmed that the choice of a similarity measure for registration was application dependent. Correlation-based

similarity measures (e.g., normalized cross-correlation and gradient correlation) tend to have large CR but poor DO, which indicates that an optimization process is less likely to be trapped in a local optimum, but stricter stopping criteria must be used to prevent premature termination. Normalized mutual information (NMI) is the measure of choice for many registration applications because of its good behavior in terms of both accuracy and robustness. However, its performances also depend on the size of the region of interest and the down-sampling ratio if a multiresolution scheme is used. NMI behaves poorly when the number of voxels used for NMI calculation is decreased.

8.5 Deformable Image Registration

Deformable registration represents a particularly difficult challenge in medical imaging. There are many orders of magnitude degrees of freedom (i.e., approximately 10^7 voxels each with three degrees of freedom) than for rigid registration. Unlike rigid registration, the transformation model typically requires regularization. Also, the establishment of an appropriate termination criterion based on the goodness of a registration remains a challenge. Numerous deformation models have been proposed. Here, we consider three models used in medical imaging.

8.5.1 B-Spline Method

The transformation model for deformable image registration can be represented by a B-spline model, where the mesh of control points can be adjusted to model a deformation field. In a B-spline model, the deformation field can be described as

$$T(x,y,z) = \sum_{l=0}^{3} \sum_{m=0}^{3} \sum_{n=0}^{3} B_l(u)B_m(v)B_n(w)\phi_{i+l,j+m,k+n}, \quad (8.23)$$

where $\phi_{i,j,k}$ is the control point in a set of control points of mesh size of $n_x \times n_y \times n_z$ and uniform spacing δ; $i = [x/n_x] - 1, j = [y/n_y] - 1,$ $k = [z/n_z] - 1; u = x/n_x - [x/n_x], v = y/n_y - [y/n_y], w = z/n_z - [z/n_z];$ and B_l represents the *l*th B-spline basis function, for which a partial list is given below:

$$\begin{aligned}
B_0(u) &= (1-u)^3/6 \\
B_1(u) &= (3u^3 - 6u^2 + 4)/6 \\
B_2(u) &= (-3u^3 + 3u^2 + 3u + 1)/6 \\
B_3(u) &= u^3/6
\end{aligned} \quad (8.24)$$

Rueckert et al. (1999) modeled the cost function of a B-spline–based deformable image registration as

$$C(\phi) = C_{\text{similarity}}(S, M(T)) + \lambda C_{\text{smooth}}(T), \quad (8.25)$$

where $C_{\text{similarity}}(S, M(T))$ and $C_{\text{smooth}}(T)$ are, respectively, the similarity measure and smoothing term with respect to transformation *T*. Here, ϕ is the set of control points that need to be optimized. In other words, the goal of solving Equation 8.25 is to find the optimal set φ to minimize the cost function $C(\phi)$. A gradient descent strategy can be used to update the control points φ:

$$\phi(i + 1) = \phi(i) + \mu\nabla C/\nabla C. \qquad (8.26)$$

The advantage of B-spline models is that they are locally controlled (i.e., change in the location of a control point only affects the transformation of the neighborhood of the control point), which makes B-spline computationally efficient (Rueckert et al. 1999). However, certain measures have to be used to prevent folding of the deformation field (Crum et al. 2004). The application of B-spline models includes real-time motion (Rohlfing et al. 2004), positron emission tomography–CT image fusion (Mattes et al. 2003), cardiac modeling (Frangi et al. 2002; McLeish et al. 2002), and breast modeling (Rueckert et al. 1999).

8.5.2 Demons Algorithm

Proposed by Thirion (1998), the Demons algorithm is a very popular registration algorithm in adaptive radiation therapy. The concept of the Demons algorithm is that the voxels in the static or reference image *S* act as local forces that move the voxels in the moving or target image. The moving image is iteratively deformed by applying a displacement vector *u*. The Demons algorithm has multiple schemes or implementations, which depend on the choice of:

(1) Location of the force (whole images or contour points)
(2) Transformation model (rigid, affine, free form, etc.)
(3) Interpolation method (linear, spline, etc.)
(4) Source of the force (optical flow, gradient-based, etc.)

A commonly used Demons scheme includes nonzero gradient points for the location of the force, free-form deformation for the transformation model, a linear interpolation method, and an optical flow source. The deformation field can be iteratively estimated by

$$u^{i+1} = \frac{(M^i - S)\nabla S}{(\nabla S)^2 + (M^i - S)^2}, \qquad (8.27)$$

where u^{i+1} is the displacement at $i + 1$ iteration, *S* is the static image, M^i is the moving image at the *i*th iteration, and ∇*S* is the gradient of the static image *S*. There are two forces in Equation 8.27: (i) the internal image gradient-based force ∇*S* and (ii) external force $(M^i - S)$. The internal force does not change during the iterations, whereas the external force changes after each iteration. The term $(M^i - S)^2$ is added to make the deformation field computation more stable. For example, without adding the $(M^i - S)^2$ term, the solution will not be stable for small values of ∇*S*. Before the next iteration, the displacement is convolved with a Gaussian kernel, as the Gaussian convolution removes noise and improves geometric continuity. Carefully tuning the smoothness parameters of the Gaussian kernel, such as the standard deviation and the

width of the Gaussian filter, is important to obtaining a good registration result. Equation 8.27 can also be considered as a special case of the multiscale, iterative version of optical flow (Thirion 1998). The optical flow method will be discussed in the next section. It should be noted that optical flow is not the only choice of force term in Demons algorithm. Other sources of forces, such as binary force, can also be applied. However, optical flow seems to provide better results than others (Thirion 1998). From the optimization point of view, the gradient descent strategy is implicitly used in Equation 8.27, where the deformation field is updated along the gradient of the static image.

Equation 8.27 indicates that Demons algorithm will not work well for intermodal image registration because the voxel intensity for the corresponding voxels of different modality imaging may not be similar. For example, in magnetic resonance brain imaging, cerebrospinal fluid (CSF) appears dark for a T1-weighted image but bright for a T2-weighted image. If one directly applies Demons algorithm to register CSF from MRI T1 images to MRI T2 images, convergence in Equation 8.27 will not occur. The iterations will continue updating even in the presence of a good registration because the voxel intensity difference $(M^i - S)$ is not zero. One solution to making the Demons algorithm applicable for intermodal image registration is to find an intensity transformation to map the voxel intensity from one imaging modality to another, making T1 images appear similar to T2 images (Guimond et al. 2001) and making CT images similar to CBCT (Nithiananthan et al. 2011).

8.5.3 Optical Flow Method

The optical flow was introduced by Horn and Schunck (1981) to estimate the motion between frames in an image sequence. The fundamental assumption is that the brightness of the images is preserved (i.e., the voxel intensity of the same object does not change within two image frames for the same anatomical object). Given two images $I(x,y,z,t)$ and $I(x + \delta x, y + \delta y, z + \delta z, t + \delta t)$, the optical flow velocity *u* can be described as

$$I(x,y,z,t) = I(x + \delta x, y + \delta y, z + \delta z, t + \delta t). \qquad (8.28)$$

Equation 8.28 can be rewritten as

$$\frac{\partial I(x,y,z,t)}{\partial t} = 0 \Rightarrow \frac{\partial I}{\partial x}\frac{dx}{dt} + \frac{\partial I}{\partial y}\frac{dy}{dt} + \frac{\partial I}{\partial z}\frac{dz}{dt} + \frac{\partial I}{\partial t} = 0. \qquad (8.29)$$

Equation 8.29 can be simplified as

$$\nabla I u = -\frac{\partial I}{\partial t}, \qquad (8.30)$$

where optical flow velocity $u = [dx/dt, dy/dt, dz/dt]$, and $\nabla I = [\nabla I_x, \nabla I_y, \nabla I_z]$ is the image intensity gradient. Here, *u* is the physical displacement vector, and *t*, conceptually equivalent to a temporal variable, is a mathematical convenience with no physical analogue. Equation 8.30 is underconstrained because one equation is not sufficient for solving three unknown components of u. To address this problem, smoothness constraints have been proposed. Carefully choosing the smoothing parameters, such as the standard deviation and the width of the Gaussian filter, is important to obtain good registration results. Horn and Schunck (1981) proposed a constraint by minimizing the square of the magnitude of the gradient of the optical flow velocity u. In terms of a similarity measure and regularization, this can be rewritten as

$$E(u) = E_{\text{sim}}(u) + \lambda E_{\text{reg}}(u)$$

$$E_{\text{sim}}(u) = \left(\nabla I u + \frac{\partial I}{\partial t} \right)^2 , \qquad (8.31)$$

$$E_{\text{reg}}(u) = \left\| \nabla u \right\|^2$$

where E_{sim} and E_{reg} are the similarity and regularization terms, respectively, and λ is the weighting factor for the similarity and regularization terms. Using calculus of variation, Equation 8.31 can be solved by the following iterative scheme based on gradient descent optimization:

$$\begin{cases} u_x^{i+1} = u_x^i - \dfrac{\nabla I_x \left(\nabla I_x u_x^i + \nabla I_y u_y^i + \nabla I_z u_z^i + I_t \right)}{(\lambda + \nabla I_x^2 + \nabla I_y^2 + \nabla I_z^2)} \\[2ex] u_y^{i+1} = u_y^i - \dfrac{\nabla I_y \left(\nabla I_x u_x^i + \nabla I_y u_y^i + \nabla I_z u_z^i + I_t \right)}{(\lambda + \nabla I_x^2 + \nabla I_y^2 + \nabla I_z^2)} , \\[2ex] u_z^{i+1} = u_z^i - \dfrac{\nabla I_z \left(\nabla I_x u_x^i + \nabla I_y u_y^i + \nabla I_z u_z^i + I_t \right)}{(\lambda + \nabla I_x^2 + \nabla I_y^2 + \nabla I_z^2)} \end{cases} \quad (8.32)$$

where u_x^i, u_y^i, and u_z^i are the optical flow in x, y, and z directions, respectively, at iteration i and ∇I_x, ∇I_y, ∇I_z are the image gradient in x, y, and z directions, respectively.

8.6 Summary

Image registration, whether rigid or nonrigid, is an optimization problem, whereby the mapping of the target image onto the reference image is computed as an optimization of the associated cost function, which measures the similarity of two data sets. We have presented a few commonly used optimization algorithms in rigid and deformable registration, and have attempted to provide a general framework for the role of optimization in image registration. As in most computational problems, the "devil" lies in the details of implementation, for which a small sampling of references is provided here. The reader is encouraged to use these references as a starting point. Furthermore, it should be noted that, in medical image registration, the selection of the optimization method is usually empirically determined. The user must consider computational efficiency, robustness, and complexity of the method involved, all of which will be affected by the registration algorithm and type of image data sets involved.

References

Bajcsy, R. and Kovačič, S. 1989. Multiresolution elastic matching. *Computer Vision, Graphics, and Image Processing, 46*, 1–21.

Bookstein, F. L. 1991. Thin-plate splines and the atlas problem for biomedical images. *In:* Colchester, A. C. F. & Hawkes, D. J. (eds.) *Information Processing in Medical Imaging.* Wye: Springer-Verlag.

Bro-Nielsen, M. and Gramkow, C. 1996. Fast fluid registration of medical images. *Visualization in Biomedical Computing. Computer Science, 1131*, 267–276.

Christensen, G. E. and Johnson, H. J. 2001. Consistent image registration. *IEEE Transactions on Medical Imaging, 20*, 568–82.

Crum, W. R., Hartkens, T. and Hill, D. L. 2004. Non-rigid image registration: theory and practice. *British Journal of Radiology, 77*(Spec No 2), S140–S153.

Fletcher, R. and Reeves, C. M. 1964. Function minimization by conjugate gradients. *Computer Journal, 7*, 149–154.

Frangi, A. F., Rueckert, D., Schnabel, J. A. and Niessen, W. J. 2002. Automatic construction of multiple-object three-dimensional statistical shape models: Application to cardiac modeling. *IEEE Transactions on Medical Imaging, 21*, 1151–1166.

Freeborough, P. A. and Fox, N. C. 1998. Modeling brain deformations in Alzheimer disease by fluid registration of serial 3D MR images. *Journal of Computer Assisted Tomography, 22*, 838–843.

Guimond, A., Roche, A., Ayache, N. and Meunier, J. 2001. Three-dimensional multimodal brain warping using the demons algorithm and adaptive intensity corrections. *IEEE Transactions on Medical Imaging, 20*, 58–69.

Horn, B. and Schunck, B. G. 1981. Determining optical flow. *Artificial Intelligence, 17*, 185–203.

Maes, F., Vandermeulen, D. and Suetens, P. 1999. Comparative evaluation of multiresolution optimization strategies for multimodality image registration by maximization of mutual information. *Medical Image Analysis, 3*, 373–386.

Mattes, D., Haynor, D. R., Vesselle, H., Lewellen, T. K. and Eubank, W. 2003. PET-CT image registration in the chest using free-form deformations. *IEEE Transactions on Medical Imaging, 22*, 120–128.

Mcleish, K., Hill, D. L., Atkinson, D., Blackall, J. M. and Razavi, R. 2002. A study of the motion and deformation of the heart due to respiration. *IEEE Transactions on Medical Imaging, 21*, 1142–1150.

Nelder, J. A. and Mead, R. 1965. A simplex method for function minimization. *The Computer Journal, 7*, 308–313.

Nithiananthan, S., Schafer, S., Uneri, A., Mirota, D. J., Stayman, J. W., Zbijewski, W., Brock, K. K. et al. 2011. Demons deformable registration of CT and cone-beam CT using an iterative intensity matching approach. *Medical Physics, 38*, 1785–1798.

Nocedal, J. and Wright, S. J. 1999. *Numerical Optimization,* New York: Springer-Verlag.

Polak, E. 1971. *Computational Methods in Optimization,* New York: Academic Press.

Press, W. H., Flannery, B. P., Teukolsky, S. A. and Vetterling, W. T. 1992. *Numerical Recipes in Pascal: The Art of Scientific Computing*, New York: Cambridge University Press.

Rohlfing, T., Maurer, C. R., Jr., O'Dell, W. G. and Zhong, J. 2004. Modeling liver motion and deformation during the respiratory cycle using intensity-based nonrigid registration of gated MR images. *Medical Physics, 31*, 427–432.

Rueckert, D., Sonoda, L. I., Hayes, C., Hill, D. L., Leach, M. O. and Hawkes, D. J. 1999. Nonrigid registration using free-form deformations: application to breast MR images. *IEEE Transactions on Medical Imaging, 18*, 712–721.

Schonemann, P. H. 1966. A generalized solution of the orthogonal procrustes problem. *Psychometrika, 31*, 1–10.

Skerl, D., Likar, B. and Pernus, F. 2006. A protocol for evaluation of similarity measures for rigid registration. *IEEE Transactions on Medical Imaging, 25*, 779–791.

Terzopoulos, D. 1986. Regularization of inverse visual problems involving discontinuities. *IEEE Transactions on Pattern Analysis and Machine Intelligence, 8*, 413–424.

Thirion, J.-P. 1998. Image matching as a diffusion process: an analogy with Maxwell's demons. *Medical Image Analysis, 2*, 243–260.

Umeyama, S. 1991. Least-squares estimation of transformation parameters between two point patterns. *IEEE Transactions on Pattern Analysis and Machine Intelligence, 13*, 376–380.

Wu, J., Kim, M., Peters, J., Chung, H. and Samant, S. S. 2009. Evaluation of similarity measures for use in the intensity-based rigid 2D-3D registration for patient positioning in radiotherapy. *Medical Physics, 36*, 5391–5403.

Xie, Z. and Farin, G. E. 2004. Image registration using hierarchical B-splines. *IEEE Transactions on Visualization and Computer Graphics, 10*, 85–94.

III

Segmentation

9

Basic Segmentation

Todd R. McNutt
Johns Hopkins University
School of Medicine

Image segmentation is the process of defining regions in a 3D image set. Segmentation is used in radiotherapy to identify the normal patient anatomy and target volumes and other regions of interest for density specification or fiducial marking. With the increased utilization of intensity-modulated radiotherapy, the demands on image segmentation are magnified, where all critical structures and target volumes need to be identified on the computed tomography scan acquired during patient simulation. This increased demand warranted easier, more efficient manual segmentation tools as well as robust automatic segmentation strategies.

In this chapter, we focus on the basic image segmentation tools. Before discussing the various types of segmentation tools, it is important to first describe the various ways in which regions of interest are stored and manipulated within the computer system.

9.1 Region of Interest Definitions

9.1.1 Image Set Definition

The 3D image data set has been typically thought of as a series of 2D images making up a volume of data (Figure 9.1). X and Y resolutions are the property of the tomographic reconstruction and are typically on the order of 1 mm, and the Z resolution is the spacing between tomographic slices and typically ranges from 2 to 5 mm. In most radiotherapy applications, the data are stored as a set of voxels in a 3D rectilinear volume. The data set also has a coordinate system associated with it that defines the x, y, and z coordinates of any point in the volume. Typically, the coordinate frame of reference is defined by the x, y, and z coordinates of the first voxel (upper left corner of first image), the voxel size in

each direction, and the number of voxels in each direction. This allows the system to determine the coordinates of each voxel in the data set and which voxel an arbitrary location is positioned in through the relationship

$$x = x_{start} + i * \Delta x$$
$$i = \text{ROUND}((x - x_{start})/\Delta x),$$

where x is the position, i is the voxel index, Δx is the voxel size, and x_{start} is the position of the first voxel in the x direction. This basic conversion from index to position is adapted to support the conversion of mouse tracking coordinates into the image indices.

9.1.2 Contours on Slices

The most intuitive definition of a region of interest is contours defined on the individual images making up the 3D volume. 2D images and the idea of identifying a region by drawing on a picture or film are very intuitive. Additionally, the contours on slices are used in the DICOM-RT Structure Set standard.

An individual contour on a 2D slice is defined as a set of ordered vertices with line segments between each vertex in the order they are defined. The contour display draws lines between points (Figure 9.2a). Because all vertices on an image share the same z-coordinate, the viewing display must be on the image, or z-coordinate, to visualize it. When only a single closed contour exists, it is clear that the area enclosed by the contour is part of the region of interest and the area outside is not. This, however, is complicated when the contours are not closed or when multiple contours are defined on the same image or z-coordinate. In these

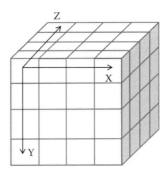

FIGURE 9.1 Tomographic images are stacked to form a 3D volumetric representation of the patient. The voxel sizes in X and Y are determined by the image reconstruction resolution and the voxel size in the Z direction is determined by the tomographic slice spacing.

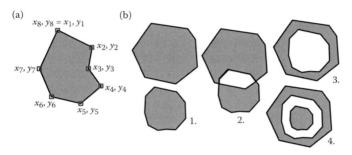

FIGURE 9.2 (a) Closed single contour on a tomographic image (slice) sharing a common z-coordinate. (b) Use of "exclusive or" logic when determining inside and outside of regions when multiple contours exist. (1) Two separate contours. (2) Contours overlap—region of intersection is considered outside of region ("exclusive or" is false). (3) A ring is defined when one contour completely encompasses the other. (4) Concentric rings are produced with three contours encompassing successive ones.

scenarios, the system must determine what is outside and what is inside the contour and what to do with regions where contours may overlap. The example in Figure 9.2b shows what happens when a logical "exclusive or" is used to combine multiple contours. When the two contours are completely separate, the inside of each is considered to be part of the overall structure as indicated by the colorwash area. When the two contours intersect, the region of intersection is considered outside, and when one contour encloses the other, the area inside the enclosed contour is considered outside of the overall region. Last, when there are three concentric contours, the "exclusive or" combination leaves a target pattern. Other methods are used to combine structures, and one must be sure they understand how the systems they use are performing this task to be sure they know what is being considered inside and outside of the region of interest. Figure 9.3 shows the relationships between the various digital representations of regions of interest.

9.1.3 Binary Mask

The binary mask is perhaps the most primitive form of a region of interest. A binary mask is simply a voxel grid matching the size and resolution of the 3D image volume with a value of one inside the region and zero outside the region. This mask is what is used in the colorwash display and identifies the inside and outside of the region by shading the voxels appropriately. There are various ways to store this information in a computer. Perhaps the cleanest way is to have a separate mask for each region of interest. Some systems store a single voxel grid and assign a value identifying the region of interest to each voxel. This model, however, does not allow multiple regions of interest to share the same voxel and thus prevents overlap of regions of interest in the representation. It is challenging to produce a colorwash display in regions where the binary masks of two regions of interest overlap. Some systems have a priority or order that the colorwash is applied and thus will use the color of the last region of interest painted as the color in the overlap regions. It is also possible to blend the colors of the regions of interest in the display; however, this can be difficult to decipher depending on the number of overlapping regions and the assigned colors of each.

9.1.4 Polygonal Surface

A polygonal surface representation is somewhat of an extension of 2D contouring; however, it is typically created from the binary mask (Figure 9.3). The polygonal surface is made up of a set of 3D vertices and a set of polygons defined by specifying the subset of vertices comprising each polygon in the order defining how the vertices are connected to form the polygon. Typically, triangles are used as the polygons, but any number of vertices can be used, provided the set of vertices for each polygon remains on a common plane. Additionally, it is possible to define which side of an individual polygon is inside the region and which is outside by the order of the vertices in the polygon definition. The direction of normal vector through the center of the polygon is defined by the clockwise ordering of vertices in the polygon representation. The normal vector typically points to the outside of the regions. Additionally, polygonal surfaces are the basis for the 3D rendering of surfaces with shading. The polygonal surface also has advantages for 2D displays. The intersection of any arbitrary 2D viewing plane and the polygonal surface can be determined and displayed on the viewing plane (Figure 9.4). This technique resolves the display problem of 2D contours having to be viewed on the same z-coordinate as they are defined, in that the intersection of the polygonal mesh is seen as if it were simply a 2D contour on any arbitrary viewing plane.

9.1.5 Conversion between Types

When working with regions of interest, it is important to understand what representation is being used for which type of task.

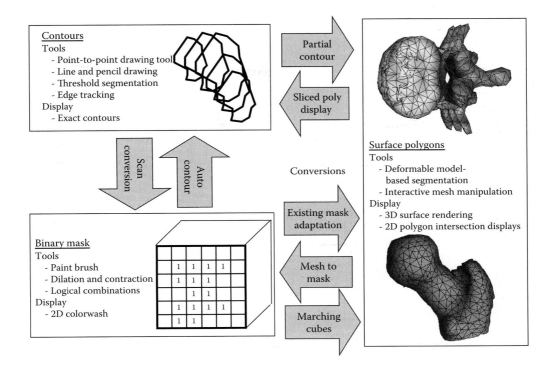

FIGURE 9.3 Region of interest representations, uses, and conversions. Various tools that manipulate regions operate on different data representations. There are also many alternatives in converting a region from one representation to another. In many cases, these conversions are not perfect and are not able to exactly recover the originally defined region.

Figure 9.3 indicates the example tools and displays used on each type. In many situations, the conversion from one representation to another is required. For example, if you are drawing a shape with a pencil but want to see the colorwash display, it is necessary to convert from the contour on a slice to the binary mask representation. With each conversion, some information is lost—or approximated.

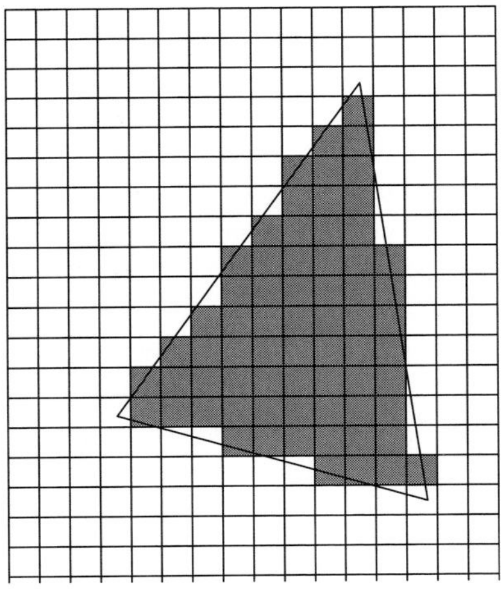

9.1.5.1 Scan Conversion

Scan conversion is the process of converting a contour on a single image slice to a binary mask (Figure 9.4). This is done by scanning the pixel mask and determining if each pixel is inside or outside of the contour. Typically, the center of the voxel is used to represent the voxel position. When multiple overlapping contours exist, the system must make a determination of whether or not the pixel is inside or not. As shown in Figure 9.2, an "exclusive or" can be used. There are other methods as well.

9.1.5.2 Edge Traversal of Mask

It is also important to be able to convert from a binary mask back to a contour on a slice. Basic autocontouring can be performed on the mask to accomplish this. These methods follow the edge of the mask and use the coordinates of the center of each edge pixel as the vertex coordinates required to form the contour. This technique generates a large number of vertices, and in some cases, it is desirable to reduce the number of vertices after this operation. There are many algorithms to do this, which will preserve the curvature of the contour while reducing the number of vertices. It is important to note that if one generates a contour (or set of contours), then converts it to a binary mask. Then, convert that binary mask back to contours, the original vertices making up the contour will not be retrieved.

FIGURE 9.4 Scan conversion is the process of converting a contour into a binary mask. In this case, every pixel of the image whose center is inside the contour (triangle) is identified as being in the region of interest (gray).

(a)
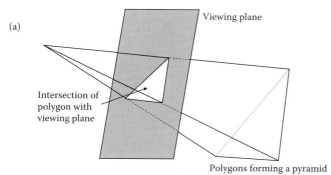

Viewing plane

Intersection of
polygon with
viewing plane

Polygons forming a pyramid

(b)
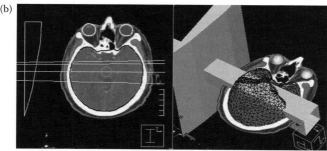

FIGURE 9.5 Example of 2D view of polygon intersection. (a) Graphical depiction of where the viewing plane intersects the 3D surface polygons of a pyramid. (b) Left image shows the 2D display of the CT with the polygon intersections of the beam, wedge and eyes in white, and the brain in black. Right image shows the 3D nature of the same polygons.

9.1.5.3 Marching Cubes

Marching cubes (Lorensen and Cline 1987) is the de facto method of converting a 3D binary mask into a polygonal surface mesh. This method basically forms a triangle at each voxel on the periphery of the mask. The combination of all these triangles forms the polygonal surface mesh. This method tends to create far more polygons than are necessary to represent typical regions of interest. For this reason, there is a great body of work dedicated to reducing the number of polygons while maintaining accurate representations of the region of interest.

9.1.5.4 Polygon Intersection

Polygonal surface meshes can be used to display the contour on a slice (Figure 9.5). This is done by drawing the line segment that represents the intersection of display plane with each polygon it intersects. This process is the basis for generating contours on slices from polygonal surface meshes. Special care has to be taken to maintain what is considered inside or outside of the contours when the surface mesh intersects the plane in complex ways.

9.2 Manual Contouring Tools

There are a variety of manual contouring tools available in the various treatment planning systems used in radiation oncology.

The type of contouring tool dictates which representation of the contour will be modified as you use the tool. The basic premise of all manual tools is to track mouse coordinates or other pointing device and convert those motions into modifications of the region of interest. The various algorithms convert these motions into the regions of interest.

9.2.1 Mouse and Other Pointing Devices

It is important to first understand the basic operation of a mouse or pointing device before discussing how their use can be translated into regions of interest on an image set. A mouse typically has one to three buttons and a mechanism for determining the position of the device on the computer screen. Software that uses the mouse receives events related to the state of the mouse through the operating system. Mouse events include "button down," "button up," "button click," and "button double-click." A mouse also sends repetitive tracking events when in motion. With each event, the state of the mouse buttons and the position information is sent to the program for processing. The first step in all contouring methods is to identify a mouse down event, or start position, and convert the screen coordinates of the event into the coordinates of the image set you are working with. From there, the system repeatedly updates tracking coordinates of the mouse position and finally will end in the mouse up event. These sets of events are then translated through various algorithms into modifications of the region of interest.

9.2.2 Operations on Contours

The most basic tool to draw contours on an image is to simply point and click the mouse on the location of each desired vertex and draw a line from the prior vertex to the newly created one. With each mouse click, a new vertex is added. When completed, the last vertex can be connected to the first one to close to contour. When contours are generated in this fashion, there will typically be a reasonably low number of vertices. It is thus possible to edit the vertices by grabbing an individual vertex with the mouse and dragging it to a new location. In this case, the mouse down event selects the existing vertex to move, and the tracking of the mouse directs where to move the vertex. The existing line segments to the neighboring vertices are then redrawn with the new location as the mouse tracks.

Slightly more complex is the pencil drawing tools. In this case, the initial mouse down event defines the starting point; then, for each new tracking update from the mouse motion, a new vertex is added and a line is drawn between the prior vertex and the new one. Because mouse tracking events are frequent, this results in significantly more vertices than the point-and-click method above. Thus, editing the vertices after drawing with the pencil tool is more difficult.

Other manual methods that operate on contours are tools such as the oval or rectangle drawing tools found in common applications such as Microsoft PowerPoint or Adobe Photoshop. In general, these tools for basic shapes are of limited use in radiation therapy applications because very few anatomical structures are well represented by these geometrical shapes.

Because we are forming a 3D region of interest, contours must be defined on each image plane where the region exists. Several shortcut tools can be used: a simple copy to next slice tool can copy the contour from one slice to the z-coordinate of the next slice, leaving only minor modification to be done by the user. Also, tools to interpolate contours on images that lie between existing contours can also assist in improving the efficiency of contouring the 3D region of interest.

9.2.3 Operations on Binary Mask

9.2.3.1 Paintbrush Tool

The primary manual tool that operates on the binary mask representation is the paintbrush tool. Paintbrush tools take a mouse down event and from that location change the pixel values in the binary mask to one in a small region around that location. The region size and shape are user-defined by a size and shape of the brush. As the mouse tracks, each new event is used to apply the brush shape and change the pixel values to 1. Thus, as you move the mouse, the system is painting the region. For ease of use, some systems combine the basic paintbrush and an eraser. For example, if you are editing an existing contour with a paintbrush tool, it is useful to initialize the function of the paintbrush with the initial mouse down event. If the initial mouse down event is outside the existing region, it is set to erase—or to change the values of the binary mask to zero in the small region around the mouse. This way, as the mouse is moved into the existing region, it erases parts of the structure with the brush shape. Conversely, if the initial mouse down event is inside the existing region, then you would set it to paint the ones into the image allowing you to expand the original region. This technique provides a convenience to the user in that they do not have to select separate tools while editing with a paintbrush.

9.2.3.2 Expansion, Contraction, and Logical Combinations

In radiation therapy, it is also important to be able to expand, contract, or produce logical combinations of regions of interest. These operations are also performed on the binary mask representation of regions of interest.

Expansion and contraction (or dilation and erosion) of a region of interest is important when applying margins to

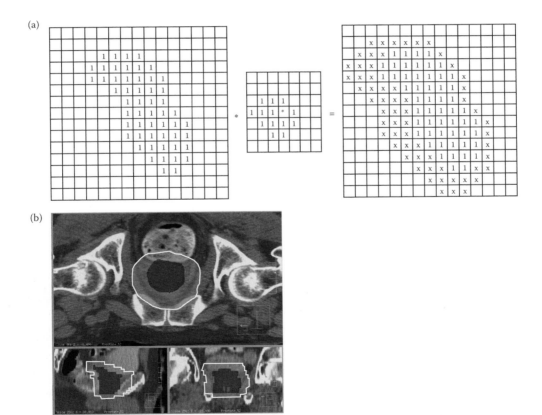

FIGURE 9.6 Dilation and contraction are operations performed on the binary mask. (a) Original mask convolved with the kernel mask, and the resulting dilated mask with "x" indicating the new portions of the region of interest. (b) 3D example of a prostate expanded nonuniformly to determine a target volume in radiation therapy.

treatment volumes. Expansions and contractions can be uniform or variable in each direction. There are many algorithms to perform these tasks. In basic terms, an expansion (or contraction) kernel is constructed representing the expansion distances in each direction. The center of the kernel is placed on the edge of the structure. Then, wherever there is a one in the kernel, a one is placed in the binary mask of the new region of interest. The resulting binary mask represents a region expanded by the amounts specified in the kernel. Figure 9.6 depicts this basic process in 2D. Conversely, contraction is performed in the same manner; however, zeros are used in the kernel, and the result is to contract the region. It is important to note that an expansion, followed by a contraction of a region, does not render the original region back. In fact, this is a smoothing operation as the expansion acts as a low-pass filter.

Binary masks are also used to produce logical combinations of regions. For example, to combine two structures (lungs or kidneys), one would create a new structure "combined lungs," which is the union of the left and right lung regions. To perform this combination with the binary mask, the algorithm simply looks at each voxel location in the left and right lungs, and if either one has a value of 1, then a 1 is placed in the corresponding voxel in the resulting "combined lungs" mask. This is a logical "or" operation. Similarly, the logical "and" operation will take the intersection of two regions and "exclusive or" will take the region where one input mask has a 1 and the other has a 0. These logical combinations can also make use of the logical "not." For example, given a planning target volume (PTV) surrounding a gross tumor volume (GTV) and a goal to define the ring-like region of the PTV, which is not in the GTV, the following logical operation could be used: ring = PTV and not GTV, and wherever PTV binary mask has a value of 1 and the GTV binary mask has a value of 0, the resulting ring mask would get a value of 1.

9.2.4 Operations on Polygons

The manual editing of polygonal surface meshes is becoming more common. As the resolution of tomographic imaging is improved, the number of axial slices contained in a typical data set is increased. This puts added burden on the conventional 2D drawing tools. The manual editing of 3D polygons, however, requires the existence of a region in the first place, so in general this method is only used to edit an existing region (Pekar et al. 2004).

When editing an existing surface mesh, the initial mouse down event will trigger the selection of either a 3D location or the closest vertex present in the mesh. As the mouse is moved, the vertices in the mesh are modified based on a 3D region of influence. For example, the vertices in the mesh can be moved based on their distance from the initial mouse down event. Each vertex in the region of influence can be moved in the same direction as the mouse motion and by a distance equal to an influence function of distance times the mouse motion distance. If a Gaussian function is used, the action will pull a Gaussian shape off the surface of the mesh.

9.3 Basic Semiautomatic Segmentation Tools

Automatic image segmentation refers to the process of using computer algorithms to detect features in an image that represent the boundaries of a structure and, from this information, define the contours of the region of interest. There are many methods of automatic segmentation in medical imaging and it remains a large field of study. The following sections briefly describe a few approaches.

9.3.1 Threshold-Based Segmentation

Threshold-based segmentation is perhaps the simplest form of autosegmentation. The user would specify a threshold image value that would represent the boundary of a structure. For example, bone is dense and would have a large CT number of approximately 600 to 1000, whereas muscle has a CT number of approximately 0 to 100. In this case, a CT number of approximately 400 would represent a threshold of a transition of muscle to bone in the image. An algorithm can search the image for two pixels where the threshold value is crossed. This is typically done starting from a seed location determined by a mouse click and then a search for the closest location where the threshold is crossed. From that crossing point, the algorithm proceeds to walk along pixels, which have neighboring pixels that cross the threshold value. The vertices of the contour are placed at each pixel boundary where the threshold is crossed and the line segments are drawn between the vertices to create the contour on a single 2D image. The process is repeated for multiple 2D images to make a full 3D segmentation. Figure 9.7 shows an example of both a lung and bone threshold-based segmentation and the corresponding seed locations.

The threshold-based method has several variants. For example, it can be set to contour a single closed contour. In this case, the algorithm would track a continuous line along the threshold boundary until it reached the starting seed point and then stop, creating a single closed contour. Alternatively, it could be used to find all contours in the image along that threshold value. Here, the algorithm would continue on with a new contour after the completion of the first one.

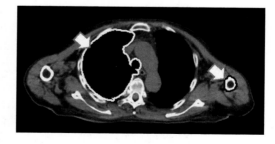

FIGURE 9.7 Images autosegmented with a threshold technique for lung (white) with a threshold CT# of –200 and bone (black) with a threshold CT# of 100. Arrows indicate the seed location for each contour.

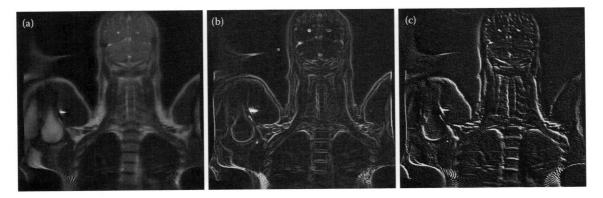

FIGURE 9.8 Example of a magnetic resonance image (a) with edge detection (b) and gradient filtering (c) to assist in the threshold-based segmentation.

In many cases, a simple threshold is not sufficient for identifying the boundary between tissues. The variations in brightness across the image, or tissues that are bounded by fatty tissue on one side and bone on the other, can cause the threshold-based methods to fail to accurately identify tissue boundaries. Other more sophisticated methods include edge detection and gradient-based tracking of boundaries. There are several algorithms for doing this, but perhaps a simple way to understand it is to consider a second data set perfectly aligned with the original images where we calculate the values used for segmentation into it. Figure 9.8 shows processed images for both edge detection and gradients. The autosegmentation can proceed on the features of these processed images in the background to improve the quality of the segmentation.

9.4 Summary

It is important to understand the fundamental structure of how different types of structures are stored and processed in the computer. Understanding these fundamental principles can assist the user while segmenting images. This chapter highlights the basic concepts. Image segmentation is a very large field of study in computer science and medical imaging. With computers getting faster and image quality improving, the realization of full autosegmentation of patient anatomy is perhaps possible, but there will always be a need to manually adjust or identify nonanatomical regions with the basic tools described above.

References

Lorensen, W. E., and Cline, H. E. (1987) Marching cubes: A high resolution 3D surface construction algorithm. *Computer Graphics, 21*(4).

Pekar, V., McNutt, T. R., and Kaus, M. R. (2004) Automated model-based organ delineation for radiation therapy planning in the prostate region. *International Journal of Radiation Oncology Biology Physics, 60*(3), 973–980.

10

Deformable Shape Models for Image Segmentation

Edward L. Chaney
University of North Carolina

Stephen M. Pizer
University of North Carolina

10.1 Overview of Model-Based Segmentation

10.1.1 Introduction

This chapter discusses the principles and uses of deformable shape models (DSMs) for segmentation of anatomy in volume images for radiotherapy applications. DSMs are being actively investigated in many research laboratories, and advances continuously appear in the literature. Although a good bit of research has been aimed at radiation therapy applications, translation into commercial clinical systems has been modest. This chapter focuses on methods that have reached clinical or near-clinical use and mentions others in translation (Section 10.4). Other approaches in commercial products, particularly nonrigid registration, are fundamentally different. Registration methods rely on contours of the target organ, which may be thought of as an organ model, drawn in a reference or atlas image. The model, however, is embedded in the atlas image and deformed by the nonrigid registration process; it cannot be independently deformed to match the target organ and is of no use without the corresponding atlas image. Level sets have some properties in common with boundary representations discussed below and in Section 10.2. However, level sets are discussed in Chapter 14 and not mentioned further.

A DSM is an explicit geometric representation of an anatomical organ or structure. The shape and size of a DSM are controlled by a set of parameters that are properties of the model and independent of image data. Some reasons for using DSMs are as follows: (1) DSMs can be trained to learn the range of credible shapes, known as the shape space, of a target organ for the population of interest. Training offers the advantage that segmentation algorithms search only within the shape space and therefore are more likely to produce clinically useful results. (2) Compared with the actual high-dimensional shape space, the learned shape space has reduced dimensionality determined by the relatively small number of model parameters. This reduction results in significant improvement in computational efficiency, albeit with theoretical sacrifice in shape fidelity. In practice, this loss occurs largely at small spatial scales that are overshadowed by the noise inherent in clinical contouring. (3) Another consideration related more to diagnostic applications, but with potential use in radiation oncology, is that certain DSMs can be used to measure shape, for example, to differentiate between normal and abnormal anatomy (Zhao et al. 2008) and to monitor shape in longitudinal studies (Gerig et al. 2001a; Cevidanes et al. 2007).

The three general steps for segmentation with a DSM are (1) initializing the starting model in the target image, (2) deforming the starting model to closely estimate the shape of the target object, and (3) editing the estimated shape to obtain a clinically acceptable segmentation. The first step can be automatic or user assisted, whereas the second step is fully automatic. Editing is interactive by nature and is not often discussed in the context of autosegmentation. However, editing is necessary in practice to reach the ultimate goal of a clinically acceptable result. Automation of steps 1 and 2 is of little clinical value if the estimated shape has errors that are difficult or time-consuming to correct. The mathematical underpinning of a particular type of DSM greatly influences editing efficiency, and thus editing is an important consideration in model selection.

DSMs are likely to play a larger role in autosegmentation than now; however, they are not regarded as a complete solution. It is widely acknowledged that general solutions will need to incorporate combinations of approaches, including hybrids tailored to each structure (Styner and Gerig 2000). Current clinically useful DSMs are for organs that have relatively simple shapes. Geometrically complex organs and some other anatomical structures are more challenging to model. The four important categories are (1) multifigure organs with lobes or prominent indentations or protrusions, (2) structures with essentially random shapes from person to person, (3) one or more objects that change shape and/or relative positions over time, and (4) interstitial tissues (e.g., connective tissues, fat, and lymph node chains). Organs relevant to radiation therapy in the first category include the lungs and liver. Most tumors are in the second category, and structures in the head and neck over a course of radiation therapy are in the third category. Some DSMs including medial representations (m-reps) discussed below show promise in research labs for modeling structures in the first (Pizer et al. 2005b; Jeong 2009) and third categories.

Two general classes of DSMs are in current clinical use: boundary representations (b-reps) and skeletal representations (s-reps), respectively (Figure 10.1). B-reps model an organ as an exterior surface with a hollow interior, and s-reps use an internal skeletal framework. M-reps are a class of s-reps with a curved grid (medial grid) that is midway between the two opposing surfaces of the organ (Pizer et al. 2005a). Pairs of equilength spokes point in opposite directions everywhere on the grid, implying the organ surface by their tips. Unlike b-reps, s-reps represent the interior volume as well as the exterior surface and thus

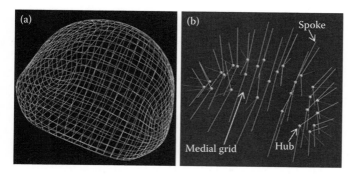

FIGURE 10.1 (a) B-rep (boundary mesh) for a prostate. (b) M-rep skeletal framework for the same prostate.

are solid models. This is an important distinction when using models for calculation of delivered dose from intratreatment images for example. In particular, solid models provide a convenient means to establish correspondence between tissue voxels, as opposed to image voxels, from one intratreatment image to another for the same patient.

10.1.2 Geometry and Image Intensity Patterns

DSMs can be classified as geometrically trained or untrained. The main distinction is that the deformation properties of an untrained DSM are assigned by the model builder, usually in terms of mechanical attributes. Trained DSMs, on the other hand, go through a training process to learn the shape space of the target organ across a statistical sample from the population of interest.

There are many possible choices for the assigned attributes for an untrained DSM. In general, they are cleverly defined and tailored to the particular model and driving problem. For example, the model surface can be assigned mechanical properties, such as elasticity and stiffness. In this case, the so-called internal energy of the DSM would represent the amount of energy required to stretch and bend the surface of the starting model to fit the target organ. Assuming a good initialization of the starting model, the optimization process would resist stretching to avoid large changes in surface area and to avoid sharp bends and bumpiness to control smoothness. The most significant disadvantage of this approach is that the shape space defined by the mechanical properties and energy function does not match the shape space of the target organ; therefore, noncredible segmentations are highly likely.

Statistically trainable DSMs (SDSMs) can learn shape properties of the target organ from images with contours of the target organ carefully drawn by one or more human experts. Training involves fitting a stock model to each set of human-drawn contours in each training case and statistically analyzing the resulting set of fit models to yield a shape space that represents the range of credible shapes for the target organ (Merck et al. 2008). This approach captures the humans' understanding of organ geometry in the context of the associated image data. During segmentation, deformations are constrained to be within the shape space inferred from expert human contours; thus, the final segmentation is much more likely to be credible. From a Bayesian point of view, the optimization process that controls model deformation is more principled if the shape space is in the form of a probability distribution on organ shape consisting of a mean shape, the modes of shape variation from the mean, and a variance for each mode (Section 10.3).

Organ geometry is understood by humans in the context of image intensity patterns in the immediate vicinity of a target organ. To mimic human understanding during contouring, it is essential to take intensity patterns into account during model-based segmentation (MBS). The term "intensity pattern" is general and refers to any quantity derived from image intensity numbers.

For untrained DSMs, intensity patterns are expressed in terms of a so-called external energy function. As with internal energy, there are many choices for the external energy function and the challenge is to craft one that works robustly. External energy could include, for example, a measurement of edge strength computed from the target image data. This is a reasonable choice when the target object has good contrast. The measurement of edge strength is only relevant near the model surface; thus, it is summed over the surface after each incremental deformation and input to an objective function. The objective function is formulated to reward high sums; hence, optimization drives the model surface closer and closer to local image regions with high edge strength. Another form of external energy could involve attractive or repulsive forces assigned to image features that are automatically computed or interactively defined by a user, for example, a nearby structure touched by the organ. An exemplar template that represents ideal intensity patterns near the target organ can serve as an implicit energy function. During deformation, the optimization process seeks a good match between the target image data near the model and the template. The disadvantage of untrained external energy functions is that they are too simplistic to fully characterize the complex relationships between image data and target organs.

Intensity training of SDSMs involves analysis of training image data in the immediate vicinities of models that have been fit to training cases. The analysis yields a quantitative description of image intensity patterns relative to the organ (and model) for each case. Examples of patterns include edge strength (Pekar et al. 2004), intensity profiles along line segments perpendicular to the organ surface (Cootes et al. 2001; Rao et al. 2005), intensity histograms (Freedman et al. 2005), quantile functions (Broadhurst et al. 2006), and feature vectors (Zhan and Shen 2006). The set of patterns from all cases is analyzed to create the so-called "appearance" that characterizes image intensities in and around the model. The appearance can be represented in a number of forms, but, as with the shape space, the ideal form is a probability distribution comprising a mean pattern, modes of variation from the mean, and variances associated with each mode (Broadhurst et al. 2006).

Unlike conventional diagnostic computed tomography (CT) imaging, other volume imaging modalities lack industry standards for defining uniform image intensity numbers. Moreover, intensity numbers and ranges for megavoltage and kilovoltage cone-beam CT (CBCT) systems for intratreatment imaging do not match well with conventional CT imaging systems across tissue types. In principle, the lack of uniformity in intensity numbers can be dealt with by measuring intensity patterns in a way that is independent of modality, for example, mutual information (Wells et al. 1996). However, this approach has not been fully investigated for clinical applications. At the present time in clinical systems, appearance is specific to an imaging modality and, with the exception of conventional CT, specific to manufacturer and perhaps machine model for the same modality and manufacturer.

Shape space is an intrinsic property of an organ, and appearance is a property of image data relative to the organ. They are mathematically independent entities that can be used alone or in combination with an SDSM. When used together, the deformation process strikes a balance between the distance of the deformed model from the mean shape and the goodness of match between the appearance and target image intensity patterns. In this context model, "distance" can be expressed as a function of the Mahalanobis distance, the number of standard deviations over all the modes of variation in the shape space (Pizer et al. 2005a).

10.1.3 Initializing the Starting Model in a Target Image

Regardless of whether a DSM is trained or untrained, the starting model should be a typical representative of the target organ to minimize the amount of deformation required to match the target organ and thereby improve computational efficiency and avoid convergence to local optima far from an acceptable result. For untrained DSMs, the starting model can be defined in a variety of ways but generally has a shape that is characteristic of the target. For SDSMs, the starting model can be the mean from the shape space or a model that has been roughly deformed interactively to match the target organ.

DSMs "live" in a reference coordinate system, and target images live in image space with different coordinate systems and units. The first step is to convert the original DICOM image data into a standard 3D Cartesian coordinate system with spatial units specified in physical or world dimensions (e.g., millimeter or centimeter). The main purpose of initialization is to compute a transformation matrix that carries the model to some position, scale (physical size), and 3D pose in the image data that has been converted to world dimensions. In principle, the position, scale, and pose from this step can be arbitrary. Before the starting model can be deformed, however, it must be positioned, scaled, and posed to match the target organ as closely as possible. This step usually is considered to be part of initialization but could be regarded as part of segmentation. However, if the computationally efficient initialization algorithm does not perform this step, the optimization approach discussed next for segmentation becomes computationally expensive and is much more likely to get trapped in a local optimum with the deformed model far from the desired result.

Fully automatic initialization can be accomplished by preprocessing the target image data to compute image features that are correlated with the target anatomy. These features are then used to position and scale the starting model close to the target object. Useful image features range from simple high-contrast edges or landmarks to more complex features such as low-contrast structures, image texture, and feature vectors. In general, fully automatic initialization is challenging for organs that have one or more of the following properties: (1) ambiguous surface regions (e.g., due to poor contrast or immediately adjacent tissues with a similar intensity range), (2) large shape variability from patient to patient (or day to day for the same patient), or (3) large internal or external intensity variations from patient to patient (or day to day for the same patient). In CT images, the prostate, bladder, and rectum can have one or more of these properties, so methods for automatic initialization currently do not produce fully acceptable results on a consistent basis.

Expert humans do a better job of accurately locating useful image features than computers, especially for organs with unfavorable image properties. A simple and fast user-assisted approach involves selecting a few points on the surface or interior of a target object. If user interaction is simple and brief and editing tools are efficient, compared with automatic initialization, the net result is a clinically acceptable segmentation with much less human intervention and shorter overall time.

10.1.4 Deforming an Initialized Model

After initialization, segmentation is accomplished by deforming the model to optimize an objective function. The model is incrementally deformed by perturbing the model parameters and evaluating the objective function after every perturbation. The objective function usually is defined so that optimization seeks to minimize the sum of all the terms in the function. Based on values of the objective function perturbation after perturbation, the optimization algorithm seeks a direction that drives the perturbed model closer and closer to the target organ.

The objective function for untrained DSMs includes the internal and external energy functions discussed above. The objective function for SDSMs contains terms called "geometric typicality" and "geometry-to-image match." Geometric typicality is an intrinsic property of the model and usually is a measure of distance in the shape space of the deformed model from the mean. Large deviations from the mean are more strongly penalized. Geometry-to-image match is a measure of how well the appearance associated with the current deformed model matches the intensity patterns in the target image. The optimization seeks a balance between these two terms. In principle, this process results in a very close estimate of the desired shape. In practice, interfering factors, such as false edges, may require special consideration to achieve an acceptably close estimation.

10.1.5 Groups of Geometrically Related Objects

For normal human anatomy, organs in the same anatomical region are geometrically correlated in the sense that the position, pose, and perhaps size of one organ can be predicted by one or more nearby organs (Pizer et al. 2005b). These correlations are most noticeable in rigid anatomical regions, such as the head and brain, but exist also in pliant regions. In the male pelvis, for example, the rectum is a tube-like structure immediately adjacent to the posterior surface of the prostate; it is oriented generally along the base–apex axis. Also, the bladder is an ellipsoid-like structure that joins and may overlap the base of the prostate; it usually has a long axis that is almost perpendicular to the base–apex axis. These types of correlations can be valuable in devising autosegmentation strategies for groups of

related objects. The first structure segmented, for example, can be trained to predict the initial position, pose, and scale for one or more of the remaining structures (Jeong 2009). Moreover, this strategy has the potential to predict structures that are poorly imaged such as lymph node chains. However, implementing this general approach efficiently and robustly enough for clinical practice is difficult and may require several more years of development.

10.1.6 Segmentation Accuracy versus Clinical Acceptability

To support scientific validity, reports in the research literature often describe experiments to compare autosegmentation results with human-drawn contours. Portable software tools have been developed to perform these measurements (e.g., Gerig et al. 2001b). Perhaps the most commonly reported accuracy metrics are average distance between the surface of the automatic segmentation and a surface fit to the human-drawn contours and volume overlap (Dice coefficient; Lin 1998). Accuracy varies from organ to organ, but the better methods, including those discussed in Section 10.3, generally agree within approximately 1 to 3 mm for average surface separation and approximately 90% or better for volume overlap. From a clinical perspective, these measurements are indicators of potential performance and are of marginal overall value for the following reasons: (1) Each clinical user judges accuracy relative to his or her own opinion about the shape of the target organ. (2) Metrics based on averages and global volume overlap incompletely characterize accuracy. (3) There are no community standards for measuring and reporting accuracy; thus, it is difficult to compare results from different groups. (4) There are no shared data sets, including images with ground truth segmentations, for comparing different approaches. (5) Accuracy experiments usually are performed by researchers familiar with methodology in a manner that does not mimic clinical use. (6) Accuracy is only one, and perhaps not the most important, metric related to clinical acceptability. The ultimate measure of acceptability is widespread clinical adoption, and adoption depends on a number of complexly related factors, including (1) user needs and expectations, (2) how well the packaged technology fits into the clinical workflow, (3) clinical robustness (i.e., the percentage of autosegmentations that are clinically useful with minimal editing), (4) fast and easy-to-use initialization and editing tools, (5) the set of structures that can be automatically segmented, and (6) perceived value to patient care.

10.2 Boundary Representations

10.2.1 Historical Roots

Active contour models, also known as snakes, are among the earliest nontrainable b-reps to receive widespread attention for image analysis (Kass et al. 1987). In their original 2D form, they are closed contours made from spline curves, and the model parameters are the spline control points. Nontrainable 3D snakes (e.g., Terzoupolus et al. 1987) are difficult to initialize and have been largely abandoned for medical applications. Trainable 3D forms are useful for research (Kelemen et al. 1999) but have not yet found routine clinical application.

Active shape models (ASMs) were the first b-reps to demonstrate the promise of SDSMs for medical imaging applications (Cootes et al. 1995). ASMs represent an organ by points distributed over its surface, perhaps complemented by a few internal or external points (e.g., the center of a concavity). When building an ASM, a subset of points is first created by careful manual labeling of points that are spatially related to important shape features, which can be stably located in corresponding positions across a training set of images from the target population. A model made from such a group of points is called a "point distribution model" (PDM). Other points can be interpolated from the labeled points. Points across all training cases are automatically aligned by minimizing the variance between corresponding points. The collection of thus aligned points is statistically analyzed to compute a PDM comprising an average position for each point and modes of variation of the PDM. Active appearance models (AAMs) (Cootes et al. 2001) compute not only an average and modes of variation of a PDM, which form a shape space, but also an average and modes of variation for image gray levels, which form a space of reference images. A training step learns a linear relation (regression) that yields a new segmentation and new reference image, given the difference of the present reference image and a deformation of the target image based on the segmentation so far. During segmentation, this linear relation is iteratively applied. Image analysis systems using AAMs have been commercially developed for medical applications, but not specifically for radiation oncology, by Imorphics (Manchester, UK) and Optasia Medical (Cheadle, UK).

10.2.2 Model-Based Segmentation: Philips Pinnacle³ MBS and Nucletron Oncentra MBS

B-reps are used in optional segmentation modules for Pinnacle³ (Philips Medical Systems) and Oncentra (Nucletron). Both treatment planning systems refer to their technology as MBS. Pinnacle³ MBS is based on the work of Pekar et al. (2004). Oncentra MBS was developed by RaySearch Laboratories, where it is known as RayAnatomy. RaySearch recently introduced their own planning system called RayStation, which includes RayAnatomy (also called MBS) as the segmentation platform. Little has been published about RayAnatomy, but the approach appears to be similar to Pekar et al. (2004) described below.

In Pinnacle³ MBS, organs are represented as a surface mesh of triangular tiles computed from conventional 2D contours (Figure 10.2). To capture sharp changes in topology, tile density is proportional to local surface curvature. The model parameters are the mesh vertices and the normal vectors for the triangular tiles.

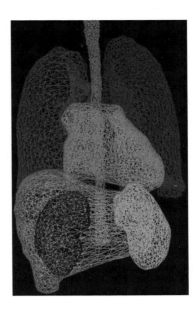

FIGURE 10.2 MBS boundary meshes for organs in the thorax and abdomen. (Courtesy of Philips Medical Systems.)

MBS comes with an inventory of previously computed mean meshes for the bladder, prostate, rectum, femurs, femoral heads, brain, brainstem, spinal cord, eyes, lenses, mandible, breasts, heart, lungs, and kidneys. Also, tools are available for users to modify existing meshes and to create their own. MBS organ models can be partially or fully trained. Partial training involves a few training cases to compute a mean mesh that has an internal energy function (Kaus et al. 2007). Only fully trained meshes are discussed here (Pekar et al. 2004).

The mesh vertices are treated like ASM surface points. The correspondence property of ASM points requires all training meshes to have the same number of tiles with vertices in corresponding locations. To accomplish this, the mesh from a random training case is rigidly and nonrigidly registered to the contours of the remaining training cases yielding a full training set of mesh models. The training models are registered using a Procrustes method that accounts for translation, rotation, and scaling differences (Goodall 1991). Principal component analysis (PCA) is performed on the aligned meshes to produce a PDM comprising a mean shape vector and principal modes of variation given by eigenvectors of the covariance matrix.

The mean shape vector $\bar{v}$ is

$$\bar{v} = \frac{1}{L}\sum_{i=1}^{L} v_i, \qquad (10.1)$$

where v_i is the shape vector for the ith training mesh ($x_{i1}, x_{i2}, \ldots x_{in}$)^T, x_{ij} is the coordinate of jth vertex for the ith training mesh, n is the number of vertices, and L is the number of training meshes.

The covariance matrix C is

$$C = \frac{1}{L} \sum_{i=1}^{L} (v_i - \overline{v})(v_i - \overline{v})^{\mathrm{T}}. \qquad (10.2)$$

The total number of eigenvectors of C is equal to $3N$. In practice, only the dominant modes are useful for reconstructing the shape of a target organ. The actual number m is determined by the model builder and usually is on the order of 10 or less. The shape S of a random target organ is approximated by

$$S \approx \overline{v} + \sum_{i=1}^{m} q_i M_i, \qquad (10.3)$$

where q_i is the weighting factor for mode M_i.

The internal energy E_{int} of the model is defined as

$$E_{\mathrm{int}} = \sum_i \sum_{j \in N(i)} \left[x_i' - x_j' - sR\left(\overline{x}_i - \overline{x}_j + \sum_{k=1}^{m} q_k \left(M_i^k - M_j^k \right) \right) \right]^2, \quad (10.4)$$

where $\overline{x}_i$, $\overline{x}_j$ are the vertices of the mean mesh, x_i', x_j' are the vertices of current deformed mesh, $N(i)$ is the set of vertices connected to the ith vertex by one triangle edge, s is the scaling factor between mean mesh and current deformed mesh, and R is the rotation matrix between mean mesh and current deformed mesh.

The scaling factors and rotation matrix R are computed for every instance of a deformed mesh by a point-based registration algorithm that minimizes the least-squares distances between the sets of vertices for the mean and deformed meshes. Using Equation 10.4 in the objective function causes optimization to favor spatial correspondence between the vertices of the mean and deformed meshes. This is analogous to a probabilistic approach that penalizes the deformed model in proportion to its distance in shape space from the mean.

The external energy E_{ext} is defined as

$$E_{\mathrm{ext}} = \sum_i w_i \left[e_{\Delta I} \left(\dot{x}_i' - p_i \right) \right]^2, \qquad (10.5)$$

where $e_{\Delta I}$ is the unit vector in the direction of the image gradient at p_i, p_i is the position of the attractor point in the target image near mesh triangle i, $\dot{x}_i'$ is the position of the center of mass of the ith triangle in the current deformed mesh, and w_i is the weighting factor.

The strength of attraction at p_i in general increases with gradient magnitude, but other considerations can come into play such as the intensity range for the target organ. Also, the user can increase the "strength" of weak attractors in the image. Using Equation 10.5 in the objective function penalizes the movement of the center of mass in the direction parallel to $e_{\Delta I}$ (i.e., toward lower image gradient) but allows it to move without penalty along the isogradient ridge, hypothetically the organ edge, intersecting p_i.

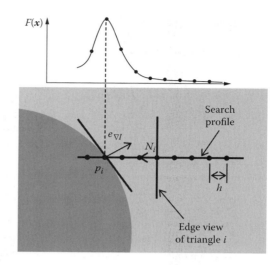

FIGURE 10.3 Search profile through the center of mass of triangle i. Values of the image feature function $F(\boldsymbol{x})$ are computed at each point.

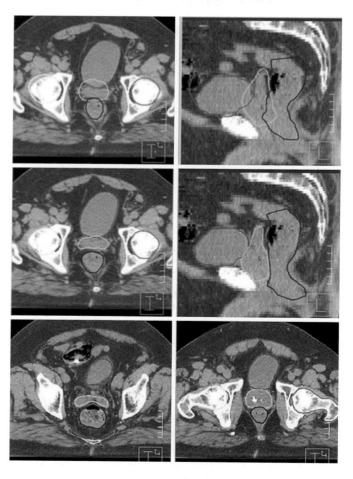

FIGURE 10.4 (See color insert.) Top row: Axial and sagittal slices showing initialized meshes for the prostate, bladder, rectum, and femoral heads. Middle row: Meshes after automatic segmentation. Bottom row: Meshes after interactive editing. (Courtesy of Philips Medical Systems.)

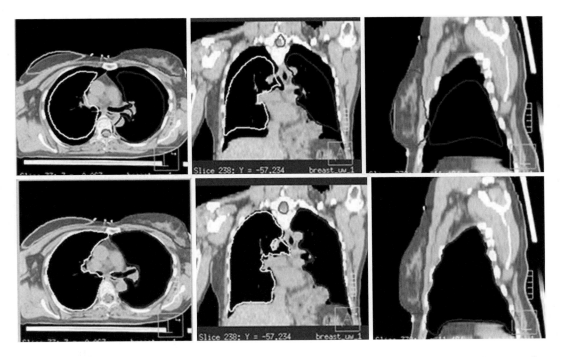

FIGURE 10.5 **(See color insert.)** Top row: Axial, sagittal, and coronal slices showing initialized meshes for the breasts and structures in the thorax. Bottom row: Unedited meshes after automatic segmentation. (Courtesy of Philips Medical Systems.)

The objective function combining the terms in Equations 10.4 and 10.5 is

$$E = E_{ext} + \alpha E_{int} + \beta E_{usr}, \qquad (10.6)$$

where E_{usr} is an optional energy term associated with user-labeled weak attractors and α and β are user adjustable weighting factors.

Segmentation begins by user-guided initialization of the starting model in the target image. The user is given tools to translate, rotate, and scale the model to closely match the target organ. Tools also are provided to nudge local regions of the model to achieve a more custom fit. When automatic segmentation begins, the starting mesh is perturbed recursively in a deterministic manner. The general approach involves a local search of the image data near each triangle to find an organ-specific attractor that pulls the triangle in its direction. For the ith triangle, the search is conducted along a search profile in the direction n_i normal to the center of mass of the triangle face (Figure 10.3). The profile has a predetermined length (~2 cm) with evenly spaced intervals h (~1 mm) between points on both sides of the triangle. Image feature values $F(x)$ are computed at each point to sample for attractor features p_i. In general, $F(x)$ is a measure of "boundariness" to attract the mesh to edges of the target organ. Rules have been developed to handle multiple attractors. When the search is completed for all triangles, the mesh is deformed within the trained shape space using Equation 10.3 to achieve a balance between matching the attractors and satisfying the constraints implied by the external energy term. This process is repeated until the objective function converges.

Segmentation results usually require editing to correct attraction to false edges (e.g., by dragging the mesh to the desired attractor and rerunning segmentation) and to label weak attractors. MBS also provides 3D morphologic shape operators that cause localized regions of the surface mesh to deform and/or shift according to user actions and the selected shape operator. Spherical and Gaussian operators have been described (Pekar et al. 2004).

Figures 10.4 and 10.5 show results for the male pelvis and female breast and thorax, respectively.

10.2.3 CMS Atlas-Based Autosegmentation

The CMS Atlas-Based Autosegmentation workstation is a vendor neutral system with a core method based on deformable registration of multiple atlas images with the target image. The head-and-neck module combines the strengths of two fundamental approaches by using the output of the deformable atlas to create a DSM that is then deformed to yield a better match to the image data for segmentation of the mandible (Figure 10.6; Han et al. 2009). The surface produced by atlas registration is converted to a triangulated mesh and parameterized to yield a b-rep in the form of a vector function

$$\vec{x}_0 = \left[x_0(r,s), y_0(r,s), z_0(r,s) \right], \qquad (10.7)$$

where $\vec{x}_0$ denotes the starting surface, x_0, y_0, z_0 are the initial mesh nodes in target image coordinates, and r,s are the node indices.

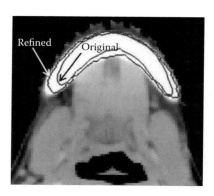

FIGURE 10.6 Thirteen axial slices through the mandible showing the segmentation from deformable atlas registration compared with the refined b-rep segmentation.

The initial surface vector $\vec{x}_0(r,s)$ is perturbed by a deformation field $\vec{d}(r,s)$ to yield a new surface $\vec{x}(r,s) = \vec{x}_0(r,s) + \vec{d}(r,s)$. The segmentation solution is formulated in terms of finding the optimal deformation field $\vec{d}_{opt}$ using a gradient-descent algorithm:

$$\vec{d}_{opt} = \arg\min \iint \exp\left(-\left\|\nabla J(\vec{x}(r,s))\right\|^2\right) drds + \iint \left\|\nabla \vec{d}(r,s)\right\|^2 drds.$$

(10.8)

In Equation 10.8, the first term on the right-hand side is the external energy and drives $\vec{x}(r,s)$ to high gradients in the image data expected at the mandible surface. The second term is the internal energy and enforces smoothness on $\vec{d}(r,s)$, which in turn is manifested as smoothness of the optimal surface $\vec{x}_{opt} = \vec{x}_0 + \vec{d}_{opt}$.

Boney structures like the mandible are good candidates for a combined approach involving an untrained b-rep because bones have high-contrast boundaries and a simple gradient-based energy function may work well. However, this idea can be extended to more challenging soft-tissue structures using trained models with strong shape and appearance components (Section 10.4.2).

10.3 Medial Representations

10.3.1 Historical Roots

Medial geometry describes objects in terms of their middles and widths. The concept of a medial skeletal structure was invented in the work of Blum (1967). In Blum's 2D work, the medial axis is implied by an object's boundary. In addition to a number of theoretical problems associated with computing Blum's axis, a very important practical consideration for image analysis applications is that Blum's axis is notoriously sensitive to small-scale details on the boundary, and real image data are notoriously noisy and unreliable for defining the boundaries of many structures of interest (Katz and Pizer 2003). Pizer was the first to recognize that medial concepts could be useful in image analysis if the relationship between the medial axis and the surface was inverted

(Nackman and Pizer 1985). In particular, Pizer took the view that a medial skeleton implies a boundary. This insight was the starting point for m-reps as discussed here. Later work by Damon (2004) showed that m-reps in their continuous form are a class of so-called skeletal structures (s-reps), around which a whole area of mathematics of geometry is evolving (Siddiqi and Pizer 2008).

10.3.2 Accuray MultiPlan

Accuray's MultiPlan Treatment Planning System incorporates a module developed by Morphormics, which uses m-reps and s-reps to segment structures in the male pelvis for treatment planning for prostate cancer.

There are two basic geometric forms of m-reps or s-reps: quasi-tubes and slabs. A quasi-tube is a structure such as the rectum, which is roughly tubular but with an axial cross section that is noncircular. Also, the scale (e.g., height and width) of the cross section can vary along the structure's length. A slab object, such as a kidney or liver, is nontubular and is analogous to a clam shell in the sense that the object is modeled as having two opposing surfaces. Some objects can be represented by either geometric form.

Both continuous (Yushkevich et al. 2003) and discrete or sampled (Pizer et al. 2005a) m-reps have been investigated. At the present state of development, the discrete form offers greater versatility and computational efficiency. Discrete slab m-reps have a curved medial sheet, a more or less regular grid with its edges aligned with the object's crest (i.e., the rounded transition zone that joins the two opposing surfaces). For some objects, a crest is not obvious and the model builder can orient the sheet in a natural plane that bisects the object. The nodes of the grid are sample points. A so-called medial atom (Figures 10.1 and 10.7)

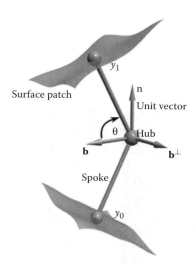

FIGURE 10.7 Interior medial atom with two equilength spokes, y_0 and y_1, which touch surface patches on opposite sides of the object. Interior atoms populate interior nodes of the medial grid illustrated in Figure 10.1. Edge atoms have a third spoke to represent the crest connecting the opposite sides of the object. Unit vectors **b**, **b**$^\perp$, and n define the coordinate system for each atom.

is located at each node, and a fine grid of atoms can be computed by interpolation (Han et al. 2007). A quasi-tube m-rep has a single chain of atoms with many spokes per atom (Figure 10.8). The number of atoms needed to adequately model an organ can be determined by a method to find the fewest atoms that capture the full range of shape variability for the target organ (Styner et al. 2003). In practice, several tens of atoms, depending on organ scale and shape variability, are usually sufficient (e.g., 28 atoms in a 4 × 7 grid for the prostate).

An atom located at an interior node of a slab grid has a hub with two equilength (or approximately equilength in the case of s-reps) spokes terminating on opposite surfaces. An edge or crest atom has a third spoke that sweeps out the crest region. A full surface is computed as a tiled skin with vertices at the spoke ends. The surface can be computed at a fine scale by subdivision surfaces (Thall 2004) or with greater accuracy by fitting a tiled surface to the spoke tips of properly interpolated atoms (Han et al. 2007). The many spokes of a quasi-tube atom are equally spaced around the hub and are allowed to have different lengths for noncircular shapes. Additional atoms and spokes can be computed by interpolation.

An interior slab atom is described by eight parameters: the coordinates of the hub (three scalars), the directions of the two spokes (two scalars each), and the common length of the two spokes. An edge atom has an additional length parameter for the third spoke. Thus, the total number of parameters for a slab m-rep is 8 × (# interior atoms) + 9 × (# edge atoms).

Atoms have individual coordinate systems; every atom "knows" the location and relative configuration of every other atom even for multiple objects. This prior knowledge allows for Markov fashion atom–neighbor predictions during segmentation (Lu et al. 2003). That is, any atom can predict the position and configuration of immediate neighbor atoms with a high level of statistical reliability. This capability improves segmentation efficiency and works against the occurrence of unlikely neighbor relations.

Single-figure objects require a single medial grid, and more complex objects require multiple grids (Han et al. 2005). Only single-figure objects are discussed here. M-reps are fit to expert human contours for shape training in a manner similar to b-reps but with some key differences. A stock m-rep model is fit by minimizing the distance between the surface of the m-rep and the surface formed from the human-drawn contours under added conditions that lead to positional correspondence across training cases (Merck et al. 2008). The added conditions include (1) a term that constrains the medial sheet to be in the same orientation relative to the organ across all training cases, (2) a term that regularizes the medial grid to avoid regions with widely spaced or tightly clustered nodes, (3) a penalty for shape deviation compared with a reference model representing a good guess of the typical shape (e.g., a good fit or mean from previous cases), and (4) a penalty that forces points on the m-rep surface into close spatial correspondence with user-identified landmarks, such as the centers of the prostate base and apex. The first three conditions impose correspondence of internal geometry across training cases, and the third condition imposes anatomical feature correspondence.

M-reps "live" in so-called symmetric space where PCA is ill-suited for statistical analysis. Instead, the collection of m-reps fit to training contours is analyzed using principal geodesic analysis (PGA; Fletcher and Joshi 2004). PGA is a form PCA that deals properly with medial atom motions. Like PCA, PGA yields a mean m-rep, principal modes of variation, and variances associated with each mode. The intrinsic shape-representation properties of m-reps lead to principal modes that match well with natural and intuitive shape changes, such as bending, twisting, rotating, filling/emptying, indenting, protruding, and magnification.

Intensity training takes advantage of the m-rep object–relative coordinate system conferred by the properties of medial atoms. This coordinate system allows the model to understand image data as a function of position relative to the object surface (Figure 10.9). For the prostate, for example, the surface regions

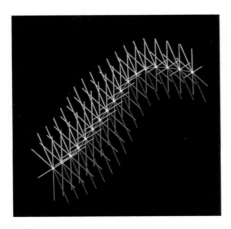

FIGURE 10.8 Chain of tube atoms for a quasi-tube m-rep of the rectum. Each atom has eight spokes.

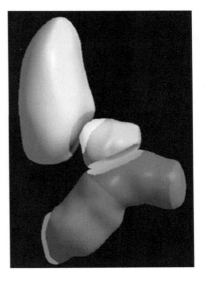

FIGURE 10.9 Organ regions for appearance training are defined in object–relative. Surface regions for the prostate, bladder, and rectum are shaded in different shades of gray.

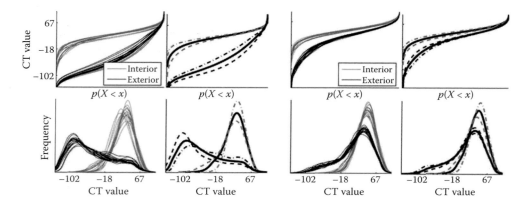

FIGURE 10.10 Top: Bladder (left) and prostate (right) training RIQFs for interior and exterior regions. For each, the left panel shows the training samples and the right panel shows the learned mean and ±2 standard deviations along the first principal direction. Bottom: Intensity histograms corresponding to the respective RIQFs.

bordering the bladder base and anterior rectal wall can be treated individually. Moreover, the image appearance can be learned at multiple spatial scales (e.g., global and region-by-region). This is useful for a multiscale segmentation approach, for example, a global appearance can be used in a first stage, and regional appearance can be applied to refine the global result.

A challenge in creating an appearance for use within a statistical optimization framework is to find an effective intensity-related metric that can be expressed in the form of a probability density function. Building on work that demonstrated the potential usefulness of intensity histograms (Freedman et al. 2005), Broadhurst et al. (2006) developed such a model that he called regional intensity quantile functions (RIQFs; Figure 10.10). The general approach is to define surface patches for an organ model, extrude each patch a specified distance outward (e.g., ~5 mm for some organs) and a separate specified distance inward, and compute and invert the cumulative intensity histograms for each region to yield internal and external appearances. The surface regions are defined in object–relative coordinates. Their shapes are arbitrary and defined by the model builder, usually to correspond with regions that have stable histogram patterns from image to image (Figure 10.9). The training image data can be preprocessed to classify voxels into broad categories (e.g., bone, soft tissue, fat, and air), so that tissue classes can be treated separately when building RIQFs.

RIQF feature space is linear in both the mean of the intensity histogram and the width (standard deviation) of the intensity distribution, suggesting that PCA analysis in this space is valid.

For each internal and external region and tissue class, PCA yields a mean RIQF, a set of principal variances, and a set of principal RIQFs, each a vector of length equal to the corresponding principal standard deviation. The principal RIQFs with dominant principal variances are chosen for the appearance.

In the male pelvis, m-reps are used to model the prostate, seminal vesicles, and rectum; the bladder and femoral head and neck are modeled with s-reps. Except for the prostate, initialization of models in a target planning image is fully automatic, although

the user can elect to use so-called initialization points for challenging cases. For the prostate, a few initialization points are interactively placed on or inside the organ surface (Figure 10.11). This strategy biases the initialization and subsequent segmentation toward the user's opinion and therefore often produces results that require minimal or no editing.

Segmentations of the prostate, seminal vesicles, and rectum (Figure 10.12) are performed in a statistical framework based around Bayes' (1764) theorem using a conjugate gradient optimization algorithm. In the context of image analysis, a Bayesian-like approach optimizes objective functions of the form

$$\arg\max_{\underline{\mathbf{m}}\in s} p(\underline{\mathbf{m}}|\underline{\mathbf{I}}) = \arg\max_{\underline{\mathbf{m}}\in s} [\log p(\underline{\mathbf{I}}|\underline{\mathbf{m}}) + K \log p(\underline{\mathbf{m}})], \tag{10.9}$$

where $\underline{\mathbf{m}}$ is the currently deformed model in the shape space s, $\underline{\mathbf{I}}$ is the target image data relative to $\underline{\mathbf{m}}$, $p(\underline{\mathbf{m}}|\underline{\mathbf{I}})$ is the probability of $\underline{\mathbf{m}}$ given $\underline{\mathbf{I}}$, $p(\underline{\mathbf{I}}|\underline{\mathbf{m}})$ is the probability of $\underline{\mathbf{I}}$ given $\underline{\mathbf{m}}$ (geometry-to-image match), $p(\underline{\mathbf{m}})$ is the probability of $\underline{\mathbf{m}}$ (geometric typicality),

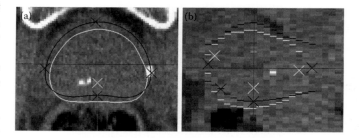

FIGURE 10.11 Mid-axial (a) and sagittal (b) slices through segmentations of the prostate resulting from two types of initialization points, interior (white Xs) and surface (black Xs). Fewer interior points (approximately 1 to 3) are required compared with surface points (approximately 14). When interior points are used, the prostate shape is completely determined from CT intensity patterns. Surface points bias the shape toward the user's opinion.

Planning CT   During treatment

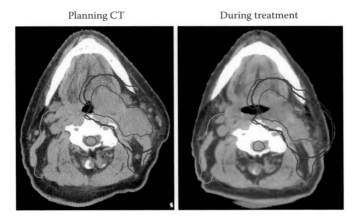

FIGURE 1.2 Significant anatomic changes can occur during radiation therapy for head-and-neck cancer patients. Left: Treatment planning CT with original target volumes overlaid on the CT images. Right: CT image acquired after 3 weeks of radiotherapy. The original target volume is no longer matching well with the patient's anatomy. The treatment will be suboptimal if a new plan is not derived from the new CT image.

Impact of tumor shrinkage for proton therapy

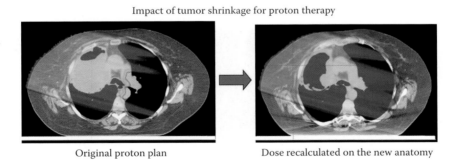

Original proton plan   Dose recalculated on the new anatomy

FIGURE 1.3 Impact of tumor shrinkage to proton dose distributions. Left: Original proton therapy plan. After about 1 month of treatment, the primary tumor shrank significantly and the originally collapsed lung tissues also expanded. As a result, the proton beam penetrated further into the contralateral (healthy) lung tissue, potentially resulting higher toxicity.

Autosegmentation by using deformable image registration

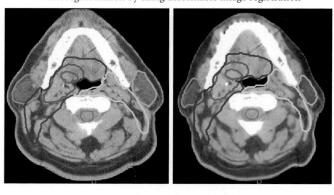

Original planning CT image and contours   Automatically segmented contours on a new CT

FIGURE 1.5 Example of autosegmentation by deformable image registration is shown for a head-and-neck cancer patient. Left: CT slice with labeled structures. The target volumes include the primary GTV (purple), the high-risk CTV (red), the intermediate-risk target volume (blue), and the low-risk target volume (yellow). The parotid glands (left and right) are also shown in blue and green contours. After deformable image registration between the new CT after 3 weeks of treatment and the original planning CT, the same contours are deformably mapped to the new CT for adaptive planning.

Nonrigid changes in patient's anatomy

Planning CT

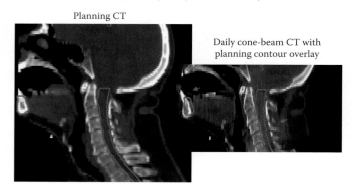

Daily cone-beam CT with planning contour overlay

FIGURE 1.6 Nonrigid variation of the patient's anatomy cannot be easily corrected by a simple couch shift even when IGRT is used. Neck curvature and chin positional variations are considered nonrigid changes. If the changes are systematic, a replanning may be the best strategy to correct such complicated shape variations.

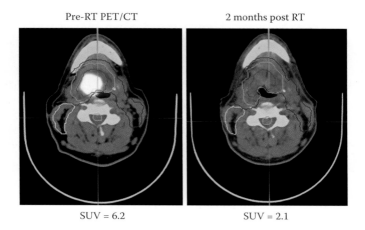

Pre-RT PET/CT   2 months post RT

SUV = 6.2   SUV = 2.1

FIGURE 1.8 Deformable image registration in quantitative assessment of functional outcome. PET/CT images acquired before and 2 months after radiation therapy are shown on the left and right, respectively. To evaluate the changes in the treated target volume, CTVs (shown in colored contours) were mapped to each PET/CT from the planning CT. A reduction of the mean SUV within the GTV (red) was measured.

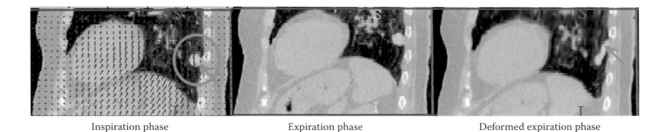

Inspiration phase   Expiration phase   Deformed expiration phase

FIGURE 1.10 Illustration of deformable registration error caused by isotropic smoothing of displacement field. Right: Deformed expiration phase image to match with the inspiration phase image. The shape of the tumor was dragged unrealistically due to the smoothing requirement inside the deformable image registration algorithm.

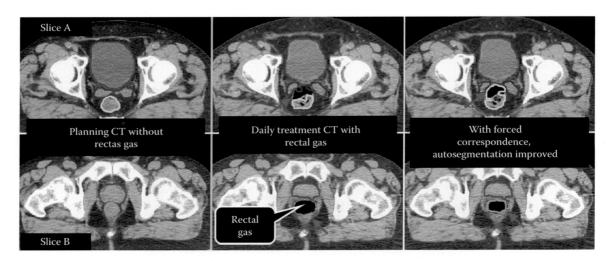

FIGURE 1.11 If the planning CT has an empty rectum (left), registration errors may occur near the area of CT images containing the rectal gas. Incorrect rectal wall is detected by the autosegmentation algorithm (middle column). One method to handle this situation is to artificially modify the planning CT images so that it will contain an air pocket in the center of the contoured (empty) rectum. This will result in a "virtual" correspondence between the gaseous regions. The improvement in autosegmentation can be seen in the CT images on the right column.

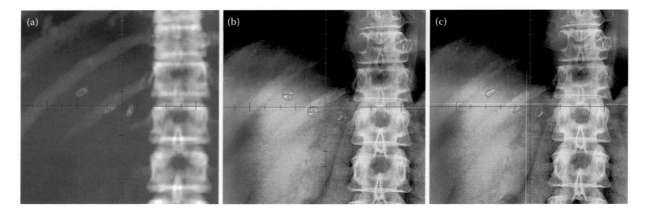

FIGURE 4.3 Fiducials as landmarks for liver tumor alignment for daily radiation treatment. (a) Reference digitally reconstructed radiography from planning CT with fiducials contoured. (b) Daily portal image with fiducials marked manually. (c) Alignment of tumor based on fiducials using sum of squared distance.

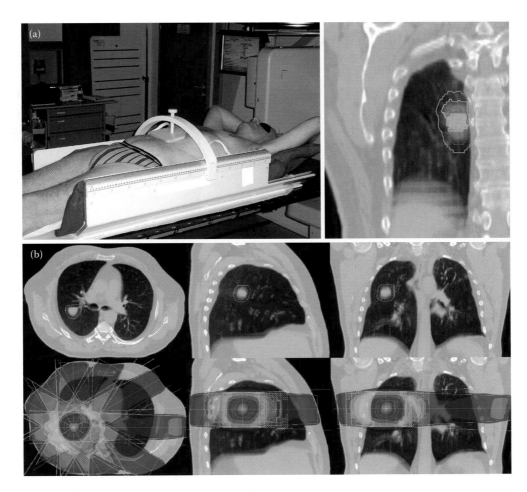

FIGURE 7.1 Top left: Example of lung cancer treatment with stereotactic body radiation therapy contention system and abdominal compression aiming at reducing motion amplitude. Top right: Coronal slice showing contoured structures representing the tumor motion (exhale, inhale positions, and ITV). Bottom: Dose distribution obtained from treatment planning system with a 12-field plan.

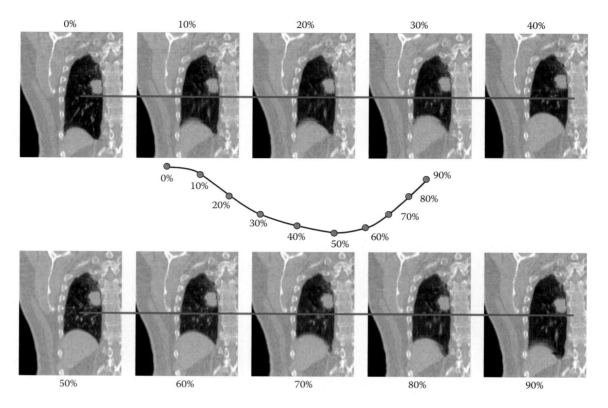

FIGURE 7.2 Example of the 10 phases that compose a 4D CT image. (Reprinted from *Cancer Radiothérapie, 15*(2), Ayadi, M., Bouilhol, G., Imbert, L., Ginestet, C., and Sarrut, D., Scan acquisition parameter optimization for the treatment of moving tumors in radiotherapy, 115–122, Copyright 2010, with permission from Elsevier.)

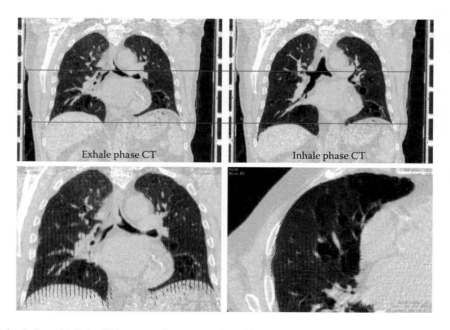

FIGURE 7.5 Top: Initial exhale and inhale CT images to be registered. Red lines help to compare the two coronal slices. Bottom: Coronal and axial slices with deformation field superimposed. The vector field is only displayed in the lung region.

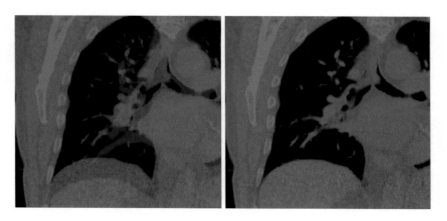

FIGURE 7.6 Green–purple differences before and after registration.

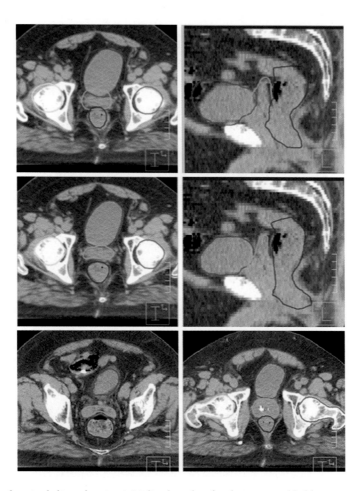

FIGURE 10.4 Top row: Axial and sagittal slices showing initialized meshes for the prostate, bladder, rectum, and femoral heads. Middle row: Meshes after automatic segmentation. Bottom row: Meshes after interactive editing. (Courtesy of Philips Medical Systems.)

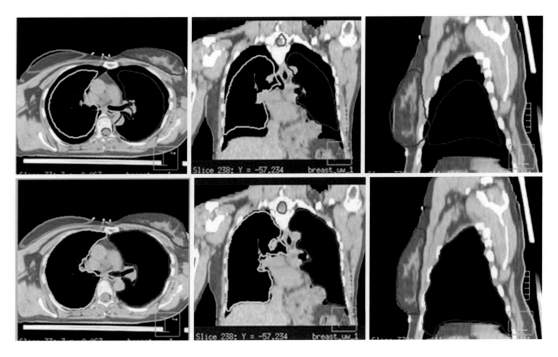

FIGURE 10.5 Top row: Axial, sagittal, and coronal slices showing initialized meshes for the breasts and structures in the thorax. Bottom row: Unedited meshes after automatic segmentation. (Courtesy of Philips Medical Systems.)

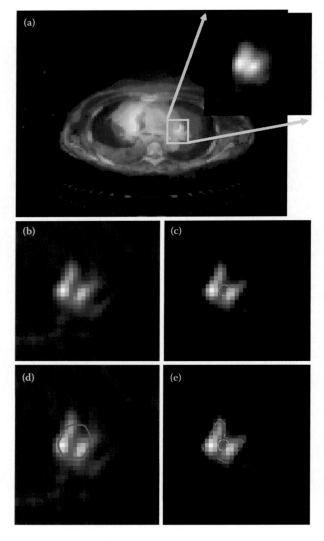

FIGURE 11.4 Application of the level set segmentation to NSCLC. (a) PET/CT lung image and ROI selected for processing. (b) Deblurred PET lesion using deconvolution. (c) Initialization of the level set with a small circle. (d) Resulting contour with level set using gradient-based approach (converged after 100 iterations). (e) Resulting contour with level set using edgeless-based approach after 40 iterations (yellow) and 100 iterations (red). Note how the edgeless-based approach has better capturing ability of the lesion boundary in this case.

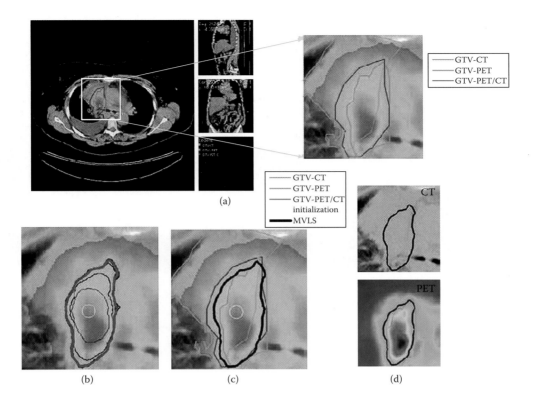

FIGURE 11.6 Analysis of lung PET/CT case. (a) Fused PET/CT displayed in CERR with manual contouring shown of the subject's right GTV. The contouring was performed separately for CT (orange), PET (green), and fused PET/CT (red) images. (b) MVLS algorithm was initialized with a circle (white) of 9.8 mm diameter, evolved contour in steps of 10 iterations (black), and the final estimated contour (thick red). The algorithm converged in 120 iterations in a few seconds. The PET/CT ratio weighting was selected as 1:1.65. (c) MVLS results shown along with manual contour results on the fused PET/CT. Note the agreement of fused PET/CT manual contour and MVLS (DSC = 0.87). (d) MVLS contour superimposed on CT (top) and PET (bottom) separately.

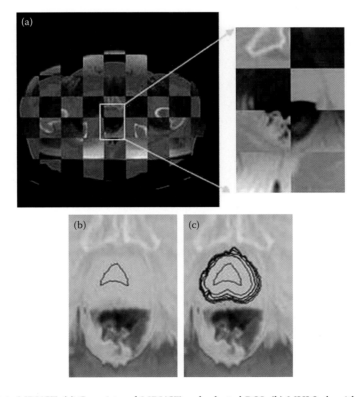

FIGURE 11.7 Analysis of prostate MRI/CT. (a) Coregistered MRI/CT and selected ROI. (b) MVLS algorithm is initialized with a shape prior that roughly resembles prostate (inside thin line). (c) Curve evolution in steps of 10 iterations and the final estimated contour (outside heavy line).

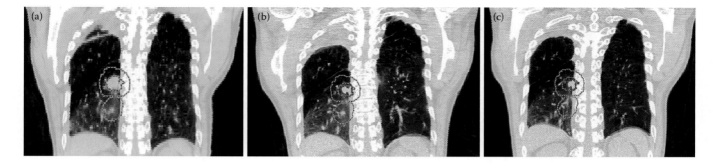

FIGURE 11.9 Example of tracking tumor changes during radiotherapy using a level set based approach. The GTV is shown in green, PTV in brown, and the active model estimate in red. (a) Pretreatment 3D scan. (b) and (c) Reconstructed mid-treatment and end-of-treatment scans, respectively, from the 4D CT data at end of exhalation phase as reference. Note the accurate capturing of tumor boundary by the level set approach at these time points by propagating the GTV.

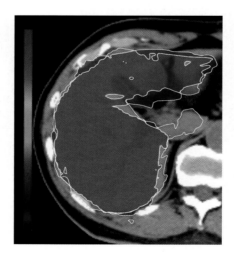

FIGURE 12.4 Deducing bulk organ shape from atlas segmentations with the STAPLE algorithm. The white lines represent liver segmentations obtained by different atlases. The color overlay is the probability map that the corresponding voxel belongs to the liver as deduced from the individual segmentations, ranging from 0 (violet) to 1 (red).

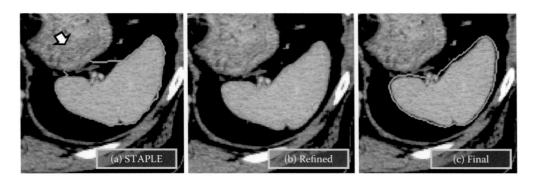

FIGURE 12.5 Steps of the refinement procedure. (a) Input spleen segmentation shown as a blue contour. (b) Regions of different Hounsfield units are eliminated using a threshold level set-based algorithm. (c) The shape is smoothed and snapped to image gradients using a geodesic level set algorithm.

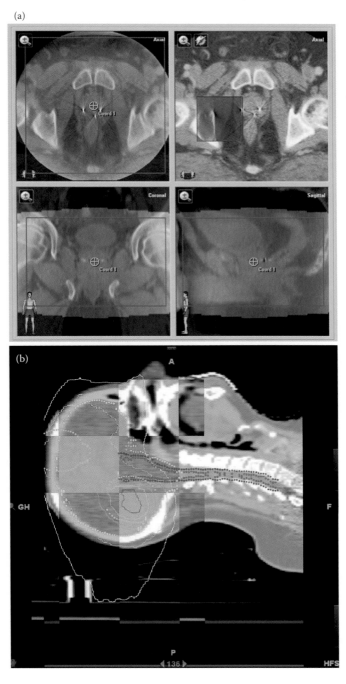

FIGURE 15.19 Illustration of various 4D CBCT registrations for SBRT of a lung cancer patient with a peak-to-peak amplitude of approximately 7 mm. Top row: Registration of the vertebrae; middle row: tumor registrations; bottom row: correction to the time-averaged tumor position. Left column: exhale phase; right column: inhale phase. (From Sonke, J.J. et al., *Int J Radiat Oncol Biol Phys*. 2009 Jun 1;74(2):567–74.)

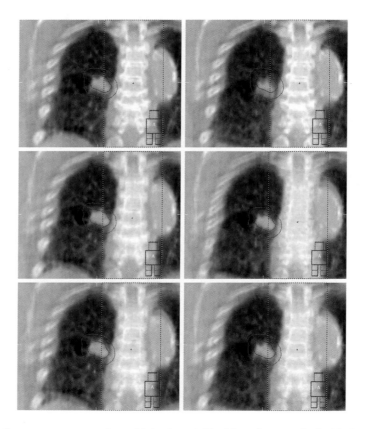

FIGURE 15.20 (a) CBCT scan of a prostate cancer patient with implanted fiducial markers acquired with the Vero system in an axial, coronal, and sagittal views and a fusion of the CBCT with the planning CT (top right). (b) Tomotherapy scan of a head-and-neck cancer patient in a sagittal view fused with the planning CT. (Image courtesy of Tom Depuydt, University Hospital Brussels, Brussels, Belgium.)

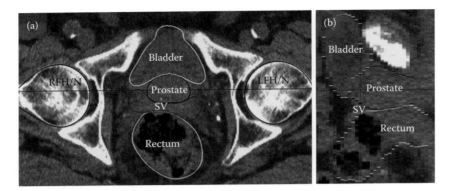

FIGURE 10.12 Mid-axial (a) and sagittal (b) slices through a planning CT image showing unedited segmentations of structures in the male pelvis. Surface initialization points were used for the prostate. All other structures were fully automatic.

and *K* accounts for the ratio of standard deviations between the two log *p* terms.

In practice, other terms can be introduced into the right-hand side of Equation 10.9 usually in the form of penalties (e.g., to keep the model surface close to the user-annotated initialization points).

The probability of **m** is computed from the shape space with the deviation of **m** from the mean expressed in terms of the Mahalanobis distance (Pizer et al. 2005a). The geometry-to-image match is computed as the sum, over the principal RIQFs, of the squares of the coefficients of the principal RIQFs plus a penalty for any residual shape that falls outside the principal RIQF space.

Due to its wide shape variability, the bladder is segmented without a trained shape space. The segmentation begins with a search along radii extending from an automatic initialization point using an appearance model to look for boundariness in the image data. The first pass produces an estimated surface and a skeletal structure that replaces the initialization point as the framework for another search. The process is repeated until convergence is achieved.

The femoral H/N has high contrast and small shape variability and can be segmented using an s-rep with geometric rules in place of a trained shape space (Figure 10.12). The rules, for example, confine the radius of the femoral head to a specified range and define the stopping length for the neck.

The m-rep architecture and object-relative coordinate system facilitate simple and efficient editing. For the bladder, for example, the user can annotate the target image with a short contour fragment to indicate the region where an edit is needed (Figure 10.13). Although the user input is 2D, the edit is performed over a 3D region in the vicinity of the annotation. The shape change is confined to be within or very close to the trained shape space to yield an edit that is credible.

10.4 Methods in Translation

This section discusses the work that is under investigation in the clinical radiation oncology setting at multiple sites. M-reps are used in these projects, but other SDSMs may have similar capabilities.

10.4.1 Adaptive Radiation Therapy Paradigm

CT images are acquired in the treatment room during the practice of image-guided radiation therapy (IGRT) and adaptive radiation therapy (ART; Yan et al. 2000). However, routine clinical implementation of ART is impeded by the lack of practical methods to segment the organs needed for calculating dose-volume metrics from delivered doses. An effective approach to deal with this problem is to use models from the planning image as patient-specific models that can be automatically initialized

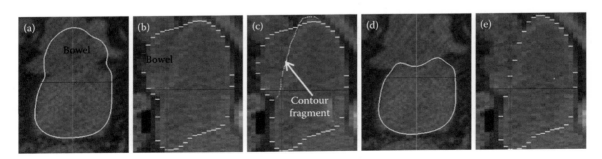

FIGURE 10.13 Mid-axial (a) and sagittal (b) slices through the bladder showing the unedited original segmentation. Bowel impinging on the bladder is included in the segmentation. A partial contour on the sagittal slice (c) results in a 3D edit that removes the unwanted bowel (d and e).

in an intratreatment CT image via rigid registration of the two images. The essential differences compared with segmenting a planning image are fully automatic initialization for all organs and the use of an intrapatient shape space trained from multiple images (approximately 10 to 15) per patient across several tens of patients (Pizer et al. 2006). Morphormics has developed a clinical prototype workstation called MxARTsuite that implements this strategy with m-reps and s-reps. Figure 10.14 shows the results for a treatment image acquired with a conventional diagnostic CT scanner, and Figure 10.15 shows the results for a CBCT image.

Delivered dose distributions computed from treatment CT images are mapped to the planning CT for each organ using the correspondence properties of m- and s-reps as given in the following equation:

$$\text{VAL}'_{\text{Mapped}}(\mathbf{m}'(i', j', k')) = \text{VAL}(\mathbf{m}(i, j, k)) \qquad (10.10)$$

where $\mathbf{m}$ and $\mathbf{m}'$ are the prostate models in treatment and planning CTs, respectively; VAL is the value of scalar in the treatment CT at $\mathbf{m}$-relative position i,j,k; $\text{VAL}'_{\text{Mapped}}$ is the value of scalar (e.g., dose, label, or intensity) mapped to $\mathbf{m}'$-relative position i',j',k'; and i,j,k and i',j',k' are corresponding positions in $\mathbf{m}$- and $\mathbf{m}'$-relative coordinates, respectively (Figure 10.16).

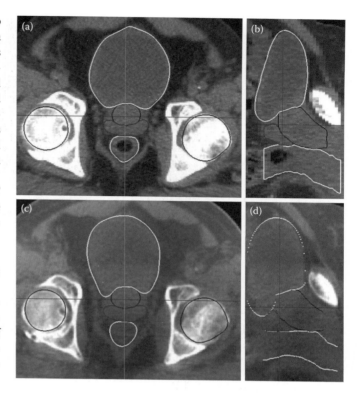

FIGURE 10.15   As in Figure 10.14, but with a kilovoltage CBCT image.

Error dose distributions are computed by subtracting the delivered dose summed over all images from the planned dose (Figure 10.17).

CBCT images are not suitable directly for the calculation of delivered dose largely because there are no industry standards for CBCT numbers equivalent to Hounsfield units. Other complicating problems include the following: (1) The x-ray tube and flat-panel receptor used for imaging have characteristics (e.g., broad beam and poor collimation) that are unfavorable for acquiring images with the same quality as conventional CT systems; (2) the electronics of the flat panel can "drift" and cause inconsistent intensity patterns for multiple images of the same patient over time; and (3) pockets of gas cause strong artifacts

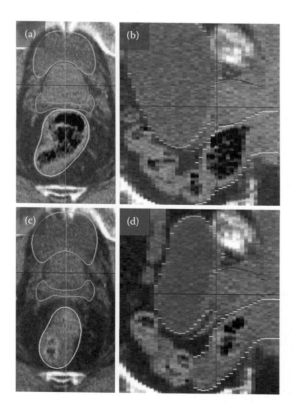

FIGURE 10.14   Mid-axial (a) and sagittal (b) slices showing patient-specific atlas organ models by segmenting the planning CT as in Figure 10.12. Unedited fully automatic segmentations from a CT image acquired with a conventional helical scanner (c and d).

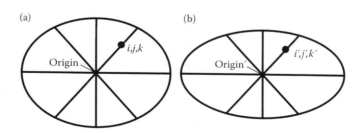

FIGURE 10.16   2D depiction of corresponding points in atlas and deformed models. (a) Atlas model with point at i,j,k. (b) Deformed model with corresponding point at i',j',k'. The points are on the same spoke at the same fractional distance from the hub origin.

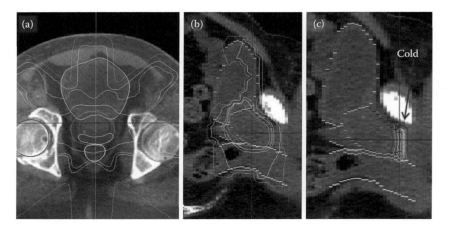

FIGURE 10.17 (a) Isodose distribution calculated from the intratreatment CBCT image with intensities mapped from the planning CT as in Figure 10.18. (b) Isodoses from the CBCT image mapped to the planning CT using m-rep correspondences alone (Equation 10.10). (c) Error dose distribution showing cold spot at prostate apex.

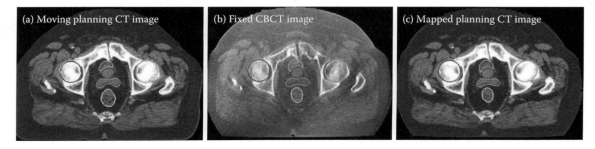

FIGURE 10.18 Planning CT image (a) is used as the moving image in a feature-constrained nonrigid registration to a cone-beam treatment image (b). The result is shown in (c).

in organ-relative intensity patterns. This problem can be overcome by mapping planning CT intensities to CBCT images of the same patient using a form of feature-constrained nonrigid registration (Montagnat and Delingette 1997). A particularly promising approach combines the correspondence properties of m-reps with the voxel-scale strengths of a nonrigid registration method based on viscoelastic flow (VEF) (Joshi and Miller 2000; Foskey et al. 2005). In particular, corresponding points sampled from MBS in a CBCT image act as constraints for VEF registration of the planning CT with the CBCT. An example case is shown in Figure 10.18.

10.4.2 Prostate-Specific Local Diffeomorphisms Based on Fiducial Marker Coordinates in Treatment Space

Systems to localize the prostate during radiation therapy by tracking markers implanted in the prostate (Murphy 1998; Balter et al. 2005) have several advantages, including ease of use and frequent sampling during each dose fraction (e.g., 10 Hz for the Calypso system). The absence of image data, however, prevents accurate calculation of delivered dose for quality assurance

purposes, the practice of ART, and, potentially, dynamic adjustment of treatment parameters. Recent work shows that marker locations recorded during treatment delivery can be used as the basis for mapping reference CT image data (e.g., the planning CT) into the treatment space to estimate treatment CT data acceptable for calculating the dose delivered to the prostate and tissues immediately adjacent to the prostate surface (Lee et al. 2010). A patient-specific m-rep is created by segmenting the prostate in the reference CT image, and the positions of the implanted markers are recorded in the m-rep coordinate system (Figure 10.19). The patient-specific m-rep is deformed in the treatment space as described in Section 10.4.1 by using the marker coordinates in treatment space as surrogate image data, that is, deformation within the shape space is driven by optimizing the match between the seed positions recorded from the planning CT and the seed coordinates recorded by the tracking system. The starting m-rep and the deformed m-rep imply a prostate-specific diffeomorphism between the planning and treatment spaces via the correspondences from Equation 10.10. The dose calculation is mapped back into the planning space by applying the reverse transformation.

Figure 10.20 shows a slice through a treatment CT with image data mapped from the planning CT as described above. This

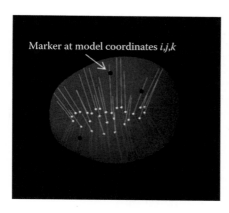

FIGURE 10.19 Patient-specific prostate atlas model created by segmenting the planning image. Markers are indicated by black dots. The markers are matched to treatment coordinates recorded by the tracking system. The m-rep is deformed within the trained shape space to yield a shape and pose for the prostate at the moment when the marker locations were recorded.

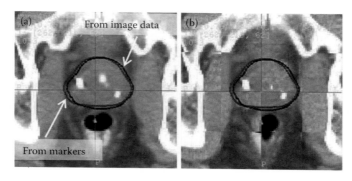

FIGURE 10.20 (a) Actual intratreatment CT image with two prostate segmentations. (b) 4 × 4 Checkerboard display of the original CT and the locally deformed original CT. Prostate contours are from the actual CT image and the Calypso marker coordinates recorded at the beginning of treatment.

figure shows that the region of interest mapped from the planning CT matches well with the surrounding image data in the treatment CT, particularly for the prostate and anterior rectal wall.

10.5 More Information

Many journals in the fields of image analysis and computer science regularly publish articles on segmentation. The collections of papers on classic deformable models can be found in the book by ter Haar Romeny (1994) and proceedings of conferences, such as CVRMed (Ayache 1995) and CVRMed-MRCAS (McInerney and Terzopoulos 1996a,b; Jones and Metaxas 1997; Troccaz et al. 1997; Vehkomäki et al. 1997). The methods for deformation within a statistical framework are discussed in papers found in the proceedings of recent meetings on *Information Processing in Medical Imaging* (IPMI) and *Medical Image Computing and Computer Assisted Intervention* (MICCAI).

References

Ayache, N., ed. 1995. In: *Proceedings of the First International Conference on Computer Vision, Virtual Reality and Robotics in Medicine*, April 3–6, 1995, Nice, France. New York: Springer.

Balter, J. M., Wright, J.N., Newell, L.J. et al. 2005. Accuracy of a wireless localization system for radiotherapy. *International Journal of Radiation Oncology Biology Physics, 61*, 933–937.

Bayes, T. 1764. An essay toward solving a problem in the doctrine of chances. *Philosophical Transactions of the Royal Society London, 53*, 370–418.

Blum, H. 1967. A transformation for extracting new descriptors of shape. In: *Models for the Perception of Speech and Visual Form*, ed. W. Wathen-Dunn, 363–380. Cambridge, MA: MIT Press.

Broadhurst, R. E., Stough, J., Pizer, S.M. et al. 2006. A statistical appearance model based on intensity quantiles. In: *Proceedings of the International Symposium on Biomedical Imaging: From Nano to Macro*, April 6–9, 2006, Arlington, VA. Los Alamitos, CA: IEEE, 422–425.

Cevidanes, L. H. S., Styner, M., Phillips, C. et al. 2007. 3D morphometric changes 1 year after jaw surgery. In: *Proceedings of the International Symposium on Biomedical Imaging: From Nano to Macro*, April 12–15, 2007, Arlington, VA. Los Alamitos, CA: IEEE, 1332–1335.

Cootes, T. F., Taylor, C., Cooper, D. et al. 1995. Active shape models—Their training and application. *Computer Vision and Image Understanding, 61*, 38–59.

Cootes, T. F., Edwards, G. J., and Taylor, C. J. 2001. Active appearance models. *IEEE Transactions on Pattern Analysis and Machine Intelligence, 23*, 681–685.

Damon, J. 2004. Smoothness and geometry of boundaries associated to skeletal structures II: Geometry in the Blum case. *Composition Mathematica, 140*, 1657–1674.

Fletcher, P. T. and Joshi, S. 2004. Principal geodesic analysis on symmetric spaces: Statistics of diffusion tensors. In: M. Sonka, ed. *International Workshop on Computer Vision Approaches to Medical Image Analysis, and Mathematical Methods in Biomedical Image Analysis*. May 15, 2004, Prague, Czech Republic. New York: Springer, 87–98.

Foskey, M., Davis, B., Chang, S. et al. 2005. Large deformation 3D image registration in image-guided radiation therapy, *Physics in Medicine and Biology, 50*, 5869–5892.

Freedman, D., Radke, R., Zhang, T. et al. 2005. Model-based segmentation of medical imagery by matching distributions. *IEEE Transactions on Medical Imaging, 24*, 281–292.

Gerig, G., Styner, M., Shenton, M. et al. 2001a. Shape versus size: Improved understanding of the morphology of brain structures. In: W.J. Niessen and M.A. Viergever, eds. *Proceedings of the 4th International Conference on Medical Image Computing and Computer Assisted Intervention*, October 14–17, 2001, Utrecht, The Netherlands. New York: Springer, 24–32.

Gerig, G., Jomier, M., and Chakos, M. 2001b. Valmet: A new validation tool for assessing and improving 3D object segmentation. In: W. Niessen and M. Viergever, eds. *Proceedings of the 1st International Conference on Medical Image Computing and Computer Assisted Intervention*, October, 14–17, 2001, Utrecht, The Netherlands. New York: Springer, 516–523.

Goodall, C. 1991. Procrustes methods in the statistical analysis of shape. *Journal of the Royal Statistical Society, 53*, 285–339.

Han, Q., Pizer, S. M., Merck, D. et al. 2005. Multi-figure anatomical objects for shape statistics. In: G. Christensen and M. Sonka eds. *Proceedings of the 19th International Conference on Information Processing in Medical Imaging*, July 10–15, 2005, Glenwood Springs, CO. New York: Springer, 701–712.

Han, Q., Merck, D., Levy, J. et al. 2007. Geometrically proper models in statistical training. In: N. Karssemeijer and B. Lelieveldt eds. *Proceedings of the 21st International Conference on Information Processing in Medical Imaging*, July 2–6, 2007, Kerkrade, The Netherlands. New York: Springer, 751–762.

Han, X., Hibbard, L. S., O'Connell, N. et al. 2009. Automatic segmentation of head and neck CT images by GPU-accelerated multi-atlas fusion. *MIDAS Journal* [online]. Available at: http://hdl.handle.net/10380/3111.

Jeong, J.-Y. 2009. Estimation of probability distributions on multiple anatomical objects and evaluation of statistical shape models. Ph.D. thesis. University of North Carolina at Chapel Hill. Available at: http://midag.cs.unc.edu.

Jones, T. N. and Metaxas, D.N. 1997. Segmentation using models with affinity-based localization. In: J. Troccaz, Grimson, E. et al., eds. *Proceedings of the First Joint Conference on Computer Vision, Virtual Reality and Robotics in Medicine and Medical Robotics and Computer-Assisted Surgery*, March 19–22, 1997, Grenoble, France. New York: Springer, 53–62.

Joshi, S. and Miller, M. I. 2000. Landmark matching via large deformation diffeomorphisms. *IEEE Transactions on Image Processing, 9(8)*, 1357–1370.

Kass, M., Witkin, A., and Terzopoulos, D. 1987. Snakes: Active contour models. *International Journal of Computer Vision, 1*, 321–331.

Katz, R. and Pizer, S. 2003. Untangling the Blum medial axis transform. *International Journal of Computer Vision, 55*, 139–154.

Kaus, M. R., Brock, K., Pekar, V. et al. 2007. Assessment of a model-based deformable image registration approach for radiation therapy planning. *International Journal of Radiation Oncology Biology Physics, 68*, 572–580.

Kelemen, A., Szekely, G., and Gerig, G. 1999. Three-dimensional model-based segmentation. *IEEE Transactions on Medical Imaging, 18*, 828–839.

Lee, H.-P., Foskey, M., Levy, J. et al. 2010. Image estimation from marker locations for dose calculation in prostate radiation therapy. In: T. Jiang, M. Navab, J. Pluim et al. eds. *Proceedings of the 13th International Conference on Medical Image Computing and Computer Assisted Intervention*, September 20–24, 2010, Beijing, China. New York: Springer, 335–342.

Lin, D., An information-theoretic definition of similarity. In: J.W. Shavlik ed. *Proceedings of the 15th International Conference on Machine Learning*, July 24–27, 1998, Madison, WI. San Francisco: Morgan Kaufmann, 296–304.

Lu, C., Pizer, S., and Joshi, S. 2003. A Markov random field approach to multi-scale shape analysis. In: *Scale Space Methods in Computer Vision*, eds. L. D. Griffin and M. Lillholm, *Lecture Notes in Computer Science 2695*, 416–431.

McInerny, T. and Terzopoulos, D. 1996a. Deformable models in medical image analysis: A survey. *Medical Image Analysis, 1*, 91–108.

McInerny, T. and Terzopoulos, D. 1996b. Deformable models in medical image analysis. In: *Proceedings of the Workshop on Mathematical Methods in Biomedical Image Analysis*, June 21–22, 1996, San Francisco, CA. Los Alamitos, CA: IEEE, 171–180.

Merck, D., Tracton, G., Saboo, R. et al. 2008. Training models of anatomic shape variability. *Medical Physics, 35*, 35–84.

Montagnat, J., and Delingette, H. 1997. A hybrid framework for surface registration and deformable models. In: *Proceedings of the Conference on Computer Vision and Pattern Recognition*, June 17–19, 1997, San Juan, Puerto Rico. Los Alamitos, CA: IEEE, 1041–1046.

Murphy, M. J. 1998. Real-time imaging for patient position alignment and tracking. In: J. Hazle and A. Boyer, eds., *Imaging in Radiotherapy*, Madison: Medical Physics Publishing, 237–258.

Nackman, L. R. and Pizer, S. M. 1985. Three-dimensional shape description using the symmetric axis transform I: Theory. *IEEE Transactions on Pattern Analysis and Machine Intelligence, 7*, 187–202.

Pekar, V., McNutt, T. R., and Kaus, M. R. 2004. Automated model-based organ delineation for radiotherapy planning in prostatic region. *International Journal of Radiation Oncology Biology Physics, 60*, 973–980.

Pizer, S. M., Fletcher, P. T., Joshi, S. et al. 2005a. A method and software for segmentation of anatomic object ensembles by deformable m-reps. *Medical Physics, 32(5)*, 1335–1345.

Pizer, S. M., Jeong, J.-Y., Lu, C. et al. 2005b. Estimating the statistics of multi-object anatomic geometry using inter-object relationships. In: Olsen O.F., Florack. L., and Kuijper, A., eds. *Proceedings of the International Workshop on Deep Structure, Singularities and Computer Vision*, June 9–10, 2005, Maastricht, The Netherlands. New York: Springer, 60–71.

Pizer, S. M., Broadhurst, R. E., Jeong, J.-Y. et al. 2006. Intra-patient anatomic statistical models for adaptive radiotherapy. In: *Proceedings of the Medical Image Computing and Computer Assisted Intervention Workshop: From Statistical Atlases to Personalized Models: Understanding Complex Diseases in Populations and Individuals*, October 1–6, 2006, Copenhagen, Denmark. New York: Springer, 43–46.

Rao, M., Stough, J., Chi, Y.-Y. et al. 2005. Comparison of human and automatic segmentations of kidneys from CT images. *International Journal of Radiation Oncology Biology Physics, 61*, 954–960.

Siddiqi, K. and Pizer, S. eds. 2008. *Medial Representations: Mathematics, Algorithms and Applications*. New York: Springer.

Styner, M. and Gerig, G. 2000. Hybrid boundary-medial shape description for biologically variable shapes. In: *Proceedings of the Workshop on Mathematical Methods in Biomedical Image Analysis*, June 11–12, 2000, Hilton Head Island, SC. Los Alamitos, CA: IEEE, 235–242.

Styner, M., Gerig, G., Pizer, S. et al. 2003. Automatic and robust computation of 3D medial models incorporating object Variability. *International Journal of Computer Vision, 55*, 107–122.

ter Haar Romeny, B.M. ed. 1994. *Geometry-Driven Diffusion in Computer Vision*. Dordrecht: Kluwer.

Terzoupolus, D., Witkin, A., and Kass, M. 1987. Symmetry-seeking models and 3D object reconstruction. *International Journal of Computer Vision, 1*, 211–221.

Thall, A. 2004. Deformable solid modeling via medial sampling and displacement subdivision. Ph.D. thesis. University of North Carolina at Chapel Hill. Available at: http://midag.cs.unc.edu.

Troccaz, J., Grimson, E., and Moesges, R., eds. 1997. *Proceedings of the 1st Joint Conference Computer Vision, Virtual Reality and Robotics in Medicine and Medical Robotics and Computer-Assisted Surgery*, March 19–22, 1997, Grenoble, France. New York: New York: Springer.

Vehkomäki, T., Gerig, G., and Székely, G.A. 1997. User-guided tool for efficient segmentation of medical image data. In: J. Troccaz, E. Grimson et al., eds. *Proceedings of the First Joint Conference on Computer Vision, Virtual Reality and Robotics in Medicine and Medical Robotics and Computer-Assisted Surgery*, March 19–22, 1997, Grenoble, France. New York: Springer, 685–694.

Wells III, W.M., Viola, P., Atsumi, H. et al. 1996. Multi-modal volume registration by maximization of mutual information. *Medical Image Analysis, 1*, 35–51.

Yan, D., Lockman, D., Brabbins, D. et al. 2000. An off-line strategy for constructing a patient-specific planning target volume in adaptive treatment process for prostate cancer. *International Journal of Radiation Oncology Biology Physics, 48*, 289–302.

Yushkevich, P., Fletcher, P. T., Joshi, S. et al. 2003. Continuous medial representations for geometric object modeling in 2D and 3D. *Image and Vision Computing, 21*, 17–27.

Zhan, Y. and Shen, D. 2006. Deformable segmentation of 3-D ultrasound prostate images using statistical texture matching method. *IEEE Transactions on Medical Imaging, 25*, 256–272.

Zhao, Z., Taylor, W.D., Styner, M. et al. 2008. Hippocampus shape analysis and late-life depression. *PLoS ONE, 3*(3), e1837. Available at http://www.ncbi.nlm.nih.gov/pmc/articles/PMC2265542/?tool = pubmed.

11

Level Set for Radiotherapy Treatment Planning

Issam El Naqa, PhD
McGill University

11.1 Introduction

Recent evolutions in radiotherapy technology and the development of intensity-modulated radiation therapy (IMRT) planning and delivery systems have led to the emergence of image-guided and adaptive radiotherapy (IGART) to meet the new requirements of precise localization and definition of targets and surrounding critical structures at the time of planning and during the course of fractionated treatment (Bortfeld et al. 2006). Although these technological advances have created new opportunities in radiotherapy treatment planning and delivery, it has been noted that robust and computationally efficient software algorithms are still lagging to achieve the optimal utilization of these recent hardware developments (Xing et al. 2007).

The ultimate goal of radiotherapy treatment of cancer is to achieve high rates of local control through tumoricidal doses, which cover all the gross and subclinical disease, while limiting toxicity to surrounding normal tissues (Perez 2004). However, one of the main obstacles to achieving better treatment outcomes is the uncertainty associated with target volumes and organs-at-risk definitions. Experts' contours have been demonstrated to suffer from significant intraobserver and interobserver variability in different cancer sites (Weiss and Hess 2003). For instance, common variations in defining the organs-at-risk in prostate cancer have been shown to have significant effects on dose-volume histogram (DVH) estimates, which are typically

used for guiding the design of treatment plans and assessing their quality (Muren et al. 2004; Boehmer et al. 2006).

In a typical treatment planning procedure, contouring (segmentation) methods are required to distinguish the gross tumor volume (GTV) from surrounding normal tissues that should be maximally spared by the radiation beam. Typically, the treatment planner would perform the cumbersome structure delineation by hand; nevertheless, in some cases, newer automated and semi-automated image segmentation techniques are applied. Formally speaking, image segmentation is defined as the process of classifying the voxels of an image into a set of distinct two or more classes (Suri et al. 2002b). For instance, in binary segmentation, the image or the region of interest (ROI) is divided into labeled classes, namely, the foreground (representing the object of interest) and the background (representing the surrounding areas). An in-depth review of segmentation methods for radiotherapy treatment planning could be found in Bondiau et al. (2004).

Deformable image analysis methods and their application in radiotherapy have garnered great interest in recent years (Brock 2009). The purpose of deformable image analysis models is to allow for natural and common deformations of tissue to be explicitly included in the image registration and/or segmentation processes. Using such deformable image maps, dose, for instance, may be accumulated for each bit of irradiated tissue despite ensuing deformations between the different intertreatment or intratreatment fractions.

Technically, deformable models are geometric representations of curves or surfaces that are defined explicitly or implicitly in the imaging domain and could be applied for both registration and segmentation problems. These models move (deform) under the influence of internal forces, which are defined within the curve or surface itself, and external forces, which are computed from the image data (Sethian 1999; Xu et al. 2002).

The level set is a variational method that belongs to the class of geometric deformable models. It is defined by sets of contour values and spatial positions that comprise the volume of interest (VOI) boundary (e.g., target volume boundary). Recently, level set methods have witnessed resurgence in their application in many different areas and are considered state-of-the-art in image processing and shape recovery applications (Sethian 1999; Suri et al. 2002a; Osher and Fedkiw 2003; Aubert 2006).

In this chapter, we will focus on the application and potential of level sets in radiotherapy treatment planning. We will provide a brief review of level sets' mathematical foundation, implementation, and application in image segmentation. Then, we will present clinical examples of applying the level sets for single-modality and multimodality image segmentation for radiotherapy treatment planning. Finally, we will discuss current challenges and future potential applications in image-guided treatment planning.

11.2 Background

11.2.1 What Are Level Sets?

Level sets are deformable models that were developed for tracking boundaries and propagating fronts in the 1980s by mathematicians Stanley Osher from the University of California at Los Angeles and James Sethian from the University of California at Berkley in a seminal work (Osher and Sethian 1988). Since their development, the level sets have been applied successfully in several disciplines, including image processing, computer vision, and computational fluid dynamics (Sethian 1999; Osher and Fedkiw 2003).

The technique is based on using the geometric concept of evolving implicit level sets by mean curvature motion as illustrated in Figure 11.1. They were originally developed in curve evolution theory to overcome the limitations encountered in parametric deformable models [e.g., snakes (Kass et al. 1987)] such as initialization requirements, generalization to higher dimensions, and topological adaptation such as splitting or merging of model parts (Xu et al. 2002).

11.2.2 Variational Methods

Variational methods are based on mathematical techniques from calculus of variations and advanced partial differential equations (PDEs). They have been applied to a wide variety of image analysis problems, including image reconstruction, restoration (e.g., denoising/deblurring), and image segmentation (Aubert 2006). In a typical scenario, the problem is modeled by

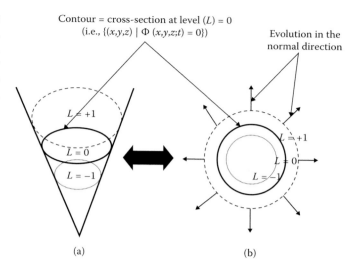

FIGURE 11.1 Deformable image segmentation by the level set method. (a) Representation of the level set surface at time t of the evolving function Φ. (b) Projected view showing the evolution direction. Typically, the function Φ evolves at a velocity proportional to the curvature of the contour and inversely proportional to the image gradient. In the present example, Φ is represented with a signed Euclidean distance transform of value L. The contour is extracted at $L = 0$, with negative values inside the contour representing the volume of interest and positive values outside representing the background.

an energy functional* and variational techniques are applied to estimate the optimal solution (minima/maxima of the functional yielding the object boundary, for instance). This solution is obtained by solving an associated Euler-Lagrange PDE using analytical or numerical techniques based on finite differences or finite element approaches.

Many frequently used deformable image modeling algorithms use variational techniques as part of their numerical machinery. For instance, the optical flow algorithm for deformable image registration is based on the continuity equation (Horn and Schunck 1981) and was applied for motion estimation in 4D computed tomography (CT) images (Yang et al. 2008) or registration of treatment planning kilovoltage CT images to daily megavoltage CT images in pelvic cases treated on tomotherapy machines (Yang et al. 2009), whereas the level set is based on the Hamilton–Jacobi formulation (Osher and Sethian 1988) and has been applied for single-modality and multimodality image segmentation problems (El Naqa et al. 2007; Li et al. 2008). These models move (deform) under the influence of internal force-like equations, which are defined within the curve or surface itself, or external forces, which are computed from the image data (Sethian 1999; Xu et al. 2002). The idea is that contours are characterized by sharp variations in the image intensity. Hence, the objective is to match deformed and reference contours by means of force equilibrium or energy minimization (Xu et al. 2002; Osher and Fedkiw 2003; Aubert 2006).

---

* A functional takes a function as its argument; in other words, it is a function of another function.

11.2.3 Snakes versus Level Sets

So-called "snakes" were among the first deformable models developed for image segmentation (Kass et al. 1987). Snakes use an explicit parametric representation of the object boundary that deforms by means of energy minimization (or dynamic force equilibrium). Mathematically, if the deformable contour/surface is represented by $C(s) = \{x(s), y(s), z(s)\}$, $s \in [0,1]$, then its movement is governed by the following objective function:

$$J(C(t)) = \int_0^1 \alpha(s) \left| \frac{\partial C(s;t)}{\partial s} \right|^2 + \beta(s) \left| \frac{\partial^2 C(s;t)}{\partial s^2} \right|^2 ds + \gamma \int_0^1 P(C(s;t)) ds,$$

(11.1)

where the first term corresponds to the internal energy and controls the tension and rigidity of the deforming contour. The first-order derivative suppresses stretching and makes the contour behave like an elastic string, whereas the second-order derivative suppresses bending and makes the model behave like a rigid rod. The second term corresponds to external energy, where P represents the potential energy given typically as a function of the image gradient ($g(|\nabla I|)$). In this case, g is selected to be a monotonically decreasing function of the image intensity (I) gradient. Using calculus of variation techniques as discussed above, the solution to Equation 11.1 is obtained by solving the associated Euler–Lagrange PDE (Xu et al. 2002; Osher and Fedkiw 2003) given by

$$\frac{\partial}{\partial s} \left(\alpha \frac{\partial C}{\partial s} \right) + \frac{\partial^2}{\partial s^2} \left(\beta \frac{\partial^2 C}{\partial s^2} \right) + \nabla P(C(s,t)) = 0.$$

(11.2)

However, the formulation in Equation 11.2 is nonconvex and suffers from several drawbacks, such as high sensitivity to contour initialization, dependency on the contour parameterization, and an inability to account for topological adaptation (e.g., delineation of a necrotic tumor). To solve some of these problems, particularly the sensitivity problem, the geodesic active contour model was proposed (Casselles et al. 1997), which, in principle, is equivalent to Equation 11.1 if the smoothness constraint is eliminated (i.e., by setting $\beta = 0$). These developments have led into further investigation of the role of this constraint, which could be shown to be redundant for most practical purposes and was one of the main motivations for the emergence of the flow or curve evolution concept as discussed below.

11.2.4 Mathematical Formulation

11.2.4.1 Curve Evolution Theory

Curve evolution is a nonparametric technique in which the contour evolution (deformation) is driven as a function of its curvature. This can be mathematically expressed as

$$\frac{\partial C}{\partial t} = \vec{V}(\kappa),$$

(11.3)

where $\vec{V}$ is the velocity function (of magnitude V) of the evolving contour in the normal direction ($\vec{N}$) as a function of time (t), and κ represents the local contour curvature. However, to resolve the main problem of parameterization and topological adaptation encountered in the snakes approach, the explicit contour representation of C is replaced by an implicit representation via the level set approach as explained below.

11.2.4.2 Level Sets with Edges

In the classic level set approach for contour representation, the curve (in Equation 11.3) is embedded into a level set function denoted as ϕ. This function defines sets of contour values and spatial positions, including the target object boundary at the zero level as shown in Figure 11.1:

$$\phi(C) = 0.$$

(11.4)

With the differentiation of Equation 11.4 with respect to the evolution time (t) and applying the chain rule, we obtain

$$\frac{\partial \phi}{\partial t} \frac{\partial t}{\partial t} + \frac{\partial \phi}{\partial C} \frac{\partial C}{\partial t} = 0.$$

(11.5)

Using the divergence operator, and substituting Equation 11.3, then

$$\begin{aligned} \frac{\partial \phi}{\partial t} &= -\nabla \phi \frac{\partial C}{\partial t} \\ &= -\nabla \phi \vec{V}(\kappa), \\ &= |\nabla \phi| V(\kappa) \end{aligned}$$

(11.6)

where the level set function (ϕ) is typically selected as a signed Euclidean distance function of the contour. V is the velocity function as before and is chosen to be proportional to the curvature (κ) and inversely proportional to the image gradient. As an example, V is given by

$$V(\kappa) = \frac{(\kappa + V_0)}{1 + |\nabla(G_\sigma \otimes I)|},$$

(11.7)

where V_0 is a constant controlling the deformation magnitude and direction similar to the balloon force used in other active contour approaches. G_σ is a Gaussian kernel with width σ, I is the image intensity, and $\otimes$ is the convolution operator. This term acts as a stopping force for the evolving contour as a function of smoothed image gradient. κ is the contour curvature and is the main driving force for contour evolution. In this case, it is given by

$$\kappa = \mathrm{div}\left(\frac{\phi}{|\phi|}\right) = \nabla \cdot \left(\frac{\phi}{|\phi|}\right). \tag{11.8}$$

More generally speaking, the evolution equation of Equation 11.6 could be rewritten as

$$\frac{\partial \phi}{\partial t} = V(\kappa)|\nabla \phi| + F(\Theta). \tag{11.9}$$

The additional term $F(\Theta)$ represents a user-defined additional constraint, where the vector parameter Θ could represent shape priors (C^+), contour string force constraints, etc. This provides further flexibility for the level set algorithm designer. Moreover, efficient solutions for level set problems were developed by using finite differences and fast marching methods as discussed below (Sethian 1999).

11.2.4.3 Level Sets without Edges

The traditional approach for level sets, mentioned above, relies on calculating image gradients for estimating image boundaries; however, this approach tends to be less robust due to the sensitivity of gradient calculations to image noise. An alternative is to use edge modeling or region-based techniques instead of gradient-based techniques. One approach is to use the Mumford–Shah model to represent edges or boundaries in the image (Chan and Vese 2001; Osher and Fedkiw 2003). In this case, the objective functional could be written as

$$\inf_C J(C,c_1,c_2) = \alpha \cdot \mathrm{length}(C) + \int^{\Omega} |I-c_1|^2 H(\phi)\mathrm{d}\mathbf{x}$$
$$+ \int^{\Omega} |I-c_2|^2 (1-H(\phi))\mathrm{d}\mathbf{x} \tag{11.10}$$

where H is the Heaviside function,* $c_1(c_2)$ corresponds to the mean intensity inside (outside) of the evolving contour, and $\inf_C$ is the infimum representing the greatest lower bound of the contours satisfying the objective functional. The first term represents a penalty function to the contour length. The next two terms represent the fitting of the foreground (F_1) and background (F_2), respectively, according to the Mumford–Shah model. This approach is illustrated in Figure 11.2, where the best fit to the boundary is obtained when both F_1 and F_2 terms are minimized.

11.2.4.4 Other Extensions

The sound mathematical foundation of the level set framework allows for its generalization to many different situations. For instance, it could be easily extended to multimodality image segmentation. In this case, assuming there are N images, then

---

* $H(x) = \begin{cases} 1, & x \geq 0 \\ 0 & \text{otherwise} \end{cases}$, in the implementation this is approximated by a smooth inverse tangent function.

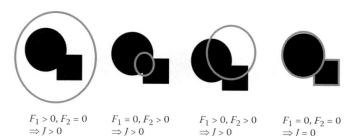

$$\begin{array}{cccc} F_1 > 0, F_2 = 0 & F_1 = 0, F_2 > 0 & F_1 > 0, F_2 > 0 & F_1 = 0, F_2 = 0 \\ \Rightarrow J > 0 & \Rightarrow J > 0 & \Rightarrow J > 0 & \Rightarrow J = 0 \end{array}$$

FIGURE 11.2 Illustration of the level set approach without edges using the Mumford–Shah model.

using the concept of multivalued level sets (MVLS) (Shah 1996; Chan et al. 2000), the objective functional in Equation 11.10 could be modified (in the case of multimodality images) to

$$\inf_C J(C,c^+,c^-) \propto \frac{1}{N} \sum_i \lambda_i^+ \int^{\Omega} |I_i - c_i^+|^2 H(\phi)\mathrm{d}\mathbf{x}$$
$$+ \lambda_i^- \int^{\Omega} |I_i - c_i^-|^2 (1-H(\phi))\mathrm{d}\mathbf{x} \tag{11.11}$$

where $(\lambda_i^+, \lambda_i^-)$ are user-defined parameter pairs providing "importance weights" for each of the imaging modalities in comparison with the other modalities.

Another example is the extension to image registration problems as in the work of Vemuri et al. (2000, 2003), in which the motion field is written as

$$\frac{\mathrm{d}V}{\mathrm{d}t} = (I_2 - I_1(V))\frac{\nabla I_1(V)}{|\nabla I_1(V)|}, \tag{11.12}$$

where I_1 is the moving image and I_2 is the reference image. In this case, the image difference becomes the velocity function for the motion field evolution (deformation). We have applied this approach for registration of 3D CT images using a multigrid implementation based on Laplace pyramid representation (Yang et al. 2007). A sample result is shown in Figure 11.3.

11.2.5 Numerical Implementation

As described earlier, the level set formalism belongs to the family of Hamilton–Jacobi PDEs. If we consider Equation 11.6, then the level set solution could be thought of as solving a heat convection problem in the normal direction of a curvature-dependent velocity function. The level set algorithm typically starts by some initial contour in the image domain; the curve then evolves under the influence of the internal forces (contour curvature and string force) and external forces (image boundaries; cf. Equation 11.7) until it reaches equilibrium.

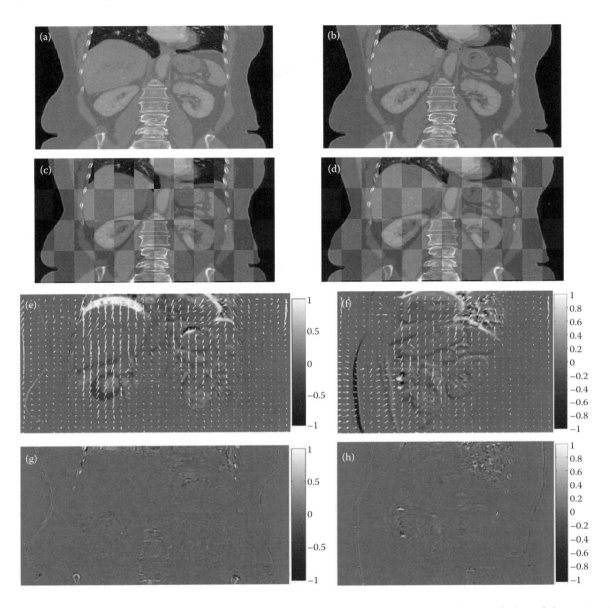

FIGURE 11.3 Results of registering the abdominal 3D CT images using the level set approach. (a) Coronal slice of the moving image. (b) Reference image. (c) Checkerboard image before registration. The brighter parts are from the moving image and the darker parts are from the reference image. (d) Checkerboard image after registration. The brighter parts are from the deformed moving image. (e) Coronal difference image before registration, overlaid with deformable vector field. (f) Sagittal difference image before registration, overlaid with deformable vector field. (g) Coronal difference image after registration. (h) Sagittal difference image after registration.

Consider the general level set formalism in Equation 11.6; the iterative solution to this PDE could be found using finite difference methods. For instance, starting on the right-hand side, the curvature could be approximated using central differences, whereas the diffusivity term $|\nabla\phi|$ could be approximated on a 2D Cartesian grid (j,k), as follows:

$$|\nabla\phi| \approx \sqrt{\frac{1}{2}[(D_+^x\phi_{jk})^2 + (D_-^x\phi_{jk})^2 + (D_+^y\phi_{jk})^2 + (D_-^y\phi_{jk})^2]}, \quad (11.13)$$

where the derivatives are approximated as follows:

$$D_\pm^x = \pm\frac{\phi_{j\pm1,k} - \phi_{jk}}{\Delta}$$

$$D_\pm^y = \pm\frac{\phi_{j,k\pm1} - \phi_{jk}}{\Delta} \qquad (11.14)$$

For time discretization, this is typically done using the Rung–Kutta approach. Osher suggests a third-order total variation diminishing the Rung–Kutta method (Osher and Fedkiw 2003). To speed up computation, further approximations based on narrow banding or fast marching could be used (Sethian 1999). In narrow banding, a short band is defined around the zero level set to restrict computational cost, and boundary conditions are updated accordingly. In the fast marching approach, the algorithm makes use of the assumption that a contour typically propagates in one direction only, and it finds the solution by propagating one voxel at a time in a systemic fashion providing significant decrease in computational time.

It is noticed in the solution of the curve evolution problem (Equation 11.6) that it does not necessarily guarantee that the evolving function $\phi(\cdot)$ will remain a valid distance function, which could potentially cause serious numerical problems. Therefore, methods for rebuilding the distance function have been proposed such that the zero level maintains its validity. One proposed approach uses the following PDE for reinitialization (Aujol and Aubert 2002):

$$\frac{\partial \phi}{\partial t} + \text{sign}(\phi)(|\nabla u| - 1) = 0, \quad (11.15)$$

where the PDE in Equation 11.14 is initialized with the current level set solution and upwinding finite difference schemes are used to approximate $|\nabla u|$ (Aubert 2006). This reinitialization process is typically repeated every 10 or 20 iterations of the level set propagation.

11.2.6 Evaluation Metrics for Level Set Application in Image Segmentation

For quantitative validation of the level set segmentation, several methods could be used. Besides measuring spatially independent metrics such as volumes, spatially dependent metrics could be also adopted such as the following (Zou et al. 2004b):

(1) Receiver operating characteristics (ROC) curve: This is a plot of the sensitivity (true classification fraction) versus 1-specificity (false-positive fraction) for a continuum of threshold values. The overall accuracy is summarized by the area under the ROC curve (AUC).

(2) Spatial overlap index: This includes the Dice similarity coefficient (DSC) defined in terms of the pixel ratio of the overlapping regions:

$$\text{DSC} = \frac{2(A \cap B)}{(A + B)}, \quad (11.16)$$

where, at any given threshold, DSC values would range from 0, indicating no spatial overlap between two sets of binary segmentation results, to 1, indicating complete overlap (Zou et al. 2004a). DSC is a special case of the κ statistics commonly used in reliability analysis to measure observers' agreement (Woolson and Clarke 2002). For purposes of statistical hypothesis

testing, a logit* transformation is applied to map DSC from the domain of [0,1] to the unbounded range $(-\infty,\infty)$. The transformed DSC values, under reasonable assumptions, follow an asymptotic normal distribution (Agresti 2002). A DSC value greater than 0.7 [i.e., logit(DSC) > 0.847] is typically selected to indicate good segmentation performance.

11.3 Level Sets for Radiotherapy Treatment Planning

11.3.1 Overview

In this section, we will demonstrate applications of the level set approach for target definition in radiotherapy planning of different cancer sites. We will first present applications to single-image modality cases and then we will show examples for generalization to multimodality image segmentation. Finally, we will present a phantom validation study of the level set approach in a multimodality image segmentation case.

11.3.2 Application to Single-Modality Images

11.3.2.1 Positron Emission Tomography Segmentation of Lung Cancer Example

Positron emission tomography (PET) imaging plays an important role in the diagnosis, staging, and target definition in radiation therapy for diverse types of tumors. For example, ^{18}F-fluorodeoxyglucose (FDG)–PET is used for staging and restaging of nonsmall cell lung carcinoma (NSCLC) and definition of the biological target (Bradley 2004). However, FDG–PET lung images are typically degraded due to respiratory motion. Therefore, it is necessary to preprocess these images to correct for motion artifacts before subsequent segmentation. As an example, we used an FDG–PET image for a patient diagnosed with NSCLC as shown in Figure 11.4a. An ROI was extracted and the image was deblurred for motion correction using a deconvolution-based approach as shown in Figure 11.4b (El Naqa 2006). In this deconvolution approach, the imaging system is modeled as

$$g(\mathbf{x}) = [h(\mathbf{x}) + TLP(\mathbf{x})] \otimes f(\mathbf{x}) + n(\mathbf{x}), \quad (11.17)$$

where g is the observed PET image and f is the ideal true PET image; the term in square brackets is the combination of two parts: h is a characteristic of the image acquisition system [referred to as the point spread function (PSF)] and TLP is the tissue localization probability that incorporates the organ motion effect; n is the additive noise; x is the spatial position; and $\otimes$ is the convolution operator. Deblurring or deconvolution in this context is defined as the process of recovering an estimate of $f(\mathbf{x})$

---

* $\text{logit}(x) = \ln\left(\dfrac{x}{1-x}\right)$.

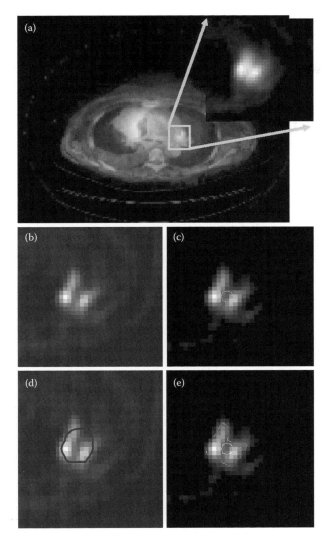

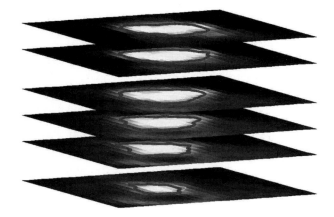

FIGURE 11.5 3D level set segmentation of an FDG–PET cervix cancer image.

11.3.2.2 PET Segmentation of Cervix Cancer Example

The level set algorithm could be easily applied to 2D or 3D images. The results of applying the level set algorithm to a 3D cervix cancer FDG–PET image are demonstrated in Figure 11.5. The algorithm converged in very short time due to the high tumor-to-background ratio in this case, independent of the initial contour shape. It should be noted that the results obtained by the level set algorithm were similar to the 40% maximum standardized uptake value (SUV) thresholding typically used with PET cervix images (Miller and Grigsby 2002).

11.3.3 Application to Multimodality Images

11.3.3.1 PET/CT Segmentation of a Lung Cancer Example

A sample case of a lung PET/CT is shown in Figure 11.6a. The PET image was corrected for motion artifacts using a deconvolution method as discussed earlier. We selected the larger tumor (right) for multimodality analysis. In Figure 11.6b, we initialized the MVLS algorithm with a circle (black) of 15.4 mm diameter. In Figure 11.6c, we show the evolved contour after 10 iterations. In Figure 11.6d, we show the curve evolution in steps of 10 iterations and the final estimated contour (thick red). The algorithm converged in 120 iterations and in less than 1 s.

11.3.3.2 Magnetic Resonance Imaging/CT Segmentation of Prostate Cancer Example

A more challenging case was the analysis of prostate magnetic resonance imaging (MRI)/CT shown in Figure 11.7a. The images were coregistered using rigid body mutual information algorithm. The normalized mutual information (NMI) improved from 1.07 to 1.11. The results of this example seemed to be more dependent on the initial shape; hence, the initial contour (white) was emphasized in the algorithm as prior knowledge as shown in Figure 11.7b. We show the MVLS evolved contour in steps of

FIGURE 11.4 **(See color insert.)** Application of the level set segmentation to NSCLC. (a) PET/CT lung image and ROI selected for processing. (b) Deblurred PET lesion using deconvolution. (c) Initialization of the level set with a small circle. (d) Resulting contour with level set using gradient-based approach (converged after 100 iterations). (e) Resulting contour with level set using edgeless-based approach after 40 iterations (yellow) and 100 iterations (red). Note how the edgeless-based approach has better capturing ability of the lesion boundary in this case.

from $g(\mathbf{x})$. Note that the effects of imaging degradation could be ignored compared with organ motion in this case. An iterative method based on the expectation maximization algorithm (EM) is used to estimate $f(\mathbf{x})$ (Katsaggelos and Lay 1990).

The level set algorithm starts with initialization by a reasonable shape prior or an arbitrary one if none is available (we chose a small circle of 7.5 mm radius). Then, the curve evolves to the boundary of the object (El Naqa et al. 2004). Figure 11.4d shows the result using the gradient-based approach, and Figure 11.4e shows the result using edgeless level set based on the Mumford–Shah model.

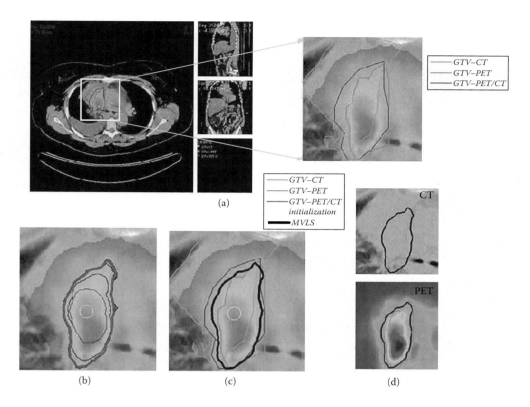

FIGURE 11.6 **(See color insert.)** Analysis of lung PET/CT case. (a) Fused PET/CT displayed in CERR with manual contouring shown of the subject's right GTV. The contouring was performed separately for CT (orange), PET (green), and fused PET/CT (red) images. (b) MVLS algorithm was initialized with a circle (white) of 9.8 mm diameter, evolved contour in steps of 10 iterations (black), and the final estimated contour (thick red). The algorithm converged in 120 iterations in a few seconds. The PET/CT ratio weighting was selected as 1:1.65. (c) MVLS results shown along with manual contour results on the fused PET/CT. Note the agreement of fused PET/CT manual contour and MVLS (DSC = 0.87). (d) MVLS contour superimposed on CT (top) and PET (bottom) separately.

10 iterations (blue) and the final estimated contour (thick red) in Figure 11.7c. The algorithm converged in 50 iterations (in less than 1 s). Note that, in this case, the delineation of the prostate on the CT image is significantly improved by incorporating the MR. On the other hand, CT also improved the convergence results on MR by offering additional external force to guide the curve evolution to the desired target's boundary.

11.3.4 Phantom Validation Study of Multimodality Level Set

We used a commercial plastic anthropomorphic head of average human size phantom. The target objects consisted of plastic spheres and rods that were placed throughout the cranium section of the phantom (Mutic et al. 2001). Tap water was used for CT imaging. However, for MRI and PET imaging, the water inside the phantom was doped with $CuNO_3$ and FDG, respectively. The cold spots spheres were considered as the segmentation targets [four spheres each at 25.4 mm in diameter (8.58 mL)]. The rods were used as landmarks to assist in alignment. The CT data were digitized at $0.94 \times 0.94 \times 3$ mm, PET data at $2.57 \times 2.57 \times 2.57$ mm, and MR at $0.98 \times 0.98 \times 2$ mm.

The phantom data were coregistered using the CT data as a reference. CT is typically used for clinical treatment planning of patients; hence, it was selected as a benchmark for comparison. The NMI between CT and MR was 1.22, and between CT and PET was 1.28 (Figure 11.8a). For validation of segmentation quality, we used the DSC metric to evaluate the algorithm performance with respect to the known phantom dimensions. The segmentation results in terms of DSC (Figure 11.8b) and estimated volume error (Figure 11.8c) of the four balls are shown.

The average DSC for PET/CT/MR was estimated to be 90%, suggesting excellent segmentation performance (22% improvement over using CT alone), and the error in estimated volume is 1.3% (74% error reduction compared with CT).

11.4 Issues, Controversies, and Problems

11.4.1 Finding a Good Initial Guess

Having a good initial guess is helpful in the case of the level set approach; however, it is not essential as in the case of parametric approaches (e.g., snakes), for instance. This is particularly

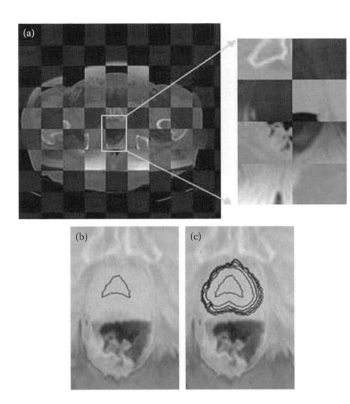

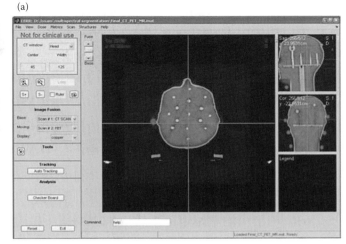

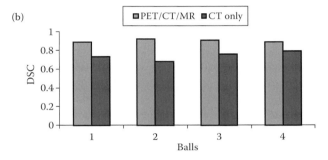

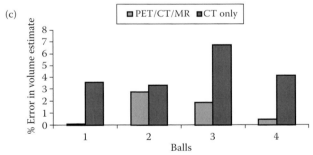

FIGURE 11.7 (See color insert.) Analysis of prostate MRI/CT. (a) Coregistered MRI/CT and selected ROI. (b) MVLS algorithm is initialized with a shape prior that roughly resembles prostate (inside thin line). (c) Curve evolution in steps of 10 iterations and the final estimated contour (outside heavy line).

true if the signal-to-noise ratio in the image is relatively high. Moreover, the initial contour could be used as a shape prior as in Figure 11.7. In this case, increasing weights toward this initialization could be embedded into the algorithm.

11.4.2 Numerical Instability

The level set PDE belongs to the family of the Hamilton–Jacobi equations, which are nonlinear versions of the traditional heat equation. The level set PDE is solved using numerical techniques as described in Section 11.2.5. These techniques are efficient and relatively stable.

To evaluate the stability and convergence characteristics of these nonlinear PDEs, one can analyze the nonlinear oscillations using the phase–plane approach. This could be done by perturbing the algorithm by different initial guesses and plotting the resulting trajectories. In addition, the Courant-Friedrichs-Levy (CFL) conditions could be used to determine optimal time steps for convergence in the numerical solution of the level set equation (Chaudhury and Ramakrishnan 2007).

FIGURE 11.8 Physical phantom validation. (a) Registration of the multimodality phantom in CERR (rods were used to evaluate alignment and the balls to evaluate the quality of segmenting the combined PET/CT/MRI data). Estimated segmentation accuracy in terms of (b) DSC and (c) percentage error in estimated volume.

11.4.3 Extension to Registration

We have presented in Section 11.2.4.4 examples of extending the level set approach to registration problems. However, an improved performance could be further achieved by combining segmentation and deformable image registration algorithms in an iterative framework. Improvement in robustness and effectiveness of coupling segmentation and registration has been recently demonstrated (Droske and Rumpf 2007). The working

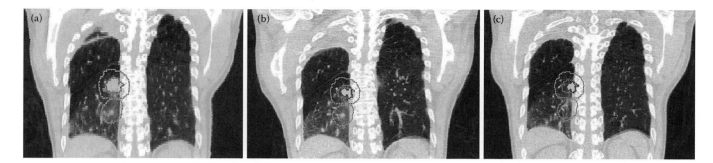

FIGURE 11.9 **(See color insert.)** Example of tracking tumor changes during radiotherapy using a level set based approach. The GTV is shown in green, PTV in brown, and the active model estimate in red. (a) Pretreatment 3D scan. (b) and (c) Reconstructed mid-treatment and end-of-treatment scans, respectively, from the 4D CT data at end of exhalation phase as reference. Note the accurate capturing of tumor boundary by the level set approach at these time points by propagating the GTV.

idea is that prior knowledge on the boundary obtained from segmentation improves the registration as in finite element registration methods (Hensel et al. 2007). At the same time, the registration can improve locally weak boundary segmentation via a correspondence to stronger edges in the other image (template) as in atlas-based segmentation (Klein et al. 2008). An example of this approach utilizing the level set approach was proposed by Yezzi et al. (2003) to jointly segment objects with registration of the active contour map using an additive energy functional of the contour and the registration mapping.

11.5 Future Research Directions

Recent years have witnessed the emergence of several commercial software packages for image registration and segmentation of structures in radiotherapy, such as VelocityAI (Velocity Medical Solutions, Atlanta, GA), Atlas-Based Autosegmentation software from CMS (St. Louis, Missouri, USA), and MiM software suite (MiMvista, Cleveland, Ohio, USA). However, despite the value of these commercial solutions, development of more robust techniques is still lagging, particularly for real-time adaptive image-guided radiotherapy (Xing et al. 2007). The level set framework combines both efficiency and robustness, which makes it a good candidate for future applications in radiotherapy. As an example, in Figure 11.9, we present an extension of the level set algorithm for tracking tumor changes during radiotherapy treatment of lung cancer. The approach uses a registration-assisted scheme, in which pretreatment contours are propagated and adapted to fraction times for selected respiratory phases. At any point of time within the course of treatment, a respiratory phase of interest is selected and the corresponding 3D CT volumes are reconstructed from the 4D acquisition using amplitude sorting based on external breathing surrogates. Next, images are globally aligned using a low-cost rigid registration algorithm. Then, necessary pretreatment contours are copied to the selected time point. For example, the GTV contour was used to initialize the algorithm in place and the PTV contour is used to narrowband the region, thus improving the algorithm's convergence. Sample results are shown in Figure 11.9.

11.6 Conclusion

In this chapter, we reviewed the basic formulation of the level set approach and its application in image segmentation. We further contrasted its two main types based on gradient or regional approaches. In addition, we discussed its extension to multiple dimensions, multiple modalities, and image registration. We have presented several examples for its application in radiotherapy for target definition based on single-modality and multiple modality images of different cancer sites. Moreover, we have demonstrated the improvement it can yield in multimodality target definition by combining complementary information using a physical phantom study. We also discussed some of the current issues associated with this technology and highlighted its potential for real-time adaptive image-guided radiotherapy.

Acknowledgments

This work was partially supported by the National Institutes of Health grant K25 CA128809 the Barnes-Jewish Hospital Foundation (6661-01), and the Natural Sciences and Engineering Research Council of Canada (NSERC-RGPIN 397711-11).

References

Agresti, A. 2002. *Categorical Data Analysis*. New York: Wiley-Interscience.

Aubert, G. 2006. *Mathematical Problems in Image Processing: Partial Differential Equations and the Calculus of Variations*. New York: Springer.

Aujol, J. and G. Aubert 2002. *Signed Distance Functions and Viscosity Solutions of Discontinuous Hamilton–Jacobi Equations*. INRIA Research Report no. 4507. France: Rocquencort.

Boehmer, D., D. Kuczer, H. Badakhshi, S. Stiefel, W. Kuschke, K. D. Wernecke et al. 2006. Influence of organ at risk definition on rectal dose-volume histograms in patients with prostate cancer undergoing external-beam radiotherapy. *Strahlentherapie und Onkologie, 182,* 277–282.

Bondiau, P. Y., G. Malandain, S. Chanalet, P. Y. Marcy, C. Foa and N. Ayache 2004. Image processing and radiotherapy. *Cancer Radiotherapy, 8,* 120–129.

Bortfeld, T., R. Schmidt-Ullrich, W. De Neve and D. Wazer, Eds. 2006. Image-Guided IMRT. Berlin: Springer-Verlag.

Bradley, J., W. L. Thorstad, S. Mutic, T. R. Miller, F. Dehdashti, B. A. Siegel, W. Bosch and R. J. Bertrand 2004. Impact of FDG–PET on radiation therapy volume delineation in non-small-cell lung cancer. *International Journal of Radiation Oncology Biology Physics, 59,* 78–86.

Brock, K. K. 2009. Results of a Multi-Institution Deformable Registration Accuracy Study (MIDRAS). *International Journal of Radiation Oncology Biology Physics, 76,* 583–596.

Casselles, V., R. Kimmel and G. Sapiro 1997. Geodesic active contours. *International Journal of Computer Vision, 22,* 61–79.

Chan, T. F., B. Y. Sandberg and L. A. Vese 2000. Active Contours without Edges for Vector-Valued Images. *Journal of Visual Communication and Image Representation, 11,* 130–141.

Chan, T. F. and L. A. Vese 2001. Active contours without edges. *IEEE Transactions on Image Processing, 10,* 266–277.

Chaudhury, K. N. and K. R. Ramakrishnan 2007. Stability and convergence of the level set method in computer vision. *Pattern Recognition Letters, 28,* 884–893.

Droske, M. and M. Rumpf 2007. Multiscale joint segmentation and registration of image morphology. *IEEE Transactions on Pattern Analysis and Machine Intelligence, 29,* 2181–2194.

El Naqa, I., J. Bradley, J. Deasy, K. Biehl, R. Laforest and D. Low 2004. Improved analysis of PET images for radiation therapy, in *14th International Conference on the Use of Computers in Radiation Therapy*, Seoul, Korea.

El Naqa, I., D. Low, J. Bradley, M. Vicic and J. Deasy 2006. Deblurring of breathing motion artifacts in thoracic PET images by deconvolution methods. *Medical Physics, 33,* 587–600.

El Naqa, I., D. Yang, A. Apte, D. Khullar, S. Mutic, J. Zheng et al. 2007. Concurrent multimodality image segmentation by active contours for radiotherapy treatment planning. *Medical Physics, 34,* 4738–4749.

Hensel, J. M., C. Menard, P. W. Chung, M. F. Milosevic, A. Kirilova, J. L. Moseley et al. 2007. Development of multiorgan finite element-based prostate deformation model enabling registration of endorectal coil magnetic resonance imaging for radiotherapy planning. *International Journal of Radiation Oncology Biology Physics, 68,* 1522–1528.

Horn, B. K. P. and B. G. Schunck 1981. Determining optical flow. *Artificial Intelligence, 17,* 185–203.

Kass, M., A. Witkin and Terzopoulos 1987. Snakes: Active contour models. *First International Conference on Computer Vision*, London, UK.

Katsaggelos, A. K. and K. T. Lay 1990. Image identification and image restoration based on the expectation–maximization algorithm. *Optical Engineering, 29,* 436–445.

Klein, S., U. A. van der Heide, I. M. Lips, M. van Vulpen, M. Staring and J. P. Pluim 2008. Automatic segmentation of the prostate in 3D MR images by atlas matching using localized mutual information. *Medical Physics, 35,* 1407–1417.

Li, H., W. L. Thorstad, K. J. Biehl, R. Laforest, Y. Su, K. I. Shoghi et al. 2008. A novel PET tumor delineation method based on adaptive region-growing and dual-front active contours. *Medical Physics, 35,* 3711–3721.

Miller, T. R. and P. W. Grigsby 2002. Measurement of tumor volume by PET to evaluate prognosis in patients with advanced cervical cancer treated by radiation therapy. *International Journal of Radiation Oncology Biology Physics, 53,* 353–359.

Muren, L. P., R. Ekerold, Y. Kvinnsland, A. Karlsdottir and O. Dahl 2004. On the use of margins for geometrical uncertainties around the rectum in radiotherapy planning. *Radiotherapy and Oncology 70,* 11–19.

Mutic, S., J. F. Dempsey, W. R. Bosch, D. A. Low, R. E. Drzymala, K. S. Chao et al. 2001. Multimodality image registration quality assurance for conformal three-dimensional treatment planning. *International Journal of Radiation Oncology Biology Physics, 51,* 255–260.

Osher, S. and R. P. Fedkiw 2003. *Level Set Methods and Dynamic Implicit Surfaces.* New York: Springer.

Osher, S. and J. A. Sethian 1988. Fronts propagating with curvature-dependent speed: Algorithms based on Hamilton–Jacobi formulations. *Journal of Computational Physics, 79,* 12–49.

Perez, C. A. 2004. *Principles and Practice of Radiation Oncology.* Philadelphia: Lippincott Williams & Wilkins.

Sethian, J. A. 1999. *Level Set Methods and Fast Marching Methods: Evolving Interfaces in Computational Geometry, Fluid Mechanics, Computer Vision, and Material Science.* Cambridge: Cambridge University Press.

Shah, J. 1996. Curve evolution and segmentation functionals: application to color images, in *Image Processing, 1996. Proceedings, International Conference, 1,* 461–464.

Suri, J. S., L. Kecheng, S. Singh, S. N. Laxminarayan, Z. Xiaolan and L. Reden 2002a. Shape recovery algorithms using level sets in 2-D/3-D medical imagery: A state-of-the-art review. *IEEE Transactions on Information Technology in Biomedicine, 6,* 8–28.

Suri, J. S., S. K. Setarehdan and S. Singh 2002b. *Advanced Algorithmic Approaches to Medical Image Segmentation: State-of-the-Art Applications in Cardiology, Neurology, Mammography, and Pathology.* New York: Springer.

Vemuri, B. C., J. Ye, Y. Chen and M. O. Leach 2000. A level-set based approach to image registration, in *Proceedings IEEE Workshop on Mathematical Methods in Biomedical Image Analysis*, Hilton Head Island, SC, USA.

Vemuri, B. C., J. Ye, Y. Chen and C. M. Leonard 2003. Image registration via level-set motion: Applications to atlas-based segmentation. *Medical Image Analysis, 7,* 1–20.

Weiss, E. and C. F. Hess 2003. The impact of gross tumor volume (GTV) and clinical target volume (CTV) definition on the total accuracy in radiotherapy. *Strahlentherapie und Onkologie, 179,* 21–30.

Woolson, R. F. and W. R. Clarke 2002. *Statistical Methods for the Analysis of Biomedical Data.* New York: Wiley-Interscience.

Xing, L., J. Siebers and P. Keall 2007. Computational challenges for image-guided radiation therapy: Framework and current research. *Seminars in Radiation Oncology, 17,* 245–257.

Xu, C., D. L. Pham and J. L. Prince 2002. Image segmentation using deformable models. Handbook of Medical Imaging: Medical Image Processing and Analysis. M. Sonka and J. M. Fitzpatrick, eds. Vol. 2: 129–174, Bellingham, WA: SPIE (The International Society for Optical Engineering).

Yang, D., S. R. Chaudhari, S. M. Goddu, D. Pratt, D. Khullar, J. O. Deasy et al. 2009. Deformable registration of abdominal kilovoltage treatment planning CT and tomotherapy daily megavoltage CT for treatment adaptation. *Medical Physics, 36,* 329–338.

Yang, D., J. Deasy, D. Low and I. El Naqa 2007. *Level set-based non-rigid 3D image registration and motion estimation.* Proceeding of SPIE, San Diego, CA.

Yang, D., W. Lu, D. Low, J. Deasy, A. Hope and I. El Naqa 2008. 4D-CT motion estimation using deformable image registration and 5D respiratory motion modeling. *Medical Physics, 35,* 4577–4590.

Yezzi, A., L. Zollei and T. Kapur 2003. A variational framework for integrating segmentation and registration through active contours. *Medical Image Analysis Mathematical Methods in Biomedical Image Analysis, 7,* 171–185.

Zou, K. H., S. K. Warfield, A. Bharatha, C. M. Tempany, M. R. Kaus, S. J. Haker et al. 2004a. Statistical validation of image segmentation quality based on a spatial overlap index. *Academic Radiology, 11,* 178–189.

Zou, K. H., W. M. Wells, 3rd, R. Kikinis and S. K. Warfield 2004b. Three validation metrics for automated probabilistic image segmentation of brain tumours. *Statistics in Medicine, 23,* 1259–1282.

Atlas-Based Segmentation: Concepts and Applications

Eduard Schreibmann
*Winship Cancer Institute
of Emory University*

Timothy H. Fox
*Winship Cancer Institute
of Emory University*

12.1 Introduction

Imaging plays an increasing role in radiation oncology as accuracy improves by taking into account anatomical changes occurring during the course of treatment as visualized through repeated imaging (Castadot et al. 2008; Lee et al. 2008; Ostergaard Noe et al. 2008; Nithiananthan et al. 2009; G. Zhang et al. 2010; Hasan et al. 2011; Velec et al. 2011). This practice, however, increases the number of images to be segmented, with critical structures having to be identified for planning, during the treatment, and in the follow-up images. In this chapter, we introduce an application of deformable registration that can automatically segment structures of interest in medical images with application to radiotherapy.

Radiotherapy aims to deliver an appropriate dose of radiation to the tumor while avoiding critical organs that, if overdosed to a level of radiation exceeding their tolerance, would cease to function properly. This is a strategic process that involves finding the right setup in terms of irradiation directions and modulation that would deposit enough energy to the treatment area while avoiding certain regions where critical organs are located. Because the prescribed dose can exceed the tolerance of critical structures even with a sharp dose gradient enclosing the target volume, it is of paramount importance to accurately define the location and shape of the critical organs to be avoided. Unfortunately, identifying and marking critical organs remains to date a time-consuming task, with significant input required from the physician to establish the shapes in a manual or semi-automatic process.

Automating the identification of critical structures, commonly called anatomical segmentation, is a cumbersome task especially for soft-tissue organs. The size of organs varies between patients and its shape changes with compression on surrounding anatomy. When surrounded by organs of similar density, the organ of interest becomes poorly differentiated from the neighboring anatomy. Algorithms based on clustering (Gao et al. 1996), thresholding (Foruzan et al. 2009), or region growing (X. Zhang et al. 2010) are currently used for clearly demarcated structures, but these algorithms commonly fail on organs with fuzzy boundaries because voxels of similar intensity belonging to different organs are indiscernible to the algorithms' mathematics (Zhou et al. 2010). Such vague boundaries are common in abdominal anatomy, where critical organs such as liver, bowel, and kidneys are grouped together and imaged with similar intensities, making their distinction cumbersome to current segmentation algorithms (Liu et al. 2005). With the noise inherent in medical images, the border between them becomes fuzzy making it hard for an automated algorithm to produce clinically significant results in a completely automated fashion.

If a structure is poorly visualized, a physician will use his or her medical knowledge to infer organ location based on the expected shape and position relative to other, well-defined, surrounding anatomy. For example, lymph nodes are structures that are too small to be visible in the computed tomography (CT) data set, with common clinical practice delineating a region containing the nodes as identified in relation to the bony anatomy. This is certainly a situation where the decision is based on previous medical knowledge rather than on voxel content. To improve autosegmentation accuracy, an algorithm has to mimic a physician's reasoning by including information on the expected organ shape when borders are fuzzy or indiscernible.

12.2 Concept

12.2.1 Overview

Atlas segmentation (Chao et al. 2006; Schreibmann et al. 2006; Zhang et al. 2006; Han et al. 2008; Wang et al. 2008; Reed et al. 2009) recently emerged as an algorithm that mimics physician judgment by incorporating prior knowledge into the segmentation process and by combining both spatial information and voxels classification schemes into the algorithm's concept. In a fundamental paradigm shift, the atlas-based approach relies on the existence of a mapping between a reference image volume (called atlas), in which structures of interest have been segmented and validated by an expert, and the image to be segmented, called subject. A point-to-point mapping obtained by a deformable image registration is used to match the atlas with the image to be segmented. Once the point-wise transformation between the images is obtained, the same transformation is applied on the structures defined on the atlas data set to warp them onto the data set to be segmented. The algorithm thus outputs a set of labels that are superimposed on the data set to be segmented. The concept is illustrated in Figure 12.1, and typical segmentation results for a head-and-neck case are shown in Figure 12.2. Compared with classic algorithms that use only voxel intensities in their mathematics to drive the segmentation process, atlas approaches use "a priori" information on shape position and size in the form of the atlas map that is presegmented manually by an expert. The accuracy depends on the presegmentation quality and on the deformable registration accuracy in transferring the predefined labels from the template to the subject data set.

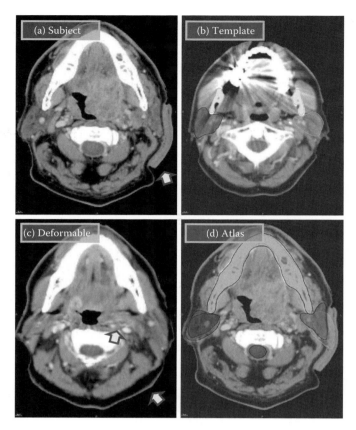

FIGURE 12.2 Sample accuracy of single atlas segmentation for head-and-neck anatomy. (a) Data set to be segmented (subject). (b) Template with the predefined segmentation. Normal interpatient anatomical differences are observable, because the subject has a different size and shape compared with the template. (c) Deformable registration applied on the template image, where the data set was warped to match, as well as possible, the subject image. Arrows show the regions of evident deformation. (d) Predefined segmentations warped with the registration result and superimposed on the subject data set. The black contours in the same display represent the manual segmentation of the same organs. Although a good match was obtained for the mandible and cord, the presence of the bolus in the subject data set and low-contrast artifacts affected the segmentation of the parotid glands.

12.2.2 Deformable Registration

In layman's terms, image registration or coregistration correlates the anatomy as visualized in two images. Initial applications of the concept correlated images of the same patient acquired with different scanners or in different conditions to superimpose data sets coming from different sources into a common reference frame. Driven by various clinical applications, the concept has been extended to matching images of different patients. At the same time, the algorithms increased in complexity driven by a need to increase accuracy, achieved by extending the allowable transform from simple translations to complex transformation matrices that track independently each point between the input images. This state-of-the-art registration approach, called

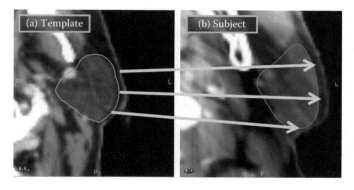

FIGURE 12.1 Concept of atlas segmentation. A displacement field matching voxel-wise a template (a) and the patient data set (b) is first obtained by a deformable registration. Then, a presegmentation of structures of interest (gray contour; a) is warped with the displacement field to adapt it to the subject data set (arrows).

deformable, elastic, or nonrigid registration, aims to find the coefficients of a transformation matrix that warps points from the moving image to their corresponding anatomical location on the fixed image using a complex mathematical optimization that iteratively modifies coefficients of the transformation matrix until anatomical discrepancies are minimized.

Atlas registration is, in essence, another application of deformable registration, where images from different patients are matched through a deformable registration, and the segmentation defined on one data set is warped with the resulting deformation matrix to segment the second data set. The critical component of the atlas-based segmentation thus becomes the accuracy of matching the template to the subject data set to be segmented. A spectrum of deformable registration algorithms has been used for the fine matching of internal anatomy, each one having its own advantages and disadvantages in terms of a trade-off between speed and accuracy. A detailed description and comparison algorithms that can be used to warp the template can be found in Chapters 7 through 16 of this book. Selection of a given algorithm is dictated by both practical and accuracy considerations but does not alter the general concept of atlas segmentation, because any deformable algorithm can be used, interchangeably, in this concept. Indeed, the output of any deformable registration algorithm is a set of vectors defining for every voxel in the template data set its corresponding location in the subject image. Independent of a particular implementation of the optimization engine used, all deformable registration algorithms have the same output format, a vector for each point in the image. Any algorithm can then be used for the atlas segmentation as long as the values of the deformation field obtained are accurate in matching the anatomical changes between the template and subject data sets. Some implementations of the deformable registration algorithms use the mutual information (MI) metric to drive the registration, allowing increased flexibility in match templates and patient data sets acquired using different imaging modalities. An example of this multimodality registration is being able to segment a magnetic resonance imaging (MRI) data set from a template defined on a CT data set. Other algorithms such as the demons approach (Thirion 1998) only match images acquired with the same modality at the advantage of a denser definition of the displacement field.

As illustrated in Figure 12.3, common steps of an atlas-based segmentation procedure are to transform the template first through a rigid registration to correct posture changes followed by a deformable registration to match the internal anatomy in terms of organ shapes and locations. For this case, the patient (1) was scanned in while lying on couch and is slightly tilted forward compared with the atlas (2) that was scanned lying flat on the couch. The rigid registration (3) did correct for differences in posture and patient positioning by tilting forward and moving down the atlas data set compared with its original scanning location, but it was not able to match shapes and volumes of internal organs such as liver or lung (shown with arrows in Figure 12.3). Results of applying the deformable registration are shown in Figure 12.3d, with the algorithm stretching the thoracic cage and liver volume in the atlas to match the appearance and shape of these organs in the patient scan.

12.2.3 Atlas Selection

Registration accuracy depends on the selected algorithms but also on the input images, because the more similar the input images are, the easier it is for the algorithm to find a suitable matching. For some anatomy, such as head and neck or brain, template-patient deviations are small, whereas, for other anatomical sites such as the thorax and abdomen, template-patient variations can be significantly larger because patients can be slim or large and short or tall, making organ location and shape change significantly. Additionally, the information within images may be significantly different, because whole organs such

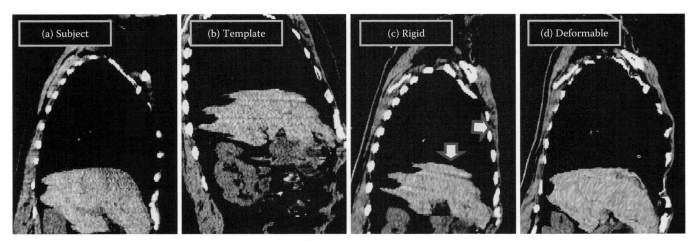

FIGURE 12.3 Steps of the atlas matching procedure. (a) Coronal slice through the subject scan. (b) Coronal slice through the template scan at the same location. For clarity, presegmentation of critical structures on the template scan is not shown. (c) Results of rigidly matching the template to the subject, where discrepancies in position and posture have been solved, but not discrepancies in organ shapes (arrows). (d) Deformable registration further matches internal organ shapes.

as kidneys may be missing, imaging artifacts are commonly present, and there are numerous soft-tissue organs that change shape and location when compressed or displaced by neighboring anatomy. Therefore, accuracy also depends on optimal selection of the template to prematch, as best as possible, the patient data set.

Different approaches have been reported in literature to select the most appropriate template for segmentation either using a representative average atlas (Ramus et al. 2010) or multiple atlases (Aljabar et al. 2009; Heckemann et al. 2010). An average atlas is usually constructed by averaging across sample patients to construct a mean, standardized data set of the expected anatomy. Another strategy is to use multiple templates where the atlas segmentation is started a few times using different templates as input, and a postprocessing step concatenating segmentations obtained by different templates into a single segmentation using an algorithm called Simultaneous Truth and Performance Level

Estimation (STAPLE) algorithm (Warfield et al. 2004). For each organ in the templates, the output of this algorithm is a probability map ranging from 0 to 1. As illustrated in Figure 12.4, using a template that is substantially different from the patient data set can result in erroneous mapping of organs into regions of similar intensity during the deformable algorithm's attempt to reconcile the anatomical differences. The figure shows liver segmentations obtained by three different atlases, with none of the atlases being able to produce a clinically acceptable segmentation because various oversegmentations or undersegmentations are visible. The STAPLE algorithm can analyze these discrepancies to produce a probability map shown as a colored overlay on the patient's CT data set, with voxels marked blue, having a low probability that the liver is present at that to yellow, marking voxels that are certainly to the algorithm inside the liver as these voxels are present in all segmentations. The probability map can be further thresholded at a value of 0.5 to obtain an improved segmentation of organ boundaries.

12.2.4 Postprocessing

Another approach to increase accuracy is to use a postrefinement step that takes the organ segmentation as obtained from the STAPLE or atlas segmentation and adapts it to local features seen in the image. Usage of the refinement stage is exemplified in Figure 12.5 on a spleen, where the initial shape is shown as a blue line superimposed on an axial slice through the patient's CT data set. Because of the proximity of the spleen to the stomach and their similar Hounsfield units, as well as a particularity in the patient's anatomy, parts of the stomach were marked as belonging to the spleen as shown in the figure. This is a common misclassification in abdominal anatomy, especially in the regions where organs attach, such as spleen–kidney or kidney–liver interfaces. The intermediary result of a refinement step based on an algorithm that preserves the appearance of the initial estimation is illustrated in Figure 12.5b, with the final result shown in Figure 12.5c, obtained after applying a filtering algorithm to smooth and include the respiratory-motion blurred voxel values at organ boundaries.

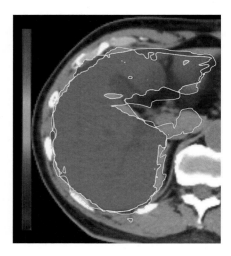

FIGURE 12.4 (**See color insert.**) Deducing bulk organ shape from atlas segmentations with the STAPLE algorithm. The white lines represent liver segmentations obtained by different atlases. The color overlay is the probability map that the corresponding voxel belongs to the liver as deduced from the individual segmentations, ranging from 0 (violet) to 1 (red).

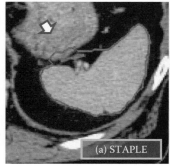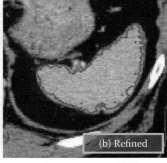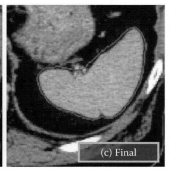

(a) STAPLE   (b) Refined   (c) Final

FIGURE 12.5 (**See color insert.**) Steps of the refinement procedure. (a) Input spleen segmentation shown as a blue contour. (b) Regions of different Hounsfield units are eliminated using a threshold level set-based algorithm. (c) The shape is smoothed and snapped to image gradients using a geodesic level set algorithm.

12.3 Applications

12.3.1 Overview

The major applications are segmenting new scans by matching the template to new acquisitions. However, any clinical scenario where a previous segmentation can be warped to new data sets is a potential application of the approach. Adaptive therapy is an example application, where the treatment plan is modified during the course of treatment to account for anatomical changes occurring during the treatment process. Although the changes are usually visualized using cone-beam CT (CBCT) data sets, the atlas segmentation method can be employed here to segment the CBCT data sets by using the planning CT and its segmentation as a template. This is an atlas segmentation procedure, with the CT used as a template and the CBCT used as a subject data set. Another application is accounting for intratreatment motion by using 4D imaging to capture anatomical changes produced by respiratory motion during the treatment process. The approach involves tracking anatomical structures across a 4D data set, images that capture the anatomy moving with the respiratory motion during the breathing cycle, typically using 10 static snapshots during the breathing cycle. To avoid laborious manual segmentation of all phases, the solution is to segment only in the first snapshot of the structures of interest, and then an atlas segmentation procedure is employed to propagate the segmentation across the whole 4D acquisition volume. For this application, the template is the first phase and the subject data set the following phases.

12.3.2 Validation Measures

For any clinical usage, one valid question is to inquire about the quality and accuracy of this automated procedure. As opposed to the deformable registration where quality assurance is hard to achieve, because there are many solutions that are indiscernible, when applied for segmentation, deformable registration accuracy can be easily measured by comparing with manually segmented contours. Ideally, the segmentations obtained by manual and automated segmentations should match. In reality, there will be regions where the two would match and regions where they would more or less disagree. This disagreement is usually measured in research projects through the classic Dice (1945) coefficient index, defined as the ratio of the overlap of two structures over their union, which is widely used in the evaluation of comparison studies. The measure is normalized, spanning from 0, when the structures to be compared do not overlap at all, to 1, representing the ideal case where the structures match perfectly. However, this perfect match is seldom encountered in clinical practice, due both to inaccuracies in the atlas segmentation algorithm and to variability in the manual segmentation that defines the ground truth. With this measure, the automatic contours were assumed to be accurate if their degree of overlap with the manual contouring was more than 0.7 (Zou et al. 2004).

In practice, the measure can be questioned in terms of utility, because the final usefulness of this approach relies on the time needed to modify the automatically generated contours to match the needed clinical accuracy. For small structures that lie into one or two slices, this editing time needed to refine the measurement is small, while for larger structures, these times are naturally larger as more slices need to be edited. The Dice measure is theoretically biased to favor larger organs, as larger organs may have more voxels in common, and thus for the same number of nonmatching voxels, the values will be higher indicating a better match.

12.3.3 Autosegmentation for Radiotherapy Planning

Due to its practicality (Chao et al. 2007), the method was recently adapted in radiotherapy for segmenting new CT data sets of the brain (Isambert et al. 2008) or head-and-neck region (Zhang et al. 2007; Commowick et al. 2008; Sims et al. 2009; Stapleford et al. 2010; Teguh et al. 2010), where interpatient anatomical variations are small; thus, an accurate mapping can be found by the deformable registration. A wealth of techniques have been reported to find the correct mapping, which include manual landmark-based registration (Qatarneh et al. 2007), B-spline (Stapleford et al. 2010; Teng et al. 2010), registration using a multimodality metric (Mattes et al. 2003), or demons approach (Wang et al. 2008). For other anatomical sites with increased interpatient variability, atlas segmentation was proposed for clearly demarcated structures. Reed et al. (2009) used atlas segmentation for detecting breast boundaries in radiotherapy planning using the demons deformable registration algorithm, and Ehrhardt et al. (2001) apply the concept for identifying bones in preplanning for hip operations using a global alignment with the aid of an affine transform followed by a local matching of corresponding structures using the classic formulation (Thirion 1998) of the demons algorithm.

When it comes to measuring the accuracy of autosegmentation of new data sets, in clinical evaluation studies (Heath et al. 2007; Isambert et al. 2008; Rodionov et al. 2009; Sims et al. 2009; Stapleford et al. 2010; Young et al. 2011), the approach was superior to classic methods on structures that do not have a clear border because their location is deduced from the atlas. Evaluating the atlas segmentation concept for brain and head-and-neck anatomy in clinical practice characterized the approach as a robust and reliable (Isambert et al. 2008) method in rapid delineation of target and normal tissues (Teguh et al. 2010). The approach provided a good trade-off between accuracy and robustness (Bondiau et al. 2005) by decreasing interobserver variability (Stapleford et al. 2010; Teguh et al. 2010), with increased consistency and time savings (Young et al. 2011). The method was found to reduce segmentation times by 35% (Stapleford et al. 2010) to 63% (Teguh et al. 2010).

In a clinical study (Young et al. 2011), the atlas pelvic nodal clinical target volume was constructed by using the data from

15 postoperative endometrial cancer patients and was compared with simulation scans from 10 additional endometrial cancer patients in terms of detecting the nodal clinical target volume autogeneration. The evaluation measure was the time needed by three radiation oncologists to correct the autocontours, and the overlap of the contours was calculated using Dice's coefficient. Results show that editing the autocontours provided a 26% time savings with the mean overlap increased from manual contouring (0.77) to correcting the autocontours (0.79), leading to increased consistency and time savings when contouring the nodal target volumes for adjuvant treatment of endometrial cancer, although the autocontours still required careful editing to ensure that the lymph nodes at risk of recurrence are properly included in the target volume.

For head-and-neck data set, Sims et al. (2009) measured accuracy at two radiotherapy centers for the brainstem, parotids, and mandible on several patients by comparing manually segmented contours with automated segmented contours in terms of their volume, sensitivity, and specificity, with the results being interpreted using the Dice similarity coefficient. Typically, automatic segmentation took approximately 7 min with an average Dice coefficient for all OAR of 0.68 to 0.82 depending on the center.

One notes that the atlas segmentation was also applied to sites with increased degree of interpatient variability. Reed et al. (2009) have recently reported on using the procedure to delineate the breast for radiotherapy. The aim is facilitated, because these are structures of high contrast, where contours were generated by an atlas segmentation procedure based on a demons deformable registration algorithm that mapped the whole-breast clinical target volume from a template. Of the eight investigated cases, oncologists modified the contours as necessary to achieve a clinical accuracy and compared with recontoured from de novo in terms of times to complete and interobserver variation. The median edit time was 12.9 min compared with 18.6 min to contour from de novo, providing a 30% time reduction, with the deformed contours achieving 94% volume overlap before correction and required editing of 5% of the contoured volume.

The atlas segmentation procedure can also be used as an educational tool. In a study by Chao et al. (2007), atlas segmentation was evaluated for reducing the variation of target delineation among physicians segmenting head-and-neck anatomy. For this study, eight radiation oncologists with varying levels of clinical experience performed target delineation by first contouring from scratch and then by modifying the contours generated in an atlas segmentation procedure. Although variation in manual delineating was significant, usage of the atlas segmentation procedure reduced volumetric variation and improved geometric consistency, with time savings of 26% to 29% for experienced physicians and 38% to 47% for the less experienced ones. The same conclusions have been observed in an independent study by Stapleford et al. (2010), which measured the efficiency of the atlas registration for lymph node delineation by concluding that a decrease in interobserver variability can be attained while maintaining clinical accuracy. The method consisted of having five physicians with head-and-neck IMRT experience use CT data from five patients to create bilateral neck clinical target volumes covering specified nodal levels, which was considered the gold standard and compared against contours created using a commercially available atlas. As with the previous cases, physicians modified the automatic contours to make them acceptable for treatment planning, with assessment using STAPLE algorithm to calculate a probabilistic estimate of the "true" segmentation. Results showed that, compared with the "true" segmentation created from manual contours, the automatic contours had a high degree of accuracy, with sensitivity, Dice similarity coefficient, and mean/max surface disagreement values comparable with the average manual contouring, but the automated group was more consistent than the manual group for multiple metrics, most notably reducing the range of contour volume and percent false positivity. Average contouring time savings with the automatic segmentation was 11.5 min per patient, which is a 35% reduction.

12.3.4 Atlas Segmentation in 4D Imaging

With the advent of imaging that allows visualization of tumor and critical organ motion during the breathing cycle, a new treatment paradigm was introduced in radiotherapy, because the tumor margin could be reduced from a static expansion to a customized margin based on the motion seen in 4D images. The approach reduces lung toxicity as a smaller volume is irradiated to higher doses. With 4D imaging, we are able to see the lung motion and customize the margins to the observed motion, but a drawback of this approach is the increased number of segmentations needed to track critical structures.

An approach to automate this segmentation is to initially delineate critical structures and target the first phase of the CT data sets and then use the atlas segmentation concept to propagate the structures to the next phases. This is, in essence, an atlas segmentation procedure that takes the segmented phase of the 4D CT as a template and applies sequentially the concept to the other phases of the data set (Figure 12.6). The concept was detailed and tested using the demons algorithm (Lu et al. 2006a), finite element model (FEM) (Brock and Dawson 2010), region matching (Chao et al. 2008), or B-spline (Schreibmann et al. 2006), with the concept validated by visual inspections and quantitative comparisons of the automatic recontours with both the gold standard segmentations and the manual contours.

Because the images of the 4D data set are acquired with the same modality and the variations between phases are minimal, caused only by respiration, this is a simple atlas segmentation problem with most deformable registration approaches being able to accurately model the anatomical changes. Consequently, the reported Dice similarity metrics are high, ranging from more than 0.94 for larger organs such as liver, kidneys, and lung and 0.81 for smaller structures such as the esophagus (Lu et al. 2006).

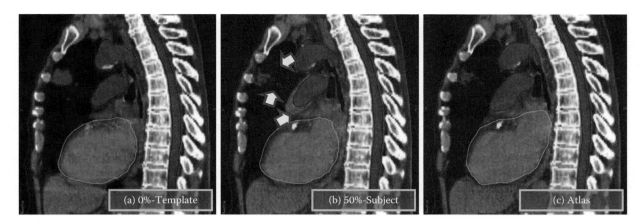

FIGURE 12.6 Application of atlas segmentation in 4D radiotherapy. (a) First phase (0%) of a 4D CT data set is segmented manually and used as template. (b) The 50% phase with the manually segmented contours overlaid with a rigid registration. The respiratory motion modified tumor, aorta and heart location and shape with respect to the original segmentation (arrows). (c) Atlas segmentation results warp the manual segmentation to match the deforming anatomy.

12.3.5 Atlas Segmentation for Adaptive Radiotherapy

Weight loss, tumor shrinkage, and tissue edema induce substantial modification of the patient's anatomy during radiotherapy or chemoradiotherapy especially that may impact upon the accuracy of the delivered dose. Adapting the plan to account for those anatomical changes, especially for head-and-neck anatomy, has drawn the interest of many clinicians because of its compelling advantages in defining tumor volume and designing treatment plans that better spare critical structures. In this approach, information obtained from online CBCT scans probe patient anatomy, and the planning dose matrix is warped with the deformable registration to compute dose volume histograms on the CBCT anatomy to determine if a replanning is needed (Lee et al. 2008; Lu et al. 2006b; Hasan et al. 2011). With the use of imaging acquired during treatment, patients may be imaged and their anatomical changes during therapy may be tracked. A bottleneck of direct application of the method in clinical settings is the time-consuming step of segmenting the additional images obtained during treatment. This can be accomplished by the use of the same concept of atlas segmentation.

Atlas segmentation based on demons (Wang et al. 2005), finite element model (Brock et al. 2008), B-spline algorithms (Lawson et al. 2007), or customized algorithms able to cope with differences in noise and contrast (Lu et al. 2006) was proposed to automate the segmentation process; the CT with the segmentation obtained with a planning time is used as a template and the CBCT acquired during treatment is the subject data set, with the deformable registration modifying the original segmentation to match the anatomy observed in the CBCT as shown in Figure 12.7.

Technically speaking, there is an intrapatient atlas segmentation problem where the anatomical changes between the input data sets are expected to be minimal. However, the deformable registration algorithm has to be able to cope with artifacts and changes Hounsfield units that are frequently present between the CT and CBCT data sets. Initial clinical evaluation of the approach was reported by Zhang et al. (2007), which uses both Dice and distances to evaluate the results. In the evaluation study, the similarity coefficient indices were approximately 0.8, with the distance between the manual and automatically segmented contours being mostly within 3 mm. A comprehensive evaluation report (Lu et al. 2006) tested 12 variants of the demons algorithm through the Dice metric (Castadot et al. 2008), which ranged more than 0.5 for all algorithms. When standard CT is used to visualize anatomical changes during therapy, an evaluation presented by Wang and colleagues reported an accuracy of 1.3 mm mean absolute surface-to-surface distance and 83% overlap index between manually generated and automated contours. Analyzed anatomy consisted of eight head-and-neck cancer patients with a total of 100 repeat CT scans, one prostate patient with 24 repeat CT scans, and nine lung cancer patients with a total of 90 4D CT images.

12.3.6 Clinical Implementation

A few companies have implemented the approach in commercial software that can be used in daily clinical practice. To date, brain and head-and-neck anatomy can be reliably segmented by commercial software, with ongoing research extending this concept to other treatment sites.

Although all atlas segmentation algorithms use the same concept, from a user perspective, small differences in implementation may significantly change their experience while using the tool and may be checked when selecting a particular implementation of this autosegmentation algorithm. The time needed for the algorithm to complete the calculations is probably the most

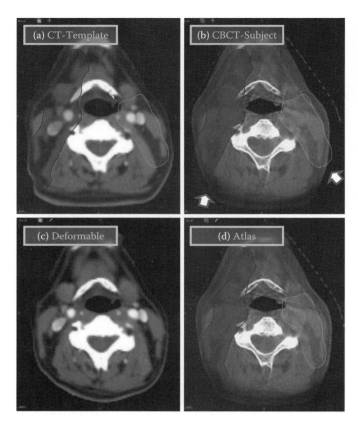

FIGURE 12.7 Application of atlas segmentation for adaptive radiotherapy. (a) CT data sets and their corresponding segmentation. (b) CBCT acquired mid-treatment, with the original segmentation overimposed. Arrows in the display show regions of anatomical changes in response to therapy. (c) Results of deforming the CT data set to account for these changes. (d) Resulting atlas segmentation of the CBCT data set.

important aspect that defines a user's experience, with current segmentation times ranging from a few minutes to between 15 to 20 min. The process can be automated by running the algorithms as an independent service in the computer's memory without freezing the user interface or by employing advanced computing techniques that take advantage of recent developments in CPU technology. However, computing time is not the only parameter to be considered when selecting an algorithm, because more complex implementations, producing in principle results of higher accuracy, need more time to complete. An algorithm's ability to segment complex anatomical regions, such as thorax and abdomen, should be considered as well. Although sites with small interpatient variability, such as the head-and-neck region and brain, can be reliably matched by current deformable registration algorithms, large interpatient anatomical variations are not easily matched, because the algorithm has to make a compromise when reconciling various anatomical differences.

Algorithms that use a multiatlas approach would need more time to complete, because the procedure has to be repeated a few times, but produce results of increased accuracy. Implementations that use a multimodality deformable algorithm to match the template to the data set have an increased flexibility in segmenting data sets, because templates and data sets acquired with different imaging modalities can be used interchangeably. In contrast, atlas segmentations using monomodality implementations of a deformable registration algorithm will be able to match templates and patient data sets acquired only with the same imaging modality (e.g., CT-to-CT or MRI-to-MRI) and will require a consistency in using the similar acquisition protocols for the template and data sets.

Results will not be perfect, because the deformable registration may have to make a compromise when matching the anatomy between the template and subject data sets and with the user having particular preferences when defining critical structures. Usually, the physician will review the results and make changes as necessary; therefore, some interaction is needed, but ideally the time needed to make the changes will be less than the time needed to manually segment. Compared with standard segmentation algorithms that need intermittent user interaction, this methodology has the advantage of being fully automated, with the user only having to approve or make changes at the end.

In terms of expectation, the atlas-based segmentation will map to the data set to be segmented any anatomical structure that is defined on the template. Apart from critical structures, the tumor can be defined on the template, as reported recently (Strassmann et al. 2010), but for this approach, the deformable registration may produce suboptimal results if the tumor location is significantly different in the template and segmented data set. A special atlas selection strategy that chooses atlases with tumor locations and sizes similar to the segmented data set may be beneficial when attempting to segment the tumor and critical structures at the same time.

Overall, the atlas segmentation procedure has recently emerged as a technique that segments medical images without much user operation at an accuracy comparable with manual segmentation but without the associated cost and time expense. The method is especially advantageous on structures with fuzzy borders—a case where the atlas interpolates from a template. Due to this advantage over classic segmentation approaches in minimizing user interaction, the method has attracted the interest of many clinicians and was implemented as part of a few treatment planning systems, being an approach system that can be used routinely in the clinic.

References

Aljabar, P., Heckemann, R. A., Hammers, A. et al. (2009). Multi-atlas based segmentation of brain images: Atlas selection and its effect on accuracy. *Neuroimage, 46,* 726–738.

Bondiau, P. Y., Malandain, G., Chanalet, S. et al. (2005). Atlas-based automatic segmentation of MR images: Validation study on the brainstem in radiotherapy context. *International Journal of Radiation Oncology Biology Physics, 61,* 289–298.

Brock, K. K., and Dawson, L. A. (2010). Adaptive management of liver cancer radiotherapy. *Seminars in Radiation Oncology, 20,* 107–115.

Brock, K. K., Hawkins, M., Eccles, C. et al. (2008). Improving image-guided target localization through deformable registration. *Acta Oncologica, 47,* 1279–1285.

Castadot, P., Lee, J. A., Parraga, A. et al. (2008). Comparison of 12 deformable registration strategies in adaptive radiation therapy for the treatment of head and neck tumors. *Radiotherapy and Oncology, 89,* 1–12.

Chao, K. S., Bhide, S., Chen, H. et al. (2007). Reduce in variation and improve efficiency of target volume delineation by a computer-assisted system using a deformable image registration approach. *International Journal of Radiation Oncology Biology Physics, 68,* 1512–1521.

Chao, M., Li, T., Schreibmann, E. et al. (2008). Automated contour mapping with a regional deformable model. *International Journal of Radiation Oncology Biology Physics, 70,* 599–608.

Chao, M., Schreibmann, E., Li, T. et al. (2006). Automatic contouring in 4D radiation therapy. *International Journal of Radiation Oncology Biology Physics, 66,* S649–S649.

Commowick, O., Gregoire, V., and Malandain, G. (2008). Atlas-based delineation of lymph node levels in head and neck computed tomography images. *Radiotherapy and Oncology, 87,* 281–289.

Dice, L. R. (1945). Measures of the amount of ecologic association between species. *Ecology, 26,* 297–302.

Ehrhardt, J., Handels, H., Malina, T. et al. (2001). Atlas-based segmentation of bone structures to support the virtual planning of hip operations. *International Journal of Medical Informatics, 64,* 439–447.

Foruzan, A. H., Aghaeizadeh Zoroofi, R., Hori, M. et al. (2009). Liver segmentation by intensity analysis and anatomical information in multi-slice CT images. *International Journal for Computer Assisted Radiology and Surgery, 4,* 287–297.

Gao, L., Heath, D. G., Kuszyk, B. S. et al. (1996). Automatic liver segmentation technique for three-dimensional visualization of CT data. *Radiology, 201,* 359–364.

Han, X., Hoogeman, M. S., Levendag, P. C. et al. (2008). Atlas-based auto-segmentation of head and neck CT images. *Medical Image Computing and Computer-Assisted Intervention, 11,* 434–441.

Hasan, Y., Kim, L., Wloch, J. et al. (2011). Comparison of planned versus actual dose delivered for external beam accelerated partial breast irradiation using cone-beam CT and deformable registration. *International Journal of Radiation Oncology Biology Physics, 80,* 1473–1476.

Heath, E., Collins, D. L., Keall, P. J. et al. (2007). Quantification of accuracy of the automated nonlinear image matching and anatomical labeling (ANIMAL) nonlinear registration algorithm for 4D CT images of lung. *Medical Physics, 34,* 4409–4421.

Heckemann, R. A., Keihaninejad, S., Aljabar, P. et al. (2010). Improving intersubject image registration using tissue-class information benefits robustness and accuracy of multi-atlas based anatomical segmentation. *Neuroimage, 51,* 221–227.

Isambert, A., Dhermain, F., Bidault, F. et al. (2008). Evaluation of an atlas-based automatic segmentation software for the delineation of brain organs at risk in a radiation therapy clinical context. *Radiotherapy and Oncology, 87,* 93–99.

Lawson, J. D., Schreibmann, E., Jani, A. B. et al. (2007). Quantitative evaluation of a cone-beam computed tomography-planning computed tomography deformable image registration method for adaptive radiation therapy. *Journal of Applied Clinical Medical Physics, 8,* 2432.

Lee, C., Langen, K. M., Lu, W. et al. (2008). Assessment of parotid gland dose changes during head and neck cancer radiotherapy using daily megavoltage computed tomography and deformable image registration. *International Journal of Radiation Oncology Biology Physics, 71,* 1563–1571.

Liu, F., Zhao, B., Kijewski, P. K. et al. (2005). Liver segmentation for CT images using GVF snake. *Medical Physics, 32,* 3699–3706.

Lu, W., Olivera, G. H., Chen, Q. et al. (2006a). Automatic re-contouring in 4D radiotherapy. *Physics in Medicine and Biology, 51,* 1077–1099.

Lu, W., Olivera, G. H., Chen, Q. et al. (2006b). Deformable registration of the planning image (kVCT) and the daily images (MVCT) for adaptive radiation therapy. *Physics in Medicine and Biology, 51,* 4357–4374.

Mattes, D., Haynor, D. R., Vesselle, H. et al. (2003). PET-CT image registration in the chest using free-form deformations. *IEEE Transactions on Medical Imaging, 22,* 120–128.

Nithiananthan, S., Brock, K. K., Daly, M. J. et al. (2009). Demons deformable registration for CBCT-guided procedures in the head and neck: convergence and accuracy. *Medical Physics, 36,* 4755–4764.

Ostergaard Noe, K., De Senneville, B. D., Elstrom, U. V. et al. (2008). Acceleration and validation of optical flow based deformable registration for image-guided radiotherapy. *Acta Oncologica, 47,* 1286–1293.

Qatarneh, S., Kiricuta, I., Brahme, A. et al. (2007). Lymphatic atlas-based target volume definition for intensity-modulated radiation therapy planning. *Nuclear Instruments and Methods in Physics Research Section A: Accelerators, Spectrometers, Detectors and Associated Equipment, 580,* 1134–1139.

Ramus, L., Commowick, O., and Malandain, G. (2010). Construction of patient specific atlases from locally most similar anatomical pieces. *Medical Image Computing and Computer-Assisted Intervention, 13,* 155–162.

Reed, V. K., Woodward, W. A., Zhang, L. et al. (2009). Automatic segmentation of whole breast using atlas approach and deformable image registration. *International Journal of Radiation Oncology Biology Physics, 73,* 1493–1500.

Rodionov, R., Chupin, M., Williams, E. et al. (2009). Evaluation of atlas-based segmentation of hippocampi in healthy humans. *Magnetic Resonance Imaging, 27,* 1104–1109.

Schreibmann, E., Chen, G. T., and Xing, L. (2006). Image interpolation in 4D CT using a B-spline deformable registration model. *International Journal of Radiation Oncology Biology Physics, 64,* 1537–1550.

Sims, R., Isambert, A., Gregoire, V. et al. (2009). A pre-clinical assessment of an atlas-based automatic segmentation tool for the head and neck. *Radiotherapy and Oncology, 93,* 474–478.

Stapleford, L. J., Lawson, J. D., Perkins, C. et al. (2010). Evaluation of automatic atlas-based lymph node segmentation for head-and-neck cancer. *International Journal of Radiation Oncology Biology Physics, 77*, 959–966.

Strassmann, G., Abdellaoui, S., Richter, D. et al. (2010). Atlas-based semiautomatic target volume definition (CTV) for head-and-neck tumors. *International Journal of Radiation Oncology Biology Physics, 78*, 1270–1276.

Teguh, D. N., Levendag, P. C., Voet, P. W. et al. (2010). Clinical validation of atlas-based auto-segmentation of multiple target volumes and normal tissue (swallowing/mastication) structures in the head and neck. *International Journal of Radiation Oncology Biology Physics, 81*, 950–957.

Teng, C. C., Shapiro, L. G., and Kalet, I. J. (2010). Head and neck lymph node region delineation with image registration. *BioMedical Engineering OnLine, 9*, 30.

Thirion, J. P. (1998). Image matching as a diffusion process: An analogy with Maxwell's demons. *Medical Image Analysis, 2*, 243–260.

Velec, M., Moseley, J. L., Eccles, C. L. et al. (2011). Effect of breathing motion on radiotherapy dose accumulation in the abdomen using deformable registration. *International Journal of Radiation Oncology Biology Physics, 80*, 265–272.

Wang, H., Dong, L., O'Daniel, J. et al. (2005). Validation of an accelerated "demons" algorithm for deformable image registration in radiation therapy. *Physics in Medicine and Biology, 50*, 2887–2905.

Wang, H., Garden, A. S., Zhang, L. et al. (2008). Performance evaluation of automatic anatomy segmentation algorithm on repeat or four-dimensional computed tomography images using deformable image registration method. *International Journal of Radiation Oncology Biology Physics, 72*, 210–219.

Warfield, S. K., Zou, K. H., and Wells, W. M. (2004). Simultaneous truth and performance level estimation (STAPLE): An algorithm for the validation of image segmentation. *IEEE Transactions on Medical Imaging, 23*, 903–921.

Young, A. V., Wortham, A., Wernick, I. et al. (2011). Atlas-based segmentation improves consistency and decreases time required for contouring postoperative endometrial cancer nodal volumes. *International Journal of Radiation Oncology Biology Physics, 79*, 943–947.

Zhang, G., Huang, T. C., Feygelman, V. et al. (2010). Generation of composite dose and biological effective dose (BED) over multiple treatment modalities and multistage planning using deformable image registration. *Medical Dosimetry, 35*, 143–150.

Zhang, L., Hoffman, E. A., and Reinhardt, J. M. (2006). Atlas-driven lung lobe segmentation in volumetric X-ray CT images. *IEEE Transactions on Medical Imaging, 25*, 1–16.

Zhang, T., Chi, Y., Meldolesi, E. et al. (2007). Automatic delineation of on-line head-and-neck computed tomography images: Toward on-line adaptive radiotherapy. *International Journal of Radiation Oncology Biology Physics, 68*, 522–530.

Zhang, X., Tian, J., Deng, K. et al. (2010). Automatic liver segmentation from CT scans based on a statistical shape model. *Conference Proceedings—IEEE Engineering in Medicine and Biology Society, 2010*, 5351–5354.

Zhou, J. Y., Wong, D. W., Ding, F. et al. (2010). Liver tumour segmentation using contrast-enhanced multi-detector CT data: Performance benchmarking of three semiautomated methods. *European Radiology, 20*, 1738–1748.

Zou, K. H., Warfield, S. K., Bharatha, A. et al. (2004). Statistical validation of image segmentation quality based on a spatial overlap index. *Academic Radiology, 11*, 178–189.

IV

Advanced Imaging Techniques

Advances in 3D Image Reconstruction

Jeffrey H. Siewerdsen
Johns Hopkins University

J. Webster Stayman
Johns Hopkins University

Frédéric Noo
University of Utah

13.1 Introduction and Overview

13.1.1 Scope of the Chapter: A Simple Primer on a Vast Space

This chapter provides a brief introduction to the topic of 3D image reconstruction in x-ray cone-beam computed tomography (CBCT). The topic represents an area of significant activity in research, commercial development, and clinical application since the first practical algorithm for 3D filtered backprojection (FBP) derived in 1984, referred to hereafter as the Feldkamp–Davis–Kress (FDK) algorithm.[1] The numerous developments in 3D image reconstruction in the ensuing three decades are as follows:

- FDK algorithm for 3D FBP[1]
- Grangeat's derivation of a well-posed inversion of the 3D radon transform[2]
- Clinical application of 3D FBP on multidetector CT (MDCT) scanners
- Use of flat-panel detectors (FPDs) for CBCT in applications ranging from image-guided radiation therapy[3,4] (IGRT) to dental/maxillofacial imaging[5]

- FBP-like reconstruction techniques for general (noncircular) trajectories[6,7]
- Katsevich's derivation of exact reconstruction in helical CBCT[8]
- A variety of iterative, statistical, and model-based reconstruction methods developed to provide higher-quality reconstruction at lower dose, from sparse projection data sets, and/or with better utilization of prior information[9–13]
- Application of iterative reconstruction techniques on clinical MDCT scanners

This cursory list only scratches the surface of a vast and vibrant area of medical imaging research, and a single chapter can do little more than provide a summary introduction of a few key principles. The sections below are organized along three broad strokes: (1) a review of the FDK algorithm (Section 13.2) and variations thereof (Sections 13.3 and 13.4); (2) an introduction to the principles and specific techniques for 3D reconstruction from noncircular trajectories (Section 13.5); and (3) a brief introduction to iterative reconstruction methods, including algebraic, likelihood-based, and compressive sensing techniques (Section 13.6). Considering the depth and breadth of such topics, emphasis is on the fundamentals, key characteristics, and distinct aspects of each reconstruction approach.

13.1.2 Definitions, Terminology, and Coordinate Systems

13.1.2.1 Is It a "Cone?" And Is It "CT?"

First, a trivial note: "cone-beam" CT is largely a misnomer. Its widespread application in combination with FPDs involves a rectangular field of view (FOV); therefore, "pyramid-beam" CT would be a more precise term, but we conform to the conventional cone-beam CT (CBCT).

A related note, perhaps less trivial from a regulatory perspective, is the question of whether CBCT is actually "CT." The point has been debated somewhat in the literature,[14,15] hinging on the question of whether fully 3D imaging (including CBCT) constitutes "tomography" (deriving etymologically from the Latin for "slice"). For CBCT, slices may be considered a choice as to how to format and view the data rather than something intrinsic to the modality—somewhat in distinction to conventional, single-slice, axial CT. Alternatives such as "volumetric imaging" have been suggested. The term carries regulatory implications (e.g., with respect to shielding requirements of "CT scanners"), cost codes for CT examination, and justification or limits on the number of CT scanners at a given institution. Here again, we conform to the conventional nomenclature—CBCT.

13.1.2.2 On Pixels, Dexels, and Voxels

Continuing with the trivial and arguably mundane, we come to the question of the "pixel." The term "pixel" refers to a "picture element," with etymology traced to a variety of sources in the early to mid 1960s and is believed to be first published in reference to digital images from space probes to the Moon and Mars.[16] In this respect, the term pertains to the basic unit or element of an image (i.e., a matrix of numbers) as displayed, for example, an element of

an array converting electrical charge to visible light on a monitor. Thus, "picture" + "element" = "pixel." Interestingly, the alternative "pel" was also proposed[17] and used interchangeably with "pixel" into the 1970s, but the latter prevailed in ubiquitous usage. The emphasis here is that "pixel" refers to a display element.

The detector presents yet another matrix, for example, a physical array of elements converting x-rays into electrical signal as in each photodiode–thin-film transistor combination on an active matrix FPD. These detector elements are commonly referred to as "pixels" in the literature, but we recognize the misappropriation of the term and the potential for confusion between "detector elements" and "picture elements." The term "del" has been alternatively suggested.[18] In light of the linguistic demise of the analogous "pel" and for phonological consistency with the well-established "pixel," we instead refer to detector elements as "dexels" (i.e., "detector" + "element" = "dexel"). The reader will be hard pressed to find this term in the literature before the time of writing, but we break with conventional terminology in this respect nonetheless.

Last is the element of a 3D matrix representing a reconstructed image (i.e., the "volume element"). Following from above, we have "volume" + "element" = "voxel." We therefore refer to the elements of a volumetric CT or CBCT image as voxels.

In usage, we have, for example, the following: "The FPD has a dexel size of 0.194 mm" or "The FPD includes a matrix of (1024 × 1024) dexels." Similarly, "Images were reconstructed at a voxel size of 0.25 mm" or "Volume images were (256 × 256 × 256) voxels in size." Finally, "Images were displayed on a monitor with (2048 × 1536) pixel format."

13.1.2.3 Geometry and Notation

Now on to the somewhat less arbitrary—geometry and notation. Figure 13.1 illustrates a typical CBCT geometry in which the source and the detector move in a circle around the patient. A summary of terms and notation is provided in Table 13.1.

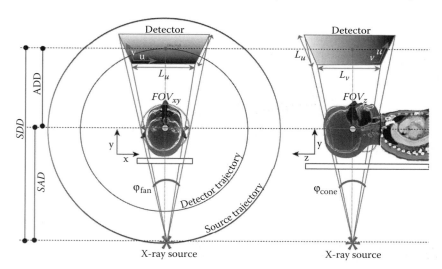

FIGURE 13.1 Illustration of CBCT system geometry and coordinate system definitions.

TABLE 13.1 Summary of Notation

Symbol	Description
g	Projection image
(u,v)	Projection image spatial domain
θ	Projection view angle
(x,y,z)	3D image reconstruction spatial domain
M	Geometric magnification = (SDD/SAD)
"dexel"	Detector element refers to an element in the 2D matrix representing the projection image domain
"voxel"	Volume element refers to an element in the 3D matrix representing the reconstructed image
a_u	Dexel size in the u direction
a_v	Dexel size in the v direction
a_{xy}	Voxel size in the axial plane (alternatively, $a_x = a_y = a_{xy}$)
a_z	Voxel size in the longitudinal (z) direction
SDD	Source-to-detector distance
SAD	Source-to-axis of rotation distance
μ	Attenuation coefficient
φ_{fan}	Fan angle (subtends the u (also x or y) dimension)
φ_{cone}	Cone angle (subtends the v (also z) dimension)
L_x	Lateral extent of the detector
L_z	Longitudinal extent of the detector
FOV_{xy}	Lateral FOV
FOV_z	Longitudinal FOV
(f_u, f_v)	Fourier domain counterpart to (u,v)
(f_x, f_y, f_z)	Fourier domain counterpart to (x,y,z)

13.1.2.3.1 World Coordinate System, Also the Object and 3D Image Coordinate System (x,y,z)

The plane of the circular orbit defines the axial (x,y) plane, and the perpendicular (z) direction corresponds to the long axis of the patient, with cardinal axes (x,y,z) chosen to obey the right-hand rule with respect to (x,y,z) as well as the direction of rotation (θ, counterclockwise in Figure 13.2). By this scheme, $+x$ points to the patient right, $+y$ to the patient anterior, and $+z$ to the patient superior. The axial plane is thus (x,y), the sagittal (y,z), and the coronal (x,z). A notable exception to this scheme is breast CBCT in which the source and the detector orbit in the coronal plane about a pendant breast.

13.1.2.3.2 Detector Coordinate System (u,v)

The detector plane is denoted by coordinates (u,v), with the origin taken at the center of the "lower-left" dexel in the right-hand system of Figure 13.3. The v coordinate is therefore parallel to z. Whether u or v corresponds to detector "rows" (e.g., FPD rows defined by a row of simultaneously switched thin-film transistors) or detector "columns" (i.e., FPD readout lines connected to an external amplifier) is arbitrary and is not specified here. (Interestingly, because detectors can exhibit distinct noise characteristics along rows or columns, orienting the detector one way or the other may have an effect on CBCT image quality; e.g., correlated structure noise between lines parallel to v imparting stronger ring artifacts.)

13.1.2.3.3 Piercing Point (u_o, v_o)

A point of interest is the so-called "piercing point" denoted (u_o, v_o) and referring to the point on the detector containing a ray from the source that is perpendicular to the detector. For a perfect circular orbit without deviation (e.g., due to mechanical flex), (u_o, v_o) is constant for all projection angles (θ), and the reader is referred to the optimal "1/4-pixel offset" (better termed the "1/4-dexel offset") in the location of the piercing point for optimal view sampling.[19] For real mechanical systems, one may expect (u_o, v_o) to vary ("move") from projection to projection. Of course, the piercing point is only one component of such variation, and more complete descriptions of geometric calibration are described below.

13.1.2.3.4 Fan Angle (φ_fan)

The fan angle is denoted φ_{fan} and refers to the angle subtended by the beam in the (x,y) domain. It refers to the full fan angle (not the half-angle) and determines the axial FOV (denoted FOV_{xy}). The fan angle also determines the minimum source–detector orbit required for complete sampling (specifically, $180° + \varphi_{fan}$). Any orbital arc extent less than this represents an incomplete sampling of the object and is technically tantamount to "tomosynthesis." Orbital extent exceeding $180° + \varphi_{fan}$ but less than $360°$ requires Parker weighting as detailed below. In these respects, the fan angle is no different in CBCT than in conventional ("fan-beam") CT. Many CBCT systems are such that the beam is symmetric in (x,y) about the central ray (i.e., the ray perpendicular to the detector), a notable exception being the "offset-detector" geometry described below, where the detector is offset laterally (i.e., in the u direction) by up to half its width to allow reconstruction of a larger FOV_{xy}.

13.1.2.3.5 Cone Angle (φ_fan)

The cone angle is denoted φ_{cone} and refers to the angle subtended by the beam in the z direction. It refers to the full angle (not the half-angle) and determines the longitudinal FOV (denoted FOV_z). Many CBCT systems are such that the beam is symmetric in z about the central ray (i.e., the ray perpendicular to the detector), a notable exception again being breast CBCT [where the angle above the central ray (proximal to the chest wall) may be only a few degrees, but the angle below the central ray (to the anterior-most aspect of the pendant breast) may be $>10°–15°$]. Alternatively, the "cone angle" may refer to the angle of a specific ray above (or below) the central axial plane (e.g., a ray at angle 5° above the central axial plane is said to have a "cone angle" of 5°), in which case we distinguish this notationally as φ'_{cone}.

The cone angle is the key factor distinguishing CBCT from conventional axial (fan-beam) CT (i.e., the divergence of the beam in the z direction). Accounting for such divergence is what distinguishes fully 3D reconstruction algorithms (the subject of this chapter) from 2D axial reconstruction. Similarly, the acquisition of a large volume from a single source–detector orbit is what distinguishes CBCT from axial and helical (fan-beam) CT (in which a volume is acquired from multiple rotations as the patient is translated on a moving bed along the z direction). Failure to account for divergence of the beam can result

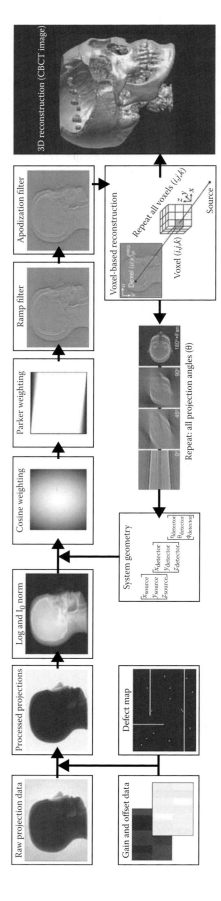

FIGURE 13.2 Flowchart diagram of 3D FBP.

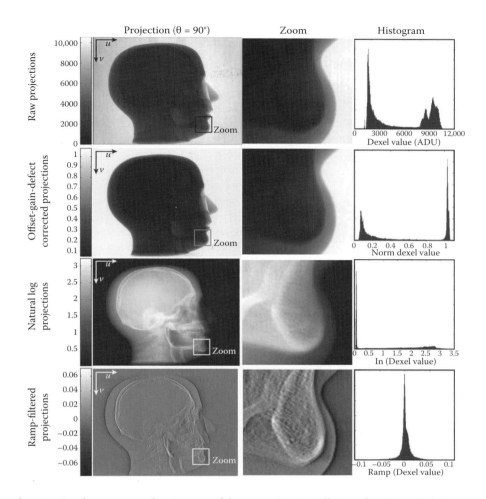

FIGURE 13.3 Example projection data corresponding to some of the processing steps illustrated in Figure 13.2. Top row: Raw projection data, $g_{raw}(u,v)$. Second row: Offset-gain-defect corrected projections, $g_{proc}(u,v)$. Third row: Log projections, $g_1(u,v)$. Bottom row: Ramp-filtered projections, $g_4(u,v)$. The zoomed-in regions give an appreciation for subtleties in signal and noise characteristics. The histograms show the manner in which the projection signal is scaled and transformed in the course of such processing.

in major image artifacts, and modern CT scanners (i.e., MDCT containing 64 rows or more and therefore a cone angle of several degrees), in fact, employ fully 3D "cone-beam" reconstruction and have done so since the introduction of MDCT scanners with approximately eight slices or more.

13.1.2.3.6 System Geometry

The system geometry is broadly determined by the source-to-detector distance (*SDD*) and the source-to-axis distance (*SAD*), with the corresponding geometric magnification (*M*) given by

$$M = \frac{SDD}{SAD}. \tag{13.1}$$

For a detector centered on the piercing point, the FOV associated with such geometry is determined by the size of the detector

(L_u and L_v along the *u* and *v* directions, respectively) and the system magnification:

$$FOV_{xy} = \frac{L_u}{M} = L_u\left(\frac{SAD}{SDD}\right) \tag{13.2}$$

$$FOV_z = \frac{L_v}{M} = L_v\left(\frac{SAD}{SDD}\right) \tag{13.3}$$

The lateral FOV may be increased in the offset-detector geometry as described below.

The voxel size (denoted a_{vox} for isotropic voxels or, alternatively, a_{xy} in the *xy*-domain and a_z in the longitudinal domain) is a free parameter that may be arbitrarily selected for a given CBCT reconstruction. Of course, the spatial resolution and noise characteristics of the resulting 3D image will depend on the choice of a_{vox}. It is important to note that the voxel size in itself is not a measure of spatial resolution (although it does affect spatial resolution, and one hears all too often the spatial resolution of a CBCT system quoted in terms of the voxel size). Rather, the spatial resolution of the system is affected by a multitude of factors ranging from the x-ray focal spot size to the dexel size and choice of reconstruction filter. Voxel size is just one factor in spatial resolution and is an arbitrary choice at that. A common choice is

$$a_{vox} = \frac{a_{dex}}{M}, \tag{13.4}$$

where a_{dex} is the size of the detector elements (dexel pitch). This choice is sometimes referred to as the "natural" voxel size. Choosing $a_{vox} < a_{dex}/M$ corresponds potentially to oversampling and can impart an antialiasing effect at the cost of increased sampling and computational load. Choosing $a_{vox} > a_{dex}/M$ requires proper binning and sampling of detector elements to avoid undersampling (aliasing) effects.

The fan angle and cone angle in these terms are therefore

$$\varphi_{fan} = 2\tan^{-1}\left[\frac{L_u}{2SDD}\right] = 2\tan^{-1}\left[\frac{FOV_u}{2SAD}\right] \tag{13.5}$$

and

$$\varphi_{cone} = 2\tan^{-1}\left[\frac{L_v}{2SDD}\right] = 2\tan^{-1}\left[\frac{FOV_v}{2SAD}\right] \tag{13.6}$$

Note the factor of 2 and the distinction from the half-fan angle (denoted γ in Ref. 19).

13.2 3D FBP

The basic processing and reconstruction steps in 3D FBP are summarized below and in Figure 13.2. The overall approach is that of the FDK algorithm. Analytical forms of the various processing and reconstruction steps are given below, but the emphasis is on the overall functional "blocks" associated with voxel-driven reconstruction rather than the analytical formalities therein. As shown in the flowchart of Figure 13.2, 3D FBP relies on a substantial number of 2D processing steps applied to the projection data, including the basic correction of detector nonuniformities and the requisite normalization, weighting, and filtering steps. The thumbnail images in Figure 13.2 are explained in greater detail in individual sections below.

13.2.1 Raw and Processed Projection Data

The Beers–Lambert law for transmission of x-rays through a medium of attenuation coefficient $\mu(x,y,z)$ gives the basic description of the raw projection data:

$$g_{raw}(u,v;\theta) = g_0 e^{-\int_0^{SDD}\mu(x,y,z)dy}, \tag{13.7}$$

where a monoenergetic beam is assumed for the sake of simplicity. The term g_0 is the mean detector signal in the unattenuated beam, and $g_{raw}(u,v;\theta)$ is the raw projection at angle θ. The process of 3D reconstruction described below is essentially a numerical method for solving for the object function $[\mu(x,y,z)]$ given a projection data set $g_{raw}(u,v;\theta)$. The first step is to correct the raw projections to account for the spatially varying nonuniformities in detector dark current, signal response, and defective detector elements. The correction is referred to as offset-gain-defect correction (or, alternatively, dark-flood-defect correction). Nonuniformities in detector offset (i.e., the detector signal in the absence of radiation) are described by $g_{dark}(u,v)$ and are typically measured as the average of a reasonable number of detector frames (~50) acquired without x-ray exposure. To the extent that the dark fields are stable in time, independent of projection angle, and described well by the mean dark image, a simple subtraction from the raw image removes nonuniformities associated with the detector offset, giving

$$g_{proc}(u,v;\theta) = K\frac{g_{raw}(u,v;\theta) - \overline{g_{dark}(u,v)}}{g'_{flood}(u,v) - g'_{dark}(u,v)}, \tag{13.8}$$

where $g_{proc}(u,v;\theta)$ is the dark-flood "processed" projection data.

Nonuniformities in detector response are corrected by pixelwise normalization of the mean flood-field $g'_{flood}(u,v)$ (i.e., the mean projection image under irradiation without an object in the field). The flood-field correction attempts to normalize nonuniform response owing to scintillator nonuniformities, detector and electronics gain variations, and even nonuniformities in the x-ray beam (e.g., Heel effect). To the extent that the flood fields are linear in detector response, independent of projection angle, and described well by the mean flood image (e.g., the average of ~50 projections acquired under uniform irradiation), the first-order correction is given by division. Note, however, the primes in the denominator, indicating that $g'_{dark}(u,v)$ is taken at a time close to $g'_{flood}(u,v)$, but $g_{dark}(u,v)$ (without the prime) is taken at a time close to $g_{raw}(u,v;\theta)$ (e.g., immediately before). The reason for this is primarily that $g_{dark}(u,v)$ is not stable in time and varies over the course of minutes or hours. The denominator (with primes) is comparatively stable, and a given flood-field calibration may hold for days or weeks. The assumption that $g'_{flood}(u,v)$ is independent of θ is a simplifying (but not necessary) assumption obeyed by many systems with reasonable mechanical and tube-output stability; however, some systems have invoked angle-dependent flood-field corrections. The factor K merely rescales the projection data to the detector signal corresponding to the (dark-subtracted) exposure level. Note that this is the simplest form of dark-flood correction, where the flood field is acquired at a single exposure level (e.g., at ~50% of sensor saturation), and linearity in detector response is assumed. Modifications include multipoint flood correction, which better handles signal nonlinearity and can reduce reconstruction artifacts.

Following offset-gain correction as in Equation 13.7, defects in individual detector elements may be corrected by median filter with a "defect map," which identifies any aberrant dexels. The defect map for modern FPDs tends to be fairly stable (months) and may be defined in terms of various characteristics by which to "flag" defective dexels. The four essential characteristics of dexel defects include (1) mean dark signal above or below given thresholds (e.g., "dead dexels"), (2) standard deviation in dark signal above or below given thresholds (e.g., "noisy dexels"), (3) mean flood signal above or below given thresholds, and (4) standard deviation in flood signal above or below given thresholds. The thresholds are user-defined and require some iteration in selecting values that accurately capture defective elements. Additional defect characteristics might include (5) linearity of response outside of a specified tolerance, (6) image lag outside of a specified tolerance, etc. Typically, the defect map identifies no more than ~1% of dexels as defects.

The types of nonuniformities typical of raw projection data are evident in the top row of Figure 13.3. Note the rectangular banks of nonuniform signal in the unattenuated beam (associated with variations in detector readout electronics). The raw dexel values (e.g., in the histogram of Figure 13.3) exhibit an offset and magnitude typical of 15- or 16-bit readout in modern FPDs. The processed projection data are illustrated in the second row of Figure 13.3. Note the relatively uniform response in the unattenuated field, the removal of line defects and nonuniformities in the zoomed-in region, and the rescaling of the detector signal value (in this case, to a bare-beam value of 1.0).

13.2.2 Logarithmic Projection Data

The next step in "solving" for the attenuation coefficient in Equation 13.7 is to take the natural logarithm of each side of the equation. The logarithm of the inverse of the normalized projection data (g_{proc}/g_0 or, alternatively, multiplying by –1) yields the "line integral"

$$g_1(u,v;\theta) = \ln\left(\frac{g_0}{g_{proc}(u,v;\theta)}\right) = \int_0^{SDD} \mu(x,y,z)\,dy. \quad (13.9)$$

Example log projection data are shown in Figure 13.3, which have the somewhat familiar appearance of a projection radiograph (owing to the logarithmic response of conventional x-ray film and/or the logarithmic processing typical of tone-scaling operations applied in digital radiography).

13.2.3 Cosine ("Feldkamp") Weighting

The next 2D projection image processing step is to weight the data according to the (inverse of the) density of rays at each pixel. These are the so-called "cosine weights" or "Feldkamp weights" and yield proper scaling for a flat detector:

$$g_2(u,v;\theta) = g_1(u,v;\theta)\left[\frac{SDD}{\sqrt{SDD^2 + u^2 + v^2}}\right]. \quad (13.10)$$

The weighting term at first glance appears to be a form of inverse-square-law factor; however, it is actually the reverse of an inverse-square correction, weighting the image less at greater

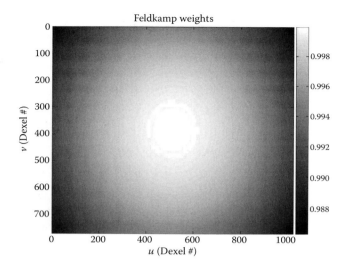

FIGURE 13.4 Cosine weights (also called Feldkamp weights) associated with the density of rays impinging on a flat detector.

distances from the central ray. (The flood-field correction in Equation 13.7 removed the nonuniformity associated with a point-source spherically symmetric field falling on a flat detector.) The cosine weights impart a reduced weighting on detector elements that received a reduced density of incident rays. The weights are illustrated in Figure 13.4.

13.2.4 Data Redundancy ("Parker") Weighting

For acquisitions exceeding total orbital extent of 180° + fan angle, there are redundant rays in the projection data that must be appropriately weighted to yield an accurate reconstruction (i.e., avoid double counting of redundant rays). A reasonable approximation to the proper weighting of redundant rays is given by the "Parker" weights (alternatively, Parker–Silver weights) described by $w_3(u; \theta)$:

$$g_3(u,v;\theta) = g_2(u,v;\theta)\,w_3(u;\theta), \quad (13.11)$$

where

$$w_3(u;\theta) =$$

$$\begin{cases} \sin^2\left(\dfrac{\pi\theta}{4\left(\dfrac{1}{2}\varphi_{fan} - \varphi(u-u_0)\right)}\right) \\[2em] 1 \\[1em] \sin^2\left(\dfrac{\pi(\pi + \varphi_{fan} - \theta)}{4\left(\dfrac{1}{2}\varphi_{fan} + \varphi(u-u_0)\right)}\right) \\[2em] \text{for } 0 \le \theta \le \varphi_{fan} - 2\varphi(u-u_0) \\ \text{for } \varphi_{fan} - 2\varphi(u-u_0) \le \theta \le \pi - 2\varphi(u-u_0) \\ \text{for } \pi - 2\varphi(u-u_0) \le \theta \le \pi + \varphi_{fan} \end{cases}$$

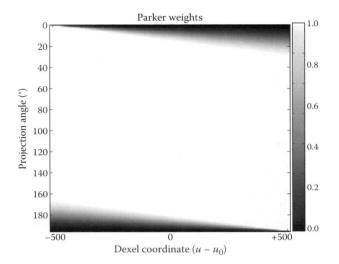

FIGURE 13.5 (a) Parker weights for proper scaling of redundant rays in FBP. Example CBCT reconstruction (b) with Parker weights applied and (c) without Parker weights applied. Note the error in the backprojection over the region of redundant views in (c).

and

$$\varphi(u) = \tan^{-1}\left(\frac{u - u_0}{SDD}\right) \quad (13.12)$$

and φ_{fan} is the full fan angle (not the half-fan angle). An example calculation of the weighting factor is shown in Figure 13.5. Example images with and without Parker weights applied across redundant views are shown in Figure 13.5b and c, where the absence of weighting terms is seen to give inaccurate attenuation values in the overlap region.

13.2.5 Ramp Filter

Exactly as in conventional fan-beam CT computed by FBP, a ramp filter is applied along the detector rows to properly account for the radial frequency dependence $(1/r)$ in backprojected rays. Without application of the ramp filter, the reconstruction is blurred by $(1/r)$, but the ramp filter exactly cancels the inverse radial dependence. Application of the ramp filter can be considered as multiplication in the Fourier domain:

$$g_4(u,v; \theta) = FT^{-1} \left[FT[g_3(u,v;\theta)]|\rho|\right], \quad (13.13)$$

where $|\rho|$ is the radial frequency and FT denotes the Fourier transform operator. Alternatively, the ramp filter may be applied as a convolution in the spatial domain, denoted by

$$g_4(u,v; \theta) = g_3(u,v;\theta) * \left(-\frac{1}{2\pi^2 u^2}\right), \quad (13.14)$$

where the spatial domain kernel corresponding to the ramp is $(1/2\pi^2 u^2)$. Note that the ramp filter is only applied along the u

direction of the projection data, corresponding to a 1D high-pass filter. The effect of this asymmetric filter on the projection data is evident in the bottom row of images in Figure 13.6, particularly in the zoomed-in region showing an enhancement of edges along the u direction.

13.2.6 Smoothing (Apodization) Filter

The ramp filter yields an accurate 3D reconstruction, but the high-pass characteristic of the filter amplifies high-frequency noise, which can reduce overall imaging performance. To counter the high-frequency noise amplification, while maintaining the proper frequency weighting associated with the $(1/r)$ effect, it is common to apply a subsequent low-pass filter, which effectively "smooths" the image and attenuates high-frequency noise. A common choice is a linear low-pass filter, such as a cosine filter:

$$T_{win}(f) = h_{win} + (1 - h_{win}) \cos(2\pi f\Delta). \quad (13.15)$$

In this case, the smoothing (apodization) filter is characterized by a parameter h_{win} varying from 0.5 (Hann filter) to 1.0 (pure ramp filter). As with the ramp filter, the low-pass filter may be applied as a product in the Fourier domain:

$$g_5(u,v; \theta) = FT^{-1} \left[FT[g_4(u,v; \theta)]T_{win}(f)\right] \quad (13.16)$$

or as a convolution in the spatial domain:

$$g_5(u,v; \theta) = g_4(u,v; \theta) \, t_{win}(u,v), \quad (13.17)$$

where t_{win} if the Fourier transform (spatial domain kernel) corresponding to T_{win}. The cosine filter mentioned above is just one example of a low-pass filter that may be applied, and the choice of filter should be considered a free parameter that may be "tuned" to the imaging task, for example, selection of smoother filters to reduce noise (recognizing a loss in spatial resolution) to better visualize low-contrast structures or, conversely, selection of sharper filters to emphasize spatial detail in high-contrast structures (e.g., bone trabeculae).

13.2.7 Definition of the 3D Reconstruction Matrix

The next step before forming a 3D reconstruction is to define the 3D matrix on which reconstruction will be computed (i.e., specify the extent and voxel size of the 3D image). This corresponds to selecting the FOV and voxel dimensions in the $x,y,$ and z directions (i.e., FOV_x, FOV_y, and FOV_z and a_x, a_y, and a_z, respectively). The axial domain typically involves $FOV_x = FOV_y \equiv FOV_{xy}$ and $a_x = a_y \equiv a_{xy}$. The axial FOV (i.e., the lateral extent of the reconstruction) is therefore FOV_{xy}, and the longitudinal FOV (i.e., the "length" of the scan) is FOV_z. Similarly, the axial voxel size is a_{xy}, and the longitudinal voxel size ("slice thickness") is a_z.

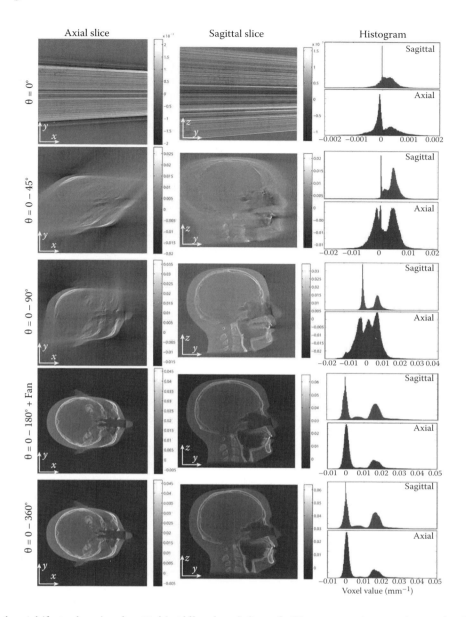

FIGURE 13.6 Example axial (first column) and sagittal (middle column) slices of a 3D reconstruction at various angles. For the axial slice, note a fairly monotonic "filling" of the axial plane from the first projection up through that at 180° + fan. For the sagittal slice, note the relatively large amount of information added to the reconstruction from the lateral view (i.e., the 90° view for a sagittal slice or the 0° slice for a coronal). Finally, note the evolution in histograms of voxel values as the reconstruction converges on a fairly accurate estimate of attenuation coefficient.

Note that FOV_{xy} and FOV_z may be freely selected, irrespective of the size of the projection data, although a nominal choice could involve the maximum size implied by the full region of support:

$$FOV_{xy} = \frac{L_u}{M} = L_u \left(\frac{SAD}{SDD} \right) \quad (13.18)$$

$$FOV_z = \frac{L_v}{M} = L_v \left(\frac{SAD}{SDD} \right). \quad (13.19)$$

However, the FOV may be selected either smaller or larger than these nominal values, the latter subject to lateral truncation and the lateral subject to incomplete sampling. A variety of clinical and image quality considerations also enter in the selection of FOV, for example, minimizing the longitudinal collimation to the length of interest reduces the volume irradiated (dose-length-product) as well as the amount of x-ray scatter; therefore, minimizing FOV_z to the region of interest is common. Similarly, minimizing FOV_{xy} to the region of interest requires methods to account for lateral truncation.

Similarly, the voxel size should be considered a free parameter, although a nominal choice is the "natural" voxel size implied by a one-to-one correspondence between dexels and voxels in a system with magnification, M:

$$a_{xy} = \frac{a_u}{M} \qquad (13.20)$$

$$a_z = \frac{a_v}{M}, \qquad (13.21)$$

where a_u and a_v are the dexel size in the u and v directions of the detector, respectively (Figure 13.1). Again, the choice of a_{xy} and a_z is arbitrary and should be "tuned" to the imaging task. Voxel size less than the "natural" size corresponds to oversampling, and selection greater than the "natural" size should carry with it appropriate binning of the projection data or be subject to undersampling. The choice $a_{xy} = a_z$ corresponds to the "isotropic" voxel size.

Furthermore, the selection of a_{xy} and a_z has a strong effect on the spatial resolution and noise in the 3D reconstruction. First, it is important to note that the voxel size is *not* tantamount to the spatial resolution of the imaging system: it is but one of many factors, and the temptation to mix the terms "voxel size" and "spatial resolution" should be avoided. The latter depends not only on the voxel size but also on the focal spot size, system geometry, x-ray detector characteristics, reconstruction filter, etc. Voxel size similarly affects image noise (in combination with numerous other system parameters, not the least of which is dose), and it is common to "tune" voxel noise in a manner consistent with the imaging task—finer voxels for high-detail structures (accepting an increase in noise) and coarser voxels for large, low-contrast structures (accepting a loss in spatial resolution).

13.2.8 Voxel-Driven Reconstruction: Interpolation

One method in forming a 3D reconstruction would involve the backprojection of the entire projection (i.e., all dexel values) from the projection domain through the 3D reconstruction volume—so-called pixel-driven reconstruction. A far more prevalent and flexible method is "voxel-driven" reconstruction in which a 3D reconstruction matrix is freely defined as in the previous section, and the location in the projection domain corresponding to the projected position of each voxel is computed. This is illustrated in the "Voxel-based reconstruction" inset of Figure 13.2. Note that this location in the projection domain is arbitrary (depending on the system geometry and definition of the 3D reconstruction matrix) and does not necessarily correspond to the center of a dexel. This location does, however, correspond to that from which the (weighted and filtered) signal value will be backprojected in the subsequent step.

The one-to-one relationship ("mapping") between voxel coordinates (x,y,z) and dexel coordinates (u,v) is described by the projection geometry relationships detailed in the following section. Because the projected location does not necessarily match

the center of a dexel, we have a choice: (1) we may interpolate the signal value at the projected location based on surrounding dexel values, or (2) we may upsample the projection to a finer grid such that the projected location more closely matches the location of a dexel center. The two choices are, of course, related and impart the effect of interpolation on the projection data. In the first case, we could simply choose nearest-neighbor interpolation (suffering potentially coarse sampling error that introduces noise in the reconstruction) or smoother interpolation filters (e.g., bilinear and bicubic). In the second case, we take advantage of the computational efficiency of nearest-neighbor interpolation but first upsample (and bilinearly interpolate) the projection data such that the projected voxel location more closely matches the location of a dexel center. This upsampling process followed by nearest-neighbor interpolation can be implemented with reasonable computational efficiency. Upsampling by 2×2 or 4×4 dexels is common.

13.2.9 Voxel-Driven Reconstruction: Voxel Values

Given the weighted, filtered projection data of Equation 13.17, the upsampling and interpolation of the projection data to an essentially continuous domain (u,v), and the projection matrix (or other) mapping of dexel locations (u,v) and voxel locations (x,y,z), we are ready to backproject domain data in the 3D domain of the image reconstruction. In "voxel-driven" reconstruction, we loop over every voxel in the 3D reconstruction matrix and over every projection in the acquired data set, at each step "sucking" the weighted, filtered, interpolated value from (u,v) to the specific voxel location (x,y,z) and adding it to the current voxel value (initialized as a 3D matrix of zeros). The process is illustrated in Figure 13.6.

Note the evolution of information in axial and sagittal domains illustrated in Figure 13.6 as well as the evolution of the voxel histogram. The first complete backprojection (i.e., covering all voxel locations for the $\theta = 0°$ projection) adds a series of unrecognizable "streaks" in the axial and sagittal planes; interestingly, in the coronal domain, the backprojection is essentially a (weighted and filtered) radiographic image demagnified to isocenter. As the backprojection proceeds, the volumetric domain "fills in" progressively, and the voxel values evolve toward a true representation of the attenuation coefficient. At $\theta = 180°$ + fan, the reconstruction is complete, and subsequent projections serve to improve sampling and reduce noise. The fairly bimodal histogram (assuming proper application of Parker weights, depending on the total extent of the source–detector orbit) is invariant beyond $\theta = 180°$ + fan, with a peak near 0 mm^{-1} representing air and a broader peak about 0.02 mm^{-1} representing various soft and bone tissues.

We therefore have an approximate numerical solution for $\mu(x,y,z)$ in Equation 13.7. Conversion to "Hounsfield units" (HU) or "CT Number" (CT#) simply scales the attenuation coefficient relative to air and water:

$$CT\# = 1000 \times \frac{\mu - \mu_{water}}{\mu_{water}} (units: HU), \qquad (13.22)$$

where μ is the reconstructed voxel value (units of attenuation coefficient), and μ_{water} is the attenuation coefficient for water (at a specified energy). Note that the CT number is, therefore, dependent on energy. The definition yields -1000 HU for air and 0 HU for water, with no theoretical upper bound (but bone tissues typically ~200 to 1000 HU).

Some treatment planning systems may prefer to offset the entire HU scale by 1000 to avoid negative numbers, in which case

$$CT\#_{offset} = 1000 \times \frac{\mu - \mu_{water}}{\mu_{water}} + 1000, \tag{13.23}$$

so that air has a value of 0, water 1000, etc.

A variety of factors in CBCT impart inaccuracy in the reconstructed voxel values. Among these are factors imparting a variety of image artifacts discussed at length in other work, for example, x-ray scatter, metal, and the "cone-beam" artifact. In applications where the accuracy of CT# is important (e.g., calculation of dose directly on CBCT data in radiation therapy in a manner similar to dose calculation in the planning CT image), correction of such artifacts and careful calibration of CT# is accordingly important and is the subject of ongoing research and development.

13.3 System Geometry and Geometric Calibration

The voxel-driven reconstruction approach summarized above relies upon a one-to-one relationship ("mapping") of voxels [location (x,y,z) in the 3D domain of the reconstructed image] to dexels [location (u,v) in the 2D domain of the detector] for each projection angle. A description of projection geometry is provided below based on the principles of perspective projection common in computer vision and robotics techniques. Other work reported in the literature describes calibration techniques and phantoms directly applicable to CBCT.[4,20–23] Consider the basic geometry illustrated in Figure 13.7, noting the differences in coordinate systems compared with Figure 13.1 for simplicity in the analytical forms in this section. Conversion among these or other coordinate systems suggested in the literature should be done with careful recognition that conventions vary from

one CBCT system to another, owing to the diversity of physical embodiments (linear accelerators, C-arms, head scanners, breast scanners, etc.) for which, at the time of writing, there is no single standard geometry—an important caveat for anyone implementing CBCT calibration and reconstruction techniques.

In this 2D example, the point (x_{source}, z_{source}) (x_A refers to a coordinate x with respect to frame A) is related to the u coordinate on the detector by

$$\frac{u - u_0}{SDD} = \frac{x_{source}}{z_{source}}$$

$$u = \frac{SDDx_{source} + u_o z_{source}}{z_{source}} \tag{13.24}$$

and to the v coordinate similarly by

$$\frac{v - v_0}{SDD} = \frac{y_{source}}{z_{source}}$$

$$v = \frac{SDDy_{source} + v_o z_{source}}{z_{source}}. \tag{13.25}$$

The system of equations is written simply in matrix form as

$$\begin{pmatrix} u \\ v \\ 1 \end{pmatrix} \sim \begin{pmatrix} SDD & 0 & u_0 \\ 0 & SDD & v_0 \\ 0 & 0 & 1 \end{pmatrix} \begin{pmatrix} x_{source} \\ y_{source} \\ z_{source} \end{pmatrix}, \tag{13.26}$$

where the ~ symbol denotes that the left and right sides are equal to within scalar multiplication [i.e., $(a\ b\ c) \sim (A\ B\ C)$ implies $a = A/C$ and $b = B/C$]. We extend to the 3D case including an arbitrarily positioned source at $(x^s_{obj}, y^s_{obj}, z^s_{obj})$ with respect to the object frame and the detector at $(x^d_{obj}, y^d_{obj}, z^d_{obj})$ with angulation about the major axes of object frame (η, θ, ϕ), giving a source–detector "pose" described by these 9 degrees of freedom: source position, detector position, and detector tilt. The geometry is illustrated in Figure 13.8.

For voxel-driven reconstruction, we want to transform from a voxel location in the object frame $(x_{obj}, y_{obj}, z_{obj})$ to a location in the detector frame $(x_{det}, y_{det}, z_{det})$ accounting for rotations of the detector about the x-axis (η), y-axis (θ), and z-axis (ϕ). The object-to-detector rotation matrix is

$$^{obj}R_{det} = {}^{obj}R_{source} = R_x(\eta)R_y(\theta)R_z(\phi)$$

$$= \begin{pmatrix} c_\theta c_\phi & -c_\theta s_\phi & s_\theta \\ c_\eta s_\phi + c_\phi s_\eta s_\theta & c_\eta c_\phi - s_\eta s_\theta s_\phi & -c_\theta s_\eta \\ s_\eta s_\phi - c_\eta c_\phi s_\theta & c_\eta s_\theta s_\phi + c_\phi s_\eta & c_\eta c_\theta \end{pmatrix}, \tag{13.27}$$

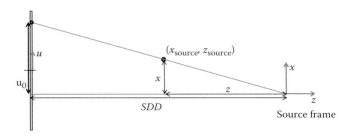

FIGURE 13.7 Illustration of perspective geometry. The subscript "source" refers to the coordinate with respect to the "source" frame. Note differences in coordinate conventions from Figure 13.1.

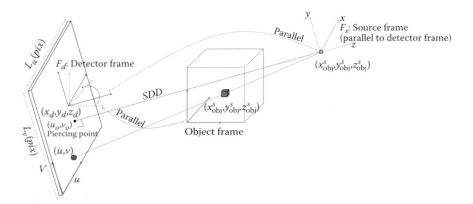

FIGURE 13.8 Illustration of projection geometry for a source—detector pose determined by 9 degrees of freedom: $(x_{obj}^s, y_{obj}^s, z_{obj}^s)$, $(x_{obj}^d, y_{obj}^d, z_{obj}^d)$, and (η, θ, φ). The source frame, F_s, is defined such that its major axes are parallel to the detector frame, F_d. The transformation scales and rotates from the $(x_{obj}, y_{obj}, z_{obj})$ coordinate system of the reconstructed image to the (u,v) coordinate system of the detector.

which is also the object-to-source rotation matrix, because detector and source frames are defined such that their major axes are parallel. The notation s_η, s_θ, and s_φ indicate $\sin(\eta)$, $\sin(\theta)$, and $\sin(\varphi)$, and c_η, c_θ, and c_φ indicate $\cos(\eta)$, $\cos(\theta)$, and $\cos(\varphi)$, respectively. The elements of the rotation matrix are denoted $r_1 - r_9$ in the row-major order. The voxel location with respect to the source frame $(x_{source}, y_{source}, z_{source})$ is therefore

$$\begin{pmatrix} x_{source} \\ y_{source} \\ z_{source} \end{pmatrix} = \begin{pmatrix} {}^{source}R_{obj} & {}^{source}t_{obj} \end{pmatrix} \begin{pmatrix} x_{obj} \\ y_{obj} \\ z_{obj} \\ 1 \end{pmatrix} = \begin{pmatrix} r_1 & r_4 & r_7 & t_x \\ r_2 & r_5 & r_8 & t_y \\ r_3 & r_6 & r_9 & t_z \end{pmatrix} \begin{pmatrix} x_{obj} \\ y_{obj} \\ z_{obj} \\ 1 \end{pmatrix}.$$

$$(13.28)$$

Here, ${}^{source}t_{obj}$ denotes a 3×1 vector (t_x, t_y, t_z) pointing from the origin of the source frame (F_s) to the origin of the object frame with respect to the source frame. The vector (t_x, t_y, t_z) can be computed by the following equation using the source location with respect to object frame $(x_{obj}^s, y_{obj}^s, z_{obj}^s)$ and the source-to-object rotation matrix ${}^{source}R_{obj}$:

$$\begin{aligned} {}^{source}t_{obj} &= \begin{pmatrix} (x^o - x^s)_{source} \\ (y^o - y^s)_{source} \\ (z^o - z^s)_{source} \end{pmatrix} = {}^{source}R_{obj} \begin{pmatrix} (x^o - x^s)_{obj} \\ (y^o - y^s)_{obj} \\ (z^o - z^s)_{obj} \end{pmatrix} \\ &= -{}^{source}R_{obj} \begin{pmatrix} x_{obj}^s \\ y_{obj}^s \\ z_{obj}^s \end{pmatrix} = \begin{pmatrix} t_x \\ t_y \\ t_z \end{pmatrix} \end{aligned}$$

$$(13.29)$$

The element $(x^o - x^s)_{source}$ denotes the x coordinate of the vector pointing from the origin of source frame to the origin of the object frame with respect to the source frame; similarly, the

element $(x^o - x^s)_{obj}$ denotes the same with respect to the object frame.

Note that various (x,y,z) coordinates up to this point are in units of length (e.g., cm), whereas (u,v) coordinates are in units of dexel coordinates (dimensionless). The (u,v) location is given by projection (by combining Equations 13.26 through 13.29) and scaling:

$$\begin{aligned} \begin{pmatrix} u \\ v \\ 1 \end{pmatrix} &\sim \begin{pmatrix} 1/a_u & 0 & L_u/2 \\ 0 & 1/a_v & L_v/2 \\ 0 & 0 & 1 \end{pmatrix} \begin{pmatrix} SDD & 0 & u_0 \\ 0 & SDD & v_0 \\ 0 & 0 & 1 \end{pmatrix} \\ &\times \begin{pmatrix} {}^{source}R_{obj} & -{}^{source}R_{obj} \begin{pmatrix} x_{obj}^s \\ y_{obj}^s \\ z_{obj}^s \end{pmatrix} \end{pmatrix} \begin{pmatrix} x_{obj} \\ y_{obj} \\ z_{obj} \\ 1 \end{pmatrix}, \quad (13.30) \end{aligned}$$

where the first matrix handles scaling according to dexel size (a_u and a_v) and detector size (L_u and L_v), and the second is the projection matrix as in Equation 13.26. The location of the piercing point (u_o, v_o) and SDD can be either directly estimated by a calibration algorithm discussed below, or if the calibration algorithm provides information of the detector position with respect to the object frame $(x_{obj}^d, y_{obj}^d, z_{obj}^d)$, we can covert those parameters into (u_0, v_0) and SDD by considering the source position with respect to the detector frame, described as follows:

$$\begin{pmatrix} u_0 \\ v_0 \\ SDD \end{pmatrix} = \begin{pmatrix} x_{det}^s \\ y_{det}^s \\ z_{det}^s \end{pmatrix} = \begin{pmatrix} {}^{det}R_{obj} & -{}^{det}R_{obj} \begin{pmatrix} x_{obj}^d \\ y_{obj}^d \\ z_{obj}^d \end{pmatrix} \end{pmatrix} \begin{pmatrix} x_{obj}^s \\ y_{obj}^s \\ z_{obj}^s \\ 1 \end{pmatrix}. \quad (13.31)$$

Here, we used a coordinate transformation similar to Equations 13.28 through 13.30 to convert the source position with respect to the object frame $(x^s_{obj}, y^s_{obj}, z^s_{obj})$ to that with respect to the detector frame $(x^s_{det}, y^s_{det}, z^s_{det})$, which is the same as (u_0, v_0, SDD).

With these transformations in hand, it becomes a matter of "calibrating" the system geometry [i.e., measuring the 9 degrees of freedom $(x^s_{obj}, y^s_{obj}, z^s_{obj})$, $(x^d_{obj}, y^d_{obj}, z^d_{obj})$, and (η, θ, φ), which usually vary as a function of projection angle for real systems subject to mechanical flex, gravity, vibration, etc.]. A variety of calibration methods have been reported in the literature—each involving a known object (e.g., an array of BBs) imaged as a function of projection angle to yield a pose determination. A single BB placed at isocenter can be used to measure approximately (u_0, v_0) as a function of projection angle,[20] which has been shown to capture the most significant aspects of geometric calibration to first order for a reasonably stable and rigid gantry (e.g., CBCT on a medical linear accelerator).[4] Among the earliest more complete calibration methods is that of Noo et al.[30] involving a line of BBs oriented parallel to the axis of rotation, projections of which trace ellipses in the projection domain. Navab et al.[21,24] showed an analogous method involving a phantom composed of a helix of BBs of various sizes—the helix minimizing overlap of BBs in projections and the various sizes providing a coding for BB recognition of individual BBs (i.e., location in the phantom) in the projection matrix analysis. An alternative approach was shown by Jain et al.[25] involving a combination of BBs, wires, and ellipses within a phantom for pose determination from a single projection view. The details of projection matrix analysis for each of these are detailed in the literature, typically involving an iterative least-squares solution to the system pose, minimizing error in pose determination across all measured projection angles, and solving for at least the most important of the 9 degrees of freedom [e.g., $(x^s_{obj}, y^s_{obj}, z^s_{obj})$ and $(x^d_{obj}, y^d_{obj}, z^d_{obj})$ but possibly only one or two of (η, θ, φ)]. The approach of Cho et al.[22] involves two circles of BBs within a cylindrical phantom, projecting as ellipses in the projection domain. The method allows a closed-form analytical (noniterative) solution for all 9 degrees of freedom in the

source–detector pose for each individual projection (repeated for all projection angles for a full CBCT calibration).

Many real CBCT systems involve large deviations from a perfect circular orbit, for example, variations in (u_0, v_0) from their mean position by >10 mm across the circular arc of a mobile "isocentric" C-arm.[23] However, the key to accurate reconstruction is that the nonideal motions are "reproducible," if not necessarily small. As illustrated in Figure 13.9 for a mobile C-arm modified to perform CBCT, departure from a semicircular orbit is significant, for example, deviations in x^d_{obj} by ±5 mm, with similar excursions exhibited by the other degrees of freedom. However, the nonidealities are reproducible—over the timescale of days, weeks, and even months—to the extent that the measurement can be used as a geometric calibration providing a mapping of the voxel domain to the dexel domain with a fairly high degree of accuracy. For example, 10 measurements overlaid in Figure 13.9a show not only reproducibility in the large-scale, low-frequency excursions but also the high-frequency "vibrations" and "jitters" that the system experiences during orbit. Figure 13.9b and c illustrates the effect of such geometric nonidealities on image quality. As shown in Figure 13.9b, failure to correct for geometric variations (i.e., assuming a semicircular orbit with the various degrees of freedom fixed at their mean value) imparts a major degradation in 3D image quality. The axial image reconstruction of a wire [point-spread function (PSF)] is seriously distorted, and fine details (e.g., temporal bone) are lost in artifact. An accurate geometric calibration, however, as shown in Figure 13.9c, properly corrects the backprojection geometry, such that the PSF properly reflects the spatial resolution limits of the system, and images (e.g., temporal bone) are largely free of geometric artifact.

13.4 "Offset-Detector" Geometry for Increased Lateral FOV

The previous sections assume a geometry in which the detector is more or less centered on the central axial ray; however,

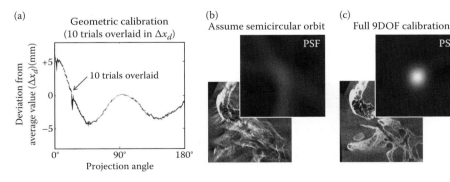

FIGURE 13.9 (a) Example variation in system geometry associated with mechanical flex and other physical nonidealities of a CBCT scanner. The example shown is for a mobile isocentric C-arm for CBCT. Variations in a single geometric parameter (x_d) are shown as a function of projection angle, demonstrating fairly large excursions from the ideal; however, the excursions are reproducible to a high degree and allow accurate image reconstruction by means of a geometric calibration. (b, c) Axial CBCT reconstructions of a wire (PSF) and cadaveric head in the region of the temporal bone. Ignoring the variations in geometry (i.e., assuming a circular orbit as in b) results in significant error. Reconstruction with geometric calibration (c) restores image quality.

this is in no way a requirement for 3D FBP, and it carries a variety of limitations. A frequent limitation is a small lateral FOV (FOV_{xy}) and significant lateral truncation artifacts, especially for body sites. An alternative to the centered-detector setup illustrated in Figure 13.1 involves a lateral shift of the detector by up to half the detector width ($L_u/2$) combined with a full 360° orbit—commonly referred to as the "offset-detector" geometry. As illustrated in Figure 13.10, the offset-detector geometry samples approximately half of the patient in any given view, but across a full 360° orbit, the system has sampled the entire volume and effected a lateral FOV up to twice that of the centered-detector setup. This does involve modifications to the weighting and filtering schemes in normal 3D FBP as summarized above, and it does not enjoy the improved sampling (e.g., 1/4 pixel offset) normally associated with a 360° orbit. It has become a fairly common CBCT setup in image-guided radiotherapy, where geometries typically involve a detector with $L_u \sim 40$ cm and magnification $M \sim 1.5$, giving $FOV_{xy} \sim 25$ cm, which may be insufficient for large body sites. A 15 cm lateral shift effectively extends this to $FOV_{xy} \sim 46$ cm.

The offset-detector geometry does involve changes to the weighting and filtering applied in 3D FBP.[26,27] As illustrated in Figure 13.10, backprojection "fills in" roughly a quarter of the volume for each 90° arc, yielding a complete reconstruction for $\theta_{tot} = 360°$—quite distinct from the filling at $\theta_{tot} = 180°$ + fan illustrated in Figure 13.6. For a lateral shift less than $L_u/2$, columns along one edge of the detector contribute redundantly in overlapping rays through the center of the reconstruction in a manner completely analogous to the redundant sampling discussed in relation to Parker weights above. The detector signal in these columns/overlapping rays must be appropriately weighted to account for such redundancy.

13.5 Beyond Feldkamp: Reconstruction from Noncircular Source–Detector Trajectories

Performing cone-beam data acquisition while moving the source–detector assembly along a noncircular orbit (trajectory) provides a means to avoid cone-beam artifacts and thereby possibly improve image quality, assuming that physical effects such as x-ray scatter and beam hardening are not limiting factors. However, the noncircular orbit may not be arbitrary: Tuy's condition[28] must be satisfied in order for the sampling of the object to be complete and the resulting 3D image reconstruction to be "exact." This condition can be expressed as follows: any plane passing through the region of interest within the scanned object must intersect the source trajectory. If Tuy's condition is not satisfied, sampling of the region of interest is incomplete, and artifact-free imaging is, in general, not possible. Furthermore, Tuy's condition is necessary and sufficient for accurate imaging only when each acquired cone-beam projection is nontruncated (i.e., the entire scanned object is visible in the projection).

Accurate image reconstruction from nontruncated projections collected on an orbit satisfying Tuy's condition is well understood and generally relies on Grangeat's formula.[29] The reconstruction can be performed indirectly by first transforming the cone-beam data into samples of the 3D radon transform of the scanned object and then inverting this transform.[30] Alternatively, it can be performed directly using an FBP formula developed by Defrise and Clack[31] and Kudo and Saito.[32]

When the projections are truncated, which is typically the case in medical imaging, performing accurate reconstruction is much more challenging and will depend on the truncation pattern. Unfortunately, there exists no simple, closed-form

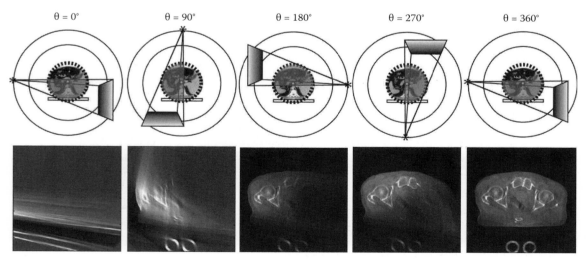

FIGURE 13.10 Illustration of offset-detector geometry. Top: Illustration of offset-detector geometry in which the detector is translated laterally by nearly half its width, effecting a larger FOV_{xy} sampled across a 360° scan. Bottom: View of the axial slice reconstruction as it evolves in FBP under offset-geometry conditions. In contrast to the centered-detector case (in which the evolution in the axial image is radially symmetric about the center of reconstruction, and complete sampling is achieved in 180° + fan), the offset-detector case results in the sampling/filling of quadrants of the axial FOV and requires a full 360° orbit.

formula to solve this problem. However, two powerful theories have been developed over the last decade: the general FBP scheme of Katsevich[33] and the differentiated backprojection (DBP) method,[7,34–38] which is also referred to as the BPF method or the two-step Hilbert method. Findings in cone-beam reconstruction are not limited to these two theories, but they define good starting points for the development of an accurate reconstruction algorithm and understanding of other published work. Both approaches heavily rely on the concept of R-lines, which are lines connecting two or more source positions. The symbol R stands for redundantly measured line. If a point P within the scanned object lies on an R-line, then the two theories offer a means to potentially achieve accurate reconstruction at P from truncated data. Otherwise, neither of the approaches is likely to work. Hence, analysis of the region covered by R-lines is critical in the selection of a noncircular orbit for imaging with truncated projections.

Practical noncircular orbits for which accurate reconstruction algorithms allow data truncation include the circle-and-line,[39,40] the circle-and-arc,[41,42] the helix,[43] and the saddle orbits.[44,45] Accurate image reconstruction from truncated cone-beam projections remains a topic of active research, particularly driven by a continuous demand for improved imaging performance, not only for diagnostic purposes but also for treatment application and monitoring in oncology and interventional radiology.

13.6 Iterative Reconstruction

There is a broad class of reconstruction algorithms that are not based on the application of a single set of analytic operations. An alternative is to successively approximate a solution to the reconstruction using iterative approaches, where one starts with an initial guess $\hat{\mu}(x, y, z)$ for the attenuation volume $\mu(x, y, z)$ and refines the values that comprise that volume with some kind of update. These approaches are often referred to as "nonanalytic" because the reconstruction estimate either cannot be written down as a closed-form solution or, even if one can write down the solution, it is too computationally cumbersome to apply a direct analytic approach.

A common property among these techniques is that the reconstruction estimate tends to be defined as the implicit maximizer of some objective function. (In some cases, the objective function exists but is not explicit.) Specifically, one adopts a mathematical model that relates a parameterized volume (i.e., image voxels) to the entire set of projection measurements collected in a tomographic acquisition. This so-called forward model can be arbitrarily complex and may include effects such as an extended x-ray source, the detector footprints (dexel size and shape), polyenergetic x-ray beams, and x-ray scatter. Similarly, the geometry specified by the forward model is general: it may represent single-slice CT, helical acquisition, cone-beam geometry, limited-angle data, noncircular orbits, etc. An objective function is formed through the selection of a particular metric that enforces a match between the predicted data, using the forward model and the reconstruction estimate, and the observed measurement

data. In many cases, the objective function is further modified (i.e., "regularized" with additional terms) to enforce desirable properties in the reconstructed image (e.g., smoothness). In general, the estimator can be written as

$$\hat{\mu} = \arg\min_{\vec{\mu}} \left\| \vec{g}, F(\vec{\mu}) \right\|, \tag{13.32}$$

where $\hat{\mu}$ represents a vector representing the current image estimate, $\vec{g}$ is a vector whose elements are the projection data measurements, $F(\vec{\mu})$ represents the predicted data vector that is a function of the image volume propagated through the forward model, and $\|\cdot,\cdot\|$ denotes some metric of similarity between the measurements and the forward model applied to the current image estimate.

A reconstruction algorithm must then be specified that iteratively approximates the solution to Equation 13.32. Such algorithms can be highly dependent on the exact form of the objective, and there are many algorithms that can be used to solve a particular objective function. Thus, an iterative approach is typically defined by those two components: (1) an objective function that includes a forward model and some metric to enforce a fit to the observed data, and (2) a specific algorithm that finds an estimate that is optimal for that particular objective.

13.6.1 Algebraic Reconstruction Techniques

A simple monoenergetic forward model was previously specified in Equation 13.7; however, this formulation presumes a continuous domain object, whereas the model and objective function in Equation 13.32 are discrete. An analogous model for a discretely parameterized volume can be written as

$$\vec{g} = g_0 \exp(-\mathbf{A}\vec{\mu}) = F(\vec{\mu}), \tag{13.33}$$

where $\mathbf{A}$ is a matrix that operates on the image vector $\vec{\mu}$. Elements of the so-called system matrix, a_{ij}, relate the contribution of each voxel to the projection integral associated with each measurement. As mentioned previously, there are no constraints on the particular projection geometry represented by $\mathbf{A}$. Recognizing that Equation 13.33 represents a nonlinear system of equations, one can potentially try to solve for $\vec{\mu}$ directly. Logarithmic transformation of the data yields

$$\vec{l} \equiv -\ln\left(\frac{\vec{g}}{g_0}\right) = \mathbf{A}\vec{\mu}, \tag{13.34}$$

which is a linear relation between the transformed data, $\vec{l}$, and the projected volume. The linear system of equations represented by Equation 13.34 can be solved using a number of different iterative approaches.[46–48] For example, one particular approach,[49] uses the update

$$\vec{\mu}^{(k+1)} = \vec{\mu}^{(k)} + \frac{\mathbf{A}^{\mathrm{T}}\left(\dfrac{\vec{l} - \mathbf{A}\vec{\mu}}{\mathbf{A}\vec{1}}\right)}{\mathbf{A}^{\mathrm{T}}\vec{1}}, \qquad (13.35)$$

where $\mathbf{A}^{\mathrm{T}}$ is the transpose of the system matrix and functionally represents the backprojection of a measurement domain vector. (Note that all divisions are element-by-element and $\vec{1}$ represents the all-ones vector.) One can see that Equation 13.35 applies a correction to the image based on a (normalized) backprojection of the residual error between $\vec{l}$ and $\mathbf{A}\vec{\mu}$. Methods following this class of method are typically referred to as algebraic reconstruction techniques (ARTs).

Under certain conditions (e.g., consistent data), ART methods are equivalent to solving the objective function

$$\hat{\vec{\mu}} = \arg\min_{\vec{\mu}} \left\| \vec{p} - \mathbf{A}\vec{\mu} \right\|^2 = \left(\mathbf{A}^T \mathbf{A} \right)^{-1} \mathbf{A}\vec{p}, \qquad (13.36)$$

which is a least-squares objective function that minimizes the variance between the predicted and modeled transformed measurements;[47] however, many ART-based approaches are not strictly convergent and would be difficult to express as the optimizer of a specific objective function.

13.6.2 Statistical Reconstruction Approaches

Although ART-based methods are attractive because they handle arbitrary geometries, they rely on linearization of the forward model, and all measurements are treated equally, regardless of their signal-to-noise ratio (SNR). Statistical reconstruction approaches are a class of image estimators that incorporate both the forward model as described above and the noise model for individual measurements.

13.6.2.1 Maximum Likelihood Reconstruction

In estimation theory,[50] there are many possible metrics that take the noise model into account. One metric that has found widespread use in tomographic reconstruction is the likelihood function. Specifically, maximizing a likelihood function over all possible $\vec{\mu}$ will produce the image volume that was most likely to be responsible for the observed measurements. Using such an objective will implicitly balance the relative information content between different measurements—essentially increasing the effect that high SNR data impart on the reconstruction and decreasing the relative importance of low SNR data.

Likelihood-based approaches require adoption of a noise model for the measurements. Typically, Poisson[51] or Gaussian (for log data)[52] distributions are used; however, one could potentially adopt a more sophisticated compound Poisson or Gaussian–Poisson mixture model to better match the physics of tomographic system—or potentially even a cascaded systems

model[53,54] that provides a fairly complete description of the imaging chain characteristics (e.g., detector blur and additive noise). Any particular model tends to be an approximation of the composition of the underlying noise processes of the physics of photon generation and detection. The simple Poisson model presumes that noise is dominated by the photon statistics. In this case, each measurement has a probability given by

$$p(g_i \mid \vec{\mu}) = \exp\left[-F_i(\vec{\mu})\right] \frac{\left[F_i(\vec{\mu})\right]^{g_i}}{g_i!}. \qquad (13.37)$$

The likelihood function is equal to the joint probability of all measurements given the object, $p(\vec{g} \mid \vec{\mu})$, which for independent observations is the product of the marginal probabilities in Equation 13.37. Although one may try to maximize the likelihood directly, it is equivalent to maximizing the log likelihood, which yields the following estimator:

$$\hat{\vec{\mu}} = \arg\max_{\vec{\mu}} L(\vec{g}, \vec{\mu}) = \arg\min_{\vec{\mu}} (-L(\vec{g}, \vec{\mu}))$$

$$L(\vec{g}, \vec{\mu}) = \ln p(\vec{g} \mid \vec{\mu}) = \ln\left(\prod_i p(g_i \mid \vec{\mu})\right)$$

$$= \sum_i \ln p(g_i \mid \vec{\mu}) = \sum_i g_i \ln\left[F_i(\vec{\mu})\right] - F_i(\vec{\mu}) - \ln g_i!. \qquad (13.38)$$

Numerous algorithms exist to iteratively solve this form of objective function, including the expectation–maximization (EM) approach,[51] ordered subsets,[55] EM variations,[56] paraboloidal surrogates approaches,[10] etc.

Both ART and maximum likelihood (ML) approaches are unregularized estimation methods. That is, the objective function does not include any control to trade-off between noise and resolution in the reconstruction. Because tomographic reconstruction is ill-conditioned and highly noise-amplifying, the solutions to these objectives can be very noisy. One simple control method that relies on the particular convergence properties of the reconstruction algorithm is to stop iterations before the estimate becomes too noisy. This technique tends to work because high spatial frequencies typically take longer to converge than low spatial frequencies. Other noise control options include postfiltering[57] and the method of sieves.[58]

13.6.2.2 Penalized Likelihood/Maximum A Posteriori Reconstruction

Another option for controlling the trade-off between noise and spatial resolution is to modify the objective function to enforce desirable image properties. Such "regularization" tends to improve the convergence rate of iterative algorithms and improve image quality in the solution of the objective. One general form for such a modified objective is

$$\hat{\vec{\mu}} = \arg\max_{\vec{\mu}} L(\vec{g}, \vec{\mu}) - \beta R(\vec{\mu}), \qquad (13.39)$$

where a regularization term, R, is subtracted from the likelihood term. This term acts as a penalty on undesirable (e.g., noisy) images and includes the regularization parameter, β, to control the balance between the data fidelity and penalty terms. Such approaches are often referred to as penalized-likelihood (PL) estimators; however, one may also derive an objective of the form in Equation 13.39 by defining a probabilistic distribution on the image volume. Specifically, if $p(\mu)$ is known, one may employ Bayes' rule to form the maximum a posteriori (MAP) estimate, which is mathematically equivalent to Equation 13.39 for a general class of prior distributions on μ.

Although many kinds of penalty and prior exist, it is common to presume the pairwise form:

$$R(\vec{\mu}) = \sum_j \sum_{k \in N_j} \psi(\mu_j - \mu_k), \qquad (13.40)$$

where the pairwise difference of voxels within some neighborhood penalizes according to some function, $\psi(\cdot)$. In effect, neighboring pixels are not allowed to deviate too much from another, thereby enforcing local smoothness (and discouraging noise). The simplest penalty function is to use the squared difference between voxels (i.e., $\psi = (\mu_j - \mu_k)^2$). However, this quadratic penalty tends to smooth out both noise and anatomical edges in the image. Many other penalty functions have been used that reduce the relative penalty (compared with the quadratic case) for increasing differences to preserve anatomical edges. Such penalties include those introduced by Lange[56,59] and Elbakri,[12] and the so-called "total variation" (TV) penalty[60,61] based on the 1-norm. The latter methods are closely related to constrained total variation minimizations.[62,63] A comparison of two different PL approaches and traditional FBP are shown in Figure 13.11.

Algorithms for PL and MAP approaches are similar to ML approaches (ML being a special case in which $\beta = 0$). Numerous algorithms have been developed, and those developed for the penalized case usually apply to the unregularized case as well.

13.6.3 Compressive Sensing Techniques

Although the iterative approaches discussed thus far do not constrain the geometry represented by A, the results provided by those methods from highly limited or sparse acquisitions will have reduced image quality due to missing data (i.e., undersampling). Traditional theories on image sampling suggest that one cannot reconstruct spatial frequencies that are not observed in the data. Recent work in the area of compressive sensing suggests otherwise. Specifically, if one possesses a certain kind of prior knowledge about the object, much less data are required for reconstruction. Compressed sensing objectives tend to have the form

$$\hat{\mu} = \arg\min_{\vec{\mu}} \left\| \Psi(\vec{\mu}) \right\| \quad \text{s.t.} \quad F(\vec{\mu}) = \vec{g}, \qquad (13.41)$$

where the implicit estimator is represented by a constrained optimization. That is, one minimizes a function, Ψ, of the volume subject to the constraint that the predicted data (via the forward model) match the measurements. In this case, it is critical that Ψ represents a so-called sparsifying operator that converts the volume into a domain that is largely composed of zeros and some nonzero elements. Some objects (i.e., images such as contrast-enhanced vasculature and background-subtracted images) are naturally sparse, whereas others might consider the spatial derivative of an image to be sparse, implying

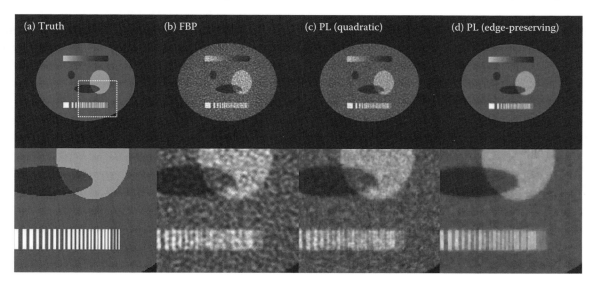

FIGURE 13.11 Comparison of FBP with two different PL estimators. (a) Simple digital phantom used to generate synthetic data. (b) FBP reconstruction. (c) PL reconstruction with a quadratic penalty. (d) PL reconstruction with an edge-preserving penalty.[59] Although all reconstructions use the same raw data, we see that both the noise decreases "and" the spatial resolution improve in moving from FBP to PL (quadratic) to PL (edge-preserving). The lower row of the images represents zoomed-in versions of the top row (zoomed region indicated by the dotted line in a).

that the image is largely piecewise constant. The norm used in Equation 13.41 is also important. The 1-norm is commonly used (although general p-norms with $p \leq 1$ are also used) because it encourages sparse estimates. Iterative algorithms exist for solving such objectives and can provide impressive results when an appropriate sparse domain can be found.

A notable example of such an approach is the prior image constrained compressed sensing (PICCS)[64,65] method, which has the following objective function:

$$\hat{\vec{\mu}} = \arg\min_{\vec{\mu}} \left[\alpha \left\| \Psi(\vec{\mu} - \vec{\mu}_P) \right\|_1 + (1 - \alpha) \left\| \Psi(\vec{\mu}) \right\|_1 \right] \text{ s.t. } \mathbf{A}\vec{\mu} = -\ln\frac{\vec{g}}{g_0},$$

(13.42)

where the objective is composed of two terms: the first being a sparsified difference between the estimate and a prior (previously acquired) image, μ_P, and the second being the sparsified image as in Equation 13.41. A parameter α controls the relative importance of the two terms. A linear constraint is applied and is easily recognized as Equation 13.34. This objective enforces similarity between the estimate and a prior image and would have particular application in situations where sequences of volumes are acquired (e.g., dynamic scans or follow-up scan sequences). Algorithms for solving Equation 13.42 often involve updates that minimize the unconstrained objective followed by enforcement of the constraint. However, because the constraint also involves an inversion, many approaches apply the constraint through the use of ART-type updates.

13.6.4 Current and Future Directions

Recent development of iterative reconstruction techniques shows increasing focus on forming highly accurate forward models that include realistic projection models including detector size effects,[66,67] incorporating a polyenergetic source and energy-dependent attenuation coefficients,[12,68,69] and scattered radiation.[70] Similarly, object models have also been refined to incorporate a moving object (e.g., respiratory[71] or cardiac motion[72]) and additional prior information such as object boundaries. Using an accurate forward model and all of the available prior information is critical to providing the best image quality for a given set of data. With the continued drive to improve image quality and/or reduce radiation dose, we expect that trends to increase the realism of the forward model and incorporate prior knowledge will continue. Gains over traditional methods are likely to have the greatest potential in situations where the quality of the projection data is limited either by very low-dose acquisitions (i.e., high quantum noise) or by sparse collections (e.g., limited-angle tomography and region-of-interest tomography).

Similarly, new technologies and specific applications will also drive the development of new iterative approaches. Technology examples include algorithms for photon-counting devices and multienergy acquisitions,[73] where new forward models are being developed. Specific applications include perfusion imaging, contrast-enhanced imaging, and interventional imaging in the presence of tools and implants, which will require improved object models and priors.

13.7 Summary and Future Directions

The first decade of the twenty-first century witnessed a revolution in CT. First and foremost was the widespread development and availability of MDCT, extending from 4-slice detectors to 8-, 16-, and 64-slice scanners now common throughout diagnostic CT as well as 320-slice scanners for fully volumetric (e.g., whole organ) imaging. When this evolution proceeded beyond approximately eight slices, CT, as we know it in broad diagnostic imaging contexts, in fact, became CBCT, and account of the divergent beam and nonnegligible cone angle became essential and intrinsic to 3D reconstruction techniques. In parallel with this major development in diagnostic CT was an evolution in the application of FPDs for CBCT. Early applications included image-guided radiotherapy and image-guided surgery, and CBCT rapidly proliferated in a variety of contexts throughout angiography and interventional guidance. In specialty applications of diagnostic imaging as well, CBCT has emerged as a promising technology for areas ranging from dental, ENT, and maxillofacial imaging to breast imaging as well as musculoskeletal, orthopedics, and rheumatology imaging. This explosion in the variety of CBCT imaging platforms and applications has benefited from the ongoing ingenuity in FPDs, reconstruction methods, and collaboration between physicists, engineers, and physicians in a broad spectrum of disciplines.

In parallel with such development of technology and clinical application have come major strides in 3D reconstruction techniques, many of which are touched upon in this chapter. Although 3D FBP approaches stemming directly from the FDK algorithm represent by far the majority of work reported and translated to clinical use, the new era in which advanced statistical and iterative reconstruction techniques is now upon us. In late 2010, iterative reconstruction was released on commercially available diagnostic CT scanners, signaling a wave of advances and greater proliferation of iterative reconstruction in real clinical applications. Fueled by high-speed computing technology and parallel programming methods, computationally intense iterative reconstruction methods such as model-based reconstruction and compressive sensing techniques have become more practical, such that high-quality reconstructions can be obtained within reasonable timescales and with affordable computing hardware. The benefits of such advances include higher image quality, lower radiation dose, faster scanning, artifact reduction, and potential application of CT in new areas previously challenged by artifacts or low-contrast limits.

Acknowledgments

The authors extend their appreciation to numerous individuals with whom conversations on 3D image reconstruction contributed to this chapter. Dr. Wojciech Zbijewski (Johns Hopkins University) helped with numerous aspects of the chapter, including generation of real and simulated data included in the figures. Dr. Yoshito Otake (Johns Hopkins University) assisted with nomenclature and analysis pertaining to system geometry and geometric calibration. Valuable discussions with Dr. John Boone (University of California-Davis), Dr. Norbert Pelc (Stanford University), Dr. Guang-Hong Chen (University of Wisconsin), Dr. Xiaochuan Pan (University of Chicago), and Dr. Jeff Fessler (University of Michigan) are gratefully acknowledged. Some the material included in this chapter was generated under research conducted with support from the National Institutes of Health (2R01-CA-112163 and R01-CA-127444), Elekta Oncology Systems (Atlanta, Georgia), and Siemens Healthcare (Erlangen, Germany).

References

1. L. A. Feldkamp, L. C. Davis and J. W. Kress, "Practical cone-beam algorithm," *Journal of the Optical Society of America A* **1** (6), 612–619 (1984).

2. P. Grangeat, A. Koenig, T. Rodet and S. Bonnet, "Theoretical framework for a dynamic cone-beam reconstruction algorithm based on a dynamic particle model," *Physics in Medicine and Biology* **47** (15), 2611–2625 (2002).

3. D. A. Jaffray and J. H. Siewerdsen, "Cone-beam computed tomography with a flat-panel imager: Initial performance characterization," *Medical Physics* **27** (6), 1311–1323 (2000).

4. D. A. Jaffray, J. H. Siewerdsen, J. W. Wong and A. A. Martinez, "Flat-panel cone-beam computed tomography for image-guided radiation therapy," *International Journal of Radiation Oncology Biology Physics* **53** (5), 1337–1349 (2002).

5. A. C. Miracle and S. K. Mukherji, "Conebeam CT of the head and neck, part 2: Clinical applications," *American Journal of Neuroradiology* **30** (7), 1285–1292 (2009).

6. S. Hoppe, J. Hornegger, F. Dennerlein, G. Lauritsch and F. Noo, "Accurate image reconstruction using real C-arm data from a Circle-plus-arc trajectory," *International Journal of Computer Assisted Radiology and Surgery* **7** (1), 73–86 (2012).

7. Y. Zou, X. Pan, D. Xia and G. Wang, "PI-line-based image reconstruction in helical cone-beam computed tomography with a variable pitch," *Medical Physics* **32** (8), 2639–2648 (2005).

8. A. Katsevich, "Analysis of an exact inversion algorithm for spiral cone-beam CT," *Physics in Medicine and Biology* **47** (15), 2583–2597 (2002).

9. H. Erdogan and J. A. Fessler, "Ordered subsets algorithms for transmission tomography," *Physics in Medicine and Biology* **44** (11), 2835–2851 (1999).

10. H. Erdogan and J. A. Fessler, "Monotonic algorithms for transmission tomography," *IEEE Transactions on Medical Imaging* **18** (9), 801–814 (1999).

11. J. A. Fessler, I. A. Elbakri, P. Sukovic and N. H. Clinthorne, "Maximum-likelihood dual-energy tomographic image reconstruction," *Physics of Medical Imaging, SPIE Proceedings* **4684** (1), 38–49 (2002).

12. I. A. Elbakri and J. A. Fessler, "Statistical image reconstruction for polyenergetic X-ray computed tomography," *IEEE Transactions on Medical Imaging* **21** (2), 89–99 (2002).

13. D. F. Yu and J. A. Fessler, "Edge-preserving tomographic reconstruction with nonlocal regularization," *IEEE Transactions on Medical Imaging* **21** (2), 159–173 (2002).

14. S. C. Bushong, S. Balter and C. G. Orton, "Point/counterpoint. Office-based cone-beam and digital tomosynthesis systems using flat-panel technology should not be referred to as CT units," *Medical Physics* **38** (1), 1–4 (2011).

15. R. Molteni, "The so-called cone beam computed tomography technology (or CB3D, rather!)," *Dentomaxillofacial Radiology* **37** (8), 477–478 (2008).

16. F. C. Billingsley, "Digital video processing at JPL," *Electronic Imaging Techniques I, SPIE Proceedings* **3** XV-1-19 (1965).

17. W. F. Schreiber, "Picture coding," *SPIE Proceedings* **55** (1967).

18. M. J. Yaffe, in *Digital Mammography*, edited by U. Bick and F. Diekman (Springer, New York, 2009).

19. A. C. Kak and M. Slaney, *Principles of Computerized Tomographic Imaging* (IEEE Press, 1988).

20. R. Fahrig and D. W. Holdsworth, "Three-dimensional computed tomographic reconstruction using a C-arm mounted XRII: Image-based correction of gantry motion nonidealities," *Medical Physics* **27** (1), 30–38 (2000).

21. N. Nassir, R. B.-H. Ali, M. M. Matthias, W. H. David, F. Rebecca, J. F. Allan and R. Graumann, "Dynamic geometrical calibration for 3D cerebral angiography," *Physics of Medical Imaging, SPIE Proceedings* **2708** (1), 361–370 (1996).

22. Y. B. Cho, D. J. Moseley, J. H. Siewerdsen and D. A. Jaffray, "Geometric calibration of cone-beam computerized tomography system and medical linear accelerator," *Proceedings of the XVIth International Conference on the Use of Computers in Radiation Therapy (ICCR)*, 482–485 (2004).

23. M. J. Daly, J. H. Siewerdsen, Y. B. Cho, D. A. Jaffray and J. C. Irish, "Geometric calibration of a mobile C-arm for intraoperative cone-beam CT," *Medical Physics* **35** (5), 2124–2136 (2008).

24. N. Navab, A. Bani-Hashemi, M. Nadar, K. Wiesent, P. Durlak, T. Brunner, K. Barth and R. Graumann, in *Medical Image Computing and Computer-Assisted Intervention—MICCAI'98*, edited by W. Wells, A. Colchester and S. Delp (Springer, Berlin, 1998), Vol. 1496, pp. 119–129.

25. A. K. Jain, T. Mustafa, Y. Zhou, C. Burdette, G. S. Chirikjian and G. Fichtinger, "FTRAC—A robust fluoroscope tracking fiducial," *Medical Physics* **32** (10), 3185–3198 (2005).

26. G. Wang, "X-ray micro-CT with a displaced detector array," *Medical Physics* **29** (7), 1634–1636 (2002).

27. V. Liu, N. R. Lariviere and G. Wang, "X-ray micro-CT with a displaced detector array: Application to helical cone-beam reconstruction," *Medical Physics* **30** (10), 2758–2761 (2003).

28. H. Tuy, "An inversion formula for cone-beam reconstruction," *SIAM Journal on Applied Mathematics* **43**, 546–552 (1983).

29. P. Grangeat, in *Mathematical Methods in Tomography*, edited by G. T. Herman, A. K. Louis and F. Natterer (Springer, New York, 1991), pp. 66–97.

30. F. Noo, R. Clack and M. Defrise, "Cone-beam reconstruction from general discrete vertex sets using Radon rebinning algorithms," *IEEE Transactions on Nuclear Science* **44**, 1309–1316 (1997).

31. M. Defrise and R. Clack, "A cone-beam reconstruction algorithm using shift-variant filtering and cone-beam backprojection," *IEEE Transactions on Medical Imaging* **13** (1), 186–195 (1994).

32. H. Kudo and T. Saito, "Fast and stable cone-beam filtered backprojection method for non-planar orbits," *Physics in Medicine and Biology* **43** (4), 747–760 (1998).

33. A. Katsevich, "A general scheme for constructing inversion algorithms for cone-beam CT," *International Journal of Mathematics and Mathematical Sciences* **21**, 1305–1321 (2003).

34. J. D. Pack, F. Noo and R. Clackdoyle, "Cone-beam reconstruction using the backprojection of locally filtered projections," *IEEE Transactions on Medical Imaging* **24** (1), 70–85 (2005).

35. Y. Ye and G. Wang, "Filtered backprojection formula for exact image reconstruction from cone-beam data along a general scanning curve," *Medical Physics* **32** (1), 42–48 (2005).

36. Y. Ye, S. Zhao, H. Yu and G. Wang, "A general exact reconstruction for cone-beam CT via backprojection-filtration," *IEEE Transactions on Medical Imaging* **24** (9), 1190–1198 (2005).

37. Y. Zou and X. Pan, "Image reconstruction on PI-lines by use of filtered backprojection in helical cone-beam CT," *Physics in Medicine and Biology* **49** (12), 2717–2731 (2004).

38. Y. Zou and X. Pan, "Exact image reconstruction on PI-lines from minimum data in helical cone-beam CT," *Physics in Medicine and Biology* **49** (6), 941–959 (2004).

39. G. L. Zeng and G. T. Gullberg, "A cone-beam tomography algorithm for orthogonal circle-and-line orbit," *Physics in Medicine and Biology* **37** (3), 563–577 (1992).

40. F. Noo, M. Defrise and R. Clack, "Direct reconstruction of cone-beam data acquired with a vertex path containing a circle," *IEEE Transactions on Image Processing* **7** (6), 854–867 (1998).

41. A. Katsevich, "Image reconstruction for the circle-and-arc trajectory," *Physics in Medicine and Biology* **50** (10), 2249–2265 (2005).

42. X. Wang and R. Ning, "A cone-beam reconstruction algorithm for circle-plus-arc data-acquisition geometry," *IEEE Transactions on Medical Imaging* **18** (9), 815–824 (1999).

43. A. Katsevich, S. Basu and J. Hsieh, "Exact filtered backprojection reconstruction for dynamic pitch helical cone beam computed tomography," *Physics in Medicine and Biology* **49** (14), 3089–3103 (2004).

44. J. D. Pack, F. Noo and H. Kudo, "Investigation of saddle trajectories for cardiac CT imaging in cone-beam geometry," *Physics in Medicine and Biology* **49** (11), 2317–2336 (2004).

45. H. Yang, M. Li, K. Koizumi and H. Kudo, "View-independent reconstruction algorithms for cone beam CT with general saddle trajectory," *Physics in Medicine and Biology* **51** (15), 3865–3884 (2006).

46. R. Gordon, R. Bender and G. T. Herman, "Algebraic reconstruction techniques (ART) for three-dimensional electron microscopy and x-ray photography," *Journal of Theoretical Biology* **29**, 471–481 (1970).

47. G. T. Herman, A. Lent and C. M. Rowland, "ART: Mathematics and applications, a report on the mathematical foundations and on the applicability to real data of the algebraic reconstruction techniques," *Journal of Theoretical Biology* **42**, 1–32 (1973).

48. M. Jiang and G. Wang, "Convergence studies on iterative algorithms for image reconstruction," *IEEE Transactions on Medical Imaging* **22** (5), 569–579 (2003).

49. A. H. Anderson and A. C. Kak, "Simultaneous algebraic reconstruction technique (SART): A superior implementation of the ART algorithm," *Ultrasonic Imaging* **6**, 81–94 (1984).

50. H. L. Van Trees, *Detection, Estimation, and Modulation Theory.* (Wiley, New York, 1968).

51. K. Lange and R. Carson, "EM reconstruction algorithms for emission and transmission tomography," *Journal of Computer Assisted Tomography* **8** (2), 306–316 (1984).

52. K. D. Sauer and C. A. Bouman, "A local update strategy for iterative reconstruction from projections," *IEEE Transactions on Signal Processing* **41** (2), 534–548 (1993).

53. J. H. Siewerdsen, L. E. Antonuk, Y. El-Mohri, J. Yorkston, W. Huang and I. A. Cunningham, "Signal, noise power spectrum, and detective quantum efficiency of indirect-detection flat-panel imagers for diagnostic radiology," *Medical Physics* **25** (5), 614–628 (1998).

54. D. J. Tward, J. H. Siewerdsen, R. A. Fahrig and A. R. Pineda, "Cascaded systems analysis of the 3D NEQ for cone-beam CT and tomosynthesis," *SPIE Medical Imaging* **6913**, 69131S (2008).

55. H. M. Hudson and R. S. Larkin, "Accelerated image reconstruction using ordered subsets of projection data," *IEEE Transactions on Medical Imaging* **13** (4), 601–609 (1994).

56. K. Lange and J. A. Fessler, "Globally convergent algorithms for maximum a posteriori transmission tomography," *IEEE Transactions on Image Processing* **4** (10), 1430–1438 (1995).

57. J. Nuyts, B. De Man, P. Dupont, M. Defrise, P. Suetens and L. Mortelmans, "Iterative reconstruction for helical CT: A simulation study," *Physics in Medicine and Biology* **43** (4), 729–737 (1998).

58. D. L. Snyder, M. I. Miller, L. J. Thomas and D. G. Politte, "Noise and edge artifacts in maximum-likelihood reconstructions for emission tomography," *IEEE Transactions on Medical Imaging* **6** (3), 228–238 (1987).

59. K. Lange, "Convergence of EM image reconstruction algorithms with Gibbs smoothing," *IEEE Transactions on Medical Imaging* **9** (4), 439–446 (1990).

60. C. R. Vogel and M. E. Oman, "Fast, robust total variation-based reconstruction of noisy, blurred images," *IEEE Transactions on Image Processing* **7** (6), 813–824 (1998).

61. C. R. Vogel and M. E. Oman, "Iterative methods for total variation denoising," *SIAM Journal on Scientific Computing* **17**, 227–238 (1996).

62. E. Y. Sidky and X. Pan, "Image reconstruction in circular cone-beam computed tomography by constrained, total-variation minimization," *Physics in Medicine and Biology* **53** (17), 4777–4807 (2008).

63. J. Bian, J. H. Siewerdsen, X. Han, E. Y. Sidky, J. L. Prince, C. A. Pelizzari and X. Pan, "Evaluation of sparse-view reconstruction from flat-panel-detector cone-beam CT," *Physics in Medicine and Biology* **55** (22), 6575–6599 (2010).

64. G. H. Chen, J. Tang and S. Leng, "Prior image constrained compressed sensing (PICCS): A method to accurately reconstruct dynamic CT images from highly undersampled projection data sets," *Medical Physics* **35** (2), 660–663 (2008).

65. G. H. Chen, J. Tang and S. Leng, "Prior Image Constrained Compressed Sensing (PICCS)," *Society of Photo-Optical Instrumentation Engineers* **6856**, 685618 (2008).

66. B. De Man and S. Basu, "Distance-driven projection and backprojection in three dimensions," *Physics in Medicine and Biology* **49** (11), 2463–2475 (2004).

67. Y. Long, J. A. Fessler and J. M. Balter, "3D forward and backprojection for X-ray CT using separable footprints," *IEEE Transactions on Medical Imaging* **29** (11), 1839–1850 (2010).

68. B. De Man, J. Nuyts, P. Dupont, G. Marchal and P. Suetens, "An iterative maximum-likelihood polychromatic algorithm for CT," *IEEE Transactions on Medical Imaging* **20** (10), 999–1008 (2001).

69. I. A. Elbakri and J. A. Fessler, "Segmentation-free statistical image reconstruction for polyenergetic x-ray computed tomography with experimental validation," *Physics in Medicine and Biology* **48** (15), 2453–2477 (2003).

70. W. Zbijewski and F. J. Beekman, "Efficient Monte Carlo based scatter artifact reduction in cone-beam micro-CT," *IEEE Transactions on Medical Imaging* **25** (7), 817–827 (2006).

71. R. Zeng, J. A. Fessler and J. M. Balter, "Estimating 3-D respiratory motion from orbiting views by tomographic image registration," *IEEE Transactions on Medical Imaging* **26** (2), 153–163 (2007).

72. G. H. Chen, J. Tang and J. Hsieh, "Temporal resolution improvement using PICCS in MDCT cardiac imaging," *Medical Physics* **36** (6), 2130–2135 (2009).

73. J. Xu, E. C. Frey, K. Taguchi and B. M. Tsui, "A Poisson likelihood iterative reconstruction algorithm for material decomposition in CT," *SPIE Proceedings* **6510** (2007).

Advancements in Respiratory Sorting: The Future of Respiratory Sorting

Daniel A. Low
*University of California
at Los Angeles*

14.1 Introduction

Breathing motion-induced artifacts plague radiation therapy by distorting images and reducing the accuracy or conformality of dose distributions. Human breathing is a quasi-voluntary function that has a period of approximately 5 s, significantly faster than the time required to deliver radiation doses and acquire cone-beam computed tomography (CT) scans, and similar to the timescale required to acquire volumetric images using diagnostic CT scanners. Breathing can be temporarily halted for up to a minute, but many cancer patients have compromised respiratory function and cannot hold their breath. Breath hold is used effectively in diagnostic imaging, especially for modern CT scanners, which can acquire a thoracic or abdominal image data set in a few seconds. Nuclear medicine image acquisition and radiation therapy delivery require longer times, so single breath-hold times are not sufficient to remove the influence of breathing motion. There are multiple breath-hold protocols, but they require that the patient can hold their breath and comply with the instructions.

Breathing motion-induced artifacts are not the only reason to manage breathing motion. Because breathing motion moves tumors during therapy, one method for assuring the tumor is not missed is to expand the radiation beam to encompass the tumor and its trajectory (Wolthaus 2006, 2008a,b). This leads to portals that are larger than the tumor cross section and consequently to the irradiation of normal tissues. Methods designed to reduce the excess radiation require that the breathing-induced motion be quantitatively measured. The primary method for this measurement is 4DCT, which is an ambiguously defined process in which multiple CT data sets are acquired throughout the breathing cycle. Because the breathing cycle is irreproducible, the definition of the breathing cycle becomes critical to understanding and applying the subsequent 4DCT image information.

This chapter reviews the state-of-the-art in respiratory sorting, describes the influence of breathing motion on imaging and therapy, discusses the definitions behind respiratory image sorting, and shows how the sorting processes are implemented in radiation therapy.

14.2 Effects of Breathing Motion on Image Quality and Quantification and Promise of Sorting

14.2.1 Computed Tomography

All volumetric diagnostic images are acquired using a temporal acquisition sequence that requires between a few seconds and a few minutes to acquire. These acquisition systems assume that there is no motion during the acquisition sequence. When motion is present, the assumption is untrue, resulting in motion-induced image artifacts.

Breathing-induced motion is relatively fast. Zhao et al. (2009) recently published that the 85th percentile of motion per inhaled tidal volume was 24 mm/L, and Low et al. (2010) recently published that the average maximum breathing rate for 35 patients was 0.443 L/s. The combination of these quantities indicates that maximum lung tissue velocities are on the order of 11 mm/s. Modern CT scanners have rotation periods of between 0.3 and 0.4 s, which still allows more than 3 mm of motion during a full-rotation scan. The motion magnitude is typically greatest in the craniocaudal direction and causes motion-induced blurring of the small bronchial structures.

Because of the needs of radiation therapy to measure breathing-induced motion, motion artifacts will continue to be a challenge. Figure 14.1 shows a simulation of the effects of in-plane motion on a circular structure. The structure is a 10-pixel-wide cylinder that moves to the right during the scan acquisition. The motion is expressed in fractions of the cylinder radius, from 0 (no motion) to 2.2 times the radius for a 360° CT gantry rotation. The impact of motion is clear in the reconstructed cylinder shape. With small motion (≤0.4 radius or 2 pixels), the shape of the cylinder is maintained, but artifact extensions appear. As the motion increases, the image becomes more distorted.

Little has been published describing breathing motion-induced artifacts for modern multislice CT scanners, presumably because, in most cases, a diagnostic CT scan can be conducted during a breath hold (Dinkel et al. 2007; Wu et al. 2010). Olsen et al. (2008) recently compared amplitude-based and phase-angle-based sorting on multislice CT helical acquisition image artifacts. They primarily looked at artifacts in the lungs, examining images reconstructed by tagging the breathing cycle measured during CT scan acquisition. They concluded that their technique could detect clinically relevant image distortions that might otherwise go unnoticed. The technique could also reduce image distortion caused by some respiratory irregularities.

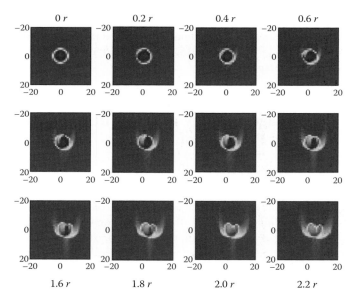

FIGURE 14.1 Motion-induced artifacts from a 10-pixel-wide circular target that moves to the right a fraction of the target radius per CT scan rotation. The motion-induced artifacts are evident with motion of as little as 40% of the radius. As the motion increases, the artifacts become more striking and include deformation of the object and projections outside the object.

14.2.2 Positron Emission Tomography

Positron emission tomography (PET) images are used in radiation therapy to assist the treatment planner in determining the tumor extent, differentiate lung atelectasis from tumor, and identify involved regional lymph nodes. An accurate image is required to use the PET image for treatment planning. PET images are acquired over many minutes per bed position. The PET data are acquired as a series of gamma coincidences and are sorted according to the detector pairs involved in the coincidence measurement. The subsequent PET image is formed by determining the lines of coincidence and the lines between the detector pairs. The greater the PET activity in a region is, the more the number of lines of coincidence will intersect that region. When breathing motion moves the tumor, the lines of coincidence will appear to come from the combination of the tumor and breathing motion envelope. The apparent PET intensity will be a function of the tumor PET intensity and the breathing motion pattern, so even a homogeneous PET intensity distribution will not appear homogeneous due to the addition of the breathing motion.

Unlike the influence of multislice CT on tumor images, the impact of breathing motion on PET images has been thoroughly studied (Intensity Modulated Radiation Therapy Collaborative Working Group 2001; Erdi et al. 2004; Hamill 2008; Boellaard 2009). The PET intensity has been correlated to important tumor characteristics, such as radioresistance and the potential for distant metastases. Breathing motion spreads the apparent intensity and therefore reduces the maximum intensity in the image. This may influence clinical staging or clinical decisions based on the maximum PET intensity.

PET intensity is often quantified by the standard uptake value (SUV) or the ratio of the amount of activity within the tumor divided by activity in nearby normal tissues (Boellaard 2009) or normalized by injected activity and patient weight. The SUVs have been correlated with outcomes (Weber 2009; Kinahan and Fletcher 2010). Because breathing motion blurs PET images, they also impact the SUVs. Therefore, a method to remove breathing motion artifacts from PET images would be extremely valuable.

The process of sorting PET images is relatively straightforward. Breathing surrogate data are acquired at the same time as the PET data. The PET data are sorted into predetermined bins, each bin corresponding to a specified breathing phase. An image is reconstructed using data from each bin, leading to images that have reduced breathing motion artifacts, but also reduced statistics. This often means that such gated images need to be acquired over a longer period of time than images without gating.

Apostolova et al. (2010) recently examined the impact of removing the motion blurring artifact from PET images. They used spherical phantoms moving in a single direction with an amplitude and period that modeled human breathing. Figure 14.2 shows an example of a coronal cross section through the different phantoms both with and without motion. Figure 14.2 shows the maximum SUV in each image without correction and with correction for motion and PET blurring. The measured

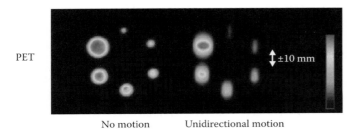

PET   ±10 mm

No motion   Unidirectional motion

FIGURE 14.2 Examples of the influence of breathing motion on apparent PET activity distributions. Left: Reconstructed PET image with no motion and spherical target shell diameters of 11.7, 14.3, 18.2, 23.3, 29.0, and 39.0 mm. Right: Apparent activity distribution of the same spheres, this time moving 20 mm total motion. The alteration of the apparent shape and activity distribution is evident if no motion compensation is employed. (With kind permission from Springer Science+Business Media: *European Radiology,* Combined correction of recovery effect and motion blur for SUV quantification of solitary pulmonary nodules in FDG PET/CT, 20, 2010, 1868–1877, Apostolova, I. Wiemker, R., Paulus, T. et al.)

versus actual SUVs are clearly significantly improved with motion correction, especially for small targets (Figure 14.3).

Bundschuh et al. (2007) proposed to detect tumor motion by using the list-mode data to correct the craniocaudal motion of the lesion. They reconstructed images with very short time bins, from 250 to 750 ms. Volume of interest was defined and used in each time bin. The center of mass of activity distribution, a quantity that essentially averages over relatively large quantities of noise in such short acquisitions, was used to determine the amount of tumor motion. They correlated the motion to a breathing surrogate and found, for patients with tumor motion, that the correlation was very good.

One of the challenges of PET image gating is that PET images require an attenuation correction to account for the body's attenuation of the PET gamma-rays. For modern PET/CT units, this is accomplished using the CT scan data set. The Hounsfield unit numbers are correlated to 511 keV gamma-ray attenuation coefficients, and these are used to determine the amount of attenuation that the gamma-rays undergo. The inverse of the attenuation is the correction factor. If the scans are acquired during free breathing, they may exhibit image artifacts. In the scans acquired during breath hold, the CT scan anatomy positions may not match the PET anatomy positions. This causes attenuation correction errors in regions such as the superior liver and inferior lung due to the diaphragm position. To reduce attenuation correction artifacts caused by this issue, quantitative 4DCT techniques that are well matched to the PET acquisition techniques are required.

Alessio et al. (2007) examined the use of ciné CT scans for attenuation correction, specifically for cardiac scans. In a phantom study, the attenuation correction from a helical CT caused an artifact presenting as a defect in the lateral heart wall. Using a CT created by examining all ciné CT scans and selecting the maximum voxel intensity for the attenuation scan, they improved the deficit by 60%. In patient studies, they again found that the best results, this time defined as the alignment between the PET image and yielded acceptable attenuation-corrected PET images, were obtained with the maximum intensity image.

14.3 Sorting Algorithms

14.3.1 Acquisition Modes

The breathing cycle is periodic but irregular. This leads to complications and challenges when designing a sorting algorithm.

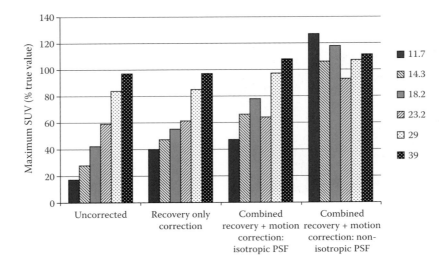

FIGURE 14.3 Quantitative assessment of the data shown in Figure 14.2. The legend indicates the spherical shell diameter in millimeters. PSF is the point-spread function used to remove nonimage-induced blurring motion. (With kind permission from Springer Science+Business Media: *European Radiology,* Combined correction of recovery effect and motion blur for SUV quantification of solitary pulmonary nodules in FDG PET/CT, 20, 2010, 1868–1877, Apostolova, I. Wiemker, R., Paulus, T. et al.)

Cardiac imaging has had to deal with involuntary motion that cannot be paused but is highly regular and has electronic surrogates, such as an electrocardiogram, which can be used to synchronize imaging and the physiologic function. The CT and PET manufacturers have taken advantage of the regular nature of the cardiac cycle to develop their sorting algorithms. Breathing is irregular, so some of the assumptions made with cardiac gating break down. If the algorithms used in cardiac imaging were used directly in breathing gating, it would lead to inaccurate sorting and an incomplete reduction of breathing motion artifacts.

With respect to imaging, there are two acquisition modes, prospective and retrospective. Prospective scanning uses the breathing cycle measurement surrogate to monitor the breathing cycle. The imaging device is maintained in a ready state until the patient has reached a user-specified breathing phase. At that time, the device is activated and the image data are acquired. Prospective scanning is typically used when an image is required at only a single breathing phase. For CT scanning, it is also used to reduce the radiation dose under conditions where the patient cannot hold their breath and must continue breathing throughout the imaging sequence.

Because of the unpredictability of the breathing cycle, there is an unknown time between acquisitions in a prospective scanning protocol. This makes synchronization challenging for acquisition protocols, such as helical scanning, where the CT couch moves in concert with the gantry rotation. Prospective protocols are typically used under ciné conditions, which is an acquisition with a couch stationary as the CT is acquired.

Retrospective acquisition techniques also employ a breathing surrogate during image acquisition. In this case, the surrogate is recorded and synchronized to the image data. These data may be reconstructed images or raw data, depending on the protocol. Unlike prospective gating, images acquired during a retrospective acquisition are not synchronized to specific gating phases.

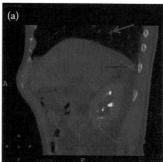
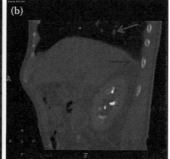

FIGURE 14.4 Sagittal reconstruction from a helical 4DCT scan with a patient that is free breathing. The patient paused their breathing while the scanner was acquiring images of their liver and breathed regularly during the remaining acquisition. (a) Peak exhalation. (b) Peak inhalation. The superior and inferior arrows point to tissues that are moving as expected. However, the liver, a section of which is being pointed to by the middle arrows, appears not to move. This yields an error in the motion measurement of the tissues near the liver and represents a major challenge to 4DCT procedures.

For ciné protocols, images are typically reconstructed and then sorted based on the surrogate information. For helical protocols, however, the data are acquired continuously and a more sophisticated sorting can be performed. The raw projection data can be sorted in time as a function of the breathing phase, and images can be generated for any breathing phase, as long as the patient spent time at that phase throughout the craniocaudal extent of the scan. Figure 14.4 shows an example of a reconstructed 4DCT helical scan when the patient paused in their breathing. In this case, the patient was being scanned and paused breathing while the scanner passed by their liver. Because the patient did not move, the scanner erroneously assigned breathing phases to the projection data and created a distorted image set.

14.3.2 Breathing Modeling

To develop a process to use CT or PET imaging to assess tumor motion, we need to first understand the characteristics of breathing motion and breathing patterns. Because breathing is less reproducible than cardiac motion, compromises and assumptions must be made that have profound impacts in the image acquisition and analysis techniques as well as the quality and robustness of the imaging results.

The general categories of breathing cycle descriptions are phase angle based and amplitude based. The phase-angle description assumes that the breathing cycle can be reproducibly subdivided in time. A specific breathing phase (e.g., peak inhalation) is selected as the start of the breathing cycle. The time at which this breathing phase occurs is identified, and the breathing cycle is then defined as the fraction of time from that time to the next time the same phase occurs. For example, if the breathing cycle starts at peak inhalation, the relative time between successive peak inhalations corresponds to a single breath. Phases in between are usually defined by the fraction of time between successive peaks. Often, these are described in terms of angles, 360° between successive peaks, hence the description phase-angle sorting. Occasionally, the breathing cycle is further subdivided between inhalation and exhalation. Peak inhalation and exhalation are designated 0° and 180°, with linear interpolation of phase angle in time between these extremes.

Phase-angle sorting works well when the patient's breathing cycle, especially their breathing amplitude, is consistent. If breathing is irregular, the algorithm breaks down. Because of the regularity of the cardiac cycle, this algorithm was the first to be commercially introduced because of the ease with which the imaging companies could transfer cardiac image gating algorithms to respiratory gating.

Seppenwoolde et al. (2002) developed a phase angle-based approach to describing tissue motion within the lungs. They examined fluoroscopic video images of implanted markers in a number of patients and determined that the motion had an elliptic behavior. They elected to parameterize the elliptic motion into components aligned with the elliptic axes. Figure 14.5 shows a schematic of the motion model for cases without (Figure 14.5b) and with (Figure 14.5d) hysteresis. The method for including

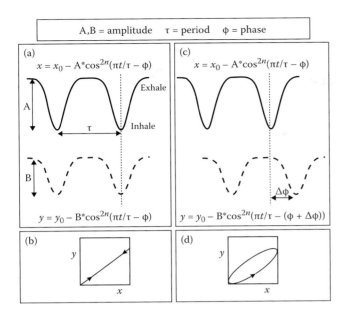

FIGURE 14.5 Breathing motion model proposed by Seppenwoolde et al. (2002) to describe tissue motion. The model describes the spatial coordinates as periodic functions in time. (Reprinted from *International Journal of Radiation Oncology Biology Physics*, 53, Seppenwoolde, Y., Shirato, H., Kitamura, K. et al., Precise and real-time measurement of 3D tumor motion in lung due to breathing and heartbeat, measured during radiotherapy, 822–834, Copyright 2002, with permission from Elsevier.)

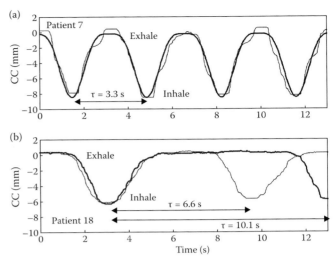

FIGURE 14.6 Example of breathing motion model results from Seppenwoolde et al. (2002) for (a) a regular-breathing patient and (b) an irregular-breathing patient. The model works well when the breathing frequency and amplitude are stable but breaks down if the breathing frequency changes. (Reprinted from *International Journal of Radiation Oncology Biology Physics*, 53, Seppenwoolde, Y., Shirato, H., Kitamura, K. et al., Precise and real-time measurement of 3D tumor motion in lung due to breathing and heartbeat, measured during radiotherapy, 822–834, Copyright 2002, with permission from Elsevier.)

hysteresis was to add a phase angle between the different directional components. The variation in time spent at specific phases, such as spending more time at exhalation–inhalation, were managed by taking the sinusoidal function to an even power. Figure 14.6 shows an example of the model for regular (Figure 14.6a) and irregular (Figure 14.6b) breathing patients. Although this model worked for regular breathing patients, it broke down for irregular breathing patients because the model could not predict changes in breathing frequency.

The second breathing cycle description category is amplitude based. Amplitude-based sorting assumes that the internal anatomic positions are related to the depth of breathing rather than the fraction of time between breaths. In cases where the patient breathes irregularly, amplitude-based sorting results in images with fewer motion-based artifacts than phase angle-based sorting. The main drawback for amplitude-based sorting is that it does not distinguish between the period of time during inhalation relative to the period of time during exhalation. Lung tissue motion is known to be different during inhalation and exhalation, a process known as hysteresis.

Lu et al. (2006) studied the impact of breathing cycle variation on phase angle-based and amplitude-based sorting. They reconstructed 3D image data sets for 12 breathing phases for 40 patients. They computed the air content, defined as the integrated amount of air in the lungs based on the CT scan, and used air content as a surrogate for tumor position. They correlated breathing phase and amplitude to air content and determined

the residual for the correlation. In most cases, the residual for amplitude gating was smaller than the residual for phase angle gating. When tidal volume was used as the comparator, the variation was always less for amplitude than phase angle gating. As an example, Figure 14.7 shows the definition of mid-inspiration for amplitude and phase angle sorting algorithms. The tidal

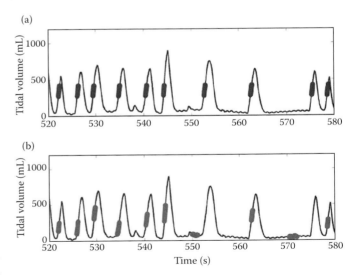

FIGURE 14.7 Example of the definition of mid-inspiration using (a) amplitude sorting and (b) phase angle sorting. The thicker lines indicate the times at which each algorithm has identified that the patient is at mid-inspiration.

volume for the points in time where amplitude sorting has determined that the patient is at mid-inspiration is shown in Figure 14.7a and would very likely correspond to consistent internal anatomy positions. The tidal volume for the points in time where phase angle sorting has determined that the patient is at mid-inspiration is shown in Figure 14.7b. Although there are many points in time where the patient's anatomy will be in a consistent location, there are three times when the patient's breathing is paused. At these times, mid-inspiration is less well defined for phase angle sorting, and the algorithm fails twice to identify a time at which the tidal volume is accurately at mid-inspiration.

14.3.3 Future of Respiratory Sorting

One of the challenges of using commercial CT sorting programs is the lack of feedback provided by the manufacturers concerning the reliability of the sorted CT data. As shown in Figure 14.4, the sorted data may contain motion artifacts caused by breathing irregularities. Also, the images may reflect the majority of breathing motion data, but breathing extremes may not be represented. If these extremes occur sufficiently often and with large amplitudes, the tumor may be underdosed, although the treatment planner uses the 4DCT data set as a guide. One way to manage this is to use amplitude-based sorting and to keep track of the relative probability for each amplitude. An example of such a histogram is shown in Figure 14.8. The breathing waveform is shown in Figure 14.8a, where the timescale is compressed so that the relative inhalation and exhalation amplitudes are easily visible. A histogram of the breathing waveform is shown in Figure 14.8b. The histogram is consistent with that of a relatively stable breathing pattern. There are peaks near exhalation and inhalation that are indicative of the increased time spent at peak inhalation and exhalation. The peak near exhalation is greater than the one near inhalation because of the more reproducible exhalation amplitude and the fact that the patient spent more time near exhalation than inhalation.

Superimposed on the histogram in Figure 14.8 are four lines indicating four percentile levels. The percentiles, labeled V_x, where x is the percentage of time that the patient spent at tidal volume V or less, provide some insight as to the relationship between breathing extremes and the rest of the breathing cycle. Typically, V_5 can represent exhalation, meaning that the patient has spent sufficient time at that amplitude and that a complete CT scan can be reliably reconstructed from projection data taken at that amplitude. Because of the characteristics of breathing, the inhalation amplitude for which a reconstructed CT scan can be reliably made is V_{85}. However, if V_5 and V_{85} are used to represent the entire breathing cycle, then the resulting image data are reflective of only 80% of the time; the remaining time has been spent at amplitudes exceeding those represented in the images. Therefore, the user must somehow extrapolate the motion data to beyond what is provided in the images. The histogram can help tremendously with this by showing the user the percentile amplitudes that lie outside those reflected in the images.

To illustrate how the percentile data could be used to assist the treatment planner, Figure 14.9 which was compares the 98th and 85th percentiles, as referenced to the 5th percentile (the definition for 0 mL tidal volume) for 32 patients. The ratio of the two percentiles was 1.39 ± 0.19, which was remarkably stable given that the inhalation tidal volumes varied by almost an order of magnitude. The consequence of these data is that, on average, using an inhalation and exhalation image, as reflected by the 85th and 5th percentiles, could result in an underestimation of the motion by 40%. In this case, V_{98} reflects the amount of motion for 93% of the time (98%–5%) rather than just 80% of the time (85%–5%). The consequence of this is not so much that the images would not reflect the motion but that the user is not provided these data to allow them to make a determination or to extrapolate the image data. This may lead to an unknown number of target volumes being underestimated by a significant amount. For phase-based images, the consequence could be even more severe, because the user has little feedback of how the images they see reflect the

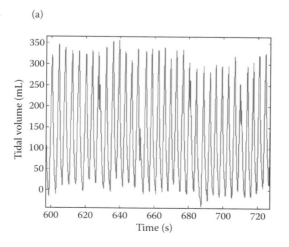
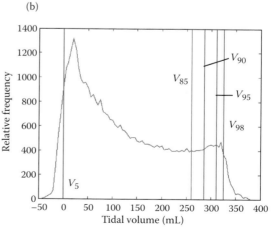

FIGURE 14.8 (a) Tidal volume as a function of time for a regularly breathing patient. The inhalation and exhalation peaks can be seen to be relatively stable. (b) Histogram for the breathing cycle shown in (a). The tidal volume percentiles are shown for reference.

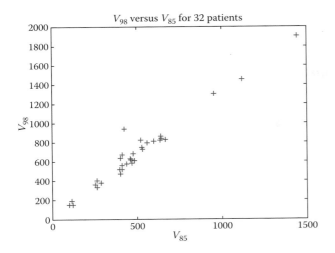

FIGURE 14.9 Comparison of V_{98} and V_{85}, as referenced to V_5, for a collection of 32 patients. Linear regression shows that the relationship between the two is 1.39 ± 0.19. The ratio indicates the relative breathing amplitude between the two percentiles.

tumor motion. The future should include such feedback to the user as well as amplitude-based gating (one of the CT simulator manufacturers has recently installed an amplitude-based sorting algorithm into their 4DCT workflow).

The drawbacks of the current clinical phase angle and amplitude approaches led Low et al. (2005, 2010) to formulate a breathing motion model based on the breathing amplitude and rate. They used tidal volume and airflow, correlating tumor positions in a linear relationship to those two variables. Using these variables allowed the model to be independent of time and therefore the irregularities of the breathing cycle. The model used a linear combination of two vectors, one proportional to tidal volume

and the other proportional to airflow. Although the motion model was linear in tidal volume and airflow, the time dependence of those two variables led to very complex motion patterns that well fit the measured breathing motion data. Figure 14.10 (Zhao et al. 2009) shows an example of the model's ability to model breathing motion in a patient that breathed regularly. The vector whose length was proportional to tidal volume and the vector whose length was proportional to airflow are shown. The crosses indicate the measured positions of a point in the lungs as imaged 15 times during a 4DCT procedure. The circles are the model's prediction of that point for the tidal volume and airflow exhibited at the 15 points in time corresponding to the 4DCT scan acquisitions. Lines connecting the circles and crosses indicate the level of error in the model. Although the model was not perfect, the mean discrepancy between the model and measured position data was less than 1 mm, approximately the size of individual image voxels.

Figure 14.11 shows a similar example, this time from an irregular breathing patient (Low et al. 2005). The measured and model predicted positions are shown as crosses and circles, respectively. The model did an excellent job of predicting the variation in tissue position for a variety of inhalation depths.

There is an excellent chance that this type of model will be used in the future for respiratory sorting and breathing motion modeling. If this algorithm is shown to be superior to phase angle-based sorting, other CT simulator manufacturers will likely follow suit. Ultimately, a more sophisticated approach to breathing gating and modeling will be needed.

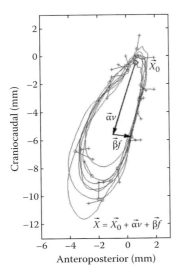

FIGURE 14.10 Breathing motion model from Low et al. (2005) and Zhao et al. (2009), which uses tidal volume and airflow as the independent variables in the motion model. α and β describe the relationship between motion in tidal volume and motion and airflow, respectively.

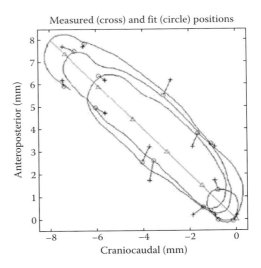

FIGURE 14.11 Breathing motion model from Low et al. (2005) showing measured positions of lung tissue (crosses) and the model prediction (circles) for an irregular breathing patient. In this case, the patient had, for inhalations, three relatively deep and one very shallow. The position data come from CT scans acquired during simultaneous spirometry measurements. In this case, the variation in breathing amplitude is well modeled. (Reprinted from *International Journal of Radiation Oncology Biology Physics*, 63, Low, D. A., Parikh, P. J., Lu, W. et al., Novel breathing motion model for radiotherapy, 921–929, Copyright 2005, with permission from Elsevier.)

References

Alessio, A. M., Kohlmyer, S., Branch, K. et al. 2007. Cine CT for attenuation correction in cardiac PET/CT. *Journal of Nuclear Medicine, 48,* 794–801.

Apostolova, I., Wiemker, R., Paulus, T. et al. 2010. Combined correction of recovery effect and motion blur for SUV quantification of solitary pulmonary nodules in FDG PET/CT. *European Radiology, 20,* 1868–1877.

Boellaard, R. 2009. Standards for PET image acquisition and quantitative data analysis. *Journal of Nuclear Medicine, 50 Suppl 1,* 11S–20S.

Bundschuh, R. A., Martinez-Moeller, A., Essler, M. et al. 2007. Postacquisition detection of tumor motion in the lung and upper abdomen using list-mode PET data: A feasibility study. *Journal of Nuclear Medicine, 48,* 758–763.

Dinkel, J., Welzel, T., Bolte, H. et al. 2007. Four-dimensional multi-slice helical CT of the lung: Qualitative comparison of retrospectively gated and static images in an ex-vivo system. *Radiotherapy and Oncology, 85,* 215–222.

Erdi, Y. E., Nehmeh, S. A., Pan, T. et al. 2004. The CT motion quantitation of lung lesions and its impact on PET-measured SUVs. *Journal of Nuclear Medicine, 45,* 1287–1292.

Hamill, J. J., Bosmans, G. and Dekker, A. 2008. Respiratory-gated CT as a tool for the simulation of breathing artifacts in PET and PET/CT. *Medical Physics, 35,* 576–585.

Intensity Modulated Radiation Therapy Collaborative Working Group. 2001. Intensity-modulated radiotherapy: Current status and issues of interest. *International Journal of Radiation Oncology Biology Physics, 51,* 880–914.

Kinahan, P. E. and Fletcher, J. W. 2010. Positron emission tomography-computed tomography standardized uptake values in clinical practice and assessing response to therapy. *Seminars in Ultrasound, CT, and MRI, 31,* 496–505.

Low, D. A., Parikh, P. J., Lu, W. et al. 2005. Novel breathing motion model for radiotherapy. *International Journal of Radiation Oncology Biology Physics, 63,* 921–929.

Low, D. A., Zhao, T. Y., White, B. et al. 2010. Application of the continuity equation to a breathing motion model. *Medical Physics, 37,* 1360–1364.

Lu, W., Parikh, P. J., Hubenschmidt, J. P., Bradley, J. D. and Low, D. A. 2006. A comparison between amplitude sorting and phase-angle sorting using external respiratory measurement for 4D CT. *Medical Physics, 33,* 2964–2974.

Olsen, J. R., Lu, W., Hubenschmidt, J. P. et al. 2008. Effect of novel amplitude/phase binning algorithm on commercial four-dimensional computed tomography quality. *International Journal of Radiation Oncology Biology Physics, 70,* 243–252.

Seppenwoolde, Y., Shirato, H., Kitamura, K. et al. 2002. Precise and real-time measurement of 3D tumor motion in lung due to breathing and heartbeat, measured during radiotherapy. *International Journal of Radiation Oncology Biology Physics, 53,* 822–834.

Weber, W. A. 2009. Assessing tumor response to therapy. *Journal of Nuclear Medicine, 50 Suppl 1,* 1S–10S.

Wolthaus, J. W. H., Schneider, C., Sonke, J. J. et al. 2006. Mid-ventilation CT scan construction from four-dimensional respiration-correlated CT scans for radiotherapy planning of lung cancer patients. *International Journal of Radiation Oncology Biology Physics, 65,* 1560–1571.

Wolthaus, J. W. H., Sonke, J. J., van Herk, M. et al. 2008a. Comparison of different strategies to use four-dimensional computed tomography in treatment planning for lung cancer patients. *International Journal of Radiation Oncology Biology Physics, 70,* 1229–1238.

Wolthaus, J. W. H., Sonke, J. J., van Herk, M. and Damen, E. M. F. 2008b. Reconstruction of a time-averaged mid-position CT scan for radiotherapy planning of lung cancer patients using deformable registration. *Medical Physics, 35,* 3998–4011.

Wu, C., Sodickson, A., Cai, T. et al. 2010. Comparison of respiratory motion artifact from craniocaudal versus caudocranial scanning with 64-MDCT pulmonary angiography. *American Journal of Roentgenology, 195,* 155–159.

Zhao, T. Y., Lu, W., Yang, D. S. et al. 2009. Characterization of free breathing patterns with 5D lung motion model. *Medical Physics, 36,* 5183–5189.

15

In-Room Imaging Techniques

Jan-Jakob Sonke
The Netherlands Cancer Institute–Antoni van Leeuwenhoek Hospital

15.1 Introduction

Considerable geometrical uncertainties such as setup error, organ motion, shape change, and treatment response limit the precision and accuracy of radiation therapy (RT; Langen and Jones 2001; van Herk 2004). Consequently, the actually delivered dose does not equal the planned dose (what you see is not what you get). Generally, generous safety margins (ICRU Report 50; ICRU Report 62) are applied around the target and optionally for organs-at-risk (OARs), such that undertreatment and overtreatment due to geometrical uncertainties can be avoided with an acceptable probability (van Herk et al. 2000). Image-guided RT (IGRT) is the process of (1) acquiring an image of the patient's anatomy in the treatment room (Figure 15.1) generally with the patient in treatment position; (2) comparing the treatment position with planned position of the tumor, OARs, or some surrogate; and (3) correcting the treatment position. This reduces the required safety margin and allows dose escalation without compromising nearby OARs. A large variety of in-room imaging modalities have been proposed for IGRT. In this chapter, an overview of these imaging modalities will be given.

15.2 In-Room Imaging Modalities

15.2.1 Portal Imaging

The first imaging system used for IGRT was the electronic portal imaging device (EPID; Boyer et al. 1992; Herman et al. 2001), a large digital x-ray camera attached to the gantry of the linear accelerator (Figure 15.2). A portal image is a transmission image obtained at the beam-exit side of the patient. It captures the beam outline and the projected patient anatomy, mostly limited to the bony anatomy (Figure 15.3).

Various types of EPIDs have been developed, such as the video camera with a scintillator screen (Strandqvist and Rosengren 1958; Shalev et al. 1989; Munro et al. 1990), fiber-optic image reducer arrays (Wong et al. 1990), scintillation crystal-photodiode detector (Morton et al. 1991), scanning liquid-filled ionization-based arrays (van Herk 1991), and large area amorphous silicon (a-Si) flat-panel imagers (FPI; Antonuk et al. 1990; Street et al. 1990; Munro and Bouius 1998). The latter systems, consisting of a phosphorus screen and thin-film transistor diode array, are currently the most widely used because of the relatively high detector quantum efficiency (Antonuk 2002; see Figure 15.4) requiring less patient dose for the same image

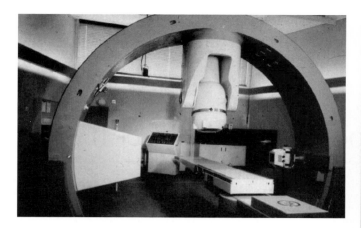

FIGURE 15.1 Early kilovoltage imaging system for patient position mounted orthogonal to a Cobalt-60 treatment developed in 1960 at The Netherlands Cancer Institute–Antoni van Leeuwenhoek Hospital (Amsterdam, The Netherlands).

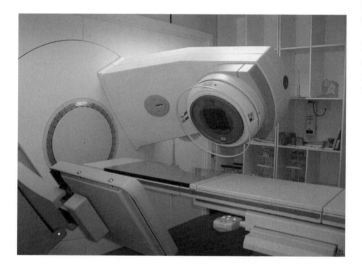

FIGURE 15.2 Example of an EPID mounted on the gantry of a linear accelerator.

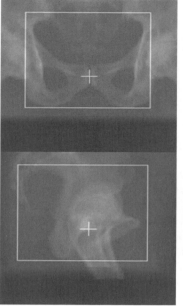
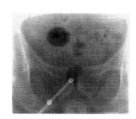
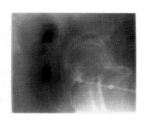

FIGURE 15.3 Coronal (top) and sagittal (bottom) digitally reconstructed radiograph and indicated field edge (left) with corresponding portal images (right) of a prostate cancer patient. The contrast information is mainly limited to the bony anatomy.

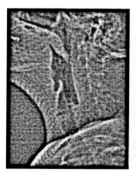
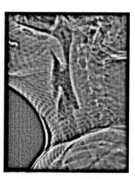

FIGURE 15.4 Examples of lateral portal images of a head-and-neck cancer patient acquired with a scanning liquid-filled ionization chamber array (left) and an a-Si FPI (right). Both images were acquired with six monitor units.

quality. All major vendors offer a-Si FPI-based portal imaging systems.

The setup of the patient is usually verified by comparing a pair of orthogonal portal images with their corresponding reference image obtained during treatment planning, either radiographs acquired during simulation or digitally reconstructed radiographs obtained from the planning computed tomography (CT) scan. Portal image analysis typically involves two steps. First, the radiation field edge in the portal image is registered to the field edge in the reference image. Second, the projected bony anatomy is registered to the bony anatomy in the image. The difference between these two registration results quantifies the patient setup error (Herman et al. 2001).

Although EPIDs are still widely used for patient position verification in RT, the limited image quality and the lack of soft-tissue contrast in portal images have limited its potential and have inspired researchers to develop more advanced in-room imaging technologies.

An alternative application of portal image devices is dosimetric verification (Essers et al. 1995; van Elmpt et al. 2008), where the delivered dose is estimated based on the signal captured by the EPID. It is expected that, in the future, this will be the most important utilization of portal imagers in modern RT.

15.2.2 Kilovoltage Planar Imaging

Planar imaging based on kilovoltage radiation has been proposed as an alternative to portal imaging. Similar to portal images, it is a transmission image obtained at the beam-exit side of the patient. Due to the smaller focal spot size of an x-ray tube compared with a linear accelerator, and the lower energy at which not only Compton scattering but also the photoelectric effect contributes to the attenuation, kilovoltage images have an improved spatial resolution and contrast compared with portal images. Hence, they are more easy to interpret and can be acquired at a lower imaging dose. Such systems facilitate fast acquisition and even real-time monitoring in fluoroscopic mode. On the contrary, additional hardware, in terms of x-ray source(s) and detector(s), are required. Kilovoltage imaging systems are fixed to the treatment room (Murphy and Cox 1996; Adler et al. 1997; Shirato et al. 1999; Yan et al. 2003) or mounted on the gantry of the treatment machine (Johns and Cunningham 1959; Biggs et al. 1985; Suit et al. 1988; Jaffray et al. 1999; Berbeco et al. 2004). Examples of room-mounted systems are the CyberKnife system (Accuray, Sunnyvale, California; see Figure 15.5a) and the ExacTrac system (BrainLab Novalis, Feldkirchen, Germany; see Figure 15.5b). Examples of gantry-mounted systems are Synergy (Elekta Oncology Systems, Crawley, West Sussex, United Kingdom; see Figure 15.6a) and OBI (Varian Medical Systems, Palo Alto, California; see Figure 15.6b).

Although these systems show significantly increased contrast of bony structures, visualization of soft-tissue detail remains problematic and correction of daily organ motion is still challenging. Investigations of organ motion (Langen and Jones 2001), however, have demonstrated that, for many sites, substantial reductions in geometric uncertainty require the visualization of soft tissue.

15.2.3 Ultrasound Imaging

An ultrasound (US) image captures differences in reflective properties within the body (Wells and Liang 2011). An US probe contains acoustic transducers that emit pulses of high-frequency sound waves that partially reflect back to the probe on each interface in the body with a different density. The microphones in the probe receive these echoes, and the time it takes to travel is measured to determine the distance to the reflecting interface. If the waves encounter bone or air, the density difference is so great that most of the acoustic energy is reflected and structures beyond these interfaces cannot be seen.

US imaging was the first modality used for in-room imaging that provides appreciable soft-tissue contrast. Additionally, US is inexpensive and noninvasive and does not rely on ionizing radiation. The first-generation US systems used for IGRT produced 2D images and were mainly used for prostate localization. 3D US combined with a navigation system to determine the orientation of the probe facilitates an easier interpretation (see Figure 15.7b).

Nevertheless, some challenges still limit the applicability of US guidance. The relatively low image quality of US images makes accurate localization challenging (Kuban et al. 2005). Uncertainties in the speed of sound between the skin and the target induce some geometrical distortions (Fontanarosa et al. 2011), and pressure applied by the probe might displace the target (Artignan et al. 2004).

15.2.4 Optical Imaging

Another nonionizing modality proposed for IGRT is optical imaging that captures a reflection of the patient's surface for setup correction (MacKay et al. 2001). Systems based on video cameras (Milliken et al. 1997), laser projections (Ploeger et al. 2003a), surface markers (Soete et al. 2003), and projected speckled patterns (Bert et al. 2005; see Figure 15.7a) have been developed producing 2D or 3D surface representation of the patient. Comparison through subtraction or registration

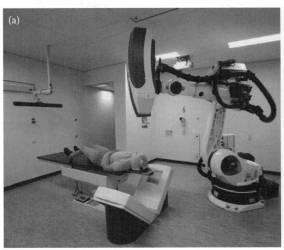 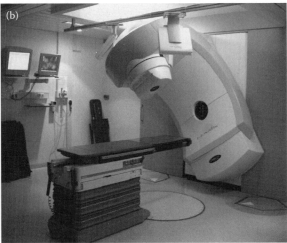

FIGURE 15.5 Examples of kilovoltage planar IGRT systems. (a) CyberKnife (Accuray, Sunnyvale, California) uses two ceiling-mounted kilovoltage sources and two floor-mounted a-Si FPI arranged to provide orthogonal views of the patient. (Image courtesy of Mischa Hoogeman, Erasmus MC, Rotterdam, The Netherlands.) (b) ExacTrac as part of the BrainLab Novalis patient positioning system (BrainLab Novalis, Feldkirchen, Germany) using two floor-mounted kilovoltage sources and two ceiling mounted a-Si FPI. (Image courtesy of Tom Depuydt, University Hospital Brussels, Brussels, Belgium.)

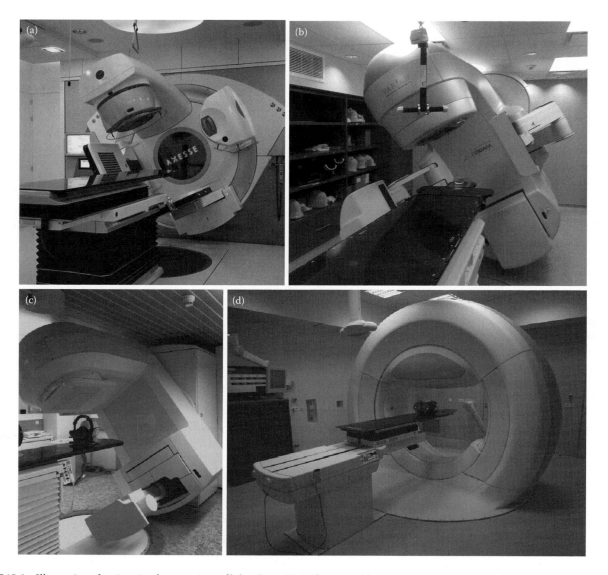

FIGURE 15.6 Illustration of various implementations of kilovoltage CBCT for IGRT. (a) Synergy (Elekta Oncology Systems, Crawley, West Sussex, United Kingdom). (Image courtesy of Patricia Fewer, The Netherlands Cancer Institute–Antoni van Leeuwenhoek Hospital, Amsterdam, The Netherlands.) (b) OBI (Varian Medical Systems, Palo Alto, California). (Image courtesy of Michael Velec, Princess Margaret Hospital, Toronto, Ontario, Canada.) (c) Siemens Artiste (Deutsches Krebsforschungszentrum Heidelberg, Heidelberg, Germany). (Image courtesy of Simeon Nill, Deutsches Krebsforschungszentrum Heidelberg, Heidelberg, Germany.) (d) Vero (Mitsubishi Heavy Industries, Tokyo, Japan). (Image courtesy of Tom Depuydt, University Hospital Brussels, Brussels, Belgium.)

quantifies setup error. Submillimeter precision has been reported for phantom experiments. Note, however, that patient setup through skin marks was found to be inadequate for most RT body sites. Only cases of a rigid relationship between the skin surface and the target optical imaging techniques have the potential to produce an accurate target alignment. Applications for intrafraction surveillance are therefore more promising than for interfraction setup verification (Noel et al. 2008). The patient body shape, however, does not always contain enough shape to uniquely determine its position (Ploeger et al. 2003b; Alderliesten et al. 2009).

15.2.5 In-Room CT

The projective nature of portal imaging and other planar imaging systems intrinsically limits the ability to discriminate different anatomical structures of such IGRT solutions. CT, on the contrary, produces cross-sectional images of tissue based on their x-ray attenuation coefficient from a large series of x-ray images taken around a single axis of rotation (Kak and Slaney 1988). Because treatment planning is based on 3D CT images, the introduction of CT imaging technology in the treatment room for treatment verification is a logical step.

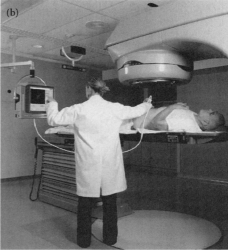

FIGURE 15.7 (a) Example of an optical IGRT system (AlignRT, Vision RT Ltd., London, United Kingdom), where three stereoscopic cameras are mounted to the ceiling and calibrated to the isocenter. (Image courtesy of Tanja Alderliesten, The Netherlands Cancer Institute–Antoni van Leeuwenhoek Hospital, Amsterdam, The Netherlands.) (b) Example of an US IGRT system installed in the treatment room. The system consists of US probe with reflector array mounted on it and a camera on ceiling to track this probe in the room. (Image courtesy of Frank Verhaegen, Maastro Clinic, Maastricht, The Netherlands.)

15.2.5.1 CT on Rails

The first IGRT system based on CT was a CT on rails (Kuriyama et al. 2003; Uematsu et al. 2001), where the CT gantry installed on rails moved across the patient. The CT gantry can be installed between 90° and 180° to the linac gantry, and rotating the couch, the patient can be positioned for either imaging or treatment. Often, additional shifts in the lateral and vertical directions are required to ensure that the couch fits through the CT bore. Installation of a "conventional" CT scanner in the treatment room benefits from all technological developments over the past decades and produces excellent image quality and fast acquisitions. On the contrary, the systems are relatively expensive, require spacious treatment rooms, and cannot image the patient in treatment position. Possible errors are introduced by the process of rotating and shifting the patient from the imaging position back to the treatment position due to patient motion, couch readouts, and controls.

15.2.5.2 Cone-Beam CT

Traditional CT is based on a fan-shaped beam and a narrow detector array that acquires one or a few thin slices per rotation. Multiple rotations are required to capture a volumetric representation of the patient's anatomy. To integrate a CT scanner with the treatment machine, such an imaging geometry is not feasible, because safety requirements limit the rotation speed of the linac to typically 1 rpm. Multiple rotations per acquisition would therefore take too long. As an alternative, adopting a 2D detector and a cone-shaped beam allows acquisition of volume in only one rotation or even a partial rotation. Such a cone-beam CT (CBCT) scanner, integrated with the linear accelerator, makes the acquisition of volumetric images feasible in the treatment position at the time of therapy. CBCT-based IGRT systems have

been explored and commercialized using x-rays in the kilovoltage as well as megavoltage range.

15.2.5.2.1 Kilovoltage CBCT

Kilovoltage CBCT scanners integrated with a linear accelerator typically have a pulsed fluoroscopic x-ray tube and large-area FPI mounted on the accelerator gantry orthogonal to the treatment beam (Jaffray et al. 2002; see Figure 15.6a and b). Alternative approaches are to mount the x-ray tube inline, that is, 180° to the treatment beam (Thilmann et al. 2006; see Figure 15.6c) or a dual kilovoltage-source imaging pair mounted at ±45° (Kamino et al. 2006; see Figure 15.6d). The source–detector orbit is circular ranging from a half-scan (180° + fan angle) or full-scan (360°) orbit. The field of view (FOV) of the circular orbit imaging geometry is limited by the size of the detector at the isocenter plane, yielding a cylinder with conical caps according to the cone angle. With detector sizes currently applied for CBCT-guided RT of 30–40 cm at a source-to-detector distance of approximately 1.6 m, FOVs are adequate for imaging the head-and-neck region but too small to image, for instance, the whole pelvis. An effective mechanism to extend the FOV of a CBCT scanner in the plane of rotation up to a factor of 2 is the application of a (partially) displaced detector position in which the detector is shifted by up to half its width (Liu et al. 2003).

The kilovoltage CBCT image quality (see Figure 15.8) is generally somewhat lower than traditional CT due to the lower efficiency and dynamic range of FPI compared with dedicated CT detectors, scatter associated with the volumetric nature of image acquisition, and motion artifacts amplified by the long rotation times. These factors are likely to improve with advancements in scatter correction (Siewerdsen et al. 2006; Reitz et al. 2009; Jin et al. 2010; Lazos and Williamson 2010) and reconstruction

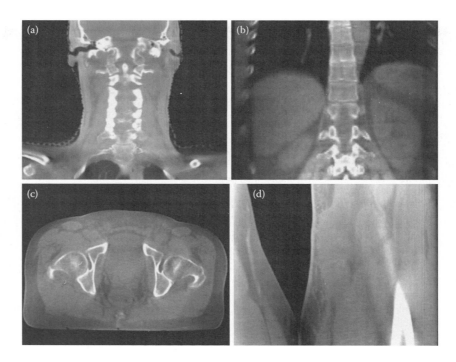

FIGURE 15.8   Examples of CBCT images for (a) head-and-neck cancer, (b) liver cancer, (c) prostate cancer, and (d) extremity soft-tissue sarcoma.

techniques (Sidky and Pang 2008; Wang et al. 2009a; Choi et al. 2010).

Motion during CBCT acquisition violates the principles of CT inducing motion artifacts. For (semi)periodic motion that is fast relative to the CBCT acquisition time, these artifacts are mainly a blurring of the moving structures over their trajectory of motion. To reduce motion artifacts, time-resolved CBCT imaging was developed to correlate respiratory motion with image acquisition (Sonke et al. 2005; Dietrich et al. 2006; Li et al. 2006). By sorting the projections into bins depending on their respective respiratory phase and subsequently reconstructing each bin separately into a 3D CBCT image or frame representing one phase of the respiratory cycle, a 4D CBCT image of the respiratory cycle is generated (see Figure 15.9). To correlate the projection images to the respiratory motion, a breathing signal is required that can be extracted directly from the series of projection images (Zijp et al. 2004). To generate 4D CBCT with adequate image quality, the gantry rotation speed is generally reduced, thereby increasing the number of respiratory cycles captured during the acquisition and thus reducing the angular gap between corresponding phases of adjacent respiratory cycles. Note that the imaging dose can be kept constant by decreasing the dose per projection in proportion to the gantry rotation speed.

15.2.5.2.2 Megavoltage CBCT

A simple and elegant alternative to kilovoltage CBCT is to use the treatment beam and an FPI-based EPID to perform megavoltage CBCT (Sidhu et al. 2003; Pouliot et al. 2005; see Figure 15.10a). This provides easy access to the patient by the therapists. To keep

the imaging dose at clinically acceptable levels, the dose-pulse rate of the accelerator beam needs to be modified to allow the delivery of a small fraction of a monitor unit (MU) per image. Typically, a 6 MV beam is used, but a dedicated imaging beam line using a carbon target and a 4 MV beam improves the image quality (Faddegon et al. 2008). Moreover, the imaging beam is

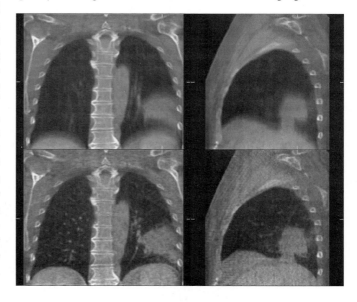

FIGURE 15.9   (See color insert.) Coronal (left column) and sagittal slices (right column) depicting a 3D CBCT (top row) and the exhale phase of a 4D CBCT scan (bottom row) reconstructed from a retrospectively sorted subset of the acquired fluoroscopic projection images (75 out of a total of 737).

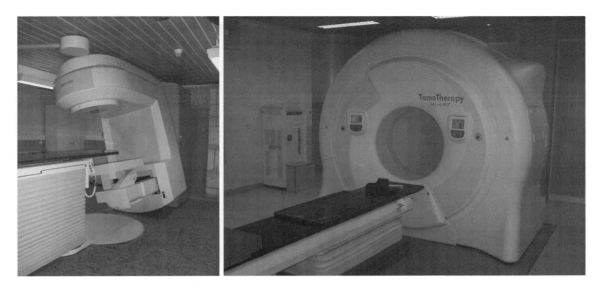

FIGURE 15.10 Illustration of various implementations of megavoltage (CB)CT for IGRT. (a) MVision (Siemens Medical Solutions, Erlangen, Germany). (Image courtesy of Simeon Nill, Deutsches Krebsforschungszentrum Heidelberg, Heidelberg, Germany.) (b) Helical TomoTherapy Hi-Art system (TomoTherapy, Inc., Madison, Wisconsin). (Image courtesy of Tom Depuydt, University Hospital Brussels, Brussels, Belgium.)

in the megavoltage range, thus rendering the images immune to artifacts caused by high-Z materials such a dental fillings. Nevertheless, the megavoltage beam for which the Compton effect is the dominant interaction process inherently causes poorer subject contrast. Additionally, the poor detection efficiency of current EPIDs in the megavoltage energy range results in poor signal-to-noise performance for clinically acceptable doses. Consequently, the CNR of megavoltage CBCT images is generally lower compared with kilovoltage CT and CBCT (Groh et al. 2002).

15.2.5.3 TomoTherapy

The TomoTherapy system is a dedicated IGRT system that can acquire megavoltage CT scans of the patient in treatment position (Figures 15.10b and 15.11). Similar to megavoltage CBCT, it uses the same accelerator to generate the imaging beam that is used to generate the treatment beam. The imaging beam has a reduced energy of 3.5 MV, whereas the image acquisition is analogous to that of helical CT imaging. Images are typically acquired with a pitch of 1, 2, or 3, a rotation time of approximately

10 s, and a beam that is typically collimated to 4 mm at the isocenter. The TomoTherapy system uses a particularly high efficient arc-shaped xenon detector that has a considerably higher detector quantum efficiency than current state-of-the-art FPIs. The TomoTherapy system therefore outperforms the megavoltage CBCT system in terms of image quality such as contrast-to-noise while exhibiting somewhat lower image quality than kilovoltage CBCT (Stützel et al. 2008). The low contrast resolution, however, is sufficient to identify some soft tissues (Meeks et al. 2005).

15.2.6 Magnetic Resonance Imaging

Although CT is still the imaging modality most widely used in RT planning, magnetic resonance imaging (MRI) is increasingly adopted as an additional imaging modality. An MRI scanner aligns the magnetization of atoms using a powerful magnetic field and subsequently alters this alignment using radiofrequency fields. This causes the materials to produce a rotating magnetic field that is detected by the scanner and reconstructed into a

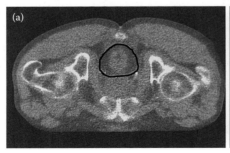
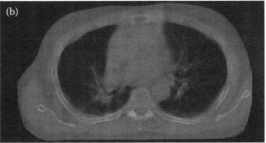

FIGURE 15.11 Examples of Tomotherapy images for (a) rectum cancer and (b) lung cancer patients.

volumetric image of the body tissues (Hendrick 1994; Duerk 1999). MRI has superior soft-tissue contrast compared with CT, which facilitates a more easy discrimination of different tissues, organs, and targets. Therefore, several groups have proposed to use MRI for IGRT as well. Lagendijk et al. (2008) are developing an MRI accelerator integrating a 6 MV treatment beam mounted on a ring around a 1.5 T MRI scanner (see Figure 15.12). Active magnetic shielding is used to create a zero magnetic field at the accelerator gun and to minimize the magnetic field at the location of the accelerator tube to decouple the accelerator from the MRI. The radiation travels through the bore of a dedicated MRI scanner before it reaches the patient. Dempsey et al. (2005) proposed an open split solenoid 0.3 T MRI scanner integrated with three equiangular spaced Cobalt-60 sources combined with MLCs. Similarly, Kron et al. (2006) proposed to combine a 0.25 T MRI with a Cobalt-60-based tomotherapy unit. An even more technologically challenging approach is to mount a 6 MV accelerator onto the open end of a biplanar MRI scanner where the linac and the MRI magnet rotate simultaneously.

Only one of these systems is clinically available at the time of writing (ViewRay Inc., Bedford, US), while proof of concepts start to emerge for alternative designs (Fallone et al. 2009; Raaymakers et al. 2009). Work continues on the mechanical integration of rotating gantries and MRI scanners and incorporation of the electron return effect in the continuously present magnetic field into the treatment planning process, especially for the higher field strengths (Raaijmakers et al. 2008; Green 2009).

15.3 Image Quality

The image quality of both radiographic as well as CT-based in-room imaging systems is, similar to their "out-of-the-room" equivalents, quantified in terms of essential metrics of imaging performance (Bushberg et al. 2002; Hendee and Ritenour 2002), such as contrast, noise, contrast-to-noise ratio (CNR), spatial resolution, and the modulation transfer function (MTF).

The measured contrast refers to the difference in mean values between two objects that are large compared with the voxel size of correlation length. As for digital imaging modalities, the window and level can be adjusted to display arbitrarily low contrast, and the contrast is primarily of interest in proportion to the magnitude of image noise. The image noise is most simply characterized as the standard deviation in voxel values over one or both ("large") objects/regions.

The spatial resolution of an imaging system can be described in terms of the minimum resolvable feature size often quantified by the number of resolvable line pairs per centimeter. A more advanced characterization of spatial resolution is the MTF describing the spatial resolution in the Fourier domain. Specifically, the MTF describes the factor by which change in image signal at a given spatial frequency (e.g., a set of line pairs or a sinusoidally varying pattern in the image) is modulated in its mean signal at the output of the imaging system. Most in-room imaging systems clinically operate at a spatial resolution that is substantially lower than their best performance. For

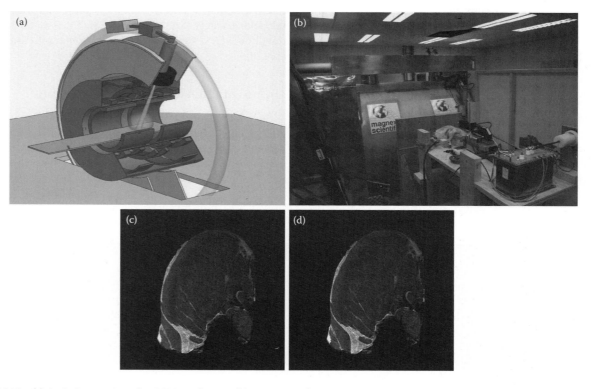

FIGURE 15.12   (a) Artist impression of an MRI accelerator. (b) Prototype of an MRI accelerator with a static beam. MR image of a pork chop with (c) and without (d) the irradiation beam on. (Images courtesy of B. Raaymakers, University Medical Centre Utrecht, Utrecht, The Netherlands.)

a given imaging dose, there is an intrinsic trade-off between spatial resolution and low-contrast detectability. Because the latter is more important for IGRT, spatial resolution is often somewhat compromised.

Most image quality parameters can be quantified using commercially available phantoms containing multiple inserts tailored to various aspects of image quality for planar imaging, such as the QC-3 phantom (Standard Imaging, Middleton, Wisconsin), the PTW EPID phantom (PTW-Freiburg, Freiburg, Germany), and the Leeds phantom TOR 18FG (Leeds Test Objects Ltd., York, UK), as well as for volumetric imaging, such as the CatPhan 500 phantom (The Phantom Laboratory, Salem, New York) or the AAPM CT performance phantom (Computerized Imaging Reference Systems, Norfolk, Virginia). Similar phantoms are being adapted for megavoltage imaging purposes (Siemens Medical Solutions, Concord, California).

Additional image quality parameters for volumetric-based systems are image uniformity and image linearity. An image of a uniform density phantom easily reveals nonuniformities and artifacts. Ring artifacts are likely caused by bad pixels. Cupping artifacts are mainly caused by scattered radiation and beam hardening. Uniformity is often characterized by the variability of the average signal over several small regions

of interest (ROIs) placed on the image of a uniform density phantom. Image linearity and CT number accuracy become important when such scans are used for dose calculations. CT number accuracy is measured by scanning a phantom containing inserts with a wide range of electron densities and comparing the CT numbers in the image with the calibrated values of the phantom.

A large range of values has been reported for the various image quality parameters depending on the type of system and the imaging protocol used for the image acquisition. Typically, a vendor of a specific in-room imaging system provides a set of thresholds these parameters should meet. Stützel et al. (2008) described a quantitative image quality comparison of kilovoltage CBCT, megavoltage CT, and megavoltage CBCT. They reported a spatial resolution of 0.55, 0.35, and 0.27 lp/mm for the Siemens Artiste kilovoltage CBCT, TomoTherapy HI-ART II, and the Siemens MVision megavoltage CBCT, respectively. Similarly, the CNR of the megavoltage CT and megavoltage CBCT were, respectively, a factor of 2 and 5 lower compared with the kilovoltage CBCT (see Figure 15.13).

15.4 Geometrical Calibration

Portal image-based IGRT protocols generally determine the position of the (bony) anatomy relative to the field edge. Such an approach is attractive, because the imaging system is intrinsically calibrated to the treatment isocenter. For all other IGRT systems, the imaging coordinate system needs to be calibrated to the treatment isocenter. The standard approach is to align a phantom to the treatment isocenter using the lasers, the light field, or portal image analysis. Subsequently, an image of the phantom is acquired to characterize the imaging geometry and account for any misalignments between the imaging and treatment isocenter (Bouchet et al. 2001; Jaffray et al. 2002; Verellen et al. 2003; Lachaine et al. 2005). Because any residual misalignments in this procedure propagate to any subsequent IGRT procedure, high accuracy is crucial and portal image analysis is the preferred method.

For gantry-mounted x-ray systems, the trajectories of the kilovoltage source and corresponding detector deviate from a perfect circle due to gravity-induced flex in the support arms (Jaffray et al. 2002). Such geometric nonidealities induce gantry dependent misalignments and can produce severe artifacts in the CT reconstruction process (see Figure 15.14). Numerous methods for geometric calibration have been reported, typically involving a phantom of known geometry (e.g., an array of BBs) projected in the detector domain and analyzed in a manner to deduce the system pose at each projection angle (Rougee et al. 1993; Jaffray and Siewerdsen 2000; Noo et al. 2000; Cho et al. 2005). CBCT systems developed for IGRT have been shown to provide a high degree of reproducibility in the nonidealities of each degree of freedom (Sharpe et al. 2006). The complexity of the calibration method required depends on the rigidity of the CBCT system and the spatial accuracy and resolution required.

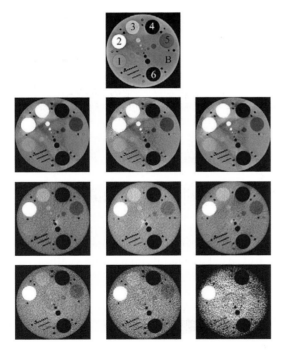

FIGURE 15.13 Central slice of the contrast-and-resolution phantom acquired with the four imaging devices at different dose levels. (Always from left to right.) Top row: Primatom. Second row: Artíste prototype kilovoltage CBCT (3.0, 1.5, and 1.0 cGy). Third row: Tomotherapy (3.0, 1.5, and 1.0 cGy). Bottom row: Artíste prototype megavoltage CBCT (17.0, 8.2, and 3.0 cGy). (Reproduced from Stützel J et al., *Radiother Oncol.* 2008 Jan;86(1):20–4.)

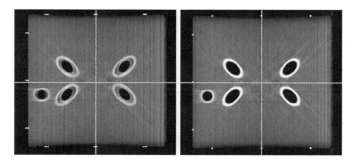

FIGURE 15.14 Geometric calibration artifacts exemplified in an axial reconstruction of a quality assurance phantom. Gantry rotation-induced geometric nonidealities induce a double image of the phantom, whereas inaccurate imaging-to-treatment isocenter calibration induces an apparent shift of the phantom (left). Proper geometric calibration restores the image quality and position (right).

15.5 Imaging Dose

Most of the currently available in-room imaging techniques, with the exception of US, optical imaging, and MRI-based systems, deliver some radiation dose to the patient. The total imaging radiation dose experienced by a patient varies significantly with modality and regimen (Murphy et al. 2007). Important factors that determine the dose to the patient are the required image quality, the attenuation characteristics of the imaged anatomy that depends on the radiation energy, and the imaging duty cycle. Many studies of imaging doses from IGRT systems have been published ranging from 2 to 10 cGy per pair of portal images using modern EPIDs to 1 to 3 cGy/scan for kilovoltage CBCT to 3 to 10 cGy/scan for megavoltage CBCT. In Table 15.1, an overview is given of the imaging doses for the different in-room imaging modalities. Note that these doses are given per scan or pair of orthogonal images. Over an entire treatment schedule, the cumulative imaging dose can in some occasions accumulate up to 380 cGy, which is above the reported threshold doses for secondary induced malignancies (Tubiana 2009). The imaging dose should therefore be carefully managed. The "as low as reasonably achievable" (ALARA) adopted by the diagnostic imaging community, however, is not generally applicable in the context of IGRT, where the imaging dose is added to an already high level of therapeutic dose. Moreover, there is an interaction between the imaging dose and its impact on the improved accuracy of the therapeutic dose delivery that suggests the possibility of optimizing rather than simply minimizing the imaging dose (Murphy et al. 2007). For example, Kron et al. (2010) demonstrated that adaptive RT for bladder cancer reduces integral dose despite daily volumetric imaging. IGRT correction strategies should therefore optimize the imaging dose with improvements in treatment delivery.

15.6 Quality Assurance

Quality assurance programs for the IGRT system are tailored to the geometrical accuracy and image quality to ensure a stable and reproducible performance.

15.6.1 Geometrical Accuracy

During the geometrical calibration of the in-room imaging system, the coincidence of the imaging isocenter and treatment isocenter is established. Because a change to this coincidence has a direct impact on the localization accuracy of this system, daily quality assurance tests are often recommended to catch change in the performance of the system. Simple integral tests have been developed to assure the geometrical accuracy of in-room imaging systems (Yoo et al. 2006; Bissonnette 2007; Mao et al. 2008). By imaging an anthropomorphic phantom and relocating it through an image guidance process, its resulting position can be checked against room lasers, light fields, and/or portal image analysis. A tolerance of ±2 mm has been demonstrated for such daily geometric accuracy checks (Bissonnette et al. 2008a). Additionally, image scale and voxel size accuracy can be tested by imaging an object of well-known sizes and subsequently compared measured with actual values. Most in-room imaging systems have displayed distance accuracy well within 1 mm (Yin et al. 2009).

15.6.2 Image Quality

An image quality assurance program generally aims to quantify relevant image quality parameters and compare them with the baseline values established during acceptance testing and/or commissioning. Typically, one or more images are acquired from an image quality phantom (see above) from which parameters such as spatial resolution, image scale, contrast, noise, and

TABLE 15.1 Overview of Typically Used Imaging Doses per Acquisition for Different IGRT Systems

Modality	Dose	References
A pair of electronic portal images	2–10 cGy	Waddington and McKensie 2004; Herman 2005
A pair of kilovoltage radiographs	0.05–0.4 cGy	Murphy et al. 2007
Kilovoltage fluoroscopy	0.1–1 cGy/min	Shirato et al. 2004
In-room CT	0.2–1 cGy	O'Daniel et al. 2004
Kilovoltage CBCT	1–2 cGy	Islam et al. 2006; Amer et al. 2007
Megavoltage CBCT	3–10 cGy	Gayou et al. 2007; Morin et al. 2007
Megavoltage CT	1–4 cGy	Shah et al. 2008

contrast-to-noise are quantified (Yoo et al. 2006; Saw et al. 2007; Bissonnette et al. 2008b; Morin et al. 2009). Such tests are often performed monthly to semiannually. Note that, for in-room CT-based systems, reductions in spatial resolution are likely caused by changes in the geometry and are therefore indicative for reduction in the geometrical accuracy of the system.

15.7 Clinical Applications of In-Room Imaging Techniques

In-room imaging techniques are widely adopted to increase the precision and accuracy of RT and are used for a large variety of treatment sites. Simpson et al. (2010) report that approximately 94% of radiation oncologists in the United States use in-room imaging for IGRT, whereas approximately 70% of the nonusers are planning to adopt IGRT in the future. The percentages using US, video, planar megavoltage, planar kilovoltage, and volumetric technologies were 22.3%, 3.2%, 62.7%, 57.7%, and 58.8%, respectively.

15.7.1 Correction Strategies

Image-guided correction strategies aim to reduce geometrical uncertainties by minimizing discrepancies between the position of target and OARs during treatment planning and during treatment delivery. Both offline and online correction strategies have been developed. In offline corrections, adjustments are based on images acquired during previous fractions, whereas in online corrections, an adjustment is made immediately following or even during imaging.

Offline correction protocols aim to correct for the mean error of a patient (systematic error) without correcting for daily variations (random error). The rationale for offline correction strategies is that margin requirements are dominated by the systematic component of the errors and much less by random errors. Therefore, offline protocols allow substantial margin reduction with limited workload. A range of statistically driven setup correction protocols have been developed by calculating the mean discrepancy over the first few fractions (Bijhold et al. 1992; Bel et al. 1996; de Boer et al. 2001b) or based on maximum likelihood estimators (Pouliot et al. 1996; Keller et al. 2004). Adaptive RT extends the idea of offline corrections and aims to not only correct for setup errors but also reduce all systematic discrepancies between treatment planning and treatment delivery (Yan et al. 2000, 2005). By building a patient model from repeat volumetric images early in the course of treatment, a better representation can be made of the average position of internal anatomy, and margins can be tailored to individual patients.

Online protocols aim to correct not only the systematic errors but also the random errors (Alasti et al. 2001; Fuss et al. 2004). The rationale for online corrections is mostly that, for hypofractionated treatment regimens (Murphy et al. 2003; Purdie et al. 2007; Cai et al. 2010), the error of each fraction has a considerable contribution to the overall uncertainty (van Herk 2004). Online image guidance applications require fast, simple, and unambiguous image analysis and correction, because otherwise the time pressure could adversely affect the accuracy of the procedure.

15.7.2 Planar Imaging

Portal imaging has been applied to correct for bony anatomy setup errors for a large range of disease sites (Herman 2005), such as head and neck (Bel et al. 1995; de Boer et al. 2001b; van Lin et al. 2003; Pehlivan et al. 2009; Mongioj et al. 2011), breast (van Tienhoven et al. 1991; Creutzberg et al. 1993; Lirette et al. 1995; Kron et al. 2004), lung (van de Steene et al. 1998; Samson et al. 1999; de Boer et al. 2001a; Erridge et al. 2003), and prostate (Bel et al. 1996; Hanley et al. 1997; van Lin et al. 2001), mostly based on offline correction strategies. The reported standard deviation (1 SD, in millimeters) of the systematic and random errors observed with portal image analysis, separately measured along the three principle axes, ranges from 1.6 to 4.6 and from 1.1 to 2.5 (head and neck), from 1.0 to 3.8 and from 1.2 to 3.5 (prostate), from 1.1 to 4.7 and from 1.1 to 4.9 (pelvis), from 1.8 to 5.1 and from 2.2 to 5.4 (lung), and from 1.0 to 4.7 and from 1.7 to 14.4 (breast), respectively (Remeijer et al. 2001). The accuracy of such portal image-based correction strategies of bony anatomy setup error, however, was difficult to quantify in the absence of a gold standard. With the introduction of volumetric imaging modalities, a comparison became feasible and demonstrated an underestimation of setup error by portal image analysis for lung (Borst et al. 2007) and breast (Topolnjak et al. 2010).

To correct for organ motion, several authors have described the use of implanted fiducial markers as a surrogate of the target, especially in prostate cancer patients (Balter et al. 1995; Crook et al. 1995; Nederveen et al. 2000; Pouliot and Lirette 2003), but also for breast cancer patients (Harris et al. 2009; Mitchell et al. 2010), lung cancer patients (Berbeco et al. 2005), and even head-and-neck cancer patients (van Asselen et al. 2004). Large-scale clinical application of portal imaging, however, has long been hampered by a lack of commercially available automatic image registration tools and decision rules for offline correction protocols.

To improve image quality, planar kilovoltage x-ray imaging systems have been implemented. In Floor/ceiling-mounted dual x-ray stereoscopic systems (Figures 15.15 and 15.16) have been employed for intracranial lesions (Hoogeman et al. 2008; Wurm et al. 2008), demonstrating comparable accuracy as obtained with frame-based systems (Ramakrishna et al. 2010). Similarly, for spinal lesions, high accuracy can be obtained (Murphy et al. 2003; Gerszten et al. 2004), especially when translations and rotations in 3D (2D–3D matching; Gilhuijs et al. 1996) were optimized for registration (Jin et al. 2006). Fiducial markers are often used as a surrogate for soft-tissue targets such as the prostate (Kitamura et al. 2002; Litzenberg et al. 2002). For gantry-mounted systems, a crossfire method has been developed using both the kilovoltage and the portal imaging system to rapidly acquire an orthogonal image pair (Mutanga et al. 2008). For liver cancer patients, the diaphragm has been used as a surrogate,

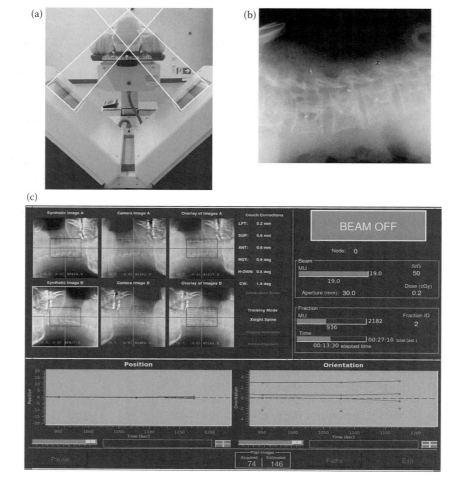

FIGURE 15.15 (a) Stereoscopic imaging system of the CyberKnife. (b) Radiograph of the neck region where the bony anatomy and fiducial markers are clearly visible. (c) Screenshot of the graphical user interface for stereoscopic guidance. (Images courtesy of Mischa Hoogeman, Erasmus MC, Rotterdam, The Netherlands.)

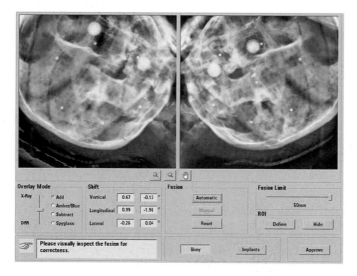

FIGURE 15.16 Screenshot of the graphical user interface of the Novalis target localization by stereoscopic x-ray imaging. (Image courtesy of Tom Depuydt, University Hospital Brussels, Brussels, Belgium.)

especially for the cranial caudal direction (Balter et al. 2002) as well as fiducial markers (Wunderink et al. 2010).

Stereoscopic imaging systems in fluoroscopic mode have the ability to quantify and correct for intrafraction motion. Shirato et al. (1999) continuously tracks the position of fiducial markers implanted in the proximity of lung tumors to gate the therapeutic beam. To reduce the imaging dose to the patient during x-ray image-based tumor tracking, an "internal–external" correlation model can be used to estimate the 3D target position (Schweikard et al. 2004; Seppenwoolde et al. 2007). A similar approach is adopted for a gantry-mounted monoscopic imaging system orthogonal to the treatment beam (Cho et al. 2009) to model the 3D motion pattern. Correlation models are also employed for combined kilovoltage and megavoltage imaging to track 3D internal marker (Wiersma et al. 2009).

Alternative to megavoltage and kilovoltage imaging, nonionizing imaging modalities have been proposed. US image guidance has been widely applied for prostate localization (Fraser et al. 2010; Lattanzi et al. 2000; see Figure 15.17a) but has been hampered with large observer errors, indicating that target

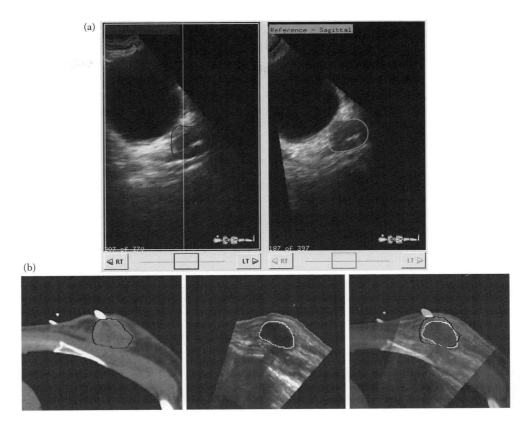

FIGURE 15.17 (a) US-guided RT for prostate cancer patients. The reference scan is contoured at a workstation (light gray), and these contours are projected (dark gray) in the daily scans. The operator can then manually register the contour to the actual position of the prostate and apply a couch correction. (b) CT tumor bed contour (black), US tumor bed contour (white), and a fused image set. (Images courtesy of Frank Verhaegen, Maastro Clinic, Maastricht, The Netherlands.)

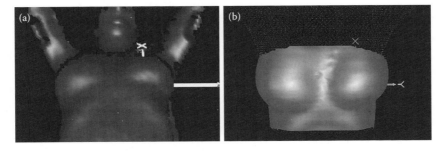

FIGURE 15.18 (a) Example of a surface capture (AlignRT, Vision RT Ltd., London, UK) of a left-sided breast cancer patient undergoing breath hold. This surface of the patient in treatment position can then be registered to a surface rendering of the planning CT (b). (Images courtesy of Tanja Alderliesten, The Netherlands Cancer Institute-Antoni van Leeuwenhoek Hospital, Amsterdam, The Netherlands.)

positioning did not improve in reference to portal imaging with implanted fiducials (Langen et al. 2003; Keller et al. 2004). With the implementation of 3D US sensors for soft-tissue localization purposes, however, these errors are reduced (Boda-Heggemann et al. 2008). Such developments extend the application of US guidance to real-time monitoring (Wu et al. 2006) and other treatment sites such as upper abdominal malignancies (Fuss et al. 2004) and breast (Kuban et al. 2005; see Figure 15.17b).

Applications of optical imaging are mainly limited to disease sites where the surface is a reasonable surrogate for the internal anatomy such as head-and-neck cancer (Cerviño et al. 2010) and breast cancer (Bert et al. 2006; Gierga et al. 2008; see Figure 15.18). Because of this restriction, clinical implementation is limited so far.

15.7.3 Volumetric Imaging

The availability of volumetric CT imaging in the treatment room has enabled a large scale of clinical protocols. The CT-on-rails system produces the same image quality as diagnostic CT scanners that are generally used for treatment planning. Several groups have used in-room CT guidance for soft-tissue targets such as lung cancer (Uematsu et al. 2001; Onishi et al. 2003; Chang et al. 2008) and prostate cancer (Fung et al. 2003; Paskalev et al. 2004; Wong et al. 2008). Also for more high-contrast targets such as spinal lesions, CT on rails has been used (Shiu et al. 2003; Yenice et al. 2003; Wang et al. 2007). Despite the excellent image quality of CT-on-rails image guidance systems, however, logistic and economic reasons limited large-scale clinical implementation of this technology.

Despite the somewhat lower image quality of kilovoltage CBCT compared with kilovoltage fan beam CT, it has been widely adopted by the RT community for IGRT. Initial clinical studies were focused on patient imaging studies to optimize the imaging techniques for different treatment sites, access usefulness of these images for clinical decision-making (Létourneau et al. 2005), and benchmark bony anatomy localization accuracy of CBCT against portal imaging (Remeijer et al. 2004; Borst et al. 2007). Subsequently, true soft-tissue guided protocols were implemented for prostate (Smitsmans et al. 2005, 2008; Moseley et al. 2007), conventionally fractionated lung RT (Bissonnette et al. 2009; Yeung et al. 2009; Higgins et al. 2011), and stereotactic body RT of pulmonary lesions (Purdie et al. 2007; Grills et al. 2008; Sonke et al. 2009; see Figure 15.19), all demonstrating an improved localization accuracy over conventional fiducial-less

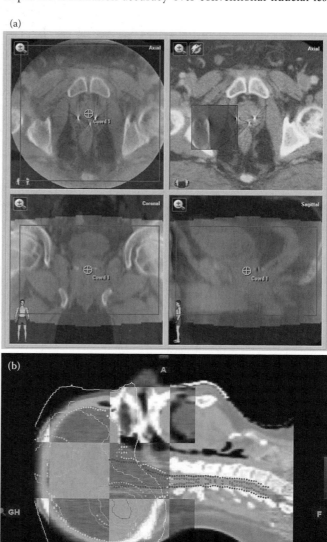

FIGURE 15.19 Illustration of various 4D CBCT registrations for SBRT of a lung cancer patient with a peak-to-peak amplitude of approximately 7 mm. Top row: Registration of the vertebrae; middle row: tumor registrations; bottom row: correction to the time-averaged tumor position. Left column: exhale phase; right column: inhale phase. (From Sonke JJ et al., *Int J Radiat Oncol Biol Phys*. 2009 Jun 1;74(2):567–74.)

FIGURE 15.20 (See color insert.) (a) CBCT scan of a prostate cancer patient with implanted fiducial markers acquired with the Vero system in an axial, coronal, and sagittal views and a fusion of the CBCT with the planning CT (top right). (b) Tomotherapy scan of a head-and-neck cancer patient in a sagittal view fused with the planning CT. (Image courtesy of Tom Depuydt, University Hospital Brussels, Brussels, Belgium.)

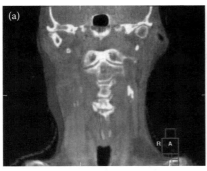 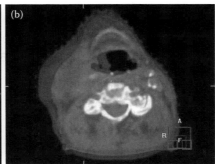

FIGURE 15.21 Illustration of anatomical change of a head-and-neck cancer patient over the course of RT. (a) Coronal slice of a fused planning CT (purple) and CBCT (green). (b) Transverse slice.

planar imaging-based methods. Kilovoltage CBCT guidance is also used extensively in other treatment sites, such as head and neck (Wang et al. 2009b; van Beek et al. 2010), breast (White et al. 2007; Fatunase et al. 2008), liver (Hawkins et al. 2006; Case et al. 2009), and bladder (Burridge et al. 2006; Lotz et al. 2006; Pos et al. 2009). Similarly, megavoltage (CB)CT alignments have been applied for prostate (Langen et al. 2005; Morin et al. 2006), head and neck (Zeidan et al. 2007), lung alignments (Ford et al. 2002; Hodge et al. 2006), breast (Langen et al. 2008), and gynecological tumors (Santanam et al. 2008). Such alignments are either based on soft-tissue targets, bony anatomy (Figure 15.20b), or implanted markers (Figure 15.20a) depending on the visibility of the target on the megavoltage (CB)CT.

In addition to patient alignment, volumetric imaging has been used to quantify anatomical variation in a range of anatomical sites. Significant tumor regression has been observed in lung cancer patients (Kupelian et al. 2005; Guckenberger et al. 2011) with a range of 0.6% to 2.3% per day. Shape changes of the mesorectal space (Tournel et al. 2008; Nijkamp et al. 2009) have been observed in up to 7 mm (1 SD). In head-and-neck cancer, volume shrinkage in both tumors and normal tissues has been observed (Figure 15.21), which also led to positional variations in the center of mass of the parotid glands and corresponding dosimetric changes (Barker et al. 2004; O'Daniel et al. 2007; Han et al. 2008). Such complex changes cannot be accounted for using simple couch shifts. Adaptive RT, in which an active feedback loop is used to build an individualized patient model to modify the treatment plan, has the potential to account for these geometrical uncertainties (Yan 2010). Such strategies are currently being explored (Lee et al. 2008; Nijkamp et al. 2008; Wu et al. 2009; Sonke and Belderbos 2010) but are awaiting large-scale clinical implementation.

15.8 Summary

The development and the large-scale adoption of in-room imaging techniques in RT have had a profound impact on the accuracy and precision of dose delivery. Initially, in-room systems were based on planar imaging, allowing the visualization of bony anatomy and implanted fiducials. The introduction of volumetric imaging technology allows the visualization of soft-tissue structures. With the high level of conformality of intensity-modulated RT and volumetric modulated arc therapy-based treatment plans, image-guided delivery is considered the standard of care. In the future, it is expected that treatment plans will be further modulated based on biological information. Developments in adaptive RT based on in-room acquired images will facilitate the accurate delivery of such plans in the presence of anatomical changes over the course of treatment.

References

Adler JR Jr, Chang SD, Murphy MJ et al. The Cyberknife: a frameless robotic system for radiosurgery, *Stereotact Funct Neurosurg.* 1997;69(1–4 Pt 2):124–8.

Alasti H, Petric MP, Catton CN, and Warde PR. Portal imaging for evaluation of daily on-line setup errors and off-line organ motion during conformal irradiation of carcinoma of the prostate, *Int J Radiat Oncol Biol Phys.* 2001 Mar 1; 49(3):869–84.

Alderliesten T, Sonke J-J, Heddes R et al. Assessment of Setup Variability for Breath-hold Radiotherapy for Breast Cancer Patients by Surface Imaging. Radiotherapy and Oncology, Proceedings of 10th Biennial ESTRO Conference on Physics and Radiation Technology for Clinical Radiotherapy, Maastricht, September 2009.

Amer A, Marchant T, Sykes J, Czajka J, and Moore C. Imaging doses from the Elekta Synergy X-ray cone beam CT system, *Br J Radiol.* 2007 Jun;80(954):476–82.

Antonuk L, Yorkston J, Boudry J et al. A development of hydrogenated amorphous silicon sensors for high energy photon radiotherapy imaging. *IEEE Trans Nucl Sci.* 1990;37: 165–70.

Antonuk LE. Electronic portal imaging devices: a review and historical perspective of contemporary technologies and research, *Phys Med Biol.* 2002 Mar 21;47(6):R31–R65.

Artignan X, Smitsmans MH, Lebesque JV et al. Online ultrasound image guidance for radiotherapy of prostate cancer: impact of image acquisition on prostate displacement, *Int J Radiat Oncol Biol Phys.* 2004 Jun 1;59(2):595–601.

Balter JM, Brock KK, Litzenberg DW et al. Daily targeting of intrahepatic tumors for radiotherapy, *Int J Radiat Oncol Biol Phys.* 2002 Jan 1;52(1):266–71.

Balter JM, Sandler HM, Lam K et al. Measurement of prostate movement over the course of routine radiotherapy using implanted markers, *Int J Radiat Oncol Biol Phys.* 1995 Jan 1;31(1):113–8.

Barker JL Jr, Garden AS, Ang KK et al. Quantification of volumetric and geometric changes occurring during fractionated radiotherapy for head-and-neck cancer using an integrated CT/linear accelerator system, *Int J Radiat Oncol Biol Phys.* 2004 Jul 15;59(4):960–70.

Bel A, Keus R, and Vijlbrief RE. Setup deviations in wedged pair irradiation of parotid gland and tonsillar tumors, measured with an electronic portal imaging device. *Radiother Oncol.* 1995;37:153–9.

Bel A, Vos PH, Rodrigus PT et al. High-precision prostate cancer irradiation by clinical application of an offline patient setup verification procedure, using portal imaging. *Int J Radiat Oncol Biol Phys.* 1996;35:321–32.

Berbeco RI, Jiang SB, Sharp GC et al. Integrated radiotherapy imaging system (IRIS): design considerations of tumour tracking with linac gantry-mounted diagnostic x-ray systems with flat-panel detectors, *Phys Med Biol.* 2004 Jan 21;49(2):243–55.

Berbeco RI, Neicu T, Rietzel E et al. A technique for respiratory-gated radiotherapy treatment verification with an EPID in cine mode. *Phys Med Biol.* 2005;50:3669–79.

Bert C, Metheany KG, Doppke KP, Taghian AG, Powell SN, and Chen GT. Clinical experience with a 3D surface patient setup system for alignment of partial-breast irradiation patients, *Int J Radiat Oncol Biol Phys.* 2006 Mar 15;64(4):1265–74.

Biggs PJ, Goitein M, and Russell MD. A diagnostic X ray field verification device for a 10 MV linear accelerator. *Int J Radiat Oncol Biol Phys.* 1985;11:635–43.

Bijhold J, Lebesque JV, Hart AAM et al. Maximizing setup accuracy using portal images as applied to a conformal boost technique for prostatic cancer. *Radiother Oncol.* 1992;24:261–71.

Bissonnette JP. Quality assurance of image-guidance technologies, *Semin Radiat Oncol.* 2007;17:278–86.

Bissonnette JP, Moseley DJ, and Jaffray DA. A quality assurance program for image quality of cone-beam CT guidance in radiation therapy, *Med Phys.* 2008a;35:1807–15.

Bissonnette JP, Moseley D, White E et al. Quality assurance for the geometric accuracy of cone-beam CT guidance in radiation therapy, *Int J Radiat Oncol Biol Phys.* 2008b;71:S57–61.

Bissonnette JP, Purdie TG, Higgins JA et al. Cone-beam computed tomographic image guidance for lung cancer radiation therapy, *Int J Radiat Oncol Biol Phys.* 2009;73:927–34.

Boda-Heggemann J, Köhler FM, Küpper B et al. Accuracy of ultrasound-based (BAT) prostate-repositioning: a three-dimensional on-line fiducial-based assessment with cone-beam computed tomography, *Int J Radiat Oncol Biol Phys.* 2008 Mar 15;70(4):1247–55.

Borst GR, Sonke JJ, Betgen A et al. Kilo-voltage cone-beam computed tomography setup measurements for lung cancer patients; first clinical results and comparison with electronic portal-imaging device, *Int J Radiat Oncol Biol Phys.* 2007 Jun 1;68(2):555–61.

Bouchet LG, Meeks SL, Goodchild G et al. Calibration of three-dimensional ultrasound images for image-guided radiation therapy, *Phys Med Biol* 2001 Feb;46(2):559–77.

Boyer AL, Antonuk L, Fenster A et al. A review of electronic portal imaging devices (EPIDs), *Med Phys.* 1992 Jan–Feb;19(1):1–16.

Burridge N, Amer A, Marchant T et al. Online adaptive radiotherapy of the bladder: small bowel irradiated-volume reduction, *Int J Radiat Oncol Biol Phys.* 2006;66:892–7.

Bushberg JT, Seibert JA, Leidholdt EM, and Boone JM. 2002. *The Essential Physics of Medical Imaging, 2.* Hagerstown MD. Lippincott Williams & Wilkins.

Cai G, Hu WG, Chen JY et al. Impact of residual and intrafractional errors on strategy of correction for image-guided accelerated partial breast irradiation, *Radiat Oncol.* 2010 Oct 26;5:96.

Case RB, Sonke JJ, Moseley DJ et al. Inter- and intrafraction variability in liver position in non-breath-hold stereotactic body radiotherapy, *Int J Radiat Oncol Biol Phys.* 2009 Sep 1;75(1):302–8.

Cerviño L, Pawlicki T, Lawson J, and Jiang S. Frame-less and mask-less cranial stereotactic radiosurgery: a feasibility study, *Phys Med Biol.* 2010;55:1863–73.

Chang JY, Dong L, Liu H et al. Image-guided radiation therapy for non-small cell lung cancer, *J Thorac Oncol.* 2008;3:177–86.

Cho B, Suh Y, Dieterich S, and Keall PJ. A monoscopic method for real-time tumour tracking using combined occasional x-ray imaging and continuous respiratory monitoring, *Phys Med Biol.* 2008 Jun 7;53(11):2837–55.

Cho Y, Moseley DJ, Siewerdsen JH, and Jaffray DA. Accurate technique for complete geometric calibration of cone-beam computed tomography systems, *Med Phys.* 2005 Apr;32(4):968–83.

Choi K, Wang J, Zhu L et al. Compressed sensing based cone-beam computed tomography reconstruction with a first-order method, *Med Phys.* 2010 Sep;37(9):5113–25.

Christoph B, Metheany KG, Doppke K, and Chen GTY. A phantom evaluation of a stereo-vision surface imaging system for radiotherapy patient setup, *Med Phys.* 2005;32(9):2753–62.

Creutzberg CL, Althof VG, and Huizenga H. Quality assurance using portal imaging: the accuracy of patient positioning in irradiation of breast cancer, *Int J Radiat Oncol Biol Phys.* 1993;25:529–39.

Crook JM, Raymond Y, Salhani D, Yang H, and Esche B. Prostate motion during standard radiotherapy as assessed by fiducial markers, *Radiother Oncol.* 1995 Oct;37(1):35–42.

Davide F, van der Meer S, Harris E, and Verhaegen F. A CT based correction method for speed of sound aberration for ultrasound based image guided radiotherapy, *Med Phys.* 2011;38:2665.

de Boer HC, van Sornsen de Koste JR, and Creutzberg CL. Electronic portal image assisted reduction of systematic set-up errors in head and neck irradiation, *Radiother Oncol.* 2001a;61:299–308.

de Boer HC, van Sornsen de Koste JR, and Senan S. Analysis and reduction of 3D systematic and random setup errors during the simulation and treatment of lung cancer patients with CT-based external beam radiotherapy dose planning. *Int J Radiat Oncol Biol Phys.* 2001b;49:857–68.

Dempsey JF, Benoit D, Fitzsimmons JR et al. A device for real-time 3D image-guided IMRT, *Int J Radiat Oncol Biol Phys.* 2005;63:S202.

Dietrich L, Jetter S, Tücking T, Nill S, and Oelfke U. Linac-integrated 4D cone beam CT: first experimental results, *Phys Med Biol.* 2006 Jun 7;51(11):2939–52. Epub 2006 May 24.

Duerk JL. Principles of MR image formation and reconstruction, *Magn Reson Imaging Clin N Am.* 1999 Nov;7(4):629–59.

Erridge SC, Seppenwoolde Y, and Muller SH. Portal imaging to assess set-up errors, tumor motion and tumor shrinkage during conformal radiotherapy of non-small cell lung cancer. *Radiother Oncol.* 2003;66:75–85.

Essers M, Hoogervorst BR, van Herk M, Lanson H, and Mijnheer BJ. Dosimetric characteristics of a liquid-filled electronic portal imaging device, *Int J Radiat Oncol Biol Phys.* 1995 Dec 1;33(5):1265–72.

Faddegon BF, Gangadharan V, Wu B, Pouliot J, and Bani-Hashemi A. Low dose megavoltage cone beam CT with an unflattened 4 MV beam from a carbon target, *Med Phys.* 2008;35(12):5777–86.

Fallone BG, Murray B, Rathee S et al. First MR images obtained during megavoltage photon irradiation from a prototype integrated linac-MR system, *Med Phys.* 2009 Jun;36(6):2084–8.

Fatunase T, Wang Z, Yoo S et al. Assessment of the residual error in soft tissue setup in patients undergoing partial breast irradiation: results of a prospective study using cone-beam computed tomography, *Int J Radiat Oncol Biol Phys.* 2008;70:1025–34.

Ford EC, Chang J, Mueller K et al. Cone-beam CT with megavoltage beams and an amorphous silicon electronic portal imaging device: potential for verification of radiotherapy of lung cancer, *Med Phys.* 2002 Dec;29(12):2913–24.

Fraser DJ, Chen Y, Poon E et al. Dosimetric consequences of misalignment and realignment in prostate 3DCRT using intramodality ultrasound image guidance, *Med Phys.* 2010 Jun;37(6):2787–95.

Fung AY, Grimm SY, Wong JR, and Uematsu M. Computed tomography localization of radiation treatment delivery versus conventional localization with bony landmarks, *J Appl Clin Med Phys.* 2003;4:112–9.

Fuss M, Salter BJ, Cavanaugh SX et al. Daily ultrasound-based image-guided targeting for radiotherapy of upper abdominal malignancies, *Int J Radiat Oncol Biol Phys.* 2004 Jul 15;59(4):1245–56.

Gayou O, Parda DS, Johnson M, and Miften M. Patient dose and image quality from mega-voltage cone beam computed tomography imaging, *Med Phys.* 2007;34:499–506.

Gerszten PC, Ozhasoglu C, Burton SA et al. CyberKnife frameless stereotactic radiosurgery for spinal lesions: clinical experience in 125 cases, *Neurosurgery.* 2004 Jul;55(1):89–98.

Gierga DP, Riboldi M, Turcotte JC et al. Comparison of target registration errors for multiple image-guided techniques in accelerated partial breast irradiation, *Int J Radiat Oncol Biol Phys.* 2008 Mar 15;70(4):1239–46.

Gilhuijs KG, van de Ven PJ, and van Herk M. Automatic three-dimensional inspection of patient setup in radiation therapy using portal images, simulator images, and computed tomography data, *Med Phys.* 1996 Mar;23(3):389–99.

Green M. Magnetic field effects on radiation dose distribution, *Med Phys.* 2009;36:2774.

Grills IS, Hugo G, Kestin LL et al. Image-guided radiotherapy via daily online cone-beam CT substantially reduces margin requirements for stereotactic lung radiotherapy, *Int J Radiat Oncol Biol Phys.* 2008 Mar 15;70(4):1045–56.

Groh BA, Siewerdsen JH, Drake DG, Wong JW, and Jaffray DA. A performance comparison of flat-panel imager-based MV and kV conebeam CT, *Med Phys.* 2002;29:967–75.

Guckenberger M, Wilbert J, Richter A, Baier K, and Flentje M. Potential of adaptive radiotherapy to escalate the radiation dose in combined radiochemotherapy for locally advanced non-small cell lung cancer, *Int J Radiat Oncol Biol Phys.* 2011 Mar 1;79(3):901–8.

Han C, Chen YJ, Liu A, Schultheiss TE, and Wong JY. Actual dose variation of parotid glands and spinal cord for nasopharyngeal cancer patients during radiotherapy, *Int J Radiat Oncol Biol Phys.* 2008;70:1256–62.

Hanley J, Lumley MA, and Mageras GS. Measurement of patient positioning errors in three-dimensional conformal radiotherapy of the prostate, *Int J Radiat Oncol Biol Phys.* 1997;37:435–44.

Harris EJ, Donovan EM, Yarnold JR, Coles CE, and Evans PM. IMPORT Trial Management Group, Characterization of target volume changes during breast radiotherapy using implanted fiducial markers and portal imaging, *Int J Radiat Oncol Biol Phys.* 2009 Mar 1;73(3):958–66.

Hawkins MA, Brock KK, Eccles C et al. Assessment of residual error in liver position using kV cone-beam computed tomography for liver cancer high-precision radiation therapy, *Int J Radiat Oncol Biol Phys.* 2006 Oct 1;66(2):610–9.

Hendee WR and Ritenour ER. 2002. *Medical Imaging Physics*, 4th Edition. Hoboken, NJ, John Wiley & Sons, Inc.

Hendrick RE. The AAPM/RSNA physics tutorial for residents. Basic physics of MR imaging: an introduction, *Radiographics.* 1994 Jul;14(4):829–46.

Herman M. Clinical use of electronic portal imaging, *Semin Radiat Oncol.* 2005;15:157–67.

Herman MG, Balter JM, Jaffray DA et al. Clinical use of electronic portal imaging: report of AAPM Radiation Therapy Committee Task Group 58, *Med Phys.* 2001 May;28(5):712–37.

Higgins J, Bezjak A, Hope A, Panzarella T, Li W, Cho JB, Craig T, Brade A, Sun A, and Bissonnette JP. Effect of image-guidance frequency on geometric accuracy and setup margins in radiotherapy for locally advanced lung cancer, *Int J Radiat Oncol Biol Phys*. 2011 Aug 1;80(5):1330–7.

Hodge W, Tome WA, Jaradat HA et al. Feasibility report of image guided stereotactic body radiotherapy (IG-SBRT) with tomotherapy for early stage medically inoperable lung cancer using extreme hypofractionation, *Acta Oncol* 2006;45:890–6.

Hoogeman MS, Nuyttens JJ, Levendag PC, and Heijmen BJ. Time dependence of intrafraction patient motion assessed by repeat stereoscopic imaging, *Int J Radiat Oncol Biol Phys*. 2008;70:609–18.

ICRU Report 50. Prescribing, Recording and Reporting Photon Beam Therapy. International Commission on Radiation Units and Measurements, Bethesda, MD, 1993.

ICRU Report 62. Prescribing, Recording and Reporting Photon Beam Therapy (supplement to ICRU report 50). Bethesda, MD: International Commission on Radiation Units and Measurements, 1999.

Islam MK, Purdie TG, Norrlinger BD et al. Patient dose from kilovoltage cone beam computed tomography imaging in radiation therapy, *Med Phys*. 2006;33:1573–82.

Jaffray DA and Siewerdsen JH. Cone-beam computed tomography with a° at-panel imager: initial performance characterization, *Med Phys*. 2000;27:1311–23.

Jaffray DA, Siewerdsen JH, Wong JW, and Martinez AA. Flat-panel cone-beam computed tomography for image-guided radiation therapy, *Int J Radiat Oncol Biol Phys*. 2002;53:1337–49.

Jaffray DA, Drake DG, Moreau M, Martinez AA, and Wong JW. A radiographic and tomographic imaging system integrated into a medical linear accelerator for localization of bone and soft-tissue targets, *Int J Radiat Oncol Biol Phys*. 1999 Oct 1;45(3):773–89.

Jin J, Ryu S, Rock J et al. Image-guided target localization for stereotactic surgery: accuracy of 6D versus 3D image fusion. In: D. Kondziolka, Editor, *Radiosurgery*, vol. 6, Karger, Basel (2006), pp. 50–9.

Jin JY, Ren L, Liu Q et al. Combining scatter reduction and correction to improve image quality in cone-beam computed tomography (CBCT), *Med Phys*. 2010 Nov;37(11):5634–44.

Johns H and Cunningham A. A precision cobalt 60 unit for fixed field and rotation therapy, *Am J Roentgenol*. 1959;4:81.

Kak AC and Slaney M. 1988. *Principles of Computerized Tomographic Imaging*. IEEE Press, ISBN: 9780780304475.

Kamino Y, Takayama K, Kokubo M et al. Development of a four-dimensional image-guided radiotherapy system with a gimbaled X-ray head, *Int J Radiat Oncol Biol Phys*. 2006;66:271–8.

Keller H, Tome W, Ritter MA, and Mackie TR. Design of adaptive treatment margins for non-negligible measurement uncertainty: application to ultrasound-guided prostate radiation therapy, *Phys Med Biol*. 2004;49: 69–86.

Kitamura K, Shirato H, and Seppenwoolde Y. Three-dimensional intrafractional movement of prostate measured during real-time tumortracking radiotherapy in supine and prone treatment positions, *Int J Radiat Oncol Biol Phys*. 2002;53:1117–23.

Kron T, Eyles D, Schreiner JL, and Battista J. Magnetic resonance imaging for adaptive cobalt tomotherapy: a proposal, *J Med Phys*. 2006;31:242–54.

Kron T, Lee C, Perera F. Evaluation of intra- and inter-fraction motion in breast radiotherapy using electronic portal Cine Imaging, *Technol Cancer Res Treat* 2004;3:443–50.

Kron T, Wong J, Rolfo A et al. Adaptive radiotherapy for bladder cancer reduces integral dose despite daily volumetric imaging, *Radiother Oncol*. 2010 Dec;97(3):485–7.

Kuban DA, Dong L, Cheung R, Strom E, and De Crevoisier R. Ultrasound-based localization, *Semin Radiat Oncol*. 2005 Jul;15(3):180–91.

Kupelian PA, Ramsey C, Meeks SL et al. Serial megavoltage CT imaging during external beam radiotherapy for non-small-cell lung cancer: observations on tumor regression during treatment, *Int J Radiat Oncol Biol Phys*. 2005;63:1024–8.

Kuriyama K, Onishi H, Sano N et al. A new irradiation unit constructed of self-moving gantry-CT and linac, *Int J Radiat Oncol Biol Phys*. 2003 Feb 1;55(2):428–35.

Lachaine M, Audet V, Huang X, and Falco T. A quick and accurate calibration method for 3D ultrasound in image guided radiotherapy, *Med Phys*. 2005;32:2154.

Lagendijk JJ, Raaymakers BW, Raaijmakers AJ et al. MRI/linac integration, *Radiother Oncol*. 2008 Jan;86(1):25–9.

Langen KM, Buchholz DJ, Burch DR et al. Investigation of accelerated partial breast patient alignment and treatment with helical tomotherapy unit, *Int J Radiat Oncol Biol Phys*. 2008;70:1272–80.

Langen KM and Jones DT. Organ motion and its management, *Int J Radiat Oncol Biol Phys*. 2001 May 1;50(1):265–78.

Langen KM, Pouliot J, Anezinos C et al. Evaluation of ultrasound-based prostate localization for image-guided radiotherapy, *Int J Radiat Oncol Biol Phys*. 2003;57:635–44.

Langen KM, Zhang Y, Andrews RD et al. Initial experience with megavoltage (MV) CT guidance for daily prostate alignments, *Int J Radiat Oncol Biol Phys*. 2005;62:1517–24.

Lattanzi J, McNeeley S, Donnelly S et al. Ultrasound-based stereotactic guidance in prostate cancer—quantification of organ motion and set-up errors in external beam radiation therapy, *Comput Aided Surg*. 2000;5(4):289–95.

Lazos D and Williamson JF. Monte Carlo evaluation of scatter mitigation strategies in cone-beam CT, *Med Phys*. 2010 Oct;37(10):5456–70.

Lee C, Langen KM, Lu W et al. Assessment of parotid gland dose changes during head and neck cancer radiotherapy using daily megavoltage computed tomography and deformable image registration, *Int J Radiat Oncol Biol Phys*. 2008 Aug 1;71(5):1563–71.

Létourneau D, Wong JW, Oldham M et al. Cone-beam-CT guided radiation therapy: technical implementation, *Radiother Oncol*. 2005 Jun;75(3):279–86.

Li T, Xing L, Munro P et al. Four-dimensional cone-beam computed tomography using an on-board imager, *Med Phys.* 2006 Oct;33(10):3825–33.

Lirette A, Pouliot J, and Aubin M. The role of electronic portal imaging in tangential breast irradiation: a prospective study. *Radiother Oncol.* 1995;37:241–5.

Litzenberg D, Dawson LA, Sandler H et al. Daily prostate targeting using implanted radiopaque markers, *Int J Radiat Oncol Biol Phys.* 2002 Mar 1;52(3):699–703.

Liu V, Lariviere NR, and Wang G. X-ray micro-CT with a displaced detector array: application to helical cone-beam reconstruction, *Med Phys.* 2003;30(10):2758–61.

Lotz HT, Pos FJ, Hulshof MC et al. Tumor motion and deformation during external radiotherapy of bladder cancer, *Int J Radiat Oncol Biol Phys.* 2006 Apr 1;64(5):1551–8.

MacKay RI, Graham PA, Logue JP, and Moore CJ. Patient positioning using detailed three-dimensional surface data for patients undergoing conformal radiation therapy for carcinoma of the prostate: a feasibility study, *Int J Radiat Oncol Biol Phys.* 2001;49:225–30.

Mao W, Lee L, and Xing L. Development of a QA phantom and automated analysis tool for geometric quality assurance of on-board MV and kV x-ray imaging systems, *Med Phys.* 2008;35:1497–506.

Meeks SL, Harmon JF, Langen KM et al. Performance characterization of megavoltage computed tomography imaging on a helical tomotherapy unit, *Med Phys.* 2005;32:2673–81.

Milliken DB, Rubin SJ, Hamilton RJ et al. Performance of a video-image-subtraction-based patient positioning, *Int J Radiat Oncol Biol Phys.* 1997;38:855–66.

Mitchell J, Formenti SC, and DeWyngaert JK. Interfraction and intrafraction setup variability for prone breast radiation therapy, *Int J Radiat Oncol Biol Phys.* 2010 Apr;76(5):1571–7.

Mongioj V, Orlandi E, Palazzi M et al. Set-up errors analyses in IMRT treatments for nasopharyngeal carcinoma to evaluate time trends, PTV and PRV margins, *Acta Oncol.* 2011 Jan;50(1):61–71.

Morin O, Aubry JF, Aubin M et al. Physical performance and image optimization of megavoltage cone-beam CT, *Med Phys.* 2009 Apr;36(4):1421–32.

Morin O, Gillis A, Chen J et al. Megavoltage cone-beam CT: system description and clinical applications, *Med Dosim.* 2006 Spring;31(1):51–61.

Morin O, Gillis A, Descovich M et al. Patient dose considerations for routine megavoltage cone-beam CT imaging, *Med Phys.* 2007;34:1819–27.

Morton EJ, Swindell W, Lewis DG, and Evans PM. A linear array, scintillation crystal-photodiode detector for megavoltage imaging, *Med Phys.* 1991 Jul–Aug;18(4):681–91.

Moseley DJ, White EA, Wiltshire KL et al. Comparison of localization performance with implanted fiducial markers and cone-beam computed tomography for on-line image-guided radiotherapy of the prostate, *Int J Radiat Oncol Biol Phys.* 2007 Mar 1;67(3):942–53.

Munro P and Bouius DC. X-ray quantum limited portal imaging using amorphous silicon flat-panel arrays, *Med Phys.* 1998 May;25(5):689–702.

Munro P, Rawlinson JA, and Fenster A. A digital fluoroscopic imaging device for radiotherapy localization, *Int J Radiat Oncol Biol Phys.* 1990 Mar;18(3):641–9.

Murphy MJ, Balter J, Balter S et al. The management of imaging dose during image-guided radiotherapy: report of the AAPM Task Group 75, *Med Phys.* 2007 Oct;34(10):4041–63.

Murphy MJ, Chang SD, Gibbs IC et al. Patterns of patient movement during frameless image-guided radiosurgery, *Int J Radiat Oncol Biol Phys.* 2003;55:1400–8.

Murphy MJ and Cox RS. The accuracy of dose localization for an image-guided frameless radiosurgery system, *Med Phys.* 1996 Dec;23(12):2043–9.

Mutanga TF, de Boer HC, van der Wielen GJ et al. Stereographic targeting in prostate radiotherapy: speed and precision by daily automatic positioning corrections using kilovoltage/megavoltage image pairs, *Int J Radiat Oncol Biol Phys.* 2008 Jul 15;71(4):1074–83.

Nederveen A, Lagendijk J, and Hofman P. Detection of fiducial gold markers for automatic on-line megavoltage position verification using a marker extraction kernel (MEK), *Int J Radiat Oncol Biol Phys.* 2000 Jul 15;47(5):1435–42.

Nijkamp J, de Jong R, Sonke JJ et al. Target volume shape variation during hypo-fractionated preoperative irradiation of rectal cancer patients, *Radiother Oncol.* 2009 Aug;92(2):202–9.

Nijkamp J, Pos FJ, Nuver TT et al. Adaptive radiotherapy for prostate cancer using kilovoltage cone-beam computed tomography: first clinical results, *Int J Radiat Oncol Biol Phys.* 2008 Jan 1;70(1):75–82.

Noel CE, Klein EE, and Moore KL. A surface-based respiratory surrogate for 4D imaging, *Med Phys.* 2008 Jun;35(6):2682.

Noo F, Clackdoyle R, Mennessier C, White TA, and Roney TJ. Analytic method based on identification of ellipse parameters for scanner calibration in cone-beam tomography, *Phys Med Biol.* 2000 Nov;45(11):3489–508.

O'Daniel J, Pan T, Mohan R, and Dong L. CT dose from daily in-room CT-guided radiotherapy, *Med Phys.* 2004;31:1876.

O'Daniel JC, Garden AS, Schwartz DL et al. Parotid gland dose in intensity-modulated radiotherapy for head and neck cancer: is what you plan what you get? *Int J Radiat Oncol Biol Phys.* 2007 Nov 15;69(4):1290–6.

Onishi H, Kuriyama K, Komiyama T et al. A new irradiation system for lung cancer combining linear accelerator, computed tomography, patient self-breath-holding, and patient-directed beam-control without respiratory monitoring devices, *Int J Radiat Oncol Biol Phys* 2003;56:14–20.

Paskalev K, Ma CM, Jacob R et al. Daily target localization for prostate patients based on 3D image correlation, *Phys Med Biol.* 2004;49:931–9.

Pehlivan B, Pichenot C, Castaing M et al. Interfractional set-up errors evaluation by daily electronic portal imaging of IMRT in head and neck cancer patients, *Acta Oncol.* 2009;48(3):440–5.

Ploeger LS, Betgen A, Gilhuijs KG, and van Herk M. Feasibility of geometrical verification of patient set-up using body contours and computed tomography data, *Radiother Oncol.* 2003a Feb;66(2):225–33.

Ploeger LS, Frenay M, Betgen A et al. Application of video imaging for improvement of patient set-up, *Radiother Oncol.* 2003b Sep;68(3):277–84.

Pos F, Bex A, Dees-Ribbers HM et al. Lipiodol injection for target volume delineation and image guidance during radiotherapy for bladder cancer, *Radiother Oncol.* 2009 Nov;93(2):364–7.

Pouliot J, Aubin M, and Langen KM. (Non)-migration of radiopaque markers used for on-line localization of the prostate with an electronic portal imaging device, *Int J Radiat Oncol Biol Phys.* 2003;56:862–6.

Pouliot J, Bani-Hashemi A, Chen J et al. Low-dose megavoltage cone-beam CT for radiation therapy, *Int J Radiat Oncol Biol Phys.* 2005;61:552–60.

Pouliot J and Lirette A. Verification and correction of setup deviations in tangential breast irradiation using EPID: gain versus workload, *Med Phys.* 1996;23:1393–8.

Purdie TG, Bissonnette JP, Franks K et al. Cone-beam computed tomography for on-line image guidance of lung stereotactic radiotherapy: localization, verification, and intrafraction tumor position, *Int J Radiat Oncol Biol Phys.* 2007 May 1;68(1):243–52.

Raaijmakers AJ, Raaymakers BW, and Lagendijk JJ. Magnetic-field-induced dose effects in MR-guided radiotherapy systems: dependence on the magnetic field strength, *Phys Med Biol.* 2008 Feb 21;53(4):909–23.

Raaymakers BW, Lagendijk JJ, Overweg J et al. Integrating a 1.5 T MRI scanner with a 6 MV accelerator: proof of concept, *Phys Med Biol.* 2009 Jun 21;54(12):N229–37.

Ramakrishna N, Rosca F, Friesen S et al. A clinical comparison of patient setup and intra-fraction motion using frame-based radiosurgery versus a frameless image-guided radiosurgery system for intracranial lesions, *Radiother Oncol.* 2010 Apr;95(1):109–15.

Reitz I, Hesse BM, Nill S, Tücking T, and Oelfke U. Enhancement of image quality with a fast iterative scatter and beam hardening correction method for kV CBCT, *Z Med Phys.* 2009;19(3):158–72.

Remeijer P, Lebesque JV, and Mijnheer BJ. Set-up verification using portal imaging; review of current clinical practice, *Radiother Oncol.* 2001;58:105–20.

Remeijer P, Sonke J-J, Betgen A, and van Herk M. First clinical experience with cone-beam CT based setup correction protocols, Proceedings 8th International Workshop on Electronic Portal Imaging—EPI2K4 Brighton, UK, 2004, pp. 92–3.

Rougee A, Picard C, Ponchut C, and Trousset Y. Geometrical calibration of x-ray imaging chains for three-dimensional reconstruction, *Comput Med Imaging Graph.* 1993;17(4–5):295–300.

Samson MJ, van Sornsen de Koste JR, and de Boer HC. An analysis of anatomic landmark mobility and setup deviations in radiotherapy for lung cancer, *Int J Radiat Oncol Biol Phys.* 1999;43:827–32.

Santanam L, Esthappan J, Mutic S et al. Estimation of setup uncertainty using planar and MVCT imaging for gynecologic malignancies, *Int J Radiat Oncol Biol Phys.* 2008;71:1511–7.

Saw CB, Yang Y, Li F et al. Performance characteristics and quality assurance aspects of kilovoltage cone-beam CT on medical linear accelerator, *Med Dosim.* 2007;32:80–5.

Schweikard A, Shiomi H, and Adler J. Respiration tracking in radiosurgery, *Med Phys.* 2004;31:2738–41.

Seppenwoolde Y, Berbeco RI, Nishioka S, Shirato H, and Heijmen B. Accuracy of tumor motion compensation algorithm from a robotic respiratory tracking system: a simulation study, *Med Phys.* 2007;34:2774–84.

Shah AP, Langen KM, Ruchala KJ et al. Patient dose from megavoltage computed tomography imaging, *Int J Radiat Oncol Biol Phys.* 2008;70:1579–87.

Shalev S, Lee T, Leszczynski K et al. Video techniques for on-line portal imaging, *Comput Med Imaging Graph.* 1989;13:217–26.

Sharpe M, Moseley D, Purdie T et al. The stability of mechanical calibration for a kV cone beam computed tomography system integrated with linear accelerator, *Med Phys.* 2006;33:136–44.

Shirato H, Shimizu S, Shimizu T, Nishioka T, and Miyasaka K. Real-time tumour-tracking radiotherapy, *Lancet.* 1999 Apr 17;353(9161):1331–2.

Shirato H, Seppenwoolde Y, Kitamura K, Onimura R, and Shimizu S. Intrafractional tumor motion: lung and liver, *Semin Radiat Oncol.* 2004;14:10–8.

Shiu AS, Chang EL, Ye JS et al. Near simultaneous computed tomography image-guided stereotactic spinal radiotherapy: an emerging paradigm for achieving true stereotaxy, *Int J Radiat Oncol Biol Phys.* 2003;57:605–13.

Sidhu K, Ford EC, Spirou S et al. Optimization of conformal thoracic radiotherapy using cone-beam CT imaging for treatment verification, *Int J Radiat Oncol Biol Phys.* 2003;55:757–67.

Sidky EY and Pan X. Image reconstruction in circular cone-beam computed tomography by constrained, total-variation minimization, *Phys Med Biol.* 2008 Sep 7;53(17):4777–807.

Siewerdsen JH, Daly MJ, Bakhtiar B et al. A simple, direct method for x-ray scatter estimation and correction in digital radiography and cone-beam CT, *Med Phys.* 2006 Jan;33(1):187–97.

Simpson DR, Lawson JD, Nath SK et al. A survey of image-guided radiation therapy use in the United States, *Cancer.* 2010 Aug 15;116(16):3953–60.

Smitsmans MH, de Bois J, Sonke JJ et al. Automatic prostate localization on cone-beam CT scans for high precision image-guided radiotherapy, *Int J Radiat Oncol Biol Phys.* 2005 Nov 15;63(4):975–84.

Smitsmans MH, Pos FJ, de Bois J et al. The influence of a dietary protocol on cone beam CT-guided radiotherapy for prostate cancer patients, *Int J Radiat Oncol Biol Phys.* 2008 Jul 15;71(4):1279–86.

Soete G, Van de Steene J, Verellen D et al. Initial clinical experience with infrared-reflecting skin markers in the positioning of patients treated by conformal radiotherapy for prostate cancer, *Int J Radiat Oncol Biol Phys.* 2002;52:694–8.

Sonke JJ and Belderbos J. Adaptive radiotherapy for lung cancer, *Semin Radiat Oncol.* 2010 Apr;20(2):94–106.

Sonke JJ, Rossi M, Wolthaus J et al. Frameless stereotactic body radiotherapy for lung cancer using four-dimensional cone beam CT guidance, *Int J Radiat Oncol Biol Phys.* 2009 Jun 1;74(2):567–74.

Sonke JJ, Zijp L, Remeijer P, and van Herk M. Respiratory correlated cone beam CT, *Med Phys.* 2005;32(4):1176–86.

Strandqvist M and Rosengren B. Television-controlled pendulum therapy, *Br J Radiol.* 1958;31:513–4.

Street R, Nelson S, Le A et al. Amorphous silicon sensor arrays for radiation imaging mater, *Res Soc Symp Proc.* 1990;192:441–52.

Stützel J, Oelfke U, and Nill S. A quantitative image quality comparison of four different image guided radiotherapy devices, *Radiother Oncol.* 2008 Jan;86(1):20–4.

Suit HD, Becht J, Leong J et al. Potential for improvement in radiation therapy. *Int J Radiat Oncol Biol Phys.* 1988;14:777–86.

Thilmann C, Nill S, Tücking T et al. Correction of patient positioning errors based on in-line cone beam CTs: clinical implementation and first experiences, *Radiat Oncol.* 2006 May 24;1:16.

Topolnjak R, Sonke JJ, Nijkamp J et al. Breast patient setup error assessment: comparison of electronic portal image devices and cone-beam computed tomography matching results, *Int J Radiat Oncol Biol Phys.* 2010 Nov 15;78(4):1235–43.

Tournel K, De Ridder M, Engels B et al. Assessment of intrafractional movement and internal motion in radiotherapy of rectal cancer using megavoltage computed tomography, *Int J Radiat Oncol Biol Phys.* 2008;71:934–9.

Tubiana M. Can we reduce the incidence of second primary malignancies occurring after radiotherapy? A critical review, *Radiother Oncol.* 2009;91:4–15; discussion 11–13.

Uematsu M, Shioda A, Suda A et al Computed tomography-guided frameless stereotactic radiotherapy for stage I non-small cell lung cancer: a 5-year experience, *Int J Radiat Oncol Biol Phys.* 2001;51:666–70.

van Asselen B, Dehnad H, Raaijmakers CP, Lagendijk JJ, and Terhaard CH. Implanted gold markers for position verification during irradiation of head-and-neck cancers: a feasibility study, *Int J Radiat Oncol Biol Phys.* 2004 Jul 15;59(4):1011–7.

van Beek S, van Kranen S, Mencarelli A et al. First clinical experience with a multiple region of interest registration and correction method in radiotherapy of head-and-neck cancer patients, *Radiother Oncol.* 2010 Feb;94(2):213–7.

van de Steene J, Van den Heuvel F, and Bel A. Electronic portal imaging with on-line correction of setup error in thoracic irradiation: clinical evaluation, *Int J Radiat Oncol Biol Phys.* 1998;40:967–76.

van Elmpt W, McDermott L, Nijsten S et al. A literature review of electronic portal imaging for radiotherapy dosimetry, *Radiother Oncol.* 2008 Sep;88(3):289–309.

van Herk M. Errors and margins in radiotherapy, *Semin Radiat Oncol.* 2004 Jan;14(1):52–64.

van Herk M. Physical aspects of a liquid-filled ionization chamber with pulsed polarizing voltage, *Med Phys.* 1991 Jul–Aug;18(4):692–702.

van Herk M, Remeijer P, Rasch C, and Lebesque JV. The probability of correct target dosage: dose-population histograms for deriving treatment margins in radiotherapy, *Int J Radiat Oncol Biol Phys.* 2000;47:1121–35.

van Lin EN, Nijenhuis E, and Huizenga H. Effectiveness of couch height based patient set-up and an off-line correction protocol in prostate cancer radiotherapy, *Int J Radiat Oncol Biol Phys.* 2001;50:569–77.

van Lin EN, van der Vight L, and Huizenga H. Set-up improvement in head and neck radiotherapy using a 3D off-line EPID-based correction protocol and a customised head and neck support, *Radiother Oncol.* 2003;68:137–48.

van Tienhoven G, Lanson JH, and Crabeels D. Accuracy in tangential breast treatment set-up: a portal imaging study, *Radiother Oncol.* 1991;22:317–22.

Verellen D, Soete G, Linthout N et al. Quality assurance of a system for improved target localization and patient set-up that combines real-time infrared tracking and stereoscopic X-ray imaging, *Radiother Oncol.* 2003 Apr;67(1):129–41.

Waddington SP and McKensie AL. Assessment of effective dose from concomitant exposures required in verification of the target volume in radiotherapy, *Br J Radiol.* 2004;77:557–61.

Wang C, Shiu A, Lii M, Woo S, and Chang EL. Automatic target localization and verification for on-line image-guided stereotactic body radiotherapy of the spine, *Technol Cancer Res Treat.* 2007;6:187–96.

Wang J, Li T, and Xing L. Iterative image reconstruction for CBCT using edge-preserving prior, *Med Phys.* 2009a Jan;36(1):252–60.

Wang J, Bai S, Chen N et al. The clinical feasibility and effect of online cone beam computer tomography-guided intensity-modulated radiotherapy for nasopharyngeal cancer, *Radiother Oncol.* 2009b;90:221–7.

Wells PN and Liang HD. Medical ultrasound: imaging of soft tissue strain and elasticity, *J R Soc Interface.* 2011 Nov 7;8(64):1521–49.

White EA, Cho J, Vallis KA et al. Cone beam computed tomography guidance for setup of patients receiving accelerated partial breast irradiation, *Int J Radiat Oncol Biol Phys.* 2007;68:547–54.

Wiersma RD, Riaz N, Dieterich S, Suh Y, and Xing L. Use of MV and kV imager correlation for maintaining continuous real-time 3D internal marker tracking during beam interruptions, *Phys Med Biol.* 2009 Jan 7;54(1):89–103.

Wong JR, Gao Z, Uematsu M et al. Interfractional prostate shifts: review of 1870 computed tomography (CT) scans obtained during image-guided radiotherapy using CT-on-rails for the treatment of prostate cancer, *Int J Radiat Oncol Biol Phys.* 2008;72:1396–401.

Wong JW, Binns WR, Cheng AY et al. On-line radiotherapy imaging with an array of fiber-optic image reducers, *Int J Radiat Oncol Biol Phys.* 1990 Jun;18(6):1477–84.

Wu J, Dandekar O, Nazareth D et al. Effect of ultrasound probe on dose delivery during real-time ultrasound-guided tumor tracking, *Conf Proc IEEE Eng Med Biol Soc.* 2006;1:3799–802.

Wu Q, Chi Y, Chen PY et al. Adaptive replanning strategies accounting for shrinkage in head and neck IMRT, *Int J Radiat Oncol Biol Phys.* 2009 Nov 1;75(3):924–32.

Wunderink W, Méndez Romero A, Seppenwoolde Y et al. Potentials and limitations of guiding liver stereotactic body radiation therapy set-up on liver-implanted fiducial markers, *Int J Radiat Oncol Biol Phys.* 2010 Aug 1;77(5):1573–83.

Wurm RE, Erbel S, Schwenkert I et al. Novalis frameless image-guided noninvasive radiosurgery: initial experience, *Neurosurgery.* 2008 May;62(5 Suppl):A11–7.

Yan D. Adaptive radiotherapy: merging principle into clinical practice, *Semin Radiat Oncol.* 2010 Apr;20(2):79–83.

Yan D, Lockman D, Brabbins D et al. An off-line strategy for constructing a patient-specific planning target volume in adaptive treatment process for prostate cancer, *Int J Radiat Oncol Biol Phys.* 2000;48:289–302.

Yan D, Lockman D, Martinez A et al. Computed tomography guided management of interfractional patient variation, *Semin Radiat Oncol.* 2005;15:168–79

Yan H, Yin FF, and Kim JH. A phantom study on the positioning accuracy of the Novalis Body system, *Med Phys.* 2003 Dec;30(12):3052–60.

Yenice KM, Lovelock DM, Hunt MA et al. CT image-guided intensity-modulated therapy for paraspinal tumors using stereotactic immobilization, *Int J Radiat Oncol Biol Phys.* 2003;55:583–93.

Yeung AR, Li JG, Shi W et al. Tumor localization using cone-beam CT reduces setup margins in conventionally fractionated radiotherapy for lung tumors, *Int J Radiat Oncol Biol Phys.* 2009 Jul 15;74(4):1100–7.

Yin F-F, Wong J, Balter J et al. The Role of In-Room kV X-Ray Imaging for Patient Setup and Target Localization: Report of AAPM Task Group 104, 2009.

Yoo S, Kim GY, Hammoud R et al. A quality assurance program for the on-board imagers, *Med Phys.* 2006;33:4431–47.

Zeidan OA, Langen KM, Meeks SL et al. Evaluation of image-guidance protocols in the treatment of head and neck cancers, *Int J Radiat Oncol Biol Phys.* 2007;67:670–7.

Zijp L, Sonke JJ, and van Herk M. Extraction of the respiratory signal from sequential thorax cone-beam X-ray images. In: International conference on the use of computers in radiation therapy (ICCR), Seoul, Republic of Korea, Jeong Publishing, 2004, p. 507.

16

Affine Medical Image Registration Using the Graphics Processing Unit

Daniel H. Adler
University of Pennsylvania

Sonny Chan
Stanford University

J. Ross Mitchell
The Mayo Clinic

16.1 Introduction

Image registration—the task of mapping images into a common coordinate space—is an important investigative tool in science (Maintz and Viergever 1998; Lester and Arridge 1999). In medicine, image registration is necessary when comparing or integrating complementary information from diagnostic scans that were acquired at different times, using different devices, or of different subjects. It is essential in many investigative and diagnostic medical workflows, such as to correct for subject motion between successive scans (Kamitani and Tong 2005), to evaluate longitudinal changes in patients (Morra et al. 2009), and to combine images of multiple individuals to create atlases of normal and abnormal variability of anatomy (Zacharaki et al. 2008), physiology (Tustison et al. 2010), and function (Gholipour et al. 2007). In routine clinical practice, registration is used to improve surgical planning and stereotactic guidance (DiMaio et al. 2006; Archip et al. 2007), electrode localization in the brain (Starreveld 2002), radiotherapy treatment planning (Kessler and Roberson 2006), identification of disease response to therapy (Jongen 2006), and many other medical procedures (Hajnal et al. 2001).

Automated, intensity-based registration is an iterative procedure that transforms a "moving image" onto a stationary, "fixed image" by maximizing a measure of similarity between the images. This form of registration is cast as an optimization problem that requires minimal user interaction, with the goal of providing reliable and repeatable results (Maes et al. 1997b). Only intrinsic image features are used, without the need for externally placed landmarks. In intensity-based registration, the similarity measures used to drive optimization are derived directly from image intensity values, whereas geometric methods are driven by shape matching. The flexibility of intensity-based registration over geometric methods is well recognized in the literature (Maintz and Viergever 1998).

Each iteration of automatic affine registration involves four steps, as shown in Figure 16.1. (1) A parameterized transformation is applied to the spatial coordinates of the moving image. (2) The transformed moving image values are obtained following resampling of the moving image into the coordinate space of the fixed image. (3) Next, a similarity metric is computed to quantify correspondence between the fixed and transformed images. (4) The metric is passed to the optimizer, which modifies the transform parameters to ultimately improve registration. The specifications and constraints of the registration problem at hand dictate the particular methods chosen to implement these four generic components.

On each iteration, the number of operations executed by the transformation, resampling, and similarity metric routines is on the order of the fixed image size. This is because these operations are performed in the coordinate space of the fixed image. This leads to high computational demand, because medical image data sets routinely contain millions of volume elements—or voxels. Factors shown to limit intensity-based registration performance include the number of interpolations following coordinate transformation (Hastreiter and Ertl 1998; Hastreiter et

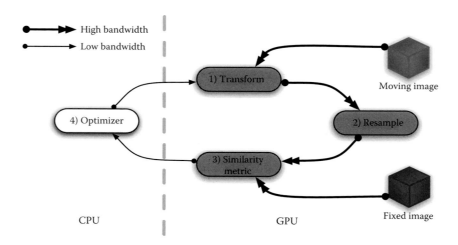

High bandwidth

Low bandwidth

1) Transform

Moving image

4) Optimizer

2) Resample

3) Similarity metric

CPU

GPU

Fixed image

FIGURE 16.1 Core components of the iterative registration cycle, with shaded components executed on the GPU in our framework.

al. 2004) and similarity metric evaluation (Rezk-Salama et al. 1999). If a derivative-free optimization scheme is used, then the logic of the optimizer typically requires relatively few operations compared to the three other components shown in Figure 16.1.

Fast and accurate automated, intensity-based medical image registration is of great utility to clinicians and researchers. However, the high computational demand of registration can lead to prohibitively lengthy execution times in imaging workflows. For instance, registration must be performed within minutes for applications in intraoperative imaging and image-guided surgery so as not to delay procedures (Hastreiter et al. 2004; Archip et al. 2007). Also, brain atlas creation (Evans et al. 1993) and clinical studies involving large cohorts of subjects (Morra et al. 2009) often require the accurate and reliable registration of hundreds or thousands of image pairs. To date, however, the enormous computational requirements of registration methods have largely precluded their use at interactive or near real-time speeds on desktop computers.

To bring down computation times, a number of groups have implemented medical image registration on mainframes and large computer clusters. Significant disadvantages of these approaches are limited accessibility to such machines in the clinical environment and high costs. We developed a novel framework that leverages recent advances in the power and flexibility of commodity desktop graphics hardware to accelerate affine registration of 3D medical images (Mitchell et al. 2008). Our framework places all computationally intensive components of the registration cycle on the graphics processing unit (GPU). In addition, all images remain on the GPU throughout the registration procedure. This eliminates bandwidth-intensive copies of data between the GPU and the host application's main memory on the central processing unit (CPU).

We accelerate several of the most commonly used intensity-based similarity metrics in our framework, including normalized mutual information (NMI) (Viola and Wells 1997; Studholme et al. 1999). NMI, which is derived from the field of

information theory, is widely regarded as one of the most reliable and accurate intensity-based metric for registering images acquired using different modalities (Pluim et al. 2003).

Our software is implemented in C++ and uses the OpenGL 2.0 graphics API (Shreiner 2009) with OpenGL Shading Language (GLSL) to access graphics hardware functionality (Rost 2004). We chose GLSL over other libraries, such as OpenCL and CUDA (NVI 2008), for its platform and device independence: some OpenCL graphics extensions are only supported on recent high-end AMD hardware and CUDA is specific to NVIDIA hardware. Our software framework is implemented modularly, with data and processes encapsulated as objects. Its structure thus generally resembles the architecture of the Insight Toolkit (ITK) (Yoo 2004; Ibáñez et al. 2005), which is an open-source, cross-platform library for performing image registration and segmentation.

16.2 Accelerated Affine Image Registration

Automated, intensity-based affine registration is cast as the iterative optimization of a similarity metric objective function. This is a single number that evaluates the quality of registration (i.e., similarity) between a moving image and a fixed image. Each iteration of the optimization cycle consists of applying a parameterized transformation to the moving image, resampling the transformed moving image into the space of the fixed image, computing the similarity metric between the two images, and generating new transformation parameters for the following iteration.

We implement the computationally intensive components of this cycle on the GPU. These components are the first three steps shown in Figure 16.1: image transformation, interpolation, and metric evaluation. We also store all 3D images in GPU video memory. This eliminates the need for repeated transfers of up to hundreds of megabytes of data over the CPU–GPU bus on each iteration. The transfer of image data has proven to be the primary bottleneck in several other GPU registration programs (Hastreiter et al. 2004). In our implementation, the amount of data sent between the CPU and GPU is negligible, consisting of the metric value and a set of transformation parameters.

16.2.1 Volume Rendering

The registration methods that we present are formulated in terms of "3D texture-based volume rendering" and are thus naturally suited to implementation on the GPU. The purpose of volume rendering is to synthesize a virtual 2D view of volumetric data, where the rendered color values represent physical interactions of light with the data (Cabral et al. 1994). This is usually achieved by a transfer function that maps scalar data intensities to light emission and absorption properties.

When performing texture-based rendering, the volumetric data set is first loaded into the GPU's video memory; hereafter, it is referred to as a 3D "texture." The next step is to map the texture's 2D slices onto a "proxy geometry" consisting of a stack of equally spaced quadrilateral polygons (or "quads") oriented parallel to the viewing plane, as shown in Figure 16.2. The number of quads generated equals the number of slices in the texture.

Finally, the quads are rendered to the display's frame buffer one by one. This is done without depth testing, so that the quads do not occlude each other. If desired, lighting effects are simulated during this stage using blending functions and additional data, such as material gradients, colors, and transparencies.

If the user changes viewing direction, then the data set must be geometrically transformed and retextured onto the proxy geometry. If the view direction is not parallel to an axis of the image, then 3D interpolation is required during the texture mapping procedure.

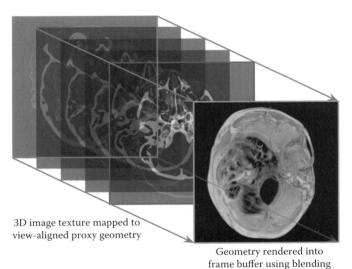

3D image texture mapped to view-aligned proxy geometry

Geometry rendered into frame buffer using blending

FIGURE 16.2 Volume rendering of a CT head data set (inferior view) using texture mapping and view-aligned proxy geometry.

Every iteration of our registration algorithm is cast as the rendering of one frame of a 3D volume—with some important modifications. To start, we load the fixed and moving volumetric medical data sets into two textures on the GPU. Next, we map both textures onto common quadrilateral proxy geometry. As discussed in the following section, the current registration transformation is applied to the moving image during mapping. The image metric between the fixed and moving images is computed by shaders as the textured quads are rendered to the frame buffer.

Most raw computed tomography (CT), magnetic resonance (MR), and positron emission tomography (PET) images have integer intensity values that span no more than a 12-bit range, whereas processed images often have floating-point values. We use GPU texture formats that match the input data format. Thus, we typically loaded raw images into 16-bit integer textures; processed images are typically loaded into 32-bit floating-point textures, which have been available on graphics hardware since 2008. Because the image data are scalar, we used textures that hold a single luminance intensity channel, as opposed to the four standard red, green, blue, and alpha channels.

Images are transferred from main memory to the GPU using the OpenGL function glTexImage3D and textures are set to exactly match the input image size. On hardware before 2006, it was necessary to pad texture dimensions to powers of two. The dimensions of the quads that we render match those of the fixed image cross-sectional slices, because the moving image is always resampled into the space of the fixed image.

16.2.2 Image Transformation

Our registration framework performs image transformations using 3D texture mapping on the GPU. Indeed, the GPU is ideally suited to performing geometric transformations, because they are called for frequently in computer graphics. The first step in the texture mapping of a polygon is to define "texture coordinates" at each of its vertices, where each coordinate serves as an index into the texture. Coordinates inside the polygon are computed using 2D interpolation during rasterization. The effect of interpolation is to create a smooth map of the texture over the entire shape. Figure 16.3 shows image slices mapped to a quad using texture coordinates at the four vertices.

We transform the moving image by modifying its mapping over the quad. This is done by multiplying its texture coordinates by a 4 × 4 homogeneous matrix. The OpenGL functions glTranslatef, glRotatef, and glScalef can be used to apply 3D translations, rotations, and scalings to the matrix. Alternatively, the components of the matrix can be set directly via glMultMatrixf. Using this method, we are able to specify arbitrary linear affine transformations. Figure 16.3 depicts a transformation **T** applied to the moving texture coordinates.

Listing 1 shows how we implement image transformations in OpenGL. In this code, the fixed and moving images are stored in textures 0 and 1, respectively. We apply an identity transformation

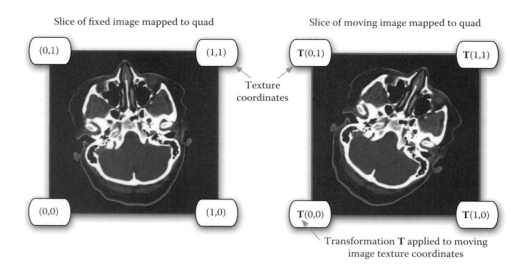

Slice of fixed image mapped to quad

Slice of moving image mapped to quad

Texture coordinates

Transformation **T** applied to moving image texture coordinates

FIGURE 16.3 Mapping the fixed and moving images to quads using texture coordinates.

to the fixed image and the current OpenGL matrix transformation to the moving image.

```
glMatrixMode(GL_TEXTURE);
glActiveTexture(GL_TEXTURE0);//fixed image
glLoadIdentity();
glActiveTexture(GL_TEXTURE1);//moving image
glLoadIdentity();
glMultMatrixf(transformationMatrix);
```

Listing 1: Applying transformations to fixed and moving image texture coordinates for registration

16.2.3 Interpolation

Image transformations in registration require interpolation of the moving image intensities. To see this, consider the image transformation, which is defined mathematically as a function **T**: $M \rightarrow F$ that maps spatial coordinates from the moving image domain M to the fixed image domain F. The map must cover the entire fixed image domain, because it is the target of registration. However, naively transforming the moving image using **T** results in a map that is not dense *onto F*.

We solve this problem by instead defining the inverse transformation $\mathbf{T}^{-1}$ that operates on fixed image coordinates. The transform is then computationally evaluated by looping over all fixed image points. Given one point $\mathbf{x} \in F$, the value of the corresponding transformed image intensity $M(\mathbf{T}^{-1}(\mathbf{x}))$ is determined by interpolation of the moving image. Figure 16.4 shows a schematic of evaluating the inverse transformation at a fixed image coordinate by interpolating the moving image.

The simplest and fastest interpolation method uses the intensity of the nearest-neighboring sample. However, this is known to result in aliasing and severe partial voluming effects, yielding poor registration (Hajnal et al. 2001). Trilinear interpolation is among the most commonly used methods, because it provides a good trade-off between accuracy and computation time and it has been shown to yield acceptable results for image registration (Levin et al. 2004). Methods using quadratic, cubic, cubic B-spline, Gaussian, and *sinc*-based interpolation kernels of different sizes have been studied in detail (Hajnal et al. 1995; Lehmann et al. 1999).

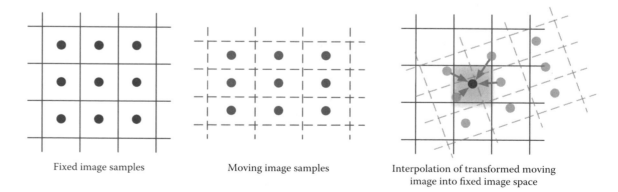

Fixed image samples

Moving image samples

Interpolation of transformed moving image into fixed image space

FIGURE 16.4 Interpolation of the moving image during transformation.

For affine registration on the CPU, it has been shown that up to approximately 90% of the computational time is spent doing image transformation and (trilinear) interpolation (Shams et al. 2010). To accelerate transformations in our GPU framework, we configure the hardware to automatically perform trilinear interpolation when reading from the moving texture. The hardware is optimized to perform fast and accurate trilinear interpolation, which is critical in real-time graphics applications, where scenes often contain multiple textures that are mapped to dynamically resizing polygons.

The GPU currently has no native support for higher-order interpolation schemes. However, efficient third-order B-spline interpolation is possible within fragment shaders using multiple texture lookups (Sigg and Hadwiger 2005). This is done by taking advantage of automatic hardware trilinear interpolation. To complete a third-order B-spline interpolation, the method makes eight trilinear texture lookups that implicitly combine the 64 neighboring intensities. The overhead compared with trilinear interpolation is thus a factor of 8.

16.2.4 Difference and Correlation-Based Similarity Metrics

Computation of the image similarity metric forms the core of our registration framework. We take advantage of the GPU's single-instruction, multiple data architecture to compute point similarity metrics, which measure correspondence between individual image samples (Rogelj et al. 2003). We implement several similarity metrics based on intensity difference and correlation: mean squared error (MSE), mean absolute error (MAE), and normalized cross-correlation (NCC). We also implement the normalized gradient field (NGF) metric (Haber and Modersitzki 2006), which has been reported to yield a good objective space for parameter optimization in multimodal registration. Information theoretic metrics are treated in the next section of this chapter.

The MSE and MAE metrics are defined between an image pair A and B as follows:

$$\text{MSE}(A,B) = \frac{1}{N} \sum_{\mathbf{x} \in \Omega} \big(A(\mathbf{x}) - B(\mathbf{x}) \big)^2, \qquad (16.1)$$

$$\text{MAE}(A,B) = \frac{1}{N} \sum_{\mathbf{x} \in \Omega} \big| A(\mathbf{x}) - B(\mathbf{x}) \big|, \qquad (16.2)$$

where Ω is the overlapping sample domain of A and B, and N is the number of samples in the domain. It has been shown that MSE is the optimal choice of metric when A and B differ only by Gaussian noise (Hajnal et al. 2001). (The assumption of pure Gaussian noise rarely holds in practice for intramodality registration.) The MSE metric is sensitive to a small number of high intensity voxels, which can be caused by injected contrast or surgical instruments in the imaging field

of view (Chisu 2005). The MAE metric reduces the effect of such outliers.

The difference metrics defined above assume direct correspondence between the intensity values of the images to be registered. This assumption does not always hold even within the same modality, especially for MR images. The NCC metric is used when a linear relationship exists between fixed and moving image intensities—a less restrictive assumption. The metric increases with better image correspondence:

$$\text{NCC}(A,B) = \frac{\sum_{\mathbf{x} \in \Omega} (A(\mathbf{x}) - \bar{A})(B(\mathbf{x}) - \bar{B})}{\sqrt{\sum_{\mathbf{x} \in \Omega} (A(\mathbf{x}) - \bar{A})^2 \sum_{\mathbf{x} \in \Omega} (B(\mathbf{x}) - \bar{B})^2}}, \qquad (16.3)$$

where $\bar{A}$ and $\bar{B}$ are the mean image intensities.

Gradient-based metrics assume that anatomical structures between the images have common boundaries, although they may exhibit different intensity and contrast characteristics. The NGF metric, introduced by Haber and Modersitzki (2006), is based on the assumption that intensity changes spatially co-occur in similar images. The metric is constructed from normalized image gradients:

$$\mathbf{n}(A, \mathbf{x}) = \frac{\nabla A(\mathbf{x})}{\sqrt{\|\nabla A(\mathbf{x})\|^2 + \epsilon^2}}, \qquad (16.4)$$

where ϵ is a regularization parameter that controls the metric's sensitivity to edges. We set it to be proportional to the estimated image noise level, which is computed as the background intensity standard deviation. Our implementation uses two-point central differences to compute the gradients.

The NGF metric accounts for structural boundaries with the same or opposing directions by maximizing the square of the normalized gradient inner products:

$$\text{NGF}(A,B) = \frac{1}{N} \sum_{\mathbf{x} \in \Omega} \langle \mathbf{n}(A,\mathbf{x}), \mathbf{n}(B,\mathbf{x}) \rangle^2. \qquad (16.5)$$

Haber and Modersitzki describe this metric as an alternative to mutual information (MI) that is easier to compute, more straightforward to implement, and more readily suitable to numerical optimization due to greater convexity over the transformation parameter space.

16.2.4.1 Metric Image Rendering

We formulate metric evaluation as a rendering operation. As we have discussed, the fixed and moving images are textured onto view-aligned quadrilaterals using the graphics hardware's fast interpolation capability. The metric evaluation takes place when we render this textured geometry using a custom fragment shader program that replaces default graphics pipeline functionality. Figure 16.5 depicts the slice-by-slice metric computation process. The process is summarized below.

1. *Texture and geometry initialization.* The fixed and transformed moving image cross-sectional slices are textured-mapped onto a stack of quadrilaterals aligned parallel to the view plane.

2. *Rendering metric images.* Each quad is rendered using a custom fragment shader that outputs an "intermediate" metric image. The shader is able to access and perform arithmetic operations on arbitrary samples of the input fixed and moving textures.

    For the MSE, MAE, and NGF similarity metrics (Equations 16.1, 16.2, and 16.5), the output fragment intensity is set to the metric summand. For the NCC metric (Equation 16.3), the three different summands are output to the fragment's red, green, and blue components.

3. *Summation of metric images.* Because they are rendered, additive alpha blending automatically sums the intermediate metric images along the slice axis into a composited metric image.

Following these steps, it remains to compute a single metric value by averaging the intensities of the composited metric image from step 3 above.

Listing 2 shows the setup of parameters in OpenGL before rendering. An orthographic view projection is used (lines 0–2) to ensure that all metric slices have the same dimensions in the frame buffer. The clipping volume is a cube with normalized

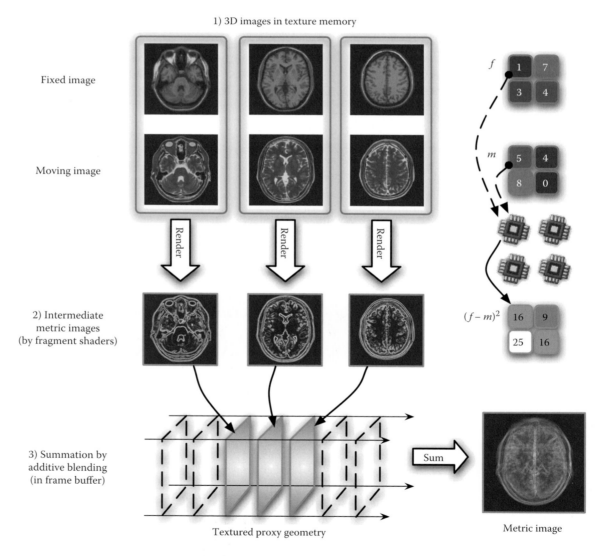

FIGURE 16.5 Computation of difference- and correlation-based metrics using the rendering pipeline. (Modified from Chan, S. Three-dimensional medical image registration on modern graphics processors, Master's thesis, University of Calgary, Calgary, Alberta, 2007. With permission.)

device coordinates $[-1,1]^3$. Depth testing is disabled (line 4), so that all fragments get rendered to the screen without occlusion. Frame buffer blending is enabled, with the blend function set to additive compositing (lines 5–7). We always render to a 32-bit floating-point frame buffer to preserve precision and to avoid overflow during blending.

```
glMatrixMode(GL_PROJECTION);
glLoadIdentity();
glOrtho(-1.0, 1.0, -1.0, 1.0, -1.0, 1.0);//
orthographic projection
glDisable(GL_DEPTH_TEST);
glEnable(GL_BLEND);
glBlendFunc(GL_ONE, GL_ONE);//blending
coefficients
glBlendEquation(GL_FUNC_ADD);//additive blending
```

Listing 2: Projection and frame buffer setup for registration rendering

As an example, Listing 3 shows the GLSL fragment shader program used to compute the metric images for the NCC metric of Equation 16.3. First, the fixed (F) and moving (M) image textures and their mean values ($\bar{F}, \bar{M}$) are declared (lines 0 and 1). The means are precomputed because they do not change.

Suppose that $\mathbf{T} : \mathbb{R}^3 \rightarrow \mathbb{R}^3$ is the current moving image transformation. At each voxel $\mathbf{x} \in \mathbb{R}^3$, the fixed image intensity $F(\mathbf{x})$ and moving intensity $M(\mathbf{T}(\mathbf{x}))$ are retrieved by sampling the textures at the original and transformed 3D texture coordinates (lines 5 and 6). The output fragment's color intensities are set to the metric summands (line 11): $(F(\mathbf{x})-\bar{F})(M(\mathbf{T}(\mathbf{x}))-\bar{M})$, $(F(\mathbf{x})-\bar{F})^2$, and $(M(\mathbf{T}(\mathbf{x}))-\bar{M})^2$. The alpha value (set to 1.0) is inconsequential.

```
uniform sampler3D fixedTexture, movingTexture;
uniform float fixedMean, movingMean;
void main(void)
{
vec4 fixed = texture3D(fixedTexture,
gl_TexCoord[0].stp);
vec4 moving = texture3D(movingTexture, gl_
TexCoord[1].stp);
float f = fixed.r - fixedMean;
float m = moving.r - movingMean;

gl_FragColor = vec4(f*m, f*f, m*m, 1.0);//NCC
metric summands
}
```

Listing 3: Computing the NCC metric images using a GLSL fragment shader program

The other metrics are computed similarly. For example, changing line 11 in the above listing to `gl_FragColor = pow(fixed - moving, 2.0)` yields a shader program that computes the MSE metric instead of the NCC metric.

The code used to render the proxy geometry is given in Listing 4. Each loop iteration creates a quad over the $[-1,1]^2$ view plane by defining its vertices with the function `glVertex3f`.

Coordinates into the fixed (GL _ TEXTURE0) and moving (GL _ TEXTURE1) textures are defined using `glMultiTexCoord3f`. Texture coordinates are normalized to the range $[0,1]^3$, with sampling done at voxel centers in the z (slice) direction.

```
glBegin(GL_QUADS);
//submit one quad per image slice
for (int i = 0; i < numSlices; i++)
{
float z = (0.5 + i)/numSlices;//slice depth
glMultiTexCoord3f(GL_TEXTURE0, 0.0, 0.0, z);
glMultiTexCoord3f(GL_TEXTURE1, 0.0, 0.0, z);
glVertex3f(-1.0, -1.0, z);//1st vertex
glMultiTexCoord3f(GL_TEXTURE0, 1.0, 0.0, z);
glMultiTexCoord3f(GL_TEXTURE1, 1.0, 0.0, z);
glVertex3f(1.0, -1.0, z);//2nd vertex
glMultiTexCoord3f(GL_TEXTURE0, 1.0, 1.0, z);
glMultiTexCoord3f(GL_TEXTURE1, 1.0, 1.0, z);
glVertex3f(1.0, 1.0, z);//3rd vertex
glMultiTexCoord3f(GL_TEXTURE0, 0.0, 1.0, z);
glMultiTexCoord3f(GL_TEXTURE1, 0.0, 1.0, z);
glVertex3f(-1.0, 1.0, z);//4th vertex
}
glEnd();
```

Listing 4: Applying transformations using texture coordinates and rendering the fixed and moving images textures in OpenGL

By default, we render directly into texture memory rather than to the on-screen frame buffer. This is done by binding a 2D texture to the frame buffer using OpenGL's "frame buffer object" specification extension (Green 2005). We do this for two reasons. First, it avoids copying from the screen into texture memory for subsequent computation of the final metric. Second, performance is improved by not rendering to the display hardware. However, there is an option to render to the display permitting visualization of intermediate metric images during registration.

Many fragment shaders compute the metric image in parallel, as illustrated schematically in Figure 16.5. In practice, however, there are more fragments to be shaded for each quad than there are available shader processors. The shaders therefore execute in parallel on subblocks of the images. As an example, a typical medical image may have $256^2 = 65,536$ pixels per slice, whereas the NVIDIA GeForce 8800 GPU has 128 shaders.

16.2.4.2 Metric Value Accumulation

At this point, the metric image has been computed and resides on the GPU. This corresponds with the last step in Figure 16.5. It remains to average the pixels of this image to compute the final similarity metric value. We do this using a sequence of parallel downsampling rendering passes (Owens 2007) on the GPU, as shown in Figure 16.6.

If the metric image has dimensions $n \times n$, then *log n* downsampling passes in a fragment shader yield the final similarity metric. The final metric value is downloaded from the GPU to the CPU optimizer using the command `glReadPixels`.

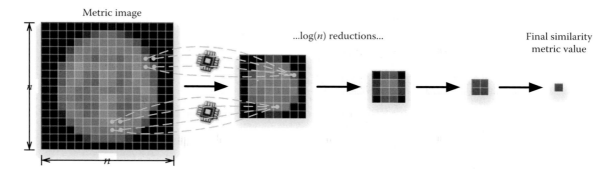

FIGURE 16.6 Parallel reduction to accumulate the final metric value by shader downsampling passes.

16.2.5 MI Similarity Metric

In this section, we describe our GPU implementation of MI and normalized MI (NMI). These are the most widely used similarity metrics for automatic registration of multimodal medical images (Maes et al. 1997a; Shekhar and Zagrodsky 2002; Pluim et al. 2003), being among the most accurate and robust metrics for retrospective studies of CT, MR, and PET images of the brain (West et al. 1997; Mattes et al. 2003).

MI measures the mutual dependence between random variables. It is based on the information theoretic quantity of entropy, which is the expected information content of a probabilistic event. As such, computation of MI requires estimation of the marginal and joint probability density functions (PDFs) of images. This is done using the marginal and joint image histograms.

MI is computed using fundamentally different GPU methods than the subtraction- and correlation-based metrics (MSE, MAE, NCC, and NGF) of Section 16.2.4. The most computationally involved component of MI evaluation is construction of the joint image histogram, because this requires iteration through the image volumes and nonsequential access to shared memory. We compute image histograms entirely on the GPU, thereby greatly accelerating registration.

Suppose that we are given two misaligned images of a subject's anatomy that may have been captured using different modalities. The images therefore contain overlapping information content with respect to the shared anatomical structures. When perfectly registered, all corresponding structures overlap and the amount of shared information is higher. In this way, we can think of registration as the maximization of the images' MI content. The discussion below formalizes this concept.

In image registration, the principle of information content is quantified by the Shannon entropy, which was originally introduced in the context of communications theory in 1948 (Pluim et al. 2003). Let us identify an image's intensity values with the random variable X and the PDF $p_X: X \to [0,1]$. This function is estimated by counting the number of occurrences of each gray value intensity and then normalizing.

Shannon defined the information content associated with image intensity $x \in X$ as $I(x) = \log \dfrac{1}{p_X(x)}$. According to this

definition, the more likely the occurrence of intensity x, the lower its associated information.* Information is therefore akin to uncertainty. The entropy of X is defined as the expected value of its information content:

$$H(X) = \sum_{x \in X} p_X(x) I(x) = -\sum_{x \in X} p_X(x) \log p_X(x). \quad (16.6)$$

We note that a completely random image, in which every intensity occurs with equal frequency, has maximal entropy for its size. An image with a single peak intensity will have a low entropy. In this sense, entropy is also a measure of dispersion of the image's probability density. The joint Shannon entropy between two discrete random variables X and Y with joint PDF p_{XY} is defined similarly:

$$H(X,Y) = -\sum_{x \in X} \sum_{y \in Y} p_{XY}(x,y) \log p_{XY}(x,y). \quad (16.7)$$

Analogous to the univariate case, this measure can be interpreted as the combined information content of the two variables (or images).

The joint image PDF changes as the degree of registration changes. With increased overlap of anatomical structures, clusters begin to appear in the PDF that correspond to structure gray values. Figure 16.7 shows the joint PDFs (estimated as the joint histogram) of a 3D T1-weighted MR image with rotated versions of itself. Histogram bin counts are shown on a natural logarithmic scale. From top left to bottom right, the rotations are 0°, 1°, 2°, 3°, 4°, 5°, 10°, and 20° in the axial plane. With increasing misalignment, the clusters become more dispersed as new combinations of co-occurring gray values emerge.

The joint entropy is a measure of this dispersion and it decreases with improved registration. However, it alone should not be used to drive image registration. This is because it is computed on the overlapping domain of the images and is therefore

---

* The logarithm function makes the definition of information content additive for independent events in a distribution, because for x_1, $x_2 \in X$, $I(x_1$ AND $x_2) = I(x_1) + I(x_2)$.

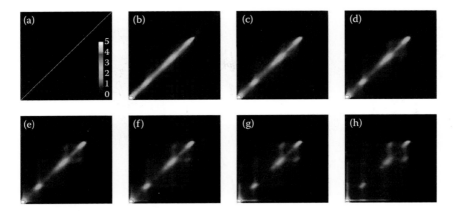

FIGURE 16.7 Joint histograms of a T1-weighted image with itself for varying degrees of misalignment by pure axial rotation (shown natural logarithmic scale). Shown are rotations of: a) 0; b) 1; c) 2; d) 3; e) 4; f) 5; g) 10; and, h) 20 degrees. Greater misalignment causes dispersion of the histogram clusters.

sensitive to the overlap size. In practice, the joint entropy may be low for complete misregistration (Studholme et al. 1999). This is because minimizing joint entropy is analogous to minimizing information content of the overlapping region, which can lead to zero overlap. Instead, we require a metric that maximizes the shared image information content over the overlap.

We solve this problem by maximizing the MI of two images A and B:

$$MI(A,B) = H(A) + H(B) - H(A,B), \qquad (16.8)$$

where we have implicitly associated the images with random variables of their intensities. MI is nonnegative, because the joint entropy of two random variables is always less than or equal to the sum of their individual entropies: $H(X,Y) \le H(X) + H(Y)$ (Maes et al. 1997a). By taking the marginal entropies into account, MI avoids the overlap problem of using joint entropy alone. This metric was first used for image registration in Viola and Wells (1997) and Maes et al. (1997a). It has been applied to the registration of most imaging modalities, including MR, CT, and PET.

In summary, MI is a measure of the shared information content and statistical dependence between images. We register two images by finding a transformation that maximizes MI. The measure is maximal when a one-to-one mapping exists between both images (i.e., one image can be completely predicted from the other). In this case, $H(A) = H(B) = H(A,B)$, so the metric equals $H(A)$. It is minimal when the images are statistically independent (i.e., contain no redundant information): $H(A,B) = H(A) + H(B)$, and the metric is zero.

A pitfall of MI is that the overlapping domain of medical images can vary considerably with their degree of alignment. In cases where the relative background and foreground areas even out, MI may incorrectly increase with greater misalignment (Pluim et al. 2003). A variant metric called NMI has been shown to be a more robust similarity metric than MI (Studholme et al. 1999). The NMI metric is invariant to changes in the marginal entropies with respect to image overlap, making it significantly

more robust than MI for automated registration. Improvements have been shown for MR-CT and MR-PET registration. It is defined as

$$NMI(A,B) = \frac{H(A) + H(B)}{H(A,B)}. \qquad (16.9)$$

MI and this closely related variant are widely accepted as two of the most robust and accurate similarity metrics for multimodal affine and deformable registration (Studholme et al. 1999; Pluim et al. 2003).

16.2.5.1 Accelerated Histogram Computation

We extend a method described by Scheuermann and Hensley (2007) for 1D histogram computation on the GPU to the case of joint (2D) histograms. The method permits creation of histograms with arbitrary size in a single rendering pass. It uses a recent extension of NVIDIA graphics hardware functionality called "vertex texture fetch," which allows the vertex shader to read from textures in video memory (Gerasimov et al. 2004). The vertex shader then uses these fetched values to "scatter" vertices to arbitrary output locations in the frame buffer.

Figure 16.8 illustrates the method used to generate a 1D histogram of an image. Corresponding GLSL vertex and fragment shaders are given in Listings 5 and 6. In this method, the histogram bins are stored as pixel intensities in a row of the frame buffer.

The method is initialized with a vertex array. One vertex is generated for each image sample, and the vertex locations are set to the image sampling coordinates (Figure 16.8, step 1). Next, the vertex shader fetches the image intensity at each vertex position (Figure 16.8, step 2; Listing 5, line 4). It sets the vertex output position to equal the fetched image intensity (Figure 16.8, step 3; Listing 5, line 5). The histogram bin counts are incremented in the frame buffer by rendering the vertices as point primitives with a color intensity of 1.0 (Figure 16.8, step 4; Listing 6, Line 2). Additive blending is enabled.

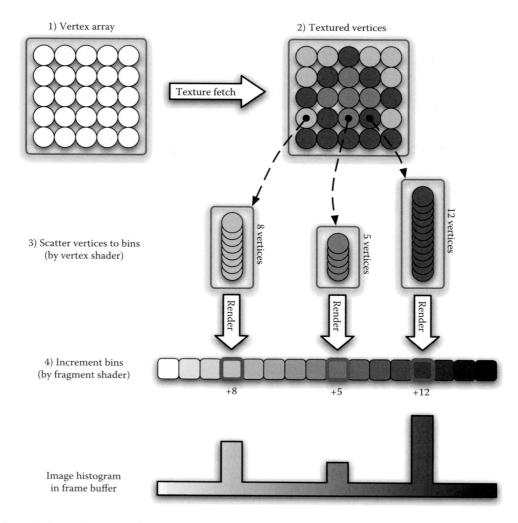

1) Vertex array

2) Textured vertices

Texture fetch

3) Scatter vertices to bins
(by vertex shader)

8 vertices

5 vertices

12 vertices

Render

Render

Render

4) Increment bins
(by fragment shader)

+8   +5   +12

Image histogram
in frame buffer

FIGURE 16.8   Computation of 1D image histograms on using vertex scattering in the rendering pipeline.

16.2.5.2 Joint Histogram Rendering

Our method for computing joint histograms is given in detail below. It is analogous to the method illustrated in Figure 16.8, except that two volumetric images are used as input and the histogram is 2D. Similar to the algorithm presented in Section 16.2.4, histograms of 3D images are computed on a "per-slice" basis and then summed in the frame buffer for all slices using additive blending.

```
uniform sampler3D imageTexture;
void main(void)
{
float intensity = texture3D(imageTexture, gl_
Vertex.xy).r;
gl_Position = vec4(intensity, 0.0, 0.0, 1.0);
}
```

Listing 5: Vertex shader using *vertex scattering* to generate an image histogram

```
void main(void)
{
gl_FragColor = vec4(1.0);
}
```

Listing 6: Trivial fragment shader for incrementing histogram bins

Our metric computations use joint histograms with 256×256 bins. This size is common among registration implementations and has been found empirically to be a good choice for most cases (Jenkinson and Smith 2001).

1. *Geometry initialization.* One vertex is created for each sample in a slice of the fixed image. The vertices are stored in an array on the GPU called a "vertex buffer object" (NVI 2004). The vertices will be rendered as point primitives, with each vertex defining one point.

   The positions of the vertices are initialized to the fixed image sampling coordinates (i,j,k), where k is the depth of the current slice being processed. (The value of k is passed to the vertex shader as a uniform variable.) We refer to each vertex by the coordinates $\mathbf{x} = (i,j,k)$. The fixed image intensities $F(\mathbf{x})$ are stored on the GPU as an array of vertex attributes.

2. *Vertex processing.* Suppose that $\mathbf{T}$ is the current affine transformation matrix. Given an input vertex $\mathbf{x}$, a custom vertex shader fetches the corresponding moving texture intensity $M(\mathbf{T}(\mathbf{x}))$. The resulting vertex is said to be "textured." Processing of the vertex is halted if the transformed coordinates $\mathbf{T}(\mathbf{x})$ are outside of the moving image domain. This ensures that the histogram is computed only using intensities in the overlap of the fixed and transformed moving images.

   The output position of the vertex is set to 2D coordinates $(F(\mathbf{x}),M(\mathbf{T}(\mathbf{x})))$, normalized to the range $[-1,1] \times [-1,1]$. These coordinates are equal to the fixed and moving image intensities.

3. *Fragment processing.* Following rasterization, a custom fragment shader sets the output intensity of each fragment to 1.0. With additive blending enabled, this results in bin $(F(\mathbf{x}),M(\mathbf{T}(\mathbf{x})))$ being incremented every time that a vertex is scattered into it during rendering.

4. *Rendering.* The vertex array corresponding to slice k is rendered. Because vertices $\mathbf{x}$ are rendered to point primitives, the resulting "image" consists of fragments at positions $(F(\mathbf{x}),M(\mathbf{T}(\mathbf{x})))$. In other words, the vertex shader "scatters" vertices into their joint histogram bin locations.

To prevent bin saturation, the joint histogram is rendered to a 32-bit floating-point buffer. Vertex array rendering initiated using the OpenGL command `glDrawArrays`.

Several optimizations in the above algorithm are worth highlighting. First, we store all vertices in a buffer on the GPU (step 1). This eliminates costly transfers of vertices from CPU to GPU on each rendering pass. Also, we render vertices for one slice of the image at a time, as we found it prohibitive to render an entire volume's worth of vertices in one pass. Because each image slice is identical in terms of sample spacing, we render the same vertex array on each pass, thereby eliminating redundancy. The only variable updated between rendering passes is a uniform corresponding to the slice number.

Because the fixed image values remain constant, they are stored in a GPU vertex buffer for fast access in the vertex shader. Before vertex processing in step 2, each vertex is assigned its corresponding fixed image value as an attribute variable. This increases vertex throughput, because the vertex shader only needs to fetch the moving image value.* Accessing vertex attributes is reported to incur less overhead than texture reads (Gerasimov et al. 2004).

Medical images often contain a large percentage of background voxels that are located outside of the subject's anatomy. It is common for up to 25% of pixels to belong to the background. In MR and PET, these voxels usually have zero intensity, because they do not contribute signal to the image. In CT, background voxels have intensities near −1000 Hounsfield units, corresponding to air.

Thus, another optimization that we implement is to discard vertices destined for histogram bin (F_b, M_b), where F_b and M_b are the fixed and moving image background intensities. For MR-to-MR registration, we therefore usually discard vertices destined for bin $(0,0)$. An "occlusion query" (Rege 2002) is issued following rendering to determine the number of fragments that were discarded, and this value is explicitly copied into bin (F_b, M_b). (In traditional graphics applications, occlusion queries are used to determine the objects obstructed from view in a scene, so as not to render them.) By eliminating the need to render a large number of vertices, this optimization significantly reduces load on the GPU without affecting the resulting histogram, yielding time savings of up to 40% in our tests on typical medical images.

16.2.5.3 Entropy Calculation

The two marginal image histograms required for MI and NMI are computed by integrating the 2D joint histogram along the fixed and moving intensity axes. This is done in a fragment shader program that uses parallel reduction along one dimension. (Figure 16.6 demonstrates parallel reduction in along two dimensions.) Because our joint histograms have 256×256 bins, eight 1D downsampling passes are required.

The marginal and joint PDFs of Equations 16.6 and 16.7 are estimated by normalizing the histogram bins by the total histogram bin counts. Next, the summands of the entropies are calculated and then accumulated using a sequence of downsampling passes to generate the marginal and joint entropy values. The downsampling passes are also executed using the parallel reduction technique demonstrated in Figure 16.6.

16.2.5.4 Partial Volume Interpolation

By default, the moving image intensity $M(\mathbf{T}(\mathbf{x}))$ is found using hardware trilinear interpolation (see Section 16.2.3). As an alternative, we also implement the method of partial volume interpolation (PVI), which has been shown to increase the robustness of NMI objective function optimization (Maes et al. 1997a). The PVI method uses fractional weights to update the joint histogram, establishing a gradual change in the bin values as the moving image is transformed. This removes undesired local optima in the similarity metric function. Compared with standard trilinear interpolation, PVI yields superior registration accuracy (Dowson and Bowden 2006; Loeckx et al. 2006).

Suppose that we are updating the histogram with intensity pair $(F(\mathbf{x}),M(\mathbf{T}(\mathbf{x})))$, where $\mathbf{T}(\mathbf{x})$ are the transformed image coordinates. Let $\mathbf{y}_i$ $(0 \le i \le 7)$ be the eight sample coordinates that neighbor $\mathbf{T}(\mathbf{x})$. The interpolated estimate of $M(\mathbf{T}(\mathbf{x}))$ using PVI is then a weighted average of the neighboring intensities:

$$\tilde{M}(\mathbf{T}(\mathbf{x})) = \sum_i w_i M(\mathbf{y}_i), \qquad (16.10)$$

---

* At the time of programming, this "vertex texture fetch" feature was available only on NVIDIA hardware. On ATI hardware, a related feature called "render to vertex buffer" is used to place image intensities directly into the vertex shader as vertex attributes.

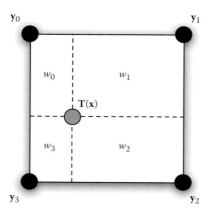

FIGURE 16.9 PVI weights (2D example).

where each weight w_i is the normalized rectilinear volume between $\mathbf{y}_i$ and $\mathbf{T}(\mathbf{x})$. Figure 16.9 illustrates the weights as areas for the 2D case corresponding to bilinear interpolation.

Rather than updating one bin, as in the standard approach described in Section 16.2.5.2, the PVI method updates the eight histogram bins that correspond to the intensity pairs $(F(\mathbf{x}), w_i M(\mathbf{y}_i))$, $0 \le i \le 7$.

Our implementation of PVI updates the joint histogram using eight rendering passes over all image samples. The first pass updates bin $(F(\mathbf{x}), M(\mathbf{y}_0))$ by the fractional amount w_0, the second pass updates bin $(F(\mathbf{x}), M(\mathbf{y}_1))$ by w_1, and so on. Thus, we scatter each vertex eight times, computing the weights explicitly in the vertex shader and sending them to the fragment shader as *varying* variables.

16.2.6 Metric Function Optimization

The goal of optimization in registration is to search for the transformation parameters that achieve the best possible similarity match between the fixed and moving images. The optimization component of the registration framework is implemented on the CPU, because it does not involve parallel operations on large data.

The objective of intensity-based registration is expressed generally as

$$\mathbf{p}^\star = \arg\min [S(F(\mathbf{x}), M(\mathbf{T_p}(\mathbf{x})))], \qquad (16.11)$$

where F and M are the fixed and moving images, S is the similarity metric function, and $\mathbf{T_p}(\mathbf{x})$ is the transformation with parameters $\mathbf{p}$ applied to the fixed image coordinates $\mathbf{x}$. Because our convention is to minimize the metric function, we multiply the NCC, NGF, and NMI metrics by -1. As described in the next section, we use a multiresolution optimization strategy to increase the likelihood and rate of convergence to the global optimum.

Full affine transforms have 12 independent parameters to optimize simultaneously: three for each rotation, translation, scaling, and shear. Rigid-body and scaling transforms have six and nine parameters, respectively. We normalize all parameters in objective space such that a unit step along any parameter axis

results in approximately the same displacement of the image in physical space (Shekhar and Zagrodsky 2002). (Displacements associated with rotation, scaling, and shearing are estimated by the movement of the image corners.)

We implement both the Powell's method and the Nelder–Mead simplex method for optimization (Press et al. 1992). These methods perform multidirectional optimization without evaluation of gradients. Both have been extensively applied to registration on the CPU and GPU (Shams et al. 2010). A comprehensive study by Maes et al. (1999) found that, compared with several common optimization strategies, Powell's method often yields the best results for multimodal image registration. They also found that the simplex method is among the fastest of the strategies studied. They recommended using the simplex method for multiresolution optimization and Powell's method for single-level optimization.

For an n-dimensional objective space, Powell's method repeatedly minimizes along a set of n directions in turn. It uses 1D line minimizations, initializing each search with the minimum found from the last direction. Powell's method ensures conjugacy of the direction set by replacing the direction of largest functional decrease with the vector between the start and end points after n minimizations. The Nelder–Mead simplex method considers all n degrees of freedom simultaneously by updating the $n + 1$ vertices of a nondegenerate simplex. The simplex follows the downhill gradient of the objective function using amoeboid-like movements until it reaches a minimum. The simplex deforms using geometric reflection, expansion, and contraction steps.

The convergence criteria for the two minimization methods are set to be as similar as possible. We stop the optimizer if $\dfrac{|\bar{f}_n - f|}{(|\bar{f}_n| + |f|)/2} \le f_{\text{tol}}$, where f denotes the current minimum function value, $\bar{f}_n$ denotes the moving average of the last n smallest function values, and f_{tol} is a specified function tolerance. The Nelder–Mead optimizer has two additional (optional) convergence criteria. Convergence can be declared if the simplex volume is below a threshold or if the relative difference between the highest and lowest simplex vertices (in terms of function value) is below a threshold. A hard limit is set for the maximum number of function evaluations in both the Powell and the Nelder–Mead optimizers.

Powell and Nelder–Mead are categorized as local optimization strategies (Press et al. 1992). This means that they search for local minima within a certain capture range of their starting point. Global optimization methods, such as "dividing rectangles" (Wachowiak and Peters 2006) and some genetic algorithms, search for the global minimum within a given parameter range. We employ a hierarchical search strategy to increase the optimization capture range.

16.2.7 Hierarchical Search

Hierarchical optimization strategies are commonly used in automated registration (Lester and Arridge 1999). Upon initialization, the input fixed and moving images are downsampled and smoothed to generate multiresolution image pyramids. Registration progresses from the lowest-resolution image pair of the pyramid to successively higher-resolution levels. Registration iterations at lower resolutions take less time, because the number of computations per iteration scales linearly with image dimension. Larger registration mismatches are recovered initially using coarser strides through search space, whereas finer details are matched as they are introduced at higher resolutions using smaller strides.

This strategy helps avoid convergence to local optima in the objective function and increases the likelihood of a global optimum match, thereby improving registration accuracy. It also accelerates optimizer convergence and increases the parameter search capture range, because relatively larger image mismatches tend to be recovered at lower resolutions (Maes et al. 1999). Multiresolution optimization thus means fewer iterations are performed in the finest pyramid level compared with a single-resolution strategy. We empirically found that using two to four resolution levels works best for most registrations.

A series of fragment shaders generate the pyramid levels by recursively blurring and downsampling the 3D textures. Blurring is done separately along the spatial dimensions using a 1D Gaussian kernel. We generally use a Gaussian with a standard deviation of 0.5 and a width of five pixels (0.06, 0.24, 0.40, 0.24, and 0.06). Filtering in the slice direction is not performed for images with relatively large slice spacing compared with in-plane spacing (Maes et al. 1999). A downsampling factor of 2 is used between pyramid levels. This scheme is depicted in Figure 16.10.

The optimizer is initialized with the identity transform at the lowest pyramid level. The registration parameters estimated at this resolution are used as the starting point for optimization at the next highest level. This process is repeated until the final registration is computed at the highest pyramid level. Other initialization choices have been reported in the literature, such as

matching image centroids and principal component analysis to estimate initial translation and rotation. However, these methods often fail for images acquired with different fields of view (Warfield et al. 1998).

16.3 Validation of GPU-Accelerated Medical Image Registration

In this section, we report several evaluations on the speed and accuracy of our GPU registration framework. Gold standard assessment of registration accuracy is based on the correspondence of homologous anatomic features between images (Klein et al. 2009). However, such quantitative evaluation of accuracy is often difficult to perform, because ground truth mappings between images are typically not known and cannot be determined exactly (Mattes et al. 2003). In addition, unique anatomic correspondences may not exist due to differing subject anatomy or image acquisition parameters. In deformable registration studies, there may exist multiple—equally valid—solutions that match image intensities between homologous structures (Wang et al. 2005).

Many registration validation studies are based on the identification and correlation of shared anatomy between images (Rogelj et al. 2002). They are usually performed by expertly identifying anatomical point landmarks or segmenting regions of interest (ROIs) in images. Landmarks and regions are defined for structures known to be shared across individuals in the test population. Klein et al. (2009) used ROI segmentations to perform the most extensive evaluation of nonlinear registration algorithms to date. In their study, source and target image pairs were manually segmented into over 50 regions. Following registration, the recovered deformations were applied to the source image segmentation labels and compared with the target image segmentation labels. They evaluated registration accuracy using various measures of source and target label overlap, similarity, and distance.

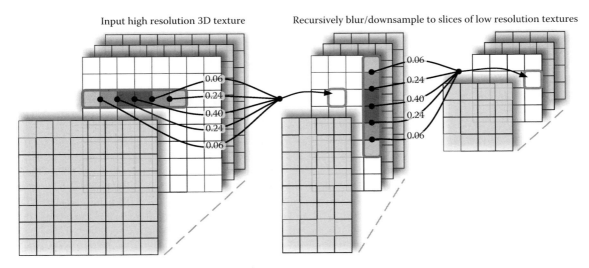

Input high resolution 3D texture   Recursively blur/downsample to slices of low resolution textures

FIGURE 16.10 Recursive Gaussian blurring and downsampling scheme to generate image pyramids.

Ardekani et al. (2005) compared the accuracy of registration algorithms using a method based on landmarks. The landmarks were manually placed on homologous structures throughout the source image brains by an operator trained in neuroanatomy. Following registration to a common reference space, accuracy was measured by the dispersion of the warped landmarks. Hellier et al. (2003) also used landmarks to measure registration accuracy. However, they focused on cortical structures, which are particularly relevant in the context of functional imaging.

Evaluation techniques based on landmarks and ROIs are often limited by the accuracy with which these features can be identified. Human judgment can be subjective and often yields results that are difficult to reproduce (Rogelj et al. 2002). In addition, such evaluation cannot detect registration mismatches that fall between landmarks or within segmented regions. It is also difficult to distinguish between registration errors and true morphologic variability among a study population. Other evaluation methods attempt to circumvent these problems by automatically extracting image features for comparison. For example, the overlap of classified tissue types (e.g., gray matter and white matter) is a frequently used measure of registration accuracy in neuroimaging (Hellier et al. 2003). Correlation between dense feature sets, such as the local curvatures of extracted surfaces, is also used.

An alternative registration evaluation method is to apply artificial transformations to the test images (Rogelj et al. 2002; Mattes et al. 2003). Following registration, the recovered transformations are compared with the artificial (ground truth) transformations. By varying the applied transformations, registration performance can be evaluated for different kinds of image mismatch. This gives the tester exquisite control for validation, although the simulated deformations may lack sufficient realism. Also, accuracy measurements may be biased by the chosen class of deformations.

We evaluate our registration framework in two ways. Our first set of tests consists of applying known artificial linear transformations to images of the human head. Our second set of tests is performed on images of brain tumor patients from a clinical database. This database is noteworthy, because gold standard rigid-body transformations are known for all patients. The true alignments were obtained from a method based on the fixation of fiducial markers to the patient skulls.

The goals of this work are to evaluate the speed and accuracy of our GPU framework for routine clinical registrations of the human head. We do not, however, aim to compare our work against other registration methods or software packages. Our tools were constructed from well-recognized methods that have been published and thoroughly evaluated in the literature (Maintz and Viergever 1998; Hellier et al. 2003; Klein et al. 2009). It is thus not necessary to reproduce prior validation work. Rather, we aim to show that our novel implementations of methods on the GPU can yield great improvements in terms of speed without sacrificing accuracy over equivalent methods implemented on the CPU.

16.3.1 Experimental Methods

We apply artificial affine transformations to synthetically constructed but realistic images of the human head. These images were obtained from the Montreal Neurological Institute (MNI) Simulated Normal Brain Database (Cocosco et al. 1997). In this section, we follow the methods of Chan (2007), who also tested his GPU registration framework by applying artificial transformations to the MNI data. A number of other registration methods have been evaluated using the MNI data (Collins et al. 1995; Rogelj et al. 2003; Rohlfing et al. 2003; Haber and Modersitzki 2006; Wachowiak and Peters 2006).

The test images were created by averaging the coregistered, intensity-normalized MRIs of 305 young, normal right-handed subjects in a common anatomical space (Collins et al. 1998). We use T1- and T2-weighted test images, which are shown in Figure 16.11. These two images are in perfect anatomic alignment and were created by averaging the same set of subjects. They both have $181 \times 217 \times 181$, 1 mm isotropic voxels, 3% noise relative to the brightest tissue, and 20% simulated radiofrequency nonuniformity. The images were converted to 16-bit unsigned integer format before processing.

16.3.1.1 Artificial Affine Transformations

Affine transformations are generated by composing random 3D translations, rotations, and scalings. The maximum translation, rotation, and scaling magnitudes along the coordinate axes are

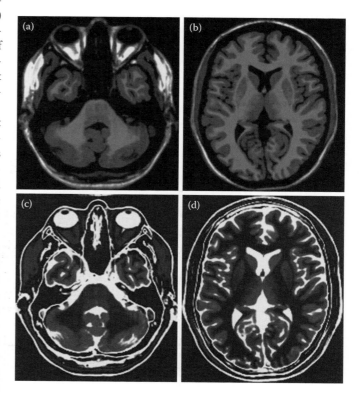

FIGURE 16.11 Slices of simulated T1-weighted (a and b) and T2-weighted (c and d) MNI images.

limited to 30 mm, 20°, and ±10%, respectively. Ten artificially transformed versions of the T1-weighted image are created. Figure 16.12 shows example affine transformations applied to the image. Before transformation, the image is zero-padded to 256^3 voxels. This is done to keep the image content in the field of view following transformation.

We evaluate our registration software by attempting to recover these artificial transformations. The monomodality MSE metric is evaluated by registering the transformed T1 image to the original T1 image. The multimodality metrics NCC, NGF, and NMI are evaluated by registering the transformed T1 image to the original T2 image. We constrain our image transformation model to nine parameters, accounting for 3D translation, rotation, and scaling.

Trilinear interpolation is used to resample the moving images, because it provides acceptable accuracy for registration and requires fewer computations than higher-order methods (Levin et al. 2004). The Nelder–Mead simplex optimizer is used for all experiments, with convergence criteria defined as follows. The function tolerance is set to $f_{tol} = 10^{-4}$, the minimum relative difference between the lowest and highest simplex vertices is set to 10^{-4}, and the maximum number of function evaluations per iteration is set to 1000. A multiresolution optimization strategy with three pyramid levels is used.

We perform all affine registrations twice. One set of runs is done using our GPU-accelerated framework. The other set of runs is done using CPU implementations of the same registration methods. The GPU and CPU registration methods perform equivalent sets of computations given the same data, although the CPU methods were written in C++ following a traditional software-based approach. The CPU methods are not multithreaded. The purpose of running all registrations twice is to compare timings between the GPU and CPU implementations.

We compute the accuracy of our affine registrations in two ways. First, we report mean errors between the applied and recovered translation, rotation, and scaling components of the artificial transformations. Translation error is measured as the vector length between the applied and recovered translations. Rotation and scaling errors are measured as the mean

absolute difference between their respective applied and recovered components.

Second, we report the root mean square (RMS) error between the applied and recovered affine transformations over a volume of interest (VOI) in the test image. The RMS error between transformations T_1 and T_2 from the source to the target image is defined by $\frac{1}{V}\sqrt{\int_{x \in VOI}\left(T_2 T_1^{-1} - I\right)(x)^2}$, where I is the identity transformation and V is the volume of the VOI. If we take the VOI to be a sphere of radius R and center x_c, then the RMS error between affine transformations T_1 and T_2 simplifies (Jenkinson 1999):

$$E_{RMS}^{affine}(T_1, T_2) = \sqrt{\frac{R^2}{5}\operatorname{trace}(A^T A) + (t + Ax_c)^T(t + Ax_c)} \quad (16.12)$$

where the 3×3 matrix A and the 3×1 vector t are components of the 4×4 matrix

$$T_2 T_1^{-1} - I = \begin{bmatrix} A & t \\ 000 & 0 \end{bmatrix} \quad (16.13)$$

We choose the VOI to be a sphere of radius 80 mm centered at the middle of the third ventricle.

We also evaluate the performance gain achieved by the specific registration components that were implemented on the GPU. To do this, we time the execution of 1000 repeated iterations of the affine transform–resample–metric cycle on both the GPU and CPU implementations of our application. These experiments are performed using the 16-bit MNI data as the moving and fixed images. To test cycle speed as a function of image size, we create versions of the images with $128^2 \times 128$, $128^2 \times 256$, $256^2 \times 128$, and $256^2 \times 256$ voxels. We apply an arbitrary affine transformation to the moving image, resample it into the space of a fixed image, and then compute the similarity metric between the two. Because no registration is performed, these experiments are independent of accuracy.

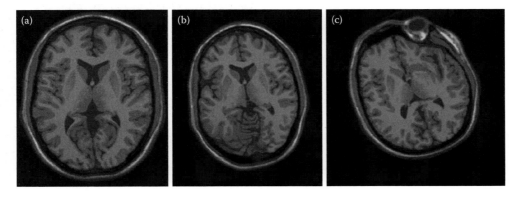

FIGURE 16.12 Slice of original T1-weighted MNI image before (a) and after linear transformations with small (b) and large (c) magnitude 3D translation, rotation, and scaling applied.

16.3.1.2 Clinical Retrospective Image Registration Evaluation

We evaluate the accuracy and speed of our GPU registration methods on a set of clinical images. The images are of brain tumor patients from Vanderbilt University's Retrospective Image Registration Evaluation (RIRE) Project (West et al. 1997). The primary objective of the RIRE Project is to evaluate the clinical accuracy of retrospective techniques for registering intermodality images of the head. Retrospective techniques, such as ours, are based on the analysis of image intensities related to anatomical features. In this study, registration results from various retrospective techniques are compared against results from a gold standard, prospective registration technique based on physical fiducial markers. The gold standard transformations remain sequestered from us and other study participants to ensure blinding in the study. Following registration of the standard data set, transformation parameters are sent to Vanderbilt, where they are compared against the gold standard. Results are reported back in terms of geometric error in millimeters, allowing for ranking of competing retrospective registration algorithms.

The authors of the RIRE study made publicly available their image database* of 18 patients, whose heads had been scanned using PET, CT, and MR imaging modalities. Table 16.1 gives the database image sizes and voxel spacings. As shown in the table, the patients are divided into two groups (A and B) based on the imaging parameters used. Each group has nine patients. One PET image, which is not represented in the table, has a voxel size of $1.94 \times 1.94 \times 8.00$ mm. Images from all modalities were acquired with contiguous slices (i.e., zero spacing between slices).

The PET images were acquired following injection of the radioactive tracer ^{18}F-fluorodeoxyglucose, which is used to assess glucose metabolism. It is characterized by elevated uptake in tissues with high metabolic activity, such as malignant brain tumors. The CT images were acquired without intravenous contrast agent; they are helpful in visualizing relatively dense structures, such as bone. The MR images were acquired for each patient using T1-weighted, T2-weighted, and proton-density (PD) spin-echo imaging sequences on a 1.5 T scanner. The MR images demonstrate soft-tissue contrast, such as between tumor and normal brain parenchyma. Figures 16.13 and 16.14 show slices of the raw PET, CT, and MR images of a sample patient in Group A of the RIRE database.

Not all patients were imaged using all modalities. Only patients of Group A were imaged using PET. In Group A, PET images were not available for two patients and CT images were not available for two (different) patients. In Group B, PD MRI was not available for five patients and there was no T2-weighted MRI for one patient.

A second set of T1, T2, and PD MR images was also included for seven patients of Group A in the database. The images in the second set were numerically corrected for geometrical distortions, which are known to decrease registration accuracy. The distortions were due to magnetic susceptibility changes induced

* The Vanderbilt RIRE database can be downloaded at http://insight-journal.org/rire/. The site includes results of alignment methods from numerous research groups.

TABLE 16.1 Size and Spacing of Images in the RIRE Database

Modality	Image Dimensions			Voxel Size (mm)		
	x	y	z	x	y	z
Group A						
PET	128	128	15	2.59	2.59	8.00
MR	256	256	20, 26	1.25–1.28	1.25–1.28	4.00–4.16
CT	512	512	27–34	0.65	0.65	4.00
Group B						
MR	256	256	52	0.78–0.86	0.78–0.86	3.00
CT	512	512	40–49	0.40–0.45	0.40–0.45	3.00

by the patient tissues within the scanner. No images in Group B were corrected for geometrical distortion.

Registrations consist of rigidly aligning the PET and CT images to the MR images of each patient. Gold standard results were obtained prospectively at Vanderbilt by attaching fiducial markers to the patients before imaging. Two types of markers were used: one bright on CT and MRI and the other bright on PET. Four fiducial markers were attached to each patient's skull by means of implanted binding posts. The study authors justified the use of this invasive procedure on the patients, stating that it was also used to aid in intraoperative guidance during subsequent neurosurgery. They used the registration results to align the preoperative images to the patients during surgery.

The gold standard rigid-body transformations were found by matching fiducial markers between image pairs. The study authors used a least-squares approach to minimize Euclidean distances between corresponding fiducials. The coordinates used for registration were defined as the centroids of the marker intensities in the images. After finding the gold standard transformations, the images were altered to remove all traces of the fiducials. These altered images were then uploaded to the RIRE database for us to download. Figure 16.15 demonstrates the removal of the fiducial markers and points on a stereotactic frame from the original images. The fiducials are circled on the original images in Figure 16.15a through c.

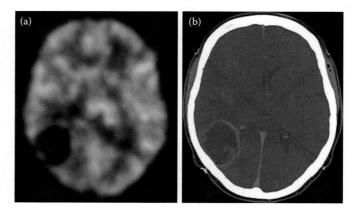

FIGURE 16.13 Sample unregistered PET (a) and CT (b) images of a patient in Group A of the RIRE database.

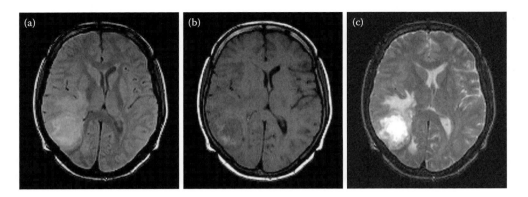

FIGURE 16.14 Sample MR images of a patient in Group A of the RIRE database: PD (a), T1 weighted (b), and T2 weighted (c).

Our PET-to-MR and CT-to-MR registration methods were evaluated against these prospectively determined, gold standard transformations. Registration error between the retrospective and prospective methods was defined as mean registration disparity between 10 target points in the images. The authors of the study defined target points to lie within regions of surgical and diagnostic importance.

Registration error is computed as follows. Let $\mathbf{x}_{MR}$ denote the coordinates of a target point in a patient's MR image. The gold standard transformation is first used to find the corresponding point $\mathbf{x}$ in the patient's CT or PET image. Next, the retrospectively determined transformation (i.e., from our software) is applied to $\mathbf{x}$, yielding the point $\mathbf{x}'_{MR}$ in the MR image. The target registration disparity for the target point is defined as $\left\| \mathbf{x}_{MR} - \mathbf{x}'_{MR} \right\|$. The overall registration error of a given method is the mean disparity computed over all target points.

We register the images using six-parameter, rigid-body transformations. Trilinear interpolation is used to resample the moving images. We evaluate the NCC metric for alignment of PET to MR, the NGF metric for alignment of CT to MR, and the NMI metric for alignment of both PET to MR and CT to MR. The NMI, NCC, and NGF metrics are defined in Equations 16.9, 16.3, and 16.5, respectively. All RIRE images are stored using 16-bit signed integer intensities.

The Nelder–Mead simplex optimizer is used for all experiments, with $f_{tol} = 10^{-4}$, the minimum relative difference between the lowest and highest simplex vertices set to 10^{-4}, and the maximum number of function evaluations per iteration set to 1000. A multiresolution optimization strategy is employed using two image resolution levels for PET-to-MR registration and three levels for CT-to-MR registration. Geometrically corrected versions of the MR images are used when available.

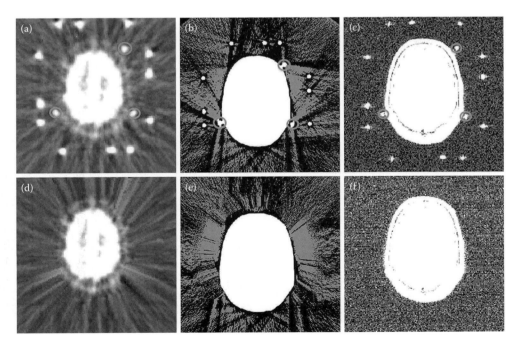

FIGURE 16.15 Sample PET, CT, and MR images from the RIRE study before (a–c) and after (d–f) removal of the fiducial markers (circled) and stereotactic frame. (From West, J. et al., *Journal of Computer Assisted Tomography*, 21(4): 554–566, 1997.)

16.3.1.3 Experimental Equipment

All experiments were performed on an Apple MacPro desktop computer, circa 2008. The system ran Mac OS X "Leopard" (version 10.5.3) and was equipped with two 3.2 GHz Quad-Core Intel Xeon processors and 2 GB of main memory. The video card used was an NVIDIA GeForce 8800 GT with 512 MB of video memory.

16.4 Results

In this section, we present the results of testing our GPU framework's accuracy and speed for registering the synthetic and clinical data sets. It is important to note that, to some extent, high registration accuracy can be achieved at the expense of convergence speed and vice versa. The precise nature of the relationship between accuracy and speed is complex, because it depends on the choice of methods and the tuning of optimization parameters. In configuring our software for these experiments, we generally opted for accuracy over speed.

16.4.1 Affine Registration Iteration Speed

The mean execution timings of the *transform–resample–metric* iteration cycle are presented in Table 16.2. Results were obtained using the MSE, NCC, NGF, and NMI similarity metrics. The experiments were run on the same workstation using both GPU- and CPU-based implementations of the methods. Each timing is reported as the mean of 1000 cycle iteration timings. The cycle was run with the MNI data as input, resampled to four different sizes.

Average cycle time performance gains using the GPU versus the CPU are 141, 143, 43, and 38 times for the MSE, NCC, NGF, and NMI metrics, respectively. Tests of the NMI metric were conducted using a 256^2 joint histogram. We also tested the NMI metric with joint histograms of size 32^2, 64^2, 128^2, 512^2, and 1024^2. No correlation was found between joint histogram computation time and histogram size.

For computing the NMI metric results in Table 16.2, we employed an optimization that was discussed in Section 16.2.5.2: joint histogram bins associated with background values from the two images are not incremented using the vertex scattering mechanism. For the MNI T1- and T2-weighted images used in these tests, our algorithm discarded all vertices destined for scattering to bin (0,0) of the 256×256 histogram. This accounted for 59% of the total vertices and resulted in an acceleration of 2.15 times over a nonoptimized version.

16.4.2 Artificially Transformed Images

In this section, we present the results of recovering transformations artificially applied to the MNI data, which has 256^3 samples and a voxel size of 1 mm³. Table 16.3 shows the

TABLE 16.3 Errors and Run Times on the GPU and CPU for Nine-Parameter, Affine Registration of the MNI Data

Method	Registration Errors				
	Translation (mm)	Rotation (°)	Scaling (%)	RMS (mm)	Timing (sec)
	MSE				
GPU	0.139	0.005	0.23	0.200	4.22
CPU	0.141	0.005	0.22	0.197	60.70
	NCC				
GPU	0.321	0.016	0.42	0.414	6.38
CPU	0.299	0.016	0.41	0.393	90.63
	NGF				
GPU	0.200	0.008	0.39	0.327	11.91
CPU	0.258	0.008	0.43	0.385	129.75
	NMI				
GPU	0.175	0.011	0.28	0.246	15.28
CPU	0.173	0.012	0.27	0.243	136.52

TABLE 16.2 Mean Run Times for the Transform–Resample–Metric Cycle on the GPU and CPU as a Function of Image Size and Similarity Metric

Method	Cycle Timing (ms)			
	Size = $128^2 \times 128$	Size = $128^2 \times 256$	Size = $256^2 \times 128$	Size = $256^2 \times 256$
	MSE			
GPU	2.5	4.3	8.9	18.0
CPU	322	642	1281	2572
	NCC			
GPU	2.9	4.5	9.1	18.8
CPU	338	678	1394	2821
	NGF			
GPU	9.1	17.4	36.6	64.3
CPU	373	687	1542	3109
	NMI			
GPU	11.5	18.4	36.9	83.0
CPU	378	746	1504	3040

registration accuracies and timings associated with recovering nine-parameter affine transformations. These are the mean values computed after complete registrations of 10 randomly transformed volumes. Accuracy is presented as the mean error between the applied and recovered transformation components as well as the overall RMS error (Equation 16.12) over a spherical VOI of radius 80 mm contained within the head. The RMS error is a measure of overall registration accuracy.

The MSE metric was used to register the transformed T1 image to the original T1 image. The other metrics were used for the multimodal registration of the transformed T1 image to the original T2 image. Registration was performed sequentially at three resolution levels, with an average of approximately 200 iterations used at the lowest resolution (64^3 samples), 150 iterations at the middle resolution (128^3), and 50 iterations at the highest resolution (256^3).

16.4.3 Retrospective Image Registration Evaluation

The results of aligning the PET and CT images to the MR images of the RIRE Project database are presented in this section. Accuracy results are given as mean and median errors with respect to the gold standard transformations. Timings for the complete registrations in our GPU-accelerated framework are also shown. Standard errors are given for all results.

Table 16.4 summarizes results of the NCC metric, which was used to align PET-to-MR images. Table 16.5 summarizes results of the NGF metric, which was used to align CT-to-MR images. Table 16.6 summarizes results of the NMI metric, which was used to align both PET and CT-to-MR images.

Figures 16.16 and 16.17 show overlaid PET, CT, and MR images of a sample brain tumor patient from Group A of the

TABLE 16.4 Registration Errors and Timing for GPU Alignment of PET-to-MR Images Using the NCC Metric

Group	Modality From	To	Mean Error (mm)	Median Error (mm)	Time (sec)	No. Cases
A	PET	PD	2.986 ± 0.043	2.821	3.49 ± 0.77	5
	PET	T1	3.824 ± 0.070	2.574	3.81 ± 0.77	4
	PET	T2	4.329 ± 0.059	3.515	3.41 ± 0.57	5

TABLE 16.5 Registration Errors and Timing for GPU Alignment of CT-to-MR Images Using the NGF Metric

Group	Modality From	To	Mean Error (mm)	Median Error (mm)	Time (sec)	No. Cases
A	CT	PD	1.391 ± 0.012	1.327	10.19 ± 0.44	7
	CT	T1	1.424 ± 0.018	1.316	11.77 ± 0.89	7
	CT	T2	1.435 ± 0.026	0.950	11.85 ± 1.02	7
B	CT	PD	2.763 ± 0.100	2.338	12.39 ± 0.36	4
	CT	T1	2.297 ± 0.039	1.895	12.24 ± 0.25	9
	CT	T2	3.044 ± 0.136	2.508	14.11 ± 0.42	8

TABLE 16.6 Registration Errors and Timing for GPU Alignment of PET and CT-to-MR Images Using the NMI Metric

Group	Modality From	To	Mean Error (mm)	Median Error (mm)	Time (sec)	No. Cases
A	PET	PD	3.021 ± 0.337	2.369	3.45 ± 0.37	5
	PET	T1	2.053 ± 0.163	1.811	3.61 ± 0.39	4
	PET	T2	2.713 ± 0.223	2.349	3.83 ± 0.17	5
	CT	PD	1.228 ± 0.084	1.058	11.19 ± 0.40	7
	CT	T1	1.040 ± 0.085	0.870	9.54 ± 0.39	7
	CT	T2	1.319 ± 0.139	1.083	8.97 ± 0.95	7
B	CT	PD	2.443 ± 0.102	2.388	12.39 ± 0.36	4
	CT	T1	1.806 ± 0.088	1.699	12.24 ± 0.25	9
	CT	T2	2.016 ± 0.048	1.965	14.11 ± 0.42	8

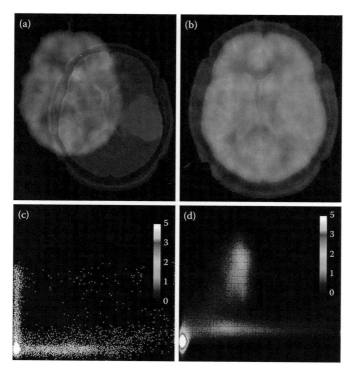

FIGURE 16.16 Overlaid PET and T2-weighted MR image slices and their joint histograms (PET vs. T2) before (a and c) and after (b and d) rigid-body alignment.

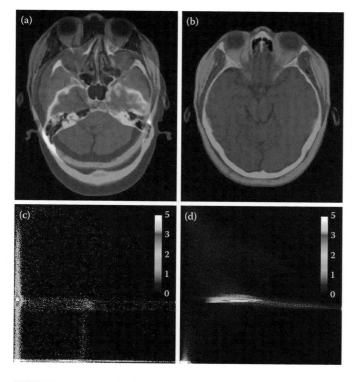

FIGURE 16.17 Overlaid CT and T1-weighted MR image slices and their joint histograms (CT vs. T1) before (a and c) and after (b and d) rigid-body alignment.

RIRE study. The representative slices are shown with their corresponding joint histograms before and after rigid-body alignment using the NMI metric. The horizontal axes of the histograms correspond to the MRI intensities. The bin values are represented on a natural logarithm scale. For this patient, the 3D alignment time for PET to T2 was 3.4 s, with a median error less than one half of the PET slice thickness. The 3D alignment time for CT to T1 was 11.8 s, with a median error equal to the in-plane voxel spacing of the MR data set or approximately one third of the CT and MR slice thicknesses.

16.5 Discussion

In the medical imaging community, certain attributes are generally demanded of a good registration tool. These include speed, accuracy, robustness, and the need for minimal user intervention (Shekhar and Zagrodsky 2002). Robustness is a broad term that we define as the tool's ability to register images of varying quality and modality, even under large initial misregistrations. As we discuss below, the results of our experiments show that our software has these desired attributes.

16.5.1 Speed and Accuracy

In Table 16.2, we compare timings between GPU- and CPU-based implementations of the image transformation, resampling, and similarity metric steps. For execution of the registration cycle using the MSE and NCC metrics, we achieve a mean acceleration of two orders of magnitude. These findings agree with those of Chan (2007), who tested his software on a high-end graphics workstation. Accelerations of one order of magnitude are found using the NGF and NMI metrics. On both the GPU and the CPU, cycle computation times roughly scale linearly with the number of voxels processed and are constant with respect to histogram size, as expected (Scheuermann and Hensley 2007).

On the GPU, the NGF metric executes approximately three times slower than the MSE and NCC metrics. This increased time complexity is mainly due to image gradient computations in the NGF routine. Such a marked slowdown was not found for the CPU implementation of NGF. This is presumably due to more sophisticated optimization of floating-point calculations by the CPU and C++ compiler than the GPU and OpenGL.

We note that accelerations determined for the transform–resample–metric cycle are not applicable to the entire image registration process. These isolated cycle timings discount relatively costly operations associated with running a complete registration, such as OpenGL environment setup, data transfers between main memory and video memory, and control logic in the optimizer. Speedups of between 10 and 20 times are commonly observed with GPU medical image registration applications (Shams et al. 2010).

To evaluate our affine registration methods, we used our software to recover affine transformations applied to synthetic test images. These results are presented in Table 16.3. The GPU and CPU versions of affine registration yielded nearly identical registration errors, all of which are subvoxel in magnitude. The slight differences between the GPU and CPU results are most likely due to a nonstandard implementation of floating-point calculations on the tested graphics hardware. These differences should not be present on later model hardware, such as the Fermi architecture, which implements IEEE standards for 32-bit floating-point operations (Glaskowsky 2009). Still, the near-perfect registration results and concordance between the GPU and CPU versions prove the validity of our affine registration methods. For the affine registration tests, we achieve a speedup of approximately 13-fold using graphics hardware.

We also showed that our affine registration methods are effective on multimodal clinical data. We aligned PET and CT images to MR images of brain tumor patients in the RIRE Project. Results are given in Tables 16.4 through 16.6. With respect to the input image voxel sizes (see Table 16.1), all RIRE mean target registration errors are subvoxel in magnitude. Larger errors are noted for images in Group B, because, unlike the Group A images, they were not corrected for geometrical distortion. All errors are reported with respect to the gold standard fiducial marker technique. In phantom studies, the gold standard has yielded mean target registration errors of 0.27 mm for CT-to-CT registration and 1.11 mm for CT-to-MR registration (Fitzpatrick et al. 1994).

16.5.2 NMI Metric

NMI has long been established as one of the most robust image similarity metrics in medical image registration. Optimization of the metric has been proven to allow fully automated affine registration of CT, MR, and PET images without the need for preprocessing, such as landmark or feature definition (Maes et al. 1999; Studholme et al. 1999). Unlike measures based on differences and correlation, MI does not assume a predefined mathematical relationship between image intensities. Results of the RIRE study have also shown that MI-based methods are among the most accurate (West et al. 1997). Indeed, a literature review reveals that either MI or NMI is used to drive linear registration in the vast majority of published neuroimaging studies (Pluim et al. 2003).

The robustness and accuracy of NMI were shown in our experiments. The metric yielded superior registration results for both PET-MR and CT-MR cases compared with the NGF and NCC metrics. The NCC metric was only applied to PET-MR cases; it was not able to successfully align CT and MR images. This is due to the inherent lack of spatial correlation between CT and MR voxel intensities. The NGF metric could not align the PET and MR images, because the PET images have weak gradient values that lack spatial coherence with MR image gradients. Thus, the NGF metric was only applied to the CT-MR cases. The NMI metric is robust and had none of these limitations. It could be employed successfully in all cases.

We use vertex scattering to generate the joint histograms required for MI: parallel threads of shaders read image intensities from texture memory and increment histogram bins in the frame buffer. With this method, we take advantage of modern hardware's ability to effectively allocate either vertex or fragment processing tasks on the fly. Heavy use of vertex processing would have been disadvantageous before the unified shader architecture, because vertex shading units had a lower level of parallelism than fragment shading units. For instance, the NVIDIA GeForce 7800 GTX GPU, released mid-2005 (just before CUDA), had 24 fragment shaders and eight vertex shaders. Modern GPUs now essentially consist of a collection of flexible floating-point engines.

We also rely on graphics hardware to serialize the bin increment operations in the frame buffer, thus guaranteeing that no two threads can simultaneously read or write a value at the same bin address. The method by which the hardware ensures atomicity and prevents memory collisions has not, to our knowledge, been made public. Regardless, we can be assured that NVIDIA's hardware implementation of memory locking is efficient. This was shown with our NMI metric cycle timings in Section 16.4.1. We compared timings with vertex scattering to bin (0,0) both enabled and disabled. When enabled, bin (0,0) is incremented by 59% of the scattered voxels or nearly 10 million times per iteration. This certainly results in thousands upon thousands of memory collisions at this bin. When disabled, none of these collisions take place. However, the resulting time savings is twofold, which is relatively low. This leads us to believe that the GPU very efficiently handles collisions at the histogram bins.

16.5.3 Real-Time Visualization

In practice, registration quality between two images is often assessed qualitatively by viewing their fused overlay or their difference image (Rogelj et al. 2002). Integration of image registration and 3D visualization was pioneered by Hastreiter and Ertl (1998), who applied their work to image-guided neurosurgery.

An advantage of our tool is that it provides such visualization of the registration process in real time (Chan 2007). Fused 2D or 3D image views, similarity metric images (as described in Figure 16.5), and the joint image density distribution can be viewed as alignment progresses. In addition, registration parameters can be initialized by manually aligning the moving and fixed images. By providing these options, our software allows the user to quickly detect and potentially correct failed registrations and to scrutinize the effects of adjusting registration parameters.

Integrated visualization was straightforward to accomplish without significant overhead, because image transformations and metrics are implemented in OpenGL. Our experiments with registration using the CUDA environment have shown that transfers of computed results to a graphics context for visualization constitute a significant bottleneck.

Figure 16.18 shows screen captures of a software application implementing our GPU registration method. Views of the images in 2D or 3D are updated in real time as registration proceeds. The T1-weighted MR images in this example were acquired of the same multiple sclerosis patient 8 months apart.

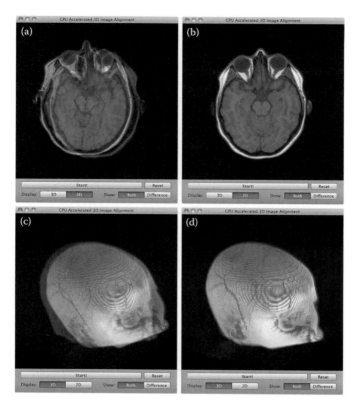

FIGURE 16.18 Screenshots of our GPU accelerated image registration tool showing MR images of the same subject at two time points before and after alignment in 2D (a and b) and rendered in 3D (c and d) (Chan 2007).

The baseline scan is loaded into the red channel; the follow-up scan is loaded into the green channel.

Before rigid-body registration in Figure 16.18a and c, it is difficult to ascertain longitudinal changes between the two scans. After registration in Figure 16.18b and d, internal brain structures are well aligned, although some flexible external soft tissues, such as the ears and ocular muscles, remain unaligned.

Other promising clinical scenarios include the detection of brain atrophy in patients with cognitive decline (Morra et al. 2009) and the assessment of tumor changes or metastatic growths in cancer patients (Zacharaki et al. 2008). Our tool could also prove valuable in time-critical settings, such as for rapidly correcting patient misalignment in emergency CT scans and for guiding radiotherapy treatment (Samant et al. 2008).

References

Archip, N., Clatz, O., Whalen, S. et al. Non-rigid alignment of pre-operative MRI, fMRI, and DT-MRI with intra-operative MRI for enhanced visualization and navigation in image-guided neurosurgery. *NeuroImage, 35*(2), 609–624, 2007.

Ardekani, B. A., Guckemus, S., Bachman, A., Hoptman, M. J., Wojtaszek, M., and Nierenberg, J. Quantitative comparison of algorithms for inter-subject registration of 3D volumetric brain MRI scans. *Journal of Neuroscience Methods, 142*(1):67–76, 2005.

Cabral, B., Cam, N., and Foran, J. Accelerated volume rendering and tomographic reconstruction using texture mapping hardware. In ACM Symposium on Volume Visualization, pages 91–98, 1994.

Chan, S. Three-dimensional medical image registration on modern graphics processors, Master's thesis, University of Calgary, Calgary, Alberta, 2007.

Chisu, R. Techniques for accelerating intensity-based rigid image registration, Master's thesis, Technische Universität München, München, 2005.

Cocosco, C. A., Kollokian, V., Kwan, R. K.-S., and Evans, A. C. BrainWeb: Online interface to a 3D MRI simulated brain database. *NeuroImage, 5*(4): 425, 1997.

Collins, D. L., Holmes, C. J., Peters, T. M., and Evans, A. C. Automatic 3-D model-based neuroanatomical segmentation. *Human Brain Mapping, 3*(3):190–208, 1995.

Collins, D. L., Zijdenbos, A. P., Kollokian, V., Sled, J. G., Kabani, N. J., Holmes, C. J., and Evans, A. C. Design and construction of a realistic digital brain phantom. *IEEE Transactions on Medical Imaging, 17*(3):463–468, 1998.

CUDA: Compute Unified Device Architecture, Programming Guide 2.0, NVIDIA Corp., 2008.

DiMaio, S. P., Archip, N., Hata, N. et al. Image-guided neurosurgery at Brigham and Women's Hospital. *IEEE Engineering in Medicine and Biology Magazine, 25*(5):67–73, 2006.

Dowson, N. and Bowden, R. A unifying framework for mutual information methods for use in non-linear optimisation. In European Conference on Computer Vision, Lecture Notes in Computer Science, volume 3951, pages 365–378. Springer-Verlag, Berlin, Germany, 2006.

Evans, A. C., Collins, D. L., Mills, S. R., Brown, E. D., Kelly, R. L., and Peters, T. M. 3D statistical neuroanatomical models from 305 MRI volumes. In *Nuclear Science Symposium and Medical Imaging Conference*, volume 3, pages 1813–1817, 1993.

Fitzpatrick, J. M., Maurer, C. R., and McCrory, J. J. Phantom testing of ACUSTAR I with comparison to stereotaxy. Technical Report CS94-04, Vanderbilt University, 1994.

Gerasimov, P., Fernando, R., and Green, S. Shader Model 3.0: Using Vertex Textures, NVIDIA Corp., June 2004. Whitepaper.

Gholipour, A., Kehtarnavaz, N., Briggs, R., Devous, M., and Gopinath, K. Brain functional localization: A survey of image registration techniques. *IEEE Transactions on Medical Imaging, 26*(4): 427–451, 2007.

Glaskowsky, P. N. NVIDIA's Fermi: The First Complete GPU Computing Architecture, NVIDIA Corp., Sept. 2009. Whitepaper.

Green, S. The OpenGL framebuffer object extension, Game Developers Conference Presentation, 2005.

Haber, E. and Modersitzki, J. Intensity gradient based registration and fusion of multi-modal images, in Medical Image Computing and Computer-Assisted Intervention (MICCAI), Volume 4191, Lecture Notes in Computer Science, pages 726–733. Springer, Heidelberg, 2006.

Hajnal, J. V., Saeed, N., Soar, E. J., Oatridge, A., Young, I. R., and Bydder, G. M. A registration and interpolation procedure for subvoxel matching of serially acquired MR images. *Journal of Computer Assisted Tomography, 19*(2): 289–296, 1995.

Hajnal, J. V., Hill, D. L. G., and Hawkes, D. J. *Medical Image Registration.* CRC Press, Boca Raton, FL, 2001.

Hastreiter, P. and Ertl, T. Integrated registration and visualization of medical image data. In *Computer Graphics International (CGI)*, pages 78–85. *IEEE Computer Society*, Hannover, Germany, 1998.

Hastreiter, P., Rezk-Salama, C., Soza, G., Bauer, M., Greiner, G., Fahlbusch, R., Ganslandt, O., and Nimsky, C. Strategies for brain shift evaluation. *Medical Image Analysis, 8*(4):447–464, 2004.

Hellier, P., Barillot, C., Corouge, I. et al. Retrospective evaluation of intersubject brain registration. *IEEE Transactions on Medical Imaging, 22*(9):1120–1130, Sept. 2003.

Ibáñez, L., Schroeder, W., Ng, L., and Cates, J. *The ITK Software Guide 2.4*, Kitware, 2005.

Jenkinson, M. Measuring transformation error by RMS deviation, Technical Report TR99MJ1, Oxford Centre for Functional Magnetic Resonance Imaging of the Brain, 1999.

Jenkinson, M. and Smith, S. M. A global optimisation method for robust affine registration of brain images. *Medical Image Analysis, 5*(2):143–156, 2001.

Jongen, C. Interpatient registration and analysis in clinical neuroimaging, Ph.D. thesis, Utrecht University, The Netherlands, March 2006.

Kamitani, Y. and Tong, F. Decoding the visual and subjective contents of the human brain. *Nature Neuroscience, 8*(5):679–685, 2005.

Kessler, M. L. and Roberson, M. Image registration and data fusion for radiotherapy treatment planning. In Schlegel, W., Bortfeld, T., and Grosu, A.-L., editors, *New Technologies in Radiation Oncology, Medical Radiology.* Springer, Heidelberg, 2006.

Klein, A., Andersson, J., Ardekani, B. A. et al. Evaluation of 14 nonlinear deformation algorithms applied to human brain MRI registration. *NeuroImage, 46*(3):786–802, 2009.

Lehmann, T. M., Gönner, C., and Spitzer, K. Survey: Interpolation methods in medical image processing. *IEEE Transactions on Medical Imaging, 18*(11):1049–1075, 1999.

Lester, H. and Arridge, S. R. A survey of hierarchical nonlinear medical image registration. *Pattern Recognition, 32*(1):129–149, 1999.

Levin, D., Dey, D., and Slomka, P. J. Acceleration of 3D, nonlinear warping using standard video graphics hardware: Implementation and initial validation. *Computerized Medical Imaging and Graphics, 28*(8):471–483, 2004.

Loeckx, D., Maes, F., Vandermeulen, D., and Suetens, P. Comparison between parzen window interpolation and generalised partial volume estimation for nonrigid image registration using mutual information, in Biomedical Image Registration, Volume 4057, Lecture Notes in Computer Science, pages 206–213. Springer-Verlag, Utrecht, The Netherlands, 2006.

Maes, F., Collignon, A., Vandermeulen, D., Marchal, G., and Suetens, P. Multimodality image registration by maximization of mutual information. *IEEE Transactions on Medical Imaging, 16*(2):187–198, 1997a.

Maes, F., Vandermeulen, F., Marchal, G., and Suetens, P. Clinical relevance of fully automated multimodality image registration by maximization of mutual information. In Proc. of the Image Registration Workshop, pages 323–330, Nov. 1997b.

Maes, F., Vandermeulen, D., and Suetens, P. Comparative evaluation of multiresolution optimization strategies for multimodality image registration by maximization of mutual information. *Medical Image Analysis, 3*(4):373–386, 1999.

Maintz, J. B. A. and Viergever, M. A. A survey of medical image registration. *Medical Image Analysis, 2*(1):1–36, 1998.

Mattes, D., Haynor, D. R., Vesselle, H., Lewellen, T. K., and Eubank, W. PET-CT image registration in the chest using free-form deformations. *IEEE Transactions on Medical Imaging, 22*(1), 2003.

Mitchell, J. R., Chan, S., and Adler, D. H. Texture-based multidimensional medical image registration. U.S. Patent No. US 2008/0143707 A1, June 2008. Filed Nov. 2007.

Morra, J. H., Tu, Z., Apostolova, L. G. et al. Automated mapping of hippocampal atrophy in 1-year repeat MRI data from 490 subjects with Alzheimer's disease, mild cognitive impairment, and elderly controls. *NeuroImage*, 45(1S):S3– 15, 2009.

Using Vertex Buffer Objects. NVIDIA Corporation, May 2004. Whitepaper.

Owens, J. Data-parallel algorithms and data structures, SIGGRAPH 2007 Presentation, Aug. 2007.

Pluim, J. P. W., Maintz, J. B. A., and Viergever, M. A. Mutual-information-based registration of medical images: A survey. *IEEE Transactions on Medical Imaging, 22*(8):986–1004, 2003.

Press, W. H., Teukolsky, S. A., Vetterling, W. T., and Flannery, B. P. *Numerical Recipes in C.* Cambridge University Press, Cambridge, 1992.

Rege, A. Occlusion (HP and NV extensions), GameDevelopers Conference Presentation, 2002.

Rezk-Salama, C., Hastreiter, P., Greiner, G., and Ertl, T. Non-linear registration of pre- and intraoperative volume data based on piecewise linear transformations. In *Vision, Modelling, and Visualization*, Erlangen, Germany, pages 365–372, 1999.

Rogelj, P., Kovačič, S., and Gee, J. C. Validation of a non-rigid registration algorithm for multi-modal data. In *SPIE Medical Imaging: Image Processing*, San Diego, CA, pages 299–307, Feb. 2002.

Rogelj, P., Kovačič, S., and Gee, J. C. Point similarity measures for non-rigid registration of multi-modal data. *Computer Vision and Image Understanding, 92*:112–140, 2003.

Rohlfing, T., Maurer, C. R., Bluemke, D. A., and Jacobs, M. A. Volume-preserving nonrigid registration of MR breast images using free-form deformation with an incompressibility constraint. *IEEE Transactions on Medical Imaging, 22*(6):730–741, 2003.

Rost, R. J. *OpenGL Shading Language*. Addison-Wesley, Boston, 2004.

Samant, S. S., Xia, J., Muyan-Özçelik, P., and Owens, J. D. High performance computing for deformable image registration: Towards a new paradigm in adaptive radiotherapy. *Medical Physics, 35*(8):3546–3553, 2008.

Scheuermann, T. and Hensley, J. Efficient histogram generation using scattering on GPUs, in Proceedings of the 2007 Symposium on Interactive 3D Graphics and Games, WA, pages 33–37. ACM, Seattle, 2007.

Shams, R., Sadeghi, P., Kennedy, R. A., and Hartley, R. I. A survey of medical image registration on multicore and the GPU. *IEEE Signal Processing Magazine*, 50, March 2010.

Shekhar, R. and Zagrodsky, V. Mutual information-based rigid and nonrigid registration of ultrasound volumes. *IEEE Transactions on Medical Imaging, 21*(1):9–22, 2002.

Shreiner, D., *OpenGL Programming Guide: The Official Guide to Learning OpenGL, Versions 3.0 and 3.1 (7th Edition)*. Addison-Wesley Professional, 2009.

Sigg, C. and Hadwiger, M. Fast third-order texture filtering. In Pharr, M., editor, GPU Gems 2, pages 313–329. Addison-Wesley, Upper Saddle River, 2005.

Starreveld, Y., Fast non-linear registration applied to stereotactic functional neurosurgery. PhD thesis, University of Western Ontario, London, Ontario, 2002.

Studholme, C., Hill, D. L. G., and Hawkes, D. J. An overlap invariant entropy measure of 3D medical image alignment. *Pattern Recognition, 32*(1):71–86, 1999.

Tustison, N. J., Awate, S. P., Cai, J., Altes, T. A., Miller, G. W., de Lange, E. E., Mugler, J. P., and Gee, J. C. Pulmonary kinematics from tagged hyperpolarized helium-3 MRI. *Journal of Magnetic Resonance Imaging, 31*(5), 2010.

Using Vertex Buffer Objects. NVIDIA Corporation, May 2004. Whitepaper.

Viola, P. and Wells, W. M. Alignment by maximization of mutual information. *International Journal of Computer Vision, 24*(2), 1997.

Wachowiak, M. P. and Peters, T. M. High-performance medical image registration using new optimization techniques. *IEEE Transactions on Information Technology in Biomedicine, 10*(2):344–353, 2006.

Wang, H., Dong, L., O'Daniel, J. et al. Validation of an accelerated "demons" algorithm for deformable image registration in radiation therapy. *Physics in Medicine and Biology, 50*(12), 2005.

Warfield, S. K., Jolesz, F. A., and Kikinis, R. A high performance computing approach to the registration of medical imaging data. *Parallel Computing, 24*(9–10):1345–1368, 1998.

West, J., Fitzpatrick, J. M., Wang, M. Y. et al. Comparison and evaluation of retrospective intermodality brain image registration techniques. *Journal of Computer Assisted Tomography, 21*(4):554–566, 1997.

Yoo, T. S., editor. *Insight into Images*. A K Peters, Wellesey, 2004.

Zacharaki, E. I., Shen, D., Lee, S.-K., and Davatzikos, C. A multiresolution framework for deformable registration of brain tumor images. *IEEE Transactions on Medical Imaging, 27*(8), 2008.

Index

Page numbers followed by f and t indicate figures and tables, respectively.